Biomarkers
of
ENVIRONMENTALLY ASSOCIATED DISEASE

Technologies, Concepts, and Perspectives

Biomarkers
of
ENVIRONMENTALLY
ASSOCIATED DISEASE

Technologies, Concepts, and Perspectives

Samuel H. Wilson, M.D.
William A. Suk, Ph.D., M.P.H.

LEWIS PUBLISHERS

A CRC Press Company
Boca Raton   London   New York   Washington, D.C.

Library of Congress Cataloging-in-Publication Data

Biomarkers of environmentally associated disease : technologies, concepts, and
perspectives / [edited by] Samuel H. Wilson, William A. Suk.
     p. cm.
     Includes bibliographical references and index.
     ISBN 156670-596-7
     1. Environmentally induced diseases—Diagnosis. 2. Environmentally induced
diseases—Molecular aspects. 3. Biochemical markers. 4. Tumor markers. I. Wilson,
Samuel H., 1939- II. Suk, William A. (William Alfred)

RB152.5 .B54 2002
616.9′8—dc21
                                                2002016076
                                                          CIP

Visit the CRC Press Web site at www.crcpress.com

© 2002 by CRC Press LLC
Lewis Publishers is an imprint of CRC Press LLC

No claim to original U.S. Government works
International Standard Book Number 156670-596-7
Library of Congress Card Number 2002016076
Printed in the United States of America 1 2 3 4 5 6 7 8 9 0
Printed on acid-free paper

Acknowledgments

We wish to thank Ms. Charle League for her editorial management and Dr. Miriam Sander for her editorial assistance. This project was supported, in part, by the National Institute of Environmental Health Sciences, National Institutes of Health.

Editors

Samuel H. Wilson, M.D., is deputy director of the National Institute of Environmental Health Sciences (NIEHS), an institute of National Institutes of Health (NIH). Dr. Wilson came to NIEHS in this capacity in 1996, where he has worked to foster basic biomedical research and to promote disease prevention research. He has helped develop NIEHS' programs in genetic susceptibility, functional genomics, children's health, risk/exposure assessment, and minority institutions' research. Dr. Wilson has strengthened partnerships between NIEHS and other federal agencies concerned with environmental health. He has also worked with the Institute of Medicine to develop broad-based discussions on issues relevant to environmental health, research, and medicine.

Dr. Wilson received his graduate and postdoctoral training in medicine and biochemistry at Harvard Medical School and the NIH. He began his career as a Principal Investigator in 1970 at the National Cancer Institute, NIH, in the Laboratory of Biochemistry. In November 1991, Dr. Wilson moved to the University of Texas Medical Branch to found the Sealy Center for Molecular Science. This large center is dedicated to interdisciplinary research in the environmental health sciences and, in particular, to the study of DNA repair and genomic stability.

Dr. Wilson's recent activities include membership on the Biochemistry Study Section at NIH and numerous other federal agency advisory groups. He has served as a scientific advisor to several private foundations involved in biomedical research. He was chair of the 2001 Mammalian DNA Repair Gordon Research Conference. He is a member of the editorial boards of the *Journal of Biological Chemistry, DNA Repair, Annual Reviews,* and *Environmental Health Perspectives.*

Dr. Wilson has had a sustained interest in eukaryotic DNA metabolism. Over the past 10 years, Dr. Wilson and his associates focused their efforts on DNA polymerase β and the mammalian base excision repair pathway. In the 1970s, Dr. Wilson helped establish that mammalian cells contain multiple DNA polymerases, and in the mid-1980s his laboratory was first to clone the DNA polymerase β gene. To assign a cellular role to DNA polymerase β, Dr. Wilson's laboratory constructed DNA polymerase b "knock-out" cell lines from a transgenic mouse strain. These cell lines, devoid of DNA polymerase β, are deficient in base excision repair and exhibit genomic instability.

Dr. Wilson and his collaborators have reported numerous DNA-protein co-crystal structures of DNA polymerases β interacting with its substrates (DNA and dNTP). His group and collaborators have solved the solution structure of the enzyme by nuclear magnetic resonance. Together, this work has improved our understanding of mechanisms of DNA synthesis and DNA repair. Dr. Wilson and his colleagues have also studied the regulation of the DNA polymerase β gene and found that expression of this gene in mammalian cells is a regulated, adaptive process responsive to certain types of environmental stress. Dr. Wilson has authored and co-authored 250 scientific publications and has been editor of four reference volumes.

William Suk, Ph.D., M.P.H., is currently director, Office of Program Development, Division of Extramural Research and Training, National Institute of Environmental Health Sciences (NIEHS), an institute of the National Institutes of Health (NIH). A primary aspect of his position has been the assessment of national and international efforts in biomedical research and its potential applications in determining adverse effects on human health resulting from exposure to environmental agents. He is responsible for designing, developing, and managing national and international programs that focus on those areas of research pertinent to the Institute's mission in experimental and molecular biology and population-based studies. To do so effectively, Dr. Suk has maintained extensive contacts with the academic and industrial scientific communities; intramural programs of the Institute as well as extramural and intramural programs of other institutes at NIH; and scientists and administrators of federal, state, professional, and private organizations, nationally and internationally.

Dr. Suk has also served as director of the NIEHS Superfund Hazardous Substances Basic Research and Training Program since it was established by Congress as part of the reauthorization of the Superfund in 1986. The goal of this university-based program is to support a wide range of research that addresses public health concerns arising from the release of hazardous substances and hazardous wastes into the environment. To this end, NIEHS sponsors coordinated multicomponent, interdisciplinary research programs that link basic biomedical research with related ecologic, hydrogeologic, and engineering studies.

Dr. Suk has had a sustained interest in linking exposures with disease etiologies and in developing research and prevention strategies to reduce risk to environmentally induced diseases and disorders. In so doing, he has helped develop NIEHS' program in children's health, genetic susceptibility, molecular medicine, risk/exposure assessment, minority institutions' research, and research on health issues related to Central and Eastern Europe, the U.S.–Mexico border environment, and the Pacific Basin.

Contributors

R. Ames
Oxford GlycoSciences
Abingdon, Oxon, U.K.

Pierre Ayotte
Quebec Public Health Centre
Beauport, Quebec, Canada

Nazzsareno Ballatori
Department of Environmental
  Medicine
University of Rochester School of
  Medicine
Rochester, New York

Åke Bergman
Department of Environmental
  Chemistry
Stockholm University
Stockholm, Sweden

Marianne Berwick
Memorial Sloan-Kettering Cancer
  Center
New York, New York

Perry J. Blackshear
Clinical Director
National Institute of Environmental
  Health Sciences
Research Triangle Park, North
  Carolina

Carole Blanchet
Public Health Research Unit
Laval University Medical Center
Beauport, Quebec, Canada

Joel D. Blum
Department of Geological Sciences
University of Michigan
Ann Arbor, Michigan

David J. Brown
Xenobiotic Detection Systems, Inc.
Durham, North Carolina

Suzanne Bruneau
Public Health Research Unit
Laval University Medical Center
Beauport, Quebec, Canada

Christof Buehler
Paul Scherrer Institut
Villigen, Switzerland

Brian Bush
School of Public Health
University of Albany
Rensselaer, New York

Mark Carey
Department of Community and
  Family Medicine
Dartmouth Medical School
Lebanon, New Hampshire

David O. Carpenter
School of Public Health
State University of New York
Albany, New York

Chia-Cheng Chang
National Food Safety and
  Toxicology Center
Department of Pediatrics and
  Human Development
Michigan State University
East Lansing, Michigan

Celia Y. Chen
Department of Biological Sciences
Superfund Basic Research Program
Dartmouth College
Hanover, New Hampshire

Yiming Chen
Laboratory of Molecular Genetics
National Institute of Environmental
  Health Sciences
Research Triangle Park, North
  Carolina

Michael Chu
Xenobiotic Detection Systems, Inc.
Durham, North Carolina

George C. Clark
Xenobiotic Detection Systems, Inc.
Durham, North Carolina

Thomas W. Clarkson
Environmental Health Science
  Center
University of Rochester Medical
  School
Rochester, New York

Max Costa
Department of Environmental
  Medicine
New York University Medical Center
New York, New York

Michael S. Denison
Environmental Toxicology
University of California
Davis, California

Mikhail F. Denissenko
Department of Biology
Beckman Research Institute of the
  City of Hope
Duarte, California
  and
Sequenom, Inc.
San Diego, California

Éric Dewailly
Quebec Public Health Centre
Beauport, Quebec, Danada

John DiGiovanni
Department of Carcinogenesis
University of Texas
M.D. Anderson Cancer Center
Smithville, Texas

Kathleen Dixon
Department of Environmental
  Health
University of Cincinnati
Cincinnati, Ohio

Chen-Yuan Dong
Department of Physics
National Taiwan University
Taipei, Taiwan

G.R. Facer
Department of Physics
Princeton University
Princeton, New Jersey

Timothy R. Fennell
CIIT Centers for Health
  Research
Research Triangle Park, North
  Carolina

Edward Fitzgerald
School of Public Health
University of Albany
Rensselaer, New York
and
New York State Department of
Health
Troy, New York

Robert A. Floyd
Free Radical Biology and Aging
Research Program
Oklahoma Medical Research
Foundation
Oklahoma City, Oklahoma

Carol L. Folt
Department of Biological Sciences
Superfund Basic Research Program
Dartmouth College
Hanover, New Hampshire

Matthew S. Forrest
Division of Environmental Health
Sciences
School of Public Health
University of California
Berkeley, California

Chris Furgal
Public Health Research Unit
Laval University Medical Center
Beauport, Quebec, Canada
and
Department of Social and
Preventive Medicine
Laval University
Sainte-Foy, Quebec, Canada

Helen C. Gallagher
Department of Pharmacology
Conway Institute
University College
Belfield, Dublin, Ireland

Patricia Ganey
Department of Pharmacology and
Toxicology
Institute for Environmental
Toxicology
Michigan State University
East Lansing, Michigan

Shirley J. Gee
Department of Entomology
University of California
Davis, California

Anne L. Golden
Mt. Sinai School of Medicine
New York, New York

Bernard D. Goldstein
Graduate School of Public Health
University of Pittsburgh
Pittsburgh, Pennsylvania

Philippe Grandjean
Institute of Public Health
University of Southern Denmark
Winsloweparken, Odense,
Denmark
and
Departments of Environmental
Health and Neurology
Boston University Schools of
Medicine and Public Health
Boston, Massachusetts

William F. Greenlee
Chemical Industry Institute of
Toxicology
Centers for Health Research
Research Triangle Park, North
Carolina

Jacques Grondin
Public Health Research Unit
Laval University Medical Center
Beauport, Quebec, Canada

Arthur Grollman
Department of Pharmacological
  Sciences
State University of New York
Stony Brook, New York

John D. Groopman
Department of Environmental
  Health Sciences
Bloomberg School of Public Health
Johns Hopkins University
Baltimore, Maryland

F. Peter Guengerich
Department of Biochemistry and
  Center in Environmental
  Toxicology
Vanderbilt University School of
  Medicine
Nashville, Tennessee

Tomás R. Guilarte
Department of Environmental
  Health Sciences
Boomberg School of Public Health
Johns Hopkins University
Baltimore, Maryland

Laura Gunn
Division of Environmental Health
  Sciences
School of Public Health
University of California
Berkeley, California

Lars Hagmar
Department of Occupational and
  Environmental Medicine
Lund University Hospital
Lund, Sweden

Bruce D. Hammock
Department of Entomology, and
  Cancer Research Center
University of California
Davis, California

Thomas D. Hancewicz
Unilever Research Inc.
Edgewater Laboratory
Edgewater, New Jersey

Peter Höglund
Department of Clinical
  Pharmacology
Lund University Hospital
Lund, Sweden

Nina T. Holland
Division of Environmental Health
  Sciences
School of Public Health
University of California
Berkeley, California

Lily Hsu
Department of Mechanical
  Engineering
Massachusetts Institute of
  Technology
Cambridge, Massachusetts

Peta E. Jackson
Department of Environmental
  Health Sciences
Bloomberg School of Public Health
Johns Hopkins University
Baltimore, Maryland

Poul J. Jørgensen
Institute of Clinical Research
Odense University Hospital
Odense, Denmark

Michael J. Kadlec
School of Public Health
University at Albany
Rensselaer, New York

Peter D. Kaplan
Unilever Research Inc.
Edgewater Laboratory
Edgewater, New Jersey

Margaret R. Karagas
Department of Community and
  Family Medicine
Dartmouth Medical School
Lebanon, New Hampshire

M.D. Kelly
Oxford GlycoSciences
Abingdon, Oxon, U.K.

Karl T. Kelsey
Department of Cancer Cell
  Biology
Harvard School of Public Health
Boston, Massachusetts

Sandy Kennedy
Oxford GlycoSciences
Abingdon, Oxon, U.K.

Thomas W. Kensler
Division of Toxicology
Bloomberg School of Public
  Health
Johns Hopkins University
Baltimore, Maryland

Ki-Hean Kim
Department of Mechnical
  Engineering
Massachusetts Institute of
  Technology
Cambridge, Massachusetts

Thomas Kluz
Department of Environmental
  Medicine
New York University School of
  Medicine
New York, New York

X. Chris Le
Department of Public Health
  Sciences University of Alberta
Edmonton, Alberta, Canada

Benoît Lévesque
Public Health Research Unit
Laval University Medical Center
Beauport, Quebec, Canada
  and
Department of Social and
  Preventive Medicine
Laval University
Sainte-Foy, Quebec, Canada

Xiufen Lu
Department of Public Health
  Sciences University of Alberta
Edmonton, Alberta, Canada

Bhaskar S. Mandavilli
Laboratory of Moloecular
  Genetics
National Institute of Environmental
  Health Sciences
Research Triangle Park, North
  Carolina

Giuseppe Matullo
I.S.I. Foundation
Institute of Scientific Interchange
Villa Gualino
  and
Department of Genetics
Biology and Biochemistry
Turin, Italy

Mario Medvedovic
Department of Environmental
  Health
University of Cincinnati College of
  Medicine
Cincinnati, Ohio

Frederick J. Miller
CIIT Centers for Health Research
Research Triangle Park, North
  Carolina

Harvey W. Hohrenweiser
Biology and Biotechnology
  Research Program
Lawrence Livermore National
  Laboratory
Livermore, California

Steven Morris
Missouri University Research
  Reactor Center
Columbia, Missouri

C. Moyses
Oxford GlycoSciences
Abingdon, Oxon, U.K.

Gina Muckle
Public Health Research Unit
Laval University Medical Center
Beauport, Quebec, Canada
  and
Department of Social and
  Preventive medicine
Laval University
Sainte-Foy, Quebec, Canada

Keith Murphy
Department of Pharmacology
Conway Institute
University College
Belfield, Dublin, Ireland

Scott R. Nagy
Department of Environmental
  Toxicology
University of California
Davis, California

Dan W. Nebert
Department of Environmental
  Health and Department of
  Pediatrics, Division of Human
  Genetics
University of Cincinnati Medical
  Center
Cincinnati, Ohio

Heather H. Nelson
Department of Cancer Cell
  Biology
Harvard School of Public Health
Boston, Massachusetts

D.A. Notterman
Department of Molecular
  Biology
Princeton University
Princeton, New Jersey

David O'Hehir
School of Public Health
University at Albany
Rensselaer, New York

R. Parekh
Oxford GlycoSciences
Abingdon, Oxon, U.K.

Wu Peng
Department of Environmental
  Medicine
New York University School of
  Medicine
New York, New York

Daria Pereg
Public Health Research Unit
Laval University Medical Center
Beauport, Quebec, Canada

Gerd P. Pfeifer
Department of Biology
Beckman Research Institute of the
  City of Hope
Duarte, California

Paul C. Pickhardt
Department of Biological Sciences
Superfund Basic Research
  Program
Dartmouth College
Hanover, New Hampshire

Ciaran M. Regan
Department of Pharmacology
University College
Belfield, Dublin, Ireland

Marc Rhainds
Public Health Research Unit
Laval University Medical Center
Beauport, Quebec, Canada
    and
Department of Social and
    Preventive Medicine
Laval University
Sainte-Foy, Quebec, Canada

Robert A. Roth
Department of Pharmacology and
    Toxicology
Institute for Environmental
    Toxicology
Michigan State University
East Lansing, Michigan

Stephen H. Safe
Department of Veterinary
    Physiology and Pharmacology
College of Veterinary Medicine
Texas A&M University
College Station, Texas

O.A. Saleh
Department of Physics
Princeton University
Princeton, New Jersey

Konstantin Salnikow
Department of Environmental
    Medicine
New York University School of
    Medicine
New York, New York

James Sanborn
Department of Entomology
University of California
Davis, California

Janine H. Santos
Laboratory of Molecular
    Genetics
National Institute of Environmental
    Health Sciences
Research Triangle Park, North
    Carolina

Guomin Shan
Department of Entomology
University of California
Davis, California

Richard R. Sharp
Division of Intramural Research
National Institute of Environmental
    Health Sciences
Research Triangle Park, North
    Carolina

Shinya Shibutani
Department of Pharmacological
    Sciences
State University of New York
Stony Brook, New York

Andreas Sjödin
Centers for Disease Control and
    Prevention
NCEH/TOX
Atlanta, Georgia

Martyn T. Smith
Division of Environmental Health
    Sciences
School of Public Health
University of California
Berkeley, California

Peter T.C. So
Department of Mechnical
    Engineering
Massachusetts Institute of
    Technology
Cambridge, Massachusetts

Lydia L. Sohn
Department of Physics
Princeton University
Princeton, New Jersey

Yukio Sugawara
Department of Entomology
University of California
Davis, California

William A. Suk
Division of Extramural Research
  and Training
National Institute of Environmental
  Health Sciences
Research Triangle Park, North
  Carolina

Jessica E. Sutherland
Department of Environmental
  Medicine
New York University School of
  Medicine
New York, New York

Naomi Suzuki
Department of Pharmacological
  Sciences
State University of New York
Stony Brook, New York

Moon-shong Tang
Department of Environmental
  Medicine
New York University School of
  Medicine
Tuxedo Park, New York

Alice Tarbell
Akwesasne Task Force on the
  Environment
Hogansburg, New York

Tor D. Tosteson
Department of Community and
  Family Medicine
Dartmouth Medical School
Lebanon, New Hampshire

James E. Trosko
National Food Safety and
  Toxicology Center
Department of Pediatrics and
  Human Development
Michigan State University
East Lansing, Michigan

Paul Turner
Epidemiology and Health Services
  Research
School of Medicine
University of Leeds
Leeds, U.K.

Brad Upham
National Food Safety and
  Toxicology Center
Department of Pediatrics and
  Human Development
Michigan State University
East Lansing, Michigan

Bennett Van Houten
Laboratory of Molecular Genetics
  and Divison of Extramural
  Research and Training
National Institute of Environmental
  Health Sciences
Research Triangle Park, North
  Carolina

Paolo Vineis
Unit of Cancer Epidemiology
University of Torino
Torino, Italy

Suryanarayana V. Vulimiri
Department of Carcinogenesis
The University of Texas
M.D. Anderson Cancer Center
Smithville, Texas

Pál Weihe
Institute of Public Health
University of Southern Denmark
Odense, Denmark
    and
Faroese Hospital System
Tórshavn, Faroe Islands

Christopher P. Wild
Epidemiology and Health Services
    Research
School of Medicine
University of Leeds
Leeds, U.K.

Kandace Williams
Biomedical Program
University of Alaska
Anchorage, Alaska
    and
Department of Biochemistry and
    Molecular Biology
Toledo, Ohio

Samuel H. Wilson
National Institute of Environmental
    Health Sciences
Research Triangle Park, North
    Carolina

Luoping Zhang
Divison of Environmental Health
    Sciences
School of Public Health
University of California
Berkeley, California

Anthony Zhitkovich
Department of Pathology and
    Laboratory Medicine
Brown University
Providence, Rhode Island

Michael Ziccardi
Department of Environmental
    Toxicology
University of California
Davis, California

Paul H. Zigas
University of North Carolina at
    Chapel Hill School of Law and
    North Carolina State University
    and
Division of Intramural Research
National Institute of Environmental
    Health Sciences
Research Triangle Park, North
    Carolina

Contents

Section VI: Nanotechniques and Biomarkers

section I

Overview

chapter one

Overview and future of molecular biomarkers of exposure and early disease in environmental health

William A. Suk and Samuel H. Wilson

Contents

1-56670-596-7/02/$0.00+$1.50
© 2002 by CRC Press LLC

Abstract   This chapter reviews the status of molecular biomarker research at the beginning of the 21st century. The characteristics of useful biomarkers are discussed and mechanisms to facilitate progress in biomarker research are proposed. These mechanisms are concerned with four important goals:

1. Developing and promoting a consistent, efficient and effective bio-marker validation process
2. Establishing and maintaining large robust relational databases to support biomarker development and use
3. Promoting expert, informed peer review of research proposals related to biomarker development
4. Improving access to biological and/or environmental samples to support biomarker research and development

These mechanisms will help bring molecular biomarkers from the laboratory environment to the clinical/population-based setting, where they can have their intended impact on reducing the burden of human disease and protecting susceptible individuals from adverse and unnecessary risk.

I. Introduction

Validated molecular biomarkers have long been recognized as invaluable tools for identifying and preventing human disease. The potential of molecular biomarkers is especially high in relation to preventing environmentally-induced disease. This is an important focus of health research at present because of significant concern over the risk of human exposure to persistent organic pollutants, heavy metals, airborne pollutants, environmental estrogens, and other environmental agents.[1] Molecular biomarkers could play a key role in facilitating advances in disease detection and prevention; however, as discussed in this chapter, much work is needed before the great potential for biomarkers in environmental health is to be realized.

This chapter is an overview of the current status of molecular biomarker research at the beginning of the 21st century. We discuss the characteristics of useful biomarkers and define criteria that can be used during biomarker development. We also focus on mechanisms that could promote progress in biomarker research. These mechanisms are concerned with four important goals:

1. Developing and promoting a consistent, efficient and effective bio-marker validation process
2. Establishing and maintaining large robust relational databases to support biomarker development and use
3. Promoting expert, informed peer review of research proposals related to biomarker development
4. Improving access to biological and/or environmental samples to support biomarker research and development

Together, efforts in these areas will facilitate more rapid development of biomarkers that can be effectively applied at the population scale. As biomarkers begin to be applied more widely, it is also important to assure that they are implemented ethically, with attention to the social and legal issues associated with their use. Future success in biomarker development and validation will undoubtedly enhance our ability to understand, treat, and prevent environmentally-induced disease. Widespread ethical application of validated biomarkers will bring great benefit to public health.

II. History and definitions

Biomarkers have been used for some time in laboratories and clinics to monitor metabolic status, disease and/or toxic responses in tissues and cells from animals and humans. They have a long history of success in the fields of analytical chemistry and have been implemented and promoted by the National Institute of Standards and Technology (NIST) and the National Center of Environmental Health at the Centers for Disease Control and Prevention (CDC). In the 1980s, rapid advances were made in molecular approaches to biology, genetics, biochemistry, and medicine, and many scientists recognized that these methods could facilitate development of molecular biomarkers. Molecular biomarkers appeared to hold the promise of transforming toxicology, epidemiology, environmental health, and clinical medicine leading to increased molecular understanding of disease and valuable applications in molecular epidemiology. With the emergence of new technologies in the last decade, including genomics-based approaches, the field of biomarker development and application has assumed even greater importance. Ultimately, biomarkers offer the promise to improve disease prevention and reduce burden of exposure and disease in the human population worldwide.

To develop molecular biomarkers in environmental health, scientists in distinct and distant fields must come together and learn to work in an interdisciplinary manner. The expertise of epidemiologists, toxicologists, statisticians, chemists, biologists, pathologists, and other scientists is needed. Because these scientists work in different ways utilizing different tools, methods, and languages, it can take considerable time for productive collaborations to develop. To address this issue, conferences have been held to promote interactions between diverse groups of scientists interested in biomarkers and to foster interdisciplinary research related to biomarker development. The National Institute of Environmental Health Sciences (NIEHS) has played a role in this process, sponsoring the landmark 1990 conference "Application of Molecular Biomarkers in Epidemiology."[2] This conference was designed to increase the visibility of research opportunities in environmental epidemiology. More recently, NIEHS sponsored the "International Conference on Arctic Development, Pollution and Biomarkers of Human Health" in Anchorage, Alaska, during which a broad assessment was made of the state of biomarker research and its impact on research on the health of the Arctic environment.[3] The present volume is also an assessment of progress in biomarker research and is in part an extension of the efforts initiated at the NIEHS-sponsored conference in Anchorage, Alaska.

A molecular paradigm or framework for biomarker development was presented in the 1983 report of the National Research Council on Risk Assessment in the Federal Government.[1] This paradigm describes progression along a continuum from exposure to disease that is influenced by the following variables: external dose, internal dose, target tissue dose, metabolic activation, detoxification and early and late biological effects (see Chapter 21 by J. Groopman, Figure 1 or Chapter 25 by Greenlee, Figure 1). Recent iterations of this paradigm[4] recognize that progression from exposure to disease can also be influenced by preventive interventions and genetic and/or environmental susceptibility of exposed individuals.

The National Academy of Sciences defines "biomarker" as an "indicator signaling events in biological systems or samples." In the context of this chapter, a biomarker is a measurement of a molecular or chemical substance or event in a biological system. Nonmolecular biomarkers have also been developed and are used extensively to measure behavioral and/or cognitive effects in humans and animals. These biomarkers are useful, but they are outside the scope of this discussion, which concerns itself with molecular biomarkers in environmental health, research and medicine.

There are three broad types of molecular biomarkers in the field of environmental health: biomarkers of exposure, biomarkers of effect, and biomarkers of susceptibility.[5] Biomarkers of exposure are methods that quantify body burden of chemicals or metabolites and they are usually applied early in the exposure-disease pathway. These markers are powerful tools for epidemiologists, allowing relatively accurate measurement of external and/or internal dose of an environmental agent; in contrast, when biomarkers of exposure are not available, epidemiologists often rely on much less quantitative approaches or mathematical models to extrapolate from external to internal dose. The kinetics of exposure complicate use of exposure biomarkers, so they should be used with attention to and awareness of this issue.

Biomarkers of effect detect functional change in the biological system under study and allow investigators to predict the outcome of exposure. However, an exposure to an environmental agent often does not have a lasting impact on a biological system. Early markers of effect can be used to identify populations at risk of disease or toxicity; late markers of effect are more tightly linked to disease progression. Ultimately, biomarkers of exposure and effect are used in a coordinated fashion to understand and prevent disease progression.

Individual susceptibility to exposure or disease is influenced by multiple genetic and environmental factors, including genetic polymorphism, age, pre-existing disease, diet, occupation, behavior and lifestyle.[1] Differential susceptibility to some diseases is also associated with ethnicity and gender. Recent advances in molecular biology, genomics, toxicology, and in understanding disease mechanisms can be used to identify susceptible subpopulations. The Human Genome Project and the Environmental Genome Project have already had a huge impact on understanding genetic susceptibility to exposure and disease, and much more information on this topic will become available in the next few years. Polymorphisms are rapidly being identified including those in human genes involved in activation and metabolism of xenobiotic agents (e.g., glutathione-S-transferase, N-acetyl transferase). These polymorphisms can be used as biomarkers of susceptibility. Gene and-protein expression technology (DNA microarray and proteomics) and high-throughput genotyping methods will facilitate development of biomarkers of susceptibility. Understanding the complex interplay between genes and environment is a tremendous challenge to scientists. Nevertheless, identification of human disease susceptibility genes is a key step towards progress in reducing the burden of human disease.

Rapid progress in genomics and genomics-associated technology has brought increased awareness of the ethical, legal, and social issues (ELSI) related to scientific research and the use or misuse of biomedical information. These issues are highly relevant to molecular biomarkers and their application, and they have received significant attention in this field. Potential misuse of biomarker data can lead to denial of employment, denial of insurance, or social marginalization, and without appropriate patient counseling, biomarker studies can adversely affect psychological welfare, long-term life goals, and family planning. Many of these have been discussed elsewhere extensively[6] (see Chapter 2 by Sharp and Zigas). Managing ELSI is an integral part of developing biomarkers for environmental health research.

III. Defining criteria for useful biomarkers

The following discussion cites many of the important characteristics of a useful molecular biomarker, many of which have been discussed previously[5,7] and are discussed elsewhere in this volume. It may be unrealistic to expect that every biomarker will fulfill all the criteria; in that case, judicious use of several biomarkers that provide complementing and confirming data is a suitable alternative. This concept of using a suite of biomarkers was pioneered by Perera and colleagues,[8–10] and it is often a beneficial approach. The strengths and/or deficiencies of each molecular biomarker should be carefully considered in experimental design.

A. Accuracy and reproducibility

A useful biomarker is highly accurate and reproducible, characteristics that may be thoroughly established during the development phase of the marker using statistical methods and laboratory studies. Internal controls are extremely valuable for ascertaining accuracy and reproducibility. Controls should also be used to characterize and quantify the dynamic range of the assay. Controls enhance data comparability between experiments in a single study, between distinct studies in a single lab, and between studies in different labs or with different models. Use of such controls can be built into the data management structure as well. Attention to accuracy and reproducibility is important as assays move from laboratory studies with model systems to clinical or population-based studies with large numbers of samples.

B.   *Specificity/sensitivity*

Biomarkers are powerful tools because they can identify and quantify exposure, effect, or susceptibility in individual members of a population. Thus, a biomarker must be sufficiently sensitive to provide an accurate measurement in a sample of limited quantity from a single individual. Sensitivity and specificity are often inversely related to one another. Specificity must be sufficient to avoid a high false positive rate, and sensitivity must be sufficient to avoid a high false negative rate. Standards and controls can be used to analyze and optimize assay sensitivity and specificity.

C.   *Database retrieval and retrospective analysis*

Data management and data comparability are extremely important in development and validation of biomarkers and for their use in population-based studies. Proper data management allows researchers to use data efficiently and to increase the data's value. Ideally, the database structure should allow the data to be reevaluated by third parties long after they are added to the database. For this reason, accessions to the database should include complete sets of raw data, and the database structure should be relational. Thus, researchers will be able to reexamine data when new insights are gained into disease processes (e.g., disease-related covariables are identified). Importantly, the statistical power of any study may increase significantly, when data from different studies are combined in a valid manner. With appropriate foresight, relational database structures can be optimized with these goals in mind.

D.   *Experimental models and biological plausibility*

Biomarkers of exposure and biomarkers of effect distribute along the continuum from exposure to disease and in some cases, a biomarker can indicate both types of events (e.g., site-specific DNA adduct in the *p53* gene and/or site-specific *p53* mutation as marker of exposure to cigarette smoke and as early marker of lung cancer[11,12]). In most cases, the link between exposure and effect is ambiguous unless it is rigorously confirmed using animal and cellular models. Animal models allow researchers to understand the mode of action of environmental chemicals and the biological mechanism of their action. Such information leads to biologically plausible biomarkers of the disease process. In addition, cell and animal models are essential for developing and refining knowledge of dose-response and exposure kinetics. Thus, the biological basis and usefulness of a molecular biomarker can be enhanced by thorough use of cell and animal models.

E. Sampling requirements

Useful molecular biomarkers for population-based studies should rely on noninvasive sampling (i.e., buccal swab, urine, blood). This requirement reflects the need for high-throughput biomarkers and for large sample size in epidemiological studies. Sample banking is also important because it enhances access to experimental materials and facilitates high-quality research studies and retrospective analyses. Retrospective analyses can take advantage of newly developed technology or allow a new hypothesis to be tested without additional field work.

F. High-throughput/feasibility on population scale

Biomarker assays need high throughput capability for application in popu-lation-based studies. Current technological advances facilitate this require-ment. DNA microarray, proteomics, immunodetection, and high-throughput sequencing or genotyping all have the potential for rapid analysis of many samples and endpoints using automated processing and data analysis.

G. Implementing biomarker criteria

It will be important to promote development of peer-reviewed guidelines that emphasize and/or require that a biomarker meets an accepted set of criteria and will be implemented in an ethical manner. These guidelines need to be implemented during the development phase of a biomarker as well as after the biomarker is implemented in a clinical or population-based setting. Mechanisms to fulfill this need have been presented previously[7,13] and are discussed below. A recent report discussing guidelines for development of biomarkers of cancer[13] recommended a phased program for biomarker development, akin to phases of drug development.

IV. Assessment of current status of biomarker research

It has been almost two decades since a molecular biomarker paradigm for study of environmental disease was advanced and discussed by the National Academy of Sciences.[1,14,15] Much research has gone into developing bio-markers of exposure and effect that help monitor, treat, and/or prevent disease. It is reasonable at this time to begin to assess progress in molecular biomarker research and to evaluate how well this field has fulfilled its prom-ise to enhance environmental health and to help identify, treat and/or pre-vent environmentally induced disease.

Only a few molecular biomarkers are currently in use in a clinical/pop-ulation setting. Among the most notable used in clinical medicine are bio-markers for exposure to aflatoxin and lead. However, in a broad sense, it is fair to say that the promise of biomarkers has not yet moved from the lab to the clinic or from development to application. For example, the recent conference "Arctic Development, Pollution, and Biomarkers of Human Health: May 1–3, 2000" produced a set of recommendations and a summary document[3] that indicate that "assessment of specific exposure using biomarker technologies holds great promise...[but] these techniques are not immediately available for widespread use." A similar conclusion was expressed in 2000 by Bennett and Waters.[16] Some researchers also express concern about the limitations of biomarkers in epidemiological studies.[17]

Nevertheless, there are exemplary results with biomarkers that can be mentioned. A recent summary highlighted the success achieved with biomarkers of exposure to aflatoxin.[7] A biologically plausible model based on adverse effects of aflatoxin bioadducts supports the strong epidemiological link between adduct level and risk of liver cancer. In both animal models and human studies, DNA adduct level correlates with known or estimated external dose of aflatoxin. A urinary metabolite of the aflatoxin DNA adduct has also been used effectively as a quantitative biomarker of exposure. In addition, large-scale epidemiological studies (cross-sectional, longitudinal, and case-control) have been carried out using these biomarkers. Prospective studies are also ongoing and biomarkers of aflatoxin exposure have been used to monitor efficacy of intervention during clinical trials of the drug oltipraz. The aflatoxin biomarker is a unique example of a well-developed and useful molecular biomarker of exposure; unfortunately, few other biomarkers have been implemented in a similar manner.

There are other promising exposure biomarkers that are based on detection of DNA and/or protein adducts (i.e., PAH-adducts including adducts from exposure to benzo-[a]pyrene, chromium-DNA adducts). However, several complications exist when using these and other DNA adducts as biomarkers. These complications include variable rates of carcinogen activation and adduct repair and the validity of using surrogate tissue in place of the target tissue in which the biological effects of a DNA lesion occur. Sufficient numbers of studies with animal models are needed to understand the dose-response curve for each environmental agent and to attempt to better define the dynamic range of the exposure-response relationship. This is especially important for evaluating chronic low-level exposure in humans.

Many studies have focused on assays for genetic instability based on the premise that genetic instability is associated with risk of cancer and certain other chronic disease and conditions. These studies propose to develop biomarkers using assays that detect DNA strand breaks, chromosomal aberrations, micronuclei, or the endogenous rate of DNA repair or mutation. In general, these assays are suboptimal for biomarker development, largely because of unresolved issues in their reproducibility, accuracy, validation and interpretation. Poor reproducibility is commonly reported in interindividual and intra-individual measurements using genomic instability assays. These assays are often labor- and cost-intensive, and are poorly suited to high-throughput applications in population-based studies. Lastly, these assays are most commonly carried out in lymphocytes as a surrogate for the target tissue at risk for disease[18,19]; the relevance of genetic stability measurements in this and other surrogate tissues has not been clearly established.

Some of the most promising areas of biomarker research involve recent advances in genomics and genomics-related fields. Toxicogenomics is an important emerging field that exploits the genomics-based technologies of DNA microarray and proteomics to solve classical problems in toxicology. One of the leading research programs in toxicogenomics is the NIEHS National Center for Toxicogenomics (NCT). Through the effort of the NCT and other groups, DNA microarray and proteomics are expected to provide useful molecular biomarkers that can be characterized and implemented rapidly. These efforts will undoubtedly also facilitate discovery of many new disease susceptibility alleles in the human genome.

Advances in human genomics have already led to identification of numerous disease-related DNA polymorphisms. Many of these polymorphisms have been characterized in clinical studies, and in some cases, human genetic polymorphism is strongly associated with disease risk and/or susceptibility to environmental agents (e.g., alleles of glutathione-S-transferase and N-acetyl transferase, mutation of *p53* and other tumor suppressor genes). As new high-throughput technologies are developed for simultaneous analysis of multiple genes, many additional disease-related polymorphisms will be discovered. The usefulness and cost effectiveness of research on human genetic polymorphism will be enhanced if human haplotypes are also identified and studied. Increased attention and effort in this area is warranted.

Other new technologies are bringing powerful capabilities to the researcher. One of these emerging technologies is nanotechnology, which is the creation of functional materials, devices, and systems at the nanometer scale (i.e., 1 to 100 nanometers) and the exploitation of novel nanoscale materials and phenomena. In the same way that miniaturization has changed the world of electronics, nanotechology has the potential to revolutionize the fields of environmental health, biomedicine, pharmaceuticals, and biotechnology. Nanoscale analytical tools could make it possible to characterize chemical and mechanical properties of cells and to discover novel processes and a wide range of tools, materials, devices, and systems with unique characteristics. Eventually, by coupling advances in the knowledge of living systems with the unique capabilities of nanostructures and materials, it may be possible to detect and intervene in disease states using biologically inspired solutions.

Nanotechnology has the potential to influence future developments in the field of molecular biomarkers. For example, nanotechnology could be important in developing biosensor technologies for detecting and analyzing molecular targets in blood, saliva, clinical specimens, and chemical or biological materials in the living body. Nanotechnology could also play a role in validation of biomarkers in population-based studies. Other important emerging capabilities include automated workstations using DNA microarrays that extract, amplify, hybridize, and detect DNA sequences, protein chips for high-throughput proteomic analyses, nano-optichemical sensors for real-time imaging of physiological processes, and nanoarrays. It is anticipated that successful development of these new approaches could facilitate rapid advances in developing sensitive, accurate, and useful molecular biomarkers. Nevertheless, these techniques and their applications are currently being developed and refined, and their anticipated impact on clinical medicine lies several years in the future.

V. Fulfilling the promise of biomarker research: instruments for future progress

Many approaches are available to enhance future progress in the process of molecular biomarker development. We have identified the following four goals as critical steps in this process:

1. Development of a consistent, efficient and effective biomarker validation process
2. Establishment of large robust relational databases to support biomarker development
3. Development of a mechanism for expert peer review of research proposals related to biomarker development
4. Establishment of large repositories of relevant biological and environmental samples

A.   The biomarker validation process

Many scientists, clinicians, and health and environmental policy makers have discussed the need for validated biomarkers and for mechanisms to promote the process of biomarker validation. Systematic efforts toward improving the biomarker validation process are critical. Biomarker validation is essential because it can assure development of biomarkers that meet specified criteria, namely, biomarkers should be accurate, reproducible, specific, sensitive, biologically plausible, high-throughput, and appropriate for use in population studies and in databases.

Because concerted effort is needed to improve biomarker validation, it is appropriate to establish validation conferences and one or more expert validation committees to work towards this goal. These groups could codify the criteria for biomarker validation and recommend approaches to standardize biomarker development. Targeted consensus development conferences for specific biomarkers would be appropriate and useful.

Existing validation committees provide a model for the process of biomarker validation. One such validation committee and process is the Interagency Coordinating Committee on the Validation of Alternative Methods (ICCVAM) established in 1997 by the NIEHS. ICCVAM is composed of representatives from 15 federal regulatory and research agencies that generate, use, or provide information relating to toxicity test methods for risk assessment. ICCVAM establishes criteria and processes that facilitate development, validation, and regulatory acceptance of new test methods. Through the National Toxicology Program (NTP) Interagency Center for the Evaluation of Alternative Toxicological Methods (NICEATM), ICCVAM also carries out peer reviews and organizes workshops on toxicological test methods. Clearly, biomarker development would be facilitated by formation of a committee whose role is similar to ICCVAM but whose focus is on biomarker assays instead of toxicology assays.

B.   Biomarker database development

As discussed above, the future of biomarker research and development will depend on the size and quality of the databases that are used to manage biomarker data. Assay standardization and comparability are essential. This can be promoted by building relational database structures that are accessible to and utilized by many research groups. A large relational database that includes the results of many different studies is important because it increases the value of existing data and saves research dollars and time. Importantly, quality control measures can also be integrated into the database structure, ensuring that data accepted into the database meet essential criteria. In a similar manner, databases can be used to establish research design standards and "best practices" and promote them throughout the research community.

Awareness of these issues is generally high in the emerging field of toxicogenomics, and the efforts underway to develop a microarray database are instructive. For example, the Microarray Gene Expression Database (MGED) group was initially established at the Microarray Gene Expression Database meeting in November 1999. As indicated on the MGED Web site (www.mged.org), "The goal of MGED is to facilitate the adoption of standards for DNA-array experiment annotation and data representation, as well as the introduction of standard experimental controls and data normalization methods. The underlying goal is to facilitate the establishment of gene expression data repositories, comparability of gene expression data from different sources and inoperability of different gene expression databases and data analysis software." Similar approaches are highly appropriate in the field of molecular biomarkers. The challenges are perhaps greater in this field because biomarkers use many different molecular techniques, each of which presents distinct issues with regard to standardization and comparability.

C.   Improved peer review of biomarker-related research
     proposals

Biomarker development is clearly an interdisciplinary endeavor that requires many types of scientific expertise. In addition, biomarker development and validation are closely linked to population-based molecular epidemiology studies. It is well recognized that this type of research does not fit well into any one of the study sections used by the National Institutes of Health (NIH) for peer-review of research grants. Therefore, these proposals have an unusually low success rate. This is a major impediment to progress in biomarker research, and it is strongly recommended that a new peer-reviewed advisory group or study section be considered for proposals in the area of molecular epidemiology and biomarker development.

D.   Sample banks and repositories to support biomarker development

Restricted sample size often limits experimental design in biomarker development and validation. If samples are archived and stored in a sample bank or repository, then sample size can potentially be increased by use of archival material. This opportunity should be exploited as much as possible in biomarker studies. Researchers should therefore be encouraged to establish and maintain sample repositories and to deposit appropriate materials in repositories whenever possible.

VI. Closing remarks

The end of the 20th century brought with it a revolution in molecular biology that culminated in advances such as completion of a preliminary draft of the human genome[20,21] and whole genome expression analyses using DNA microarray technology.[22] These advances brought an optimism to the fields of toxicology and environmental health and the anticipation that molecular biomarkers might soon come of age and have a major impact in environmental health. Optimism is justified for the future of molecular biomarkers, but current progress in biomarker application has not fulfilled the expectations of many scientists and policy makers.

   In this discussion, we present a challenge to researchers in all scientific fields relating to environmental health. This challenge is to develop mechanisms that allow the potential of molecular biomarkers to be fulfilled as rapidly as possible. These mechanisms are needed so that biomarkers are brought from the laboratory environment to the clinical- and population-based arena, where they can have an impact on health issues that affect the human population. Ultimately, ethical application of validated biomarkers has the potential to significantly reduce burden of exposure and disease and to protect susceptible individuals from adverse and unnecessary risk. Concerted and focused efforts should allow that potential to be fulfilled, with great benefit to human health.

References

  1. National Research Council, *Risk Assessment in the Federal Government: Managing the Process.* National Academy Press, Washington, D.C., 1983.

2. Smith, M. and Suk, W., Application of molecular biomarkers in epidemiology, *Environ. Health Perspect.*, 102 (Suppl. 1), 229, 1994.
3. Summary of the International Conference on Arctic Development, Pollution and Biomarkers of Human Health, National Institute of Environmental Health Sciences, 2001.
4. Olden, K. and Wilson, S., Environmental health and genomics: visions and implications, *Nat. Rev. Genet.*, 1, 149, 2000.
5. Committee on Biological Markers of National Research Council, Biological markers in environmental health research, *Environ. Health Perspect.*, 74, 3, 1987.
6. Christiani, D.C. et al., Applying genomic technologies in environmental health research: challenges and opportunities, *J. Occup. Environ. Med.*, 43, 526, 2001.
7. Groopman, J.D. and Kensler, T.W., The light at the end of the tunnel for chemical-specific biomarkers: daylight or headlight?, *Carcinogenesis*, 20, 1, 1999.
8. Perera, F. and Weinstein, I.B., Molecular epidemiology: recent advances and future directions, *Carcinogenesis*, 21, 517, 2000.
9. Perera, F., The potential usefulness of biological markers in risk assessment, *Environ. Health Perspect.*, 76, 141, 1987.
10. Perera, F. and Weinstein, I.B., Molecular epidemiology and carcinogen-DNA adduct detection: new approaches to studies of human cancer causation, *J. Chronic Dis.*, 35, 581, 1982.
11. Hainaut, P. and Pfeifer, G.P., Patterns of p53 G to T transversions in lung cancers reflect the primary mutagenic signature of DNA-damage by tobacco smoke, *Carcinogenesis*, 22, 367, 2001.
12. Greenblat, M.S. et al., Mutations in the p53 tumor suppressor gene: clues to cancer etiology and molecular pathogenesis, *Cancer Res.*, 55, 4855, 1994.
13. Sullivan, Pepe M. et al., Phases of biomarker development for early detection of cancer, *J. Natl. Cancer Inst.*, 93, 1054, 2001.
14. National Research Council, *Biologic Markers in Pulmonary Toxicology*, National Academy Press, Washington, D.C., 1989.
15. National Research Council, *Biologic Markers in Reproductive Toxicology*, National Academy Press, Washington, D.C., 1989.
16. Bennett, D.A. and Waters, M.D., Applying biomarker research, *Environ. Health Perspect.*, 108(9), 907–910, 2000.
17. Wild, C.P. et al., A critical evaluation of the application of biomarkers in epidemiological studies on diet and health, *Br. J. Nutr.*, 6, 37, 2001.
18. Bonassi, S. et al., Are chromosome aberrations in circulating lymphocytes predictive of a future cancer onset in humans? Preliminary results of an Italian cohort study, *Cancer Genet. Cytogenet.*, 79, 133, 1995.
19. Vogel, U. et al., DNA repair capacity: inconsistency between effect of over-expression of five NER genes and the correlation to mRNA levels in primary lymphocytes, *Mutat. Res.*, 461, 197, 2000.
20. Venter, J.C. et al., The sequence of the human genome, *Science*, 291, 1304, 2001.
21. McPherson, J.D. et al., A physical map of the human genome, *Nature*, 409, 934, 2001.
22. Roberts, C.J. et al., Signaling and circuitry of multiple MAPK pathways revealed by a matrix of global gene expression profiles, *Science*, 287, 873, 2000.

chapter two

Ethical and legal considerations in biological markers research

Richard R. Sharp and Paul H. Zigas

Contents

Abstract   Studies of biological markers raise a number of challenging ethical and legal issues. In this paper, we describe several of these issues. Our analysis focuses on ethical and legal considerations in studies of three different types of biological markers: biomarkers of exposure, biomarkers of early clinical effect, and biomarkers of susceptibility. We argue that studies of these three distinct types of biomarkers introduce different ethical and legal issues. We also suggest that the development of new molecular and genetic techniques for identifying biological markers will result in increased public scrutiny of biomarker research, making it all the more important for environmental health researchers to consider the social dimensions of their work.

I. Introduction

A large number of research studies in environmental health utilize biological markers. Biological markers, or biomarkers, can be defined as surrogate measures of biological or physiological processes.[1-2] These surrogate markers can be biochemical, cellular, genetic, immunological, or physiological measures used to assess some biological event or process. For example, the identification of an environmental toxicant or its metabolites in the blood can be used to estimate an individual's exposure to hazardous agents; the presence of DNA adducts can serve as a proxy for exposure to carcinogens; mutations in tumor suppressor genes can reveal early indications of cancer; and biochemical assays of enzymatic activity can indicate genomic damage.[3-4]

Since the analysis of biological markers has become common in environmental health research, it is easy to overlook the many ethical and social issues embedded in these analyses.[5-9] These issues will become of paramount importance, however, as the information derived from biomarkers is inevitably applied outside the research setting to new contexts ranging from environmental regulation to legal practice. The application of new biomarkers, in environmental health research and in a broader public context, will likely focus public attention on genetic and molecular technologies. Unfortunately, the public's response to other types of genetic research has been mixed, often characterized by gross misinterpretations of the risks and benefits of genetic analyses.[10-11] To avoid inaccurate or uninformed public perspectives on the ethical and social dimensions of environmental health research, it is important for scientists to engage the lay public in discussions regarding ethical, legal, and social issues implicit in studies of biological markers.

In this paper, we hope to help frame these discussions by highlighting several ethical and legal considerations in studies of biological markers. Our discussion is divided into three sections, in accordance with three categories of biological markers: biomarkers of exposure, biomarkers of early clinical effect, and biomarkers of susceptibility. The issues we present are characterized in general terms, with little analysis of the various subtleties of law or uncertainties regarding the present state of biomarker research. We do this with the hope of engaging a broad audience of environmental health researchers and other interested stakeholders. In addition, since many of these ethical and legal issues are just beginning to emerge as important social considerations, we believe it is important to encourage broad and creative thinking about the social issues embedded in biomarker research. Although we believe it premature to develop particular policy strategies for addressing these social issues, it is crucial to begin shaping our understanding of what those issues will ultimately entail.

II. Biomarkers of exposure

Much of the motivation for developing more sensitive molecular and genetic biomarkers is to help detect environmental exposures at lower levels than presently is possible.[12] These new biomarkers of exposure could reveal human exposures that are presently unknown or could confirm hazards whose presence is merely speculated. In addition, these biomarkers could be used to monitor individuals at risk of exposure, to assess the relationships between ambient exposure levels and internal doses, or to disentangle complex mixtures of exposures into their respective components. Ultimately, the hope is that molecular and genetic biomarkers will allow researchers to pinpoint which exposures are present in a given environment and which of these exposures play an important role in the development of human disease.

There is little doubt that the development of more sensitive biological markers of exposure will help researchers identify harmful human exposures and provide powerful tools for epidemiological research on the biological effects of environmental toxicants.[13] Despite those potential benefits, however, the development of new biomarkers of exposure will introduce a number of issues regarding how best to communicate risk-related information to individuals.[14] For example, there is a distinct possibility, given exaggerated public views regarding the significance of genetic information, that genetic and molecular biomarkers of exposure will be viewed by laypersons as more accurate or more significant than other sources of information regarding environmental risks.[15] Developing effective educational programs to help individuals interpret the significance of biological markers should thus be a key component of biomarkers research.

How environmental policy makers and regulators interpret data from studies of biological markers is also at issue. Traditionally, environmental protection efforts have focused much attention on categories of risks deemed "involuntary" and beyond individual control (e.g., clean air and water), in contrast to risks that individuals "voluntarily" impose on themselves (e.g., health risks associated with cigarette smoke). As more sensitive biological markers of exposure are developed and more environmental hazards identified, this new information is likely to significantly expand an individual's so-called "voluntary" risks by providing more accurate estimates of the harmful effects of particular environments. In other words, by revealing potential risks associated with an individual's choice to remain in a particular environment, the ability to identify individuals and subpopulations at risk from living or working in specific environments will broaden the class of risks deemed voluntary. This new information will force both individuals and environmental policymakers to reexamine the role of human agency in the development of environmentally-induced diseases.

In legal proceedings, the development of new biological markers of exposure will introduce other issues. For example, with respect to tort claims, biomarkers of exposure will likely play a role in establishing causal relationships between environmental hazards and human exposure. Establishing proof of causation in toxic tort cases is notoriously difficult.[16-17] Not only does a plaintiff have to prove the ability of a compound to cause specific disease or injury (proof of "general causation"), he or she also must prove that the wrongful exposure alleged was the specific cause of plaintiff's injury (proof of "specific causation").

Biomarkers of exposure will play an important role in toxic tort cases by providing both plaintiffs and defendants with additional tools to establish or undermine claims about cause-and-effect relationships. For example, a plaintiff might introduce evidence of a chemical-specific biomarker of exposure to prove that he or she was wrongfully exposed to a defendant's compound. Alternatively, a defendant may introduce evidence of a biomarker of exposure to prove that a plaintiff was exposed to a chemical other than the one alleged. In the second scenario, the introduction of such evidence could support an alternate theory of causation for a plaintiff's injuries that may relieve the defendant of any legal liability.

III. Biomarkers of early clinical effect

Another application of biological markers is to reveal preclinical indicators of biological processes before gross phenotypic manifestations are detected.[18] These biological markers are sometimes referred to as biomarkers of response or biomarkers of early clinical effect. In some cases, surrogate markers may allow the detection of early patterns of disease or biological conditions that frequently lead to clinical conditions.

The ability to identify early indicators of biological effect prior to other clinical manifestations of the condition raises difficult issues regarding the protection of individual privacy. For example, insurers may be interested in obtaining access to such information since it could be used to estimate an individual's likelihood of developing a specific disease. Employers may gain access to such information via worker monitoring programs, raising difficult issues regarding the appropriate balance between employer and employee interests. Biomarkers of early clinical effect may also be of much interest to other third parties such as family members. Thus, while discussions about the release of such information to third parties have not taken place, these discussions will be an important element of social conversations regarding the use of new molecular and genetic markers of early clinical effect.

A more basic set of issues surrounding the identification of biomarkers of early clinical effect is how we should think about the health status of persons with these biomarkers. For example, should we view persons with significant numbers of mutations in tumor suppressor genes as: potentially diseased, in the earliest stages of disease, or in a "preclinical" state of disease. How we resolve this conceptual issue may have important social implications for an individual who possesses such biomarkers of early clinical effect. Presence or absence or such markers may, for example, affect attitudes regarding the appropriateness of early clinical interventions.

In addition, like biomarkers of exposure, the ability to identify biomarkers of early clinical effect will have implications for toxic tort litigation and other legal allegations of negligence. For example, biomarkers of early clinical effect may play a role in determining damages in toxic tort cases. In toxic tort cases, damages normally include compensatory damages for disease or injury present as a result of plaintiff's wrongful exposure, including medical expenses arising from the exposure. To the extent that biomarkers of clinical effect are indicative of injury resulting from a toxic exposure, even where plaintiff does not yet have gross phenotypic manifestations of the injury or the disease in question, biomarkers of clinical effect could be used to support a claim for compensatory damages.

In addition to compensatory damages, damages may be sought for medical monitoring for diseases arising from the exposure that are not yet present, but may manifest in the future.[19] In deciding whether medical monitoring should be included as part of the damages, courts generally consider four factors: increased risk of contracting disease, the seriousness of the disease, toxicity of the exposure, and the diagnostic value of the medical surveillance.[20] Biomarkers of clinical effect could support an award of damages for medical monitoring by demonstrating plaintiff's increased risk for contracting a disease in the future.

Lastly, biomarkers of clinical effect could play an important role in establishing claims for an increased risk of future disease due to the toxic exposure. Traditionally, courts have placed a high standard of proof on damages claims for increased risk of disease due to their speculative nature.[21] Courts have adopted a reasonable probability standard as to these claims whereby a plaintiff cannot recover unless the prospective disease is reasonably certain to occur.[22] While the predictive value of biomarkers of clinical effect may be unclear, they could be used to support a claim for increased risk by offering evidence that a prospective disease will occur.

IV. Biomarkers of susceptibility

Biomarkers of susceptibility are a third category of biomarkers commonly studied by environmental health researchers. Susceptibility biomarkers are surrogate measures of inherited predispositions to disease or vulnerabilities to environmental agents.[23-25] Individuals do not bear the burdens of risk equally. Some individuals are more sensitive to certain environmental hazards than others because of genes they have inherited from their parents. Identifying these genetic polymorphisms associated with increased sensitivity to environmental agents thus could result in a better understanding of why people respond differently to various hazardous environments.[26] This improved understanding of genetic influences on response to environmental exposures may enable clinicians to identify those individuals who are most vulnerable to various environmental toxins. This technology may in turn facilitate more careful medical monitoring of those individuals.

Despite such potential benefits, a number of social issues are raised by the ability to identify individuals and subpopulations with heightened genetic sensitivities to environmental agents.[27] For example, identifying genetic sensitivities to environmental toxins may shift the focus of risk-management efforts away from the improvement of unhealthy environmental conditions. Employers and manufacturers, for example, may find it less costly to dismiss genetically sensitive workers rather than to eliminate workplace hazards. These concerns have been described as the "geneticization" of disease, whereby environmental and behavioral disorders are increasingly viewed as the product of genetic influences.[28] This reduction of social problems to biological problems might very well alter social priorities, for example, resulting in research funding being diverted from preventive strategies for improving public health to approaches stressing genetic causes of disease.

A different set of considerations pertain to how information on genetic susceptibilities to environmental toxins may affect how we view an individual's responsibility for his or her health. Suppose, for example, an individual has a known sensitivity to a particular occupational exposure but knowingly chooses to accept a job that will expose him or her to that very exposure. It seems plausible to maintain that this person has acted irresponsibly should he or she fall ill to the relevant disease. In this case, insurers may claim that individuals who do not minimize their exposure to these agents are responsible for any subsequent illness because they knowingly placed themselves at risk. Alternatively, individuals might argue that there was little they could do to prevent their illness since they had "bad genes." Presently, it is unclear how to resolve such disputes. The fundamental questions these disputes raise about the role human agency in the development of disease are complex and will require continued attention from all interested parties.

In the law, such disputes are likely to emerge with some frequency. For example, tort litigants might appeal to an individual's heightened genetic sensitivity to a particular agent to help establish a defendant's liability for damages. Two opposing theories define the scope of a defendant's liability in tort claims. The first theory limits defendant's liability for damages only to the foreseeable consequences that arise from defendant's negligent conduct.[29] The second theory states that a defendant who is negligent must take existing circumstances as they are and may therefore be liable for consequences brought about by the defendant's acts even though they were not reasonably foreseeable.[30] Under this second theory, a defendant is liable for the full extent of injuries suffered by a so-called "thin-skulled" or "eggshell" plaintiff whose predisposition to harm may cause severe, but unforeseeable, injury. Biomarkers of susceptibility could be used to establish the scope of damages for which a defendant is liable by identifying thin-skulled plaintiffs with genetic predispositions to injury in tort cases.

Lastly, though many molecular and genetic biomarkers are being developed by researchers for use in environmental health research, it is possible that tests for inherited sensitivities to environmental hazards may be used in other contexts before those biomarkers are well validated. In this regard, a recent EEOC claim brought forth against Burlington Northern Santa Fe Railroad (Burlington Northern) illustrates the potential for poorly validated biomarkers to be used inappropriately.[31] Prior to the EEOC claim, Burlington Northern required employees who had submitted workers compensation claims of work-related carpal tunnel syndrome to provide blood samples for testing to determine whether they had a genetic predisposition to that condition. The gene that Burlington Northern was screening for, however, is considered so rare and poorly validated as to be useless in an employment context. While the case was settled before it went to trial, the EEOC maintained that this type of practice is in violation of the Americans with Disabilities Act (ADA) of 1990.[32]

The Burlington Northern case illustrates how biomarkers of susceptibility might be used for purposes other than those for which they were intended. Although the U.S. Constitution, Title VII, and the ADA provide some protection against workplace discrimination, there is no federal anti-genetic discrimination legislation in place to regulate genetic testing by employers. Although more than half of the states have laws pertaining to genetic discrimination, a comprehensive federal statute may better protect worker's interests given the disparate nature of state protections and the still unresolved issue of how courts will apply the ADA to genetic-testing programs in the workplace.[33]

V. Conclusion

As environmental health scientists develop more sensitive molecular and genetic biomarkers, these surrogate measures will influence risk-assessment and risk-management efforts in numerous ways. Biomarkers of susceptibility, for example, will be used to identify individuals with heightened genetic sensitivities to environmental hazards. Ultimately, those most sensitive to particular environmental toxins might be able to avoid exposure based on this knowledge. Indeed, the benefits of research on biological markers are many and include: a better understanding of biological mechanisms of environmental response, more accurate assessments of exposure levels, and identification of sensitive individuals and subpopulations.

Despite these benefits, however, we should be mindful of the ethical and legal challenges that will likely stem from the development of more sensitive molecular and genetic biomarkers. The appropriate use of biological markers in workplace monitoring and genetic screening for susceptibility to diseases remains unclear, as does the use of biomarker information by insurers and environmental regulators. Furthermore, despite the potential use of biomarkers as evidence in toxic tort and product liability cases, it is uncertain whether, and to what extent, such information will be admissible under either federal or state rules of evidence. In spite of these unanswered questions, it is clear that molecular and genetic biomarkers will profoundly impact tort litigation in the near future.

In this paper we have discussed these ethical and legal considerations in broad terms, outlining many issues in need of attention. Our goal is not to suggest that research on biological markers should be abandoned. Rather, it is our hope that by highlighting a range of ethical and legal issues, we can encourage additional discussions of these issues by environmental health researchers. We believe that by proactively addressing these social issues and engaging a range of interested stakeholders in this research, the full benefits of environmental health studies are more likely to be realized.

References

1. National Research Council, Biological markers in environmental health research, *Environ. Health Persp.*, 74, 1, 1987.
2. Groopman, J.D. and Kensler, T.W., The light at the end of the tunnel for chemical specific biomarkers: daylight or headlight?, *Carcinogenesis*, 1, 20, 1999.
3. Suk, W.A. and Wilson, S.H., Overview and future of molecular biomarkers of exposure, in *Biomarkers of Environmentally Associated Disease: Technologies, Concepts, and Perspectives*, Wilson, M.A., Ed., CRC Press, Boca Raton, 2002, chap. 1.
4. Olden, K. and Guthrie, J., Genomics: implications for toxicology, *Mutat. Res.*, 473, 3, 2001.
5. Brandt-Rauf, P.W. and Brant-Rauf, S.I., Biomarkers — scientific advances and societal implications, in *Genetic Secrets: Protecting Privacy and Confidentiality in the Genetic Era*, Rothstein, M.A., Ed., Yale University Press, New Haven and London, 1997, chap. 10.
6. Schulte, P.A. et al., Ethical issues in the use of genetic markers in occupational epidemiologic research, *J. Occup. Environ. Med.*, 41, 639, 1999.
7. Schulte, P.A. et al., Ethical and social issues in the use of biomarkers in epidemiological research, *IARC Sci. Publ.*, 142, 313, 1997.
8. Hunter, D. et al., Informed consent in epidemiologic studies involving genetic markers, *Epidemiology*, 8, 596, 1997.
9. Sharp, R.R. and Barrett, J.C., The Environmental Genome Project: ethical, legal, and social implications, *Environ. Health Perspect.*, 108, 279, 2000.
10. Nelkin, D. and Lindee, M.S., *The Gene as a Cultural Icon*. New York: W.H. Freeman, 1995.
11. Condit, M.C., *The Meanings of the Gene*, University of Wisconsin Press, Madison and London, 1999, chap. 10.
12. Renwick, A.G. and Walton, K., The use of surrogate endpoints to assess potential toxicity in humans, *Toxicol. Lett.*, 120, 97, 2001.
13. Khoury, M., Genetic epidemiology and the future of disease prevention and public health, *Epidemiol. Rev.*, 19, 175, 1997.
14. Schulte, P.A. and Singal, M., Ethical issues in the interaction with subjects and disclosure of results, in *Ethics and Epidemiology*, Coughlin, S.S. and Beauchamp, T.L., Eds., Oxford University Press, New York, 1996, 178–196.
15. Hubbard, R. and Wald, E., *Exploding the Gene Myth*, Beacon Press, Boston, 1993, chap. 5.
16. Poulter, S.R., Genetic testing in toxic injury litigation: the path to scientific certainty or blind alley? *Jurimetrics* 41, 211, 2001.
17. Marchant, G.E., Genetic susceptibility and biomarkers in toxic injury litigation, *Jurimetrics*, 41, 67, 2000.
18. Bonassi, S. et al., Validation of biomarkers as early predictors of disease, *Mutat. Res.*, 480, 349, 2001.

19. See e.g., *Miranda v. Shell Oil Co.*, 17 Cal App 4th 1651, 26 Cal Rptr 2d 655, 1993.
20. See e.g., *Askey v. Occidental Chemical Corp.*, 102 App Div 2d 130, 477 NYS2d 242, 1984.
21. See e.g., *Abuan v. General Elec. Co.*, (CA9 Guam), 3 F3d 329, 93 CDOS 6360, cert den (US) 127 L Ed 2d 383, 114 S Ct 1064, 1993.
22. See e.g., *Mauro v. Raymark Industries, Inc.*, 116 NJ 126, 138, 561 A2d 257, 264, 1989.
23. Knudsen, L.E. et al., Risk assessment: the importance of genetic polymorphisms in man, *Mutat. Res.*, 482, 83, 2001.
24. Au, W.W. et al., Usefulness of genetic susceptibility and biomarkers for evaluation of environmental health risk, *Environ. Mol. Mutagen*, 37, 215, 2001.
25. Olden, K. and Wilson, S., Environmental health and genomics: visions and implications, *Nat. Rev. Genet.*, 1, 149, 2000.
26. Nerbert, D.W., Ecogenetics: genetic susceptibility to environmental adversity, in *Biomarkers of Environmentally Associated Disease: Technologies, Concepts, and Perspectives*, Wilson, M.A., Ed., CRC Press, Boca Raton, 2002, chap. 4.
27. Sharp, R.R., The evolution of predictive genetic testing: deciphering gene-environment interactions, *Jurimetrics*, 41, 145, 2001.
28. Edlin, G.J., Inappropriate use of genetic terminology in medical research: a public health issue, *Perspect. Biol. Med.*, 31, 47, 1987.
29. Keeton, W.P. et al., *Prosser and Keeton on Torts*, 5th ed. West Publishing Co., St. Paul, 1984, 281.
30. See e.g., *Snyder v. Allied Mut. Ins. Co.*, Neb. App. Lexis 190, 1994.
31. See *EEOC v. Burlington Northern Santa Fe Railroad*, N.D. Iowa, No. C01–4013, filed Feb. 9, 2001.
32. Taylor, T.S., Job gene test raises alarm: many predict discrimination by employers, *Chicago Tribune*, Sept. 3, 2001.
33. See e.g., *Mario Echazabal v. Chevron USA, Inc.*, (9th Cir Cal), 226 F.3d 1063, 2000, cert granted (U.S.) 122 S. Ct. 456, 2001.

chapter three

Scientific and policy issues affecting the future of environmental health sciences

Bernard D. Goldstein

Contents

Abstract Human development can be summarized as our species response in the last million or so years to the environmental challenges of our planet. For 5 billion years, Earth had been on automatic pilot, governed by planetary geophysical and biological processes and by feedback loops that controlled its development. In a relatively short time, our planet has moved from automatic to manual, from a natural unfolding of evolutionary processes to a significant measure of human domination of these processes. The essence of the environmental movement is recognition that there are major dangers to ourselves and our planet from this attempt at human domination, and any control of our environment must be done with a high level of respect for the natural forces that truly dominate our planet and for the cultural value and wisdom of those who have wrestled with these forces. Nowhere is this more evident than in the Arctic. The human health and ecosystem challenges there are unparalleled, and they exemplify the interaction between local and global environmental issues. Sustainable development of the Arctic requires an approach that is built upon knowledge and upon the basic concepts of prevention. Understanding sources and pathways for environmental degradation and adverse human health impacts allows eradication or replacement of the sources and interdiction of the pathways. Developing and applying biological markers of exposure, effect, and susceptibility will be central to this effort. Success in applying advanced scientific tools to the challenges of the Arctic depends upon respect for the geophysical, climatic, and biological forces that shape the Arctic environment and for the indigenous human cultures that have wrestled with its challenges. Environmental health science can and should be a major facilitator of appropriate stewardship of our planet and its biosphere. But achieving this role faces major impediments that extend beyond the limitations imposed by scientific understanding.

I. Introduction

It is easy to predict that there will be many highly significant environmental stressors in the next century. Among the many other sobering statistics that can be cited is that human population will continue to increase to at least 10 billion before we reach what is hoped to be a new equilibrium. Perhaps the best quote that exemplifies the challenges of our new century comes from a survey of school children by Beijing municipal authorities. A fifth grader said: "When I grow up, I want to be a business mogul, live in a villa in the suburbs, wear a suit by Pierre Cardin, and drive a Mercedes-Benz 600 to work in Beijing."[1] This common aspiration represents a pathway to a future that, based on current knowledge, is not environmentally sustainable.[2] It is obviously inequitable to demand that developing nations not consume as much as we do in the United States, nor would such a demand be heeded. Yet without new technology or a major adjustment in the lifestyles of the developed world, we can expect development to bring unsustainable environmental problems.

Fragile environments, like fragile humans, are particularly at risk from environmental stressors. The Arctic is rich in culture and in resources, but it requires a hardiness to survive with little room for additional stress or for mistakes. This conference has explored many of the issues that now confront the Arctic, with a particular emphasis on Alaska. We have also heard about how rapid advances in biomedical science are providing biological markers relevant to human and ecosystem health, and how they can be appropriately applied to Arctic environmental problems.

Table 3.1   Environmental Health Science Determinants
for the 21st Century

Population Growth	Population Stability
Command and Control	Stakeholder Guided
Take It or Leave It	Options
Toxin-based	Disease-Based
DNA Damage	DNA Function
Disease	Health
Powerless Epidemiology	Powerful Epidemiology
Single Agent	Multiple Agent
Unidirectional Mechanisms	Multidirectional Mechanisms
Exposure-Based Cohorts	Susceptibility and Dose-Based Cohorts
Exposure Surrogates	Dose Measures
Effluent Measurement	Indicator Measurement
Source Assumptions	Source Attribution
Secondary Prevention	Primary Prevention

I will not attempt to summarize all that we have heard. Rather, I will try to weave a coherent fabric of science, societal trends and the value of biological markers in meeting future challenges. The fabric will be interlaced with cautionary strands of some of the problems that we will or may interfere in achieving our design.

II. Environmental health science for the next century

Outlined in Table 3.1 are a series of phrases which mark the direction I suggest we are or will be moving toward. Population growth leads the list. Perhaps surprisingly, I view the issue of whether we will continue to regulate based upon a command and control philosophy as the next most important. This issue puts environmental health science squarely amidst the battle now being waged among risk managers. The battle is between those who take the more legalistic approach of focusing on regulatory command and control of toxic agents and those who take the more public health oriented approach of dealing holistically with health problems by involving stakeholders.[3-7] The Framework for Risk Assessment and Risk Management proposed by the Presidential/Congressional Commission on Risk Assessment and Risk Management attempts to put risk management in a stakeholder-based perspective.[8] It begins with an emphasis on understanding the context of the problem, something that surely is of importance in the Arctic. The framework also includes a needed focus on presenting options to stakeholders and on evaluating the consequences of our actions. In my view this public health approach is one that provides the greatest likelihood that advances in environmental health science will appropriately be of value in meeting crucial environmental health threats that cannot effectively be confronted by command and control.

As part of this new approach, we must move from toxin-based regulatory control to instead focus on diseases and work backward to their causes. For example, we must make a major attack on the causes of asthma and of birth defects; it will not be sufficient to control only specific pollutants that might be involved in asthma causation or exacerbation or in the development of birth defects.

This risk management debate has often been conducted using political, economic, or social rhetoric. Yet the key determinant is the availability of valid and robust scientific methodology pertinent to understanding the etiology of disease. This methodology must first enable the necessary definition and surveillance of diseases caused by environmental factors. Then, it must be able to determine causes. Fortunately, the recent rapid advances in molecular biology and in other biologically relevant technologies provide the necessary tools to develop indicators of cause and effect relationships by starting first with the disease and working backward to the cause. Whether by fingerprinting specific patterns of DNA codons altered by a chemical carcinogen or by understanding the subtle alterations in the immune system underlying pulmonary asthma, our new biology provides an opportunity to understand causation.

The push beyond genomics to proteomics will allow the development of endpoints related to DNA function, not just DNA damage — also in keeping with a public health model that defines health as not just the absence of disease but the total physical and mental functioning of humans. By extension, we soon should be able to ask more intelligent questions about puzzling problems of seemingly unexplained symptoms such as those in the Gulf War veterans and those who suffer from disorders labeled as multiple chemical sensitivities, chronic fatigue syndrome, and related problems. These people may or may not have a currently definable disease, but they are certainly not healthy.

Biological markers to help define susceptible populations will greatly increase the power of epidemiology to elicit cause and effect relationships.[9] An improved mechanistic understanding of how mixtures interact will permit us to disentangle the effect of multiple agents. We will also be applying advances in biometry and biostatistics to move beyond our unidirectional approach to toxicological interactions, which can be overly simplistic. For example, mathematically more complex models will be developed to understand and predict the effects of endocrine active substances around receptor sites, to assess the effects of chemicals on the plasticity of the nervous system, and to analyze the role of immune active agents in complex sensitization and desensitization processes.

We can expect that exposure assessment will make major strides forward because biological markers will estimate dose more accurately than relatively imprecise exposure surrogates. Scientific advances should permit us to develop better indicators of human health or ecosystem effects and to work backward from these indicators to identify causative factors accurately.

All of these advances, and particularly those related to better biological markers, will have a major impact on risk assessment. Not only will we be able to define susceptible populations better and to characterize exposure and dose, biological markers will provide the opportunity to explore the lower end of the dose response curve in a way that is not possible otherwise.[10] Carboxyhemoglobin is an ideal marker as it is a biomarker of carbon monoxide exposure and a direct predictor of effect. This biomarker depends upon our mechanistic understanding of the toxicity of carbon monoxide. Similar biomarkers that allow us to relate exposure to direct predictors of effect will allow characterization of the lower end of the dose response curve that is not now possible with animal toxicology or epidemiology studies.

Of course, risk assessment has problems. Currently, it is a useful tool for secondary prevention, not primary prevention.[3] At its best, risk assessment is valuable in estimating and ranking risks from existing problems. But far better tools are needed to be able to ensure problems do not occur in the first place. If we were to depend solely upon risk assessment, we would spend our time and resources chasing after an ever increasing group of new problems caused by our rapidly transforming society. Fortunately, our new biology should provide us with better tools for primary prevention and for avoiding new problems.

III. The Gaia Hypothesis: a human health model

It is particularly important to avoid new problems because the earth's biosphere is losing its resiliency. In essence, this is the Gaia Hypothesis, which has been unfairly derided because New Age advocates have fixed on it as a mystical revelation. In fact, Lovelock's basic concept of biogeophysical feedback loops governing planetary processes has been supported by the scientific evidence.[11,12] There is no question that we live on a planet dominated by humans.[13] Accelerating human impact on the Earth has altered the basic buffering capacity of the biosphere to homeostatically respond to these feedback processes. For example, much of the primeval forest of the northeast and midwestern United States was cut down rather rapidly during the colonial and postcolonial period. Presumably, the impact on global processes of the loss of this extensive forest was far less than a similar loss today, just two centuries later, of forests in Southeast Asia or the Amazon Basin. We are already undergoing global climate changes from our present level of release of carbon dioxide, methane, CFCs, and other gases which reach the higher atmosphere — and part of the problem is the loss of arboreal buffering. Again, there are equity issues involved in the response to this threat, but respond we must. Similarly, there are long-term issues that are pertinent to the availability of usable water resources, particularly with the stress of further population growth.[14] Changes in biogeophysical feedback loops are not readily reversible. Whether there will be runaway effects with major catastrophic climate change is conjecture. But a precautionary approach is needed.

In my view, this global loss in the resiliency of the biosphere is similar to the loss in resiliency that occurs with normal aging in humans. Young individuals can wall off problems in a single organ without systemic effects. For example, a kidney infection can be treated as a local problem without there being an impact on other organs. Yet, in an elderly individual, a similar minor loss in renal function can rapidly lead to serious impairment of the heart and lungs, loss of central nervous system acuity, and a rapidly progressive cycle of multiple organ failures. Fortunately, progeria of the biosphere is reversible, but only if we take a lengthier and more holistic view of the earth and its environment.

IV. Global climate research

An obvious source of long-term concern about the earth's biosphere comes from the gathering evidence of global climate change due to human activities. It is useful to consider the role of biological research in this issue. Fortunately, there has been a recent welcome change in thrust in the US research effort on global climate changes. Until recently, the focus was almost totally on modeling the impact of global carbon dioxide emissions on global climate change with very little research on the consequences of the climate change. Preventing potentially significant impacts from global climate change is likely to require major lifestyle alterations by Americans, including responding to a substantial increase in fuel prices. This is not likely to happen in our democracy unless more research focuses on answering the "so what" questions which are to a large extent the province of biological sciences. Unless the scientific community can provide the information needed by citizens to understand the adverse impacts of our climate changing activities, it is unlikely that actions will be taken to avoid these impacts. A body of information about the human health implications of global climate change is being developed.[15, 16] Providing this information will be greatly facilitated if biological markers can be developed that are useful indicators of health and environmental impacts of global climate change.

V. Biological markers as a basis for regulatory control of toxic agents

Much of this chapter has focused on the value of biological markers. It is appropriate to question whether we can in fact justify this reliance on biomarkers as a means to move toward intelligent and sustainable stewardship of the environment. There is an interesting case study comparing asbestos and lead that demonstrates the value of biological markers to the regulatory control of environmental contaminants. During the Reagan administration nearly two decades ago, the Environmental Protection Agency (EPA) developed a cost-benefit analysis demonstrating that the economic costs of lead in gasoline outweighed its economic benefits.[17] This directly resulted in the removal of all remaining lead from gasoline. Using the same approach, the EPA then tried to ban asbestos from almost all of its uses. However, the White House Office of Management and Budget (OMB) rejected this economic analysis, stating that it would only be acceptable if the argument was as strong as the one put forth against leaded gasoline.

The lead cost-benefit analysis succeeded over that of asbestos because there was a valuable biological marker indicative of low-level lead exposure; no such marker exists for asbestos. One of the key driving forces in the control of lead has been the ability to relate relatively low levels of exposure to adverse population endpoints, such as a lowering of IQ. At the relatively low blood lead levels due solely to leaded gasoline, it is not possible to point to a specific individual and state with certainty that their IQ has been affected. However, the power provided by coupling population-based research to a valid measure of exposure has permitted the recognition of subtle IQ effects occurring at relatively low blood lead levels.[18] Studies showing such effects were responsible for the Centers for Disease Control (CDC) decreasing the blood lead level for which corrective action was required. The economic costs associated with such action, including just repeating the assay or prescribing chelation therapy, were crucial in the EPA's cost-benefit analysis that resulted in lead being removed from gasoline.

In contrast, there are no biological markers for asbestos exposure at levels below those responsible for pulmonary fibrotic changes visible on a chest x-ray. There is evidence of a loss in pulmonary function occurring in asymptomatic individuals with asbestosis of the lungs that is barely detectable on a chest x-ray.[19-20] It is conceivable that even lower levels of lung asbestos cause fibrosis that is not visible on a chest x-ray yet leads to a decrement in lung function. In fact, Miller et al.[21] reported lower than expected pulmonary function in long-term insulators with no observable radiographic evidence of asbestosis. However, it is impossible to ask this question in the general population because there is no biological marker of lower level asbestos exposure that can be quantitatively related to measurements of lung function. Without such a measure, analyses of the economic and human health impacts of a slight decline in lung function in the population of individuals exposed to environmental levels of asbestos cannot be carried out.[22]

VI. Impediments to research as a mechanism to anticipate and solve 21st century problems

The recent advances in medical and biological sciences provide much grounds for optimism. Yet there are many problems as well. Those of us in science assume that our cultural mind-set of continual scientific and technical progress is shared by almost everyone. After all, society has benefitted greatly from technology and from health research, and there have been centuries of rapid advances which have seemingly imprinted us with an implicit belief in the power of science. Yet human history contains many examples of retrogression in science and technology, such as the loss of Roman science and technology during the Middle Ages.

It is conceivable that we are today facing a major cultural change in which the value placed on scientific progress will be diminished. The battle against the intrusion of creationism into the high school curriculum has degenerated to become a battle against the exclusion of Darwinian evolution as scientific fact. Animal rights activists have slowly but surely circumscribed the ability of toxicologists to perform needed and highly ethical research. It is not unwarrantedly pessimistic to say a few decades remain for animal experimentation, particularly given the increasing political sophistication of the opposition coupled with the relative unwillingness of the scientific community to engage in the debate. Controlled human experimental studies are now under siege by activists who liken such studies to the dark deeds of Nazis for whom the Nuremberg Code was adopted.

The increasingly problematic world of human and animal research is particularly an issue for biomarker studies. The classic approach to developing and validating biological markers has been to work out the relation of the marker to exposure, effect, or susceptibility in laboratory animal models and then to explore its relevance to humans by carefully controlled experimental studies at low end exposures. These include studies in humans and in animals of toxicokinetics of minuscule doses likely to be encountered in daily activities. Such studies are needed so as to extrapolate from animals to humans effectively and to understand the basic mechanisms to predict the potential for toxicity of higher doses. Without studies of exposed humans, the classic parallelogram approach that attempts to relate biomarkers of exposure in humans to biomarkers of exposure and effect in laboratory animals is not possible.

Another potential impediment to research comes from at least some of the advocates of the Precautionary Principle. The Precautionary Principle is much in vogue today. Its interpretation has been subject to considerable debate, particularly now that it is being embodied in various laws and international treaties. There are some interpretations which consider the Precautionary Principle to be antithetical to research. This is erroneous. The Precautionary Principle cannot be invoked without there being sufficient scientific information to suggest that a threat exists. Further, two corollaries are inherent in invoking the Precautionary Principle: (1) there is some finite likelihood that the proposed precautionary action is erroneous because if there were reasonable certainty of success there would be no need to speak of precautionary action; and (2) there are significant social or economic costs because if the costs were trivial there would be no need to invoke the Precautionary Principle; we would just do it. Recognition that the more precautionary we are, the more likely we will be erroneously subjecting society to significant economic or social costs, should lead us to develop a scientific agenda to test whether any given precautionary action is a success.[23] The alternative to not testing for success is to allow costly and potentially erroneous actions to go unchecked and to also develop a false sense of security that a problem has been addressed.

VII. The sure thing

I conclude with what is the only certain prediction for environmental health in the 21st century: There will be major environmental health threats that no one now predicts.[24] This argues for a strong and flexible basic environmental health science community that can be called upon to respond rapidly with the information and understanding needed for resolution.

Acknowledgments

My deepest thanks to my colleagues at the Environmental and Occupational Health Sciences Institute whose discussions and advice have been immensely helpful. Thanks also to Betty Davis and Janet Huang for their expert technical assistance with the presentation and manuscript.

References

1. Vision of Sugarplums and Big Cars, *World Press Review*, 44, 7, 1997.
2. Pedersen, D., Disease ecology at a crossroads: man-made environments, human rights and perpetual development utopias, *Soc. Sci. Med.*, 43, 745–758, 1996.
3. Goldstein, B.D., Commentary: The need to restore the public health base for environmental control, *Am. J. Pub. Health*, 85, 481–483 1995.
4. Ruckelshaus, W.D., Stopping the pendulum, *Environ. For.*, 25–29, Nov/Dec 1995.
5. Powers, C.W. and Chertow, M.R., Industrial ecology: Overcoming policy fragmentation, in *Thinking Ecologically*, Esty, D.C. and Chertow, M.R., Eds., Yale University Press, New Haven, 1997, 19–36.
6. Sussman, R.M., EPA at the crossroads, *Environ. For.*, 14–23, Mar/Apr, 1995.
7. U.S. Environmental Protection Agency, "Reinventing Environmental Protection, Annual Report, 1998," EPA 100-R-99-002, U.S. Environmental Protection Agency, Washington, D.C., March 1999.
8. The Presidential/Congressional Commission on Risk Assessment and Risk Management, "Framework for Environmental Health Risk Management," Final Report, Vol. 1, Washington, D.C., 1997.
9. Perera, F.P., Molecular epidemiology: on the path to prevention? *J. Natl. Cancer Inst.*, 92, 602–12, 2000.
10. Goldstein, B.D., Biological markers and risk assessment, *Drug Met. Rev.*, 28(1&2), 225–233, 1996.
11. Lovelock, J.E., Taking care, in *Interpreting the Precautionary Principle*, O'Riordan, T. and Cameron, C., Eds., Earthscan Publications Ltd, London, 1994, 108–116.
12. Schneider, S.H. and Boston, P.J., Eds., *Scientists on Gaia*, MIT Press, Cambridge, MA, 1991.

13. Vitousek, P.M. et al., Human domination of Earth's ecosystems, *Science*, 277, 494–499, 1997.
14. Vosamarty, C.J. et al., Global water resources: vulnerability from climate change and population growth, *Science*, 289, 284–288, 2000.
15. McMichael, A.J. and Woodward, A., Environmental health, in *Critical Issues in Global Health*, Koop, C.E., Pearson, C.E., and Schwarz, M.R., Eds., Wiley, New York, 2000, 181–188.
16. Patz, J.A. et al., The potential health impacts of climate variability and change for the United States: executive summary of the report of the health sector of the U.S. national assessment, *Environ. Health Perspect.*, 108, 367–376, 2000.
17. Federal Register, 50 FR 9386, Regulation of fuels and fuel additives: gasoline lead content, March 7, 1985.
18. Needleman, H.L. et al., Deficits in psychologic and classroom performance of children with elevated dentine lead levels, *N. Engl. J. Med.*, 300, 689–695, 1979.
19. Rosenstock, L. et al., The relation among pulmonary function, chest roentgenographic abnormalities, and smoking status in an asbestos-exposed cohort, *Am. Rev. Respir. Dis.*, 138, 272–277, 1988.
20. Schwartz, D.A. et al., Restrictive lung function and asbestos-induced pleural fibrosis, *J. Clin. Invest.*, 91, 2685–2692, 1993.
21. Miller, A. et al., Relationship of pulmonary function to radiographic interstitial fibrosis in 2,611 long-term asbestos insulators, *Am. Rev. Respir. Dis.*, 145, 263–270, 1992.
22. Goldstein, B.D. and McMenamin, M.A., Biomarkers in cost-benefit analysis and regulatory control: lead, asbestos, carbon monoxide and benzene, in *Biomarkers: Medical and Workplace Applications*, Joseph Henry Press, National Academy of Sciences, Washington, D.C., 1998, 423–434.
23. Goldstein, B.D., The precautionary principle and scientific research are not antithetical. Editorial, *Environ. Health Perspect.*, 107, 594–595, 1999.
24. Goldstein, B.D., Environmental and occupational health, in *Critical Issues in Global Health*, Koop, C.E., Pearson, C.E., and Schwarz, M.R., Eds., Wiley, New York, 2000, 170–180.

section II

Genomics-based biomarkers/genetic toxicology biomarkers

chapter four

Ecogenetics: genetic susceptibility to environmental adversity

Daniel W. Nebert

Contents

Abstract   Individual risk of toxicity or cancer, caused by environmental pollutants, fundamentally depends on two factors. One, a sufficiently high exposure to a particular environmental agent (e.g., chemicals, heavy metals, irradiation, drugs, foodstuff), or to a complex mixture, is necessary. Two, each individual's underlying genetic predisposition plays an important role. It is now clear that the human genome varies considerably from one person to another and from one ethnic group to another. If unequivocal DNA tests for genetic susceptibility to toxicity and cancer can be successfully developed, then identification of individuals at increased risk would be helpful to the fields of public health and preventive medicine. In this chapter, we will summarize our current knowledge about human genetics and genomics and describe several examples of polymorphisms involving environmentally relevant susceptibility genes.

I. Introduction

An environmental disease reflects any disorder having an environmental component. It now seems likely, however, that all human diseases — those predominantly due to a single gene (monogenic) and those involving contributions from many genes (polygenic) — include an environmental component. Why are some individuals and some families affected more severely than others? Indeed, even within families, why are some members affected whereas others are not? When taking the same dose of a prescribed medication, why do some patients and not others show therapeutic failure or experience toxicity? Why do less than 10 of every 100 cigarette smokers die of lung cancer? The answer to each of these questions involves the combination of a dose of the environmental toxicant plus an interindividual genetic predisposition.

This chapter begins with brief definitions of genetic terms and concepts, including the recent appreciation that no human disease can really be considered monogenic. Next, we will examine seven examples of polymorphisms in environmentally relevant susceptibility genes. Lastly, we will address why these human polymorphisms might exist in the first place. Most of the references cited herein include reviews in which the reader will find numerous additional studies referenced and details described.

II. Definition of genetic terms

Humans normally have 23 pairs of chromosomes: 22 autosomal pairs plus the sex chromosomal pair (XX or XY). A gene denotes the location (stretch of DNA), along each of a chromosome pair, that encodes a gene product (enzyme or other protein); any regulatory regions upstream or downstream of the coding sequence should also be considered as part of the gene.[1] A locus indicates the location of a segment of DNA on each of a pair of chromosomes that need not necessarily code for a gene product. Diploid refers to having paired chromosomes (as found in yeast, nematodes, fruit fly, and vertebrates); haploid refers to one active chromosome of each pair (as found in the mammalian sperm and egg). In diploid mammals, each gene (or locus) is made of two alleles, one from the father and one from the mother; the combination of these two alleles is called the genotype. An allele can transmit a dominantly inherited trait (e.g., Huntington disease) or a recessive trait (e.g., phenylketonuria). Another term for trait is phenotype. Homozygous and heterozygous mean having two identical alleles, and two different alleles, respectively, at the locus under study.

Ecogenetics is defined as the study of heritable variability in response to any environmental agent (e.g., chemicals, heavy metals, ionizing radiation, drugs, and foodstuff). Pharmacogenetics is a subset of ecogenetics and is the study of differences in response to pharmaceutical agents due to heredity. The recently coined term pharmacogenomics refers to the field of new drug development based on our rapidly increasing knowledge of allelic variants of all genes contained in the human genome. However, the two terms pharmacogenetics and pharmacogenomics are often used interchangeably.

A.   Polymorphisms

A polymorphism is defined as the presence of two or more subgroups in any species population. During the past seven decades, E.B. Ford who studied insects and, more recently, Harry Harris[2] stated that a polymorphism exists when the "commonest identifiable allele has a frequency no greater than 0.99...". A polymorphic variant is a minor allele with a frequency of 0.01 or greater, and a rare variant is a minor allele having a frequency of less than 0.01. Considering the Hardy-Weinberg distribution ($p + q = 1.0$ and $p^2 + 2pq + q^2 = 1$), q represents the sum of the variant alleles. If $q = 0.01$, this means that the frequency of individuals in the population being studied who are homozygous for an autosomal recessive trait would be 0.0001, or one in 10,000.

B.   Predominantly monogenic traits

In the 1860s when Mendel described the alleles in the garden pea as colored *C* and noncolored *c*, this was the prototypical example of a simple Mendelian (single gene) trait. Allele *C* (red phenotype) was dominant to allele *c* (white phenotype). The Hardy-Weinberg distribution states that, if the allelic frequencies of *C* (= p) and *c* (= q) in the population are 0.6 and 0.4, respectively, this would mean that 16% of the population [$q^2 = (0.4)^2 = 0.16$] would have the recessive white trait, i.e. homozygous for the *cc* genotype. Crossing two *Cc* heterozygous garden peas (having the red trait) would give a 1:2:1 ratio of the *CC:Cc:cc* genotypes, but a 3:1 ratio of the red to white phenotypes. In the snapdragon, Lorenz in the 1880s discovered that the inheritance of these two colors was additive (codominant, gene-dose). In other words, crossing two *Cc* heterozygous snapdragons (pink trait) gave a 1:2:1 ratio of the *CC:Cc:cc* genotypes, and a 1:2:1 ratio of the red to pink to white phenotypes.

C.   Multiplex phenotypes

A trait that is dependent on two or more genes is called polygenic, multifactorial, or a multiplex phenotype. Examples of a polygenic trait include height, obesity, blood pressure, coronary artery disease, asthma, diabetes mellitus, or the formation of the jaw during embryonic development. Multiplex phenotypes might also include some defined toxicity of an environmental agent (at a given exposure for a defined time). It is becoming increasingly clear that there is never really a simple Mendelian (single gene) human disease.[3] Diseases that had been considered monogenic, such as Gaucher disease or phenylketonuria, are known to be affected by modifier genes andenvironmental factors. For example, the same N370S mutation in the glucocerebrosidase gene can result in the death of an 8-year-old boy with severe Gaucher disease and in a reasonably normal life for the child's 80-year-old grandfather who has only a somewhat enlarged liver and spleen. Such strikingly different Gaucher disease phenotypes strongly suggest that modifier genes and/or environmental factors contribute to what was previously taught to medical students as being a monogenic disease.

For two alleles at one locus, as described above, the ratio of genotype distribution is 1:2:1. For two alleles at two loci, this distribution becomes 1:4:6:4:1. For two alleles at three loci, this genotype distribution becomes 1:6:15:20:15:6:1. One can see how quickly the genotype (and usually also the phenotype) distribution becomes complex as the number of genes increases. The number of genes contributing to the risk of coronary artery disease is estimated to be greater than 100.[3] If toxicity to an environmental chemical involves just 5 or 10 genes rather than 100, it is easy to appreciate that in most cases one will see a gradient from those most sensitive to the chemical to those most resistant to the chemical. The same can be said for pharmacogenetic disorders, and a mathematical analysis involving extreme discordant phenotype methodology has been recently offered[4] as a statistically powerful means by which a genotype can be correlated with an unequivocal phenotype. The same should also hold true when any ecogenetic or other complex disease is studied. The outliers, or individuals at the extreme ends of the spectrum of any phenotype, are the most informative patients to scientists who wish to dissect the genes responsible for any phenotype.[4]

III. Examples of gene polymorphisms

The subject of pharmacogenetics (and, in the past several years, pharmacogenomics) has been reviewed dozens of times.[4-19] The broader subject of ecogenetics has not been specifically reviewed often;[11,20] however, many of the same gene polymorphisms affecting pharmacogenetic differences are also important in ecogenetic diseases.

Variations in ecogenetics among individuals can range from 10-fold to more than 40-fold. It is now clear that all ecogenetic differences are polygenic and multifactorial, i.e., traits are caused by at least two and usually many more than two major genes and dozens if not hundreds of modifier genes plus effects of the environment. This means that, at one time or another, perhaps every gene in the human genome is capable of qualifying as a susceptibility gene; thus, an environmental chemical or heavy metal might interact directly or indirectly with its gene product as an agonist or antagonist (for receptors, transporters, and channel proteins), or as an activator or inhibitor (for enzymes, moieties in signal transduction pathways, and transcription factors). The complete sequence of the human genome, with all its genes, should be completed during 2001. Hence, the field of ecogenomics will become increasingly rich in new cutting-edge knowledge about the effects of allelic variants of all candidate genes on an environmental chemical's kinetic and dynamic pathways, thereby leading to new promises of success in preventive toxicology and public health.

There are well over 120 pharmacogenetic disorders described in the literature.[19] Table 4.1 lists just a portion of these and illustrates one possible classification of ecogenetic differences. Roughly one-third of all ecogenetic disorders represents defects in metabolism; another one-third includes defects in receptors/transporters/channel proteins; the remaining third requires further study before they can be classified. It should also be emphasized that, contrary to the belief of some, ecogenetic differences can affect pathways (toxicodynamics) in every cell type of the body and not just the liver plus the portals of entry (skin, lung, gastrointestinal tract).

A.   N-Acetylation (NAT2)

Differences in this metabolism phenotype were first identified in the late 1940s when patients who converted to a positive tuberculin test were routinely treated with isoniazid. A high incidence of peripheral neuropathy was found among those taking isoniazid. By giving isoniazid and measuring plasma levels 6 hours later (Figure 4.1), individuals could be phenotyped as slow acetylators (*r*, clearing the drug slowly) or rapid acetylators (*R*, clearing the drug quickly). The slow phenotype is inherited as an autosomal recessive trait. The frequency of the *r* allele was found to be about 0.72 in the United States, meaning that about one in every two individuals (using the Hardy-Weinberg equation, $q^2 = 0.72 \times 0.72 = 0.518$) is homozygous for *r/r* and thus shows the slow acetylator trait.

Two human *N*-acetyltransferase functional genes (*NAT1, NAT2*) were discovered, and the *NAT2* gene was found to be responsible for the rapid and slow acetylator phenotypes. It was (somewhat arbitrarily) decided to name the reference (wild-type) allele (*NAT2*4*) as the one encoding the rapid acetylator enzyme. Three major *NAT2* slow acetylator variant alleles exist: *NAT2*5B* and *NAT2*6A* are common in Caucasians; *NAT2*6A* and *NAT2*7A* are common in Asians. There are now at least 24 other, relatively rare, *NAT2* alleles that have been discovered.[22] Large ethnic differences exist in the frequency of the rapid and slow acetylator alleles. For example, the slow acetylator homozygote frequency ranges worldwide from less than 10% in Japanese populations to more than 90% in Egyptians.[8]

Although the *N*-acetylation polymorphism represents predominantly one gene, i.e. *NAT2*, Figure 4.1 shows plasma concentrations ranging from 0.3 to 11.8 µg per ml, in other words, about a 30-fold difference between the extreme individuals although the average rapid and slow acetylators showed about 1.0 and 5.0 µg per ml, respectively. This gradient, as opposed to

Table 4.1 One Possible Classification of Human
Ecogenetic Disorders

A. Less enzyme/defective protein
1. N-acetylation (*NAT2, NAT1*)
2. Glucose-6-phosphate dehydrogenase (*G6PD*)
3. P450 monooxygenases (oxidation deficiencies) debrisoquine (*CYP2D6*), *S*-mephenytoin (*CYP2C19* & *CYP2C9*), nifedipine (*CYP3A4*), coumarin and nicotine (*CYP2A6*), theophylline (*CYP1A2*), acetaminophen (*CYP2E1*), *CYP1A1, CYP2B6*
4. Null mutants of glutathione transferase, mu class (*GSTM1*); theta class (*GSTT1*)
5. Sulfotransferases (*SULT*)
6. Thiopurine methyltransferase (*TPMT*)
7. Thiol methyltransferase (*THMT*)
8. Catechol O-methyltransferase (*COMT*)
9. Paraoxonase, sarinase deficiency (*PON1*)
10. UDP glucuronosyltransferases (Gilbert's disease, *UGT1A1*; [*S*]-oxazepam, *UGT2B7*)
11. NAD(P)H:quinone oxidoreductase (*NQO1*)
12. Microsomal, soluble epoxide hydrolases (*EPHX1, EPHX2*)
13. Aldehyde dehydrogenase (*ALDH2*)
B. Alteration in receptor, transporter or channel protein
1. Inability to taste phenylthiourea
2. Coumarin anticoagulant resistance (receptor-based?)
3. Long-QT syndrome (*KVLQT1, HERG, KCBMB1, KCNMB2, SCN5A*)
4. Malignant hyperthermia/general anesthesia (defect in Ca^{++}-release channel ryanodine receptor) (*RYR1*)
5. Cyanocobalamine (vitamin B$_{12}$ malabsorption), absence of intrinsic factor
6. β-Adrenergic receptors (*ADRB1, ADRB2, ADRB3*) and sensitivity to β-agonists in asthmatics
C. Change in response due to enzyme induction, overexpression
1. Porphyrias (*esp.* cutanea tarda)
2. Aryl hydrocarbon receptor (*AHR*) (*CYP1A1, CYP1A2, CYP1B1* inducibility) cancer, immunosuppression, birth defects, chloracne, porphyria, (?)eye toxicity, (?)ovarian toxicity
D. Abnormal metal distribution
1. Iron (hereditary hemochromatosis, *HFE*)
2. Copper (Wilson disease, Menkes disease)
3. Cadmium toxicity (*CDM*)
4. Lead toxicity and δ-aminolevulinate dehydratase (*ALAD*)
E. Disorders of unknown etiology
1. Corticosteroid (eye drops)-induced glaucoma
2. Halothane-induced hepatitis
3. Chloramphenicol-induced aplastic anemia
4. Phenytoin-induced gingival overgrowth
5. Aminoglycoside antibiotic-induced deafness
6. Methotrexate-induced toxicity in juvenile rheumatoid arthritis
7. *L*-DOPA-induced dyskinesis in Parkinson disease
8. Glucosidation of amobarbital
9. Beryllium-induced lung disease
10. Bleomycin-induced pulmonary toxicity
11. Myocardial toxicity by anthracyclines (adriamycin, doxorubicin)

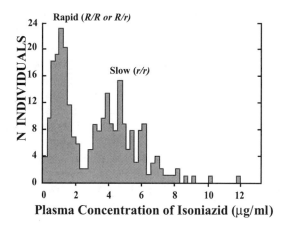

Figure 4.1 Plasma isonazid concentrations 6 hours after the drug was given. Results were obtained in 267 members of 53 complete family units. All subjects received approximately 9.8 mg isoniazid per kg body weight. (Modified from Price-Evans, D.A., Manley, K., and McKusick, V.A., *Brit. Med. J.,* 2, 485–498, 1960.)

extremely highly defined peaks and valleys, would suggest that modifier genes encoding enzymes or other proteins (or environmental factors including other drugs given to these patients) might influence plasma isoniazid concentrations.

The association of acetylation phenotypes with toxicity or cancer has received considerable attention. The slow acetylator phenotype shows a threefold lower incidence of colorectal carcinoma but a higher incidence (odds ratio = at least 16) of bladder cancer.[12] Occupational exposure to arylamines and cigarette smoking are required, however, in conjunction with the slow acetylator phenotype for bladder cancer to occur. No relationship is found between acetylator phenotype and smoking-related bladder cancer in the absence of exposure to arylamines or cigarette smoking.

B.   Glucose-6-phosphate dehydrogenase (G6PD)

Patients taking certain drugs, such as the antimalarial drug primaquine, were observed to develop painful acute hemolytic crises. This led to the discovery of low red blood cell G6PD activity associated with red cell hemolysis. G6PD activity is essential for NADPH generation and, in turn, the return of oxidized glutathione to reduced glutathione. Subsequently, it was found that about one in three Italians or Sardinians and one in ten African-Americans have the *A*-type of G6PD deficiency. Chemicals (e.g., naphthalene, methylene blue, trinitrotoluene) and more than 24 commonly prescribed drugs cause hemolytic anemia in G6PD-deficient patients.[6,12] G6PD deficiency is inherited as an X-linked recessive trait and currently affects probably more than 500 million people worldwide.

The *G6PD* gene has several hundred allelic variants, a large number of which cause one or more amino acid changes. The *G6PD* gene is located on the X chromosome, which is consistent with G6PD deficiency being transmitted as an X-linked recessive trait; this means that (similar to what is found in red-green color blindness) a carrier mother and a healthy father will have children displaying one of four possible phenotypes: a healthy female, a carrier female, a healthy male, or an afflicted male. As with the *NAT2* polymorphism, many amino acid mutations in G6PD cause lowered enzyme activity; however, in contrast to NAT2 activity, it appears that the complete absence of G6PD activity is incompatible with life. There is a more than 100-fold difference in the incidence of G6PD deficiency between Ashkenazic (0.4%) and Sephardic (53%) Jewish males.

C. *Debrisoquine/sparteine oxidation (CYP2D6)*

The debrisoquine/sparteine polymorphism reflects differences in CYP2D6 activity, a cytochrome P450. Poor metabolizers (PMs) of debrisoquine represent 6% to 10% of Caucasians, and extensive metabolizers (EMs) handle the drug 10 to more than 40 times more efficiently. PM frequencies are about 5% in African populations and less than 1% in Asians.

The PM trait represents alleles that encode a defective protein and/or incorrect splicing of the gene transcript, or even complete deletion of the gene, resulting in lowered, or completely absent, enzyme activity. An ultra-rapid metabolizer (UM) phenotype has recently been shown to be caused by amplification of the *CYP2D6* gene as many as 13 times.[7] The incidence of the UM phenotype is about 0.8% in Northern Europeans, 21% in Saudi Arabians, and 29% in Ethiopians (19);[19] the reason for these striking subethnic differences is not known.

The *CYP2D6*1* allele is the consensus sequence (wild-type, EM), and currently there are more than 70 allelic variants reported responsible for low activity and no activity as well as the extremely high activity. The most common alleles in Caucasians are the *CYP2D6*4* series (splicing defects), the *CYP2D6*5* allele (deletion of the entire gene), and the *CYP2D6*6* series (nucleotide deletions leading to codon frameshifts). All of these result in either no enzyme protein or an unstable truncated protein, i.e., no enzymic activity.[23]

CYP2D6 is now known to metabolize more than 72 drugs[19] but few environmental pollutants. Numerous epidemiological studies, associating the *CYP2D6* allelic differences with toxicity and cancer, however, have been reported. MPTP (*N*- methyl-4-phenyl-1,1,3,6- tetra-hydropyridine, formed as a byproduct during the illegal drug synthesis of a meperidine analog), and similar addictive compounds, can cause Parkinsonian tremors in humans; MPTP inhibits CYP2D6-mediated metabolism *in vitro*. Individuals having the PM phenotype also appear to have a twofold to 2.5-fold increased risk of developing Parkinson's disease. The EM phenotype appears to be correlated with an increased incidence of tumors in the bladder, liver, pharynx, and stomach, and especially in cigarette smoking-induced lung cancer. These data suggest that enhanced CYP2D6-mediated metabolism of one or more unknown dietary or other environmental agents over decades of life, i.e., formation of a reactive intermediate, might play a role in cancer initiation and/or progression in the above-named tissues.

D.   Paraoxonase (PON1)

The chemical paraoxon is the P450-mediated metabolite of parathion, an organophosphate insecticide. Paraoxon and the nerve gas sarin are substrates for PON1. Three paraoxonases (PON1, 2, 3) have also been called calcium-dependent A-esterases and are found in human plasma. For a long time, it had been believed that paraoxonase activity must exist for some reason other than detoxifying organophosphates, which were first synthesized in the 20th century. PON1 is now known to be an apoJ high-density lipoprotein (HDL)-associated enzyme that plays an important role in cardiovascular homeostasis, yet PON1 also hydrolyzes many toxic organophosphates.

Two common variant *PON1* alleles have been identified. The *PON1*2* allele (R192Q mutation) and the *PON1*3* allele (L55M) are associated with low activity of the enzyme. Frequencies of the high/high, high/low, and low/low *PON1* phenotypes are approximately 50%, 40% and 10%, respectively, in Caucasians. Five DNA sequence variants in the PON1 promoter region and four in the 3' noncoding region have recently been identified; at least some of these affect PON1 activity.

The organophosphates allegedly used as biological warfare, and pyridostigmine used as a prophylactic agent against possible nerve gas attack, have caused speculation that the *PON1* polymorphism might have some association with the Gulf War Syndrome.[19] Striking variations in illness were found in some, but not other, soldiers who fought in the 1991 war, and it is presumed that entire platoons of soldiers were equally exposed. The low-activity PON1 individual (having a slower rate of sarin hydrolysis than intermediate-activity or high-activity PON1 individuals) appears to have been more protected against neurologic symptom complexes in Gulf War veterans.[24]

E.   Aryl hydrocarbon receptor (AHR)

The AHR is a ligand-activated transcription factor that controls a number of genes, including drug-metabolizing enzymes such as CYP1A1, CYP1A2, and CYP1B1.[25] AHR ligands include dioxin, halogenated aromatic hydrocarbons, and polycyclic hydrocarbons commonly found in cigarette smoke and combustion processes. Inbred mice having a high-affinity receptor allele (*Ahr*[b1], *Ahr*[b2], *Ahr*[b3]) are more sensitive to CYP1A1/1A2/1B1 inducibility at lower doses of AHR ligands and exhibit more toxicity and cancer, compared with resistant mice having the low-affinity receptor (*Ahr*[d]).[26] More than a 12-fold difference in affinity of the AHR for dioxin, similar to that found in mice, is known to exist in human populations.[19] From numerous studies, at any given dose of an AHR ligand, the animal having the high-affinity AHR exhibits more polycyclic hydrocarbon-induced toxicity and cancer than the animal having the low-affinity AHR.[26] Thus, it seems likely that a high-affinity AHR human might develop cigarette smoke-induced lung cancer after 20 or 40 pack-years of smoking, whereas a low-affinity AHR human might never develop cancer even after more than 100 pack-years of smoking.

To date, nonsynonymous allelic variants of the human *AHR* gene include the L146P, W176R, T407I, P517S, R554K, D574N, V570I, and M786V mutations, plus the 2416delT leading to an additional 44 amino acids in the C-terminus. A small effect on AHR affinity occurs with the R554K mutation. On the other hand, the D574N mutation in the AHR affects CYP1A1 activity inducibility.[19] It is, of course, ethically difficult to determine the dioxin-inducibility phenotype in a clinical population. A convenient assay, such as one that concomitantly compares the AHR phenotype (CYP1A1 inducibility in a yeast two-hybrid system) with the *AHR* genotype (nucleotide sequencing) might prove useful for large-scale screening of human populations.[27]

F.   Porphyria cutanea tarda and CYP1A2

CYP1A2 metabolizes many environmental aromatic amine procarcinogens, including tobacco smoke-specific dibenzo[c,g]carbazole and nitrosamines such as 4-(methylnitrosamino)- 1-(3-pyridyl)-1-butanone (NNK) and numerous food-derived heterocyclic amines generated from the frying of meat and fish. There is suggestive evidence of an association between diets of fried meats and certain forms of cancer. Metabolic activation of these environmental chemicals, largely by CYP1A2, appears to be responsible for their mutagenic and tumorigenic properties. Difuranocoumarins, such as the mycotoxin aflatoxin B_1, and nitrated polycyclic aromatic hydrocarbons, such as 6-nitrochrysene and 1,3-dinitropyrene in diesel exhaust particles, are CYP1A2 substrates. Occupationally hazardous arylamines — including 4-aminobiphenyl, 2-naphthylamine, benzidine and methylene-*bis*-2-chloroaniline — are known to cause bladder tumors and are also CYP1A2 substrates. CYP1A1 and CYP1B1 metabolism in certain tissues might also contribute to the activation of some of the above-mentioned chemicals.

Endogenous substrates of CYP1A2 include estradiol and uroporphyrinogen.[19] Numerous studies have shown more than 60-fold interindividual differences in CYP1A2 protein levels and enzyme activity. Twelve *CYP1A2* mutant alleles have been described so far,[23] yet none has been experimentally proven to be responsible for these striking interindividual differences in CYP1A2 activity. Five of the variant alleles result in amino-acid changes. In conclusion, the genetic basis for the *CYP1A2* structural gene polymorphism has been extensively investigated but is not yet understood.

Essentially all types of porphyria are made worse by inducers of cytochrome P450 because the heme synthesis pathway is stimulated in order to make more heme which, when combined with the apoproteins, results in more cytochromes P450. Porphyrias can be either inherited or acquired. The

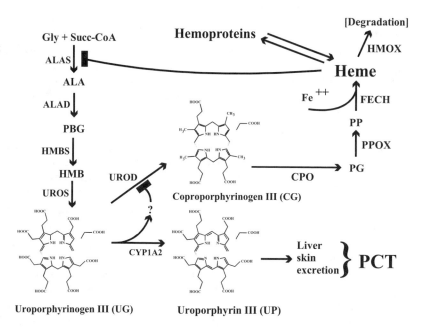

Figure 4.2 Current understanding of the heme protein biosynthetic pathway. Gly, glycine. Succ-CoA, succinyl-coenzyme A. ALAS, 5-aminolevulinic acid synthase. ALA, 5-aminolevulinic acid. ALAD, 5-aminolevulinic acid dehydratase. PBG, porphyrobilinogen. HMBS, hydroxymethylbilane synthase. HMB, hydroxymethylbilane. UROS, uroporphyrinogen synthase. UG, uroporphyrinogen. CYP1A2, cytochrome P450 1A2. UP, uroporphyrin. UROD, uroporphyrinogen decarboxylase. CG, coproporphyrinogen. CPO, coproporphyrinogen oxidase. PG, protoporphyrinogen. PPOX, protoporphyrinogen oxidase. PP, protoporphyrin IX. FECH, ferrochelatase. HMOX, heme oxygenase. Decreased levels of heme are known to stimulate ALAS activity, whereas elevated heme levels repress ALAS activity. The UROD enzyme mediates the sequential removal of four carboxylic groups of the acetic acid side chains — from UG to hepta-, hexa-, penta-, and tetra-carboxylate porphyrinogen (this latter being CG); the naturally most abundant type III isomer of UG and precursor of heme is metabolized most rapidly. (Modified from Smith, A.G. et al., *Toxicol. Appl. Pharmcol.*, 173, 89–98, 2001.)

hereditary forms of porphyria have defects (low or absent activity) in one or another of the eight enzymes of the heme biosynthetic pathway. With the exception of the first enzyme (Figure 4.2), an enzymatic defect at every step leads to tissue accumulation and excessive excretion of porphyrins and/or their precursors, such as δ-aminolevulate and porphyrobilinogen. Heme, the final product of the biosynthetic pathway, is biologically important. On the other hand, porphyrins and their precursors can sometimes be toxic.

Porphyria cutanea tarda (PCT) is the most common clinical form of porphyria. PCT can be either inherited as an autosomal dominant trait (familial or type II), or more commonly, acquired by exposure to environmental chemicals or drugs (sporadic or type I). The activity of uroporphyrinogen decarboxylase (UROD), in the heme biosynthetic pathway (Figure 4.2), is decreased more than 50% in both types of PCT, leading to the urinary excretion of uroporphyrin isomers and decarboxylated analogs as well as strikingly elevated uroporphyrin and coproporphyrin levels in the liver, skin, and feces. The onset of acquired PCT typically is either spontaneous (the etiology is unknown) or, more commonly, occurs in conjunction with known precipitating factors, especially alcohol abuse but also estrogen, drug use, viral hepatitis, or occupational exposure to halogenated environmental chemicals.

Mice exposed to polyhalogenated aromatic hydrocarbons, such as hexachlorobenzene and dioxin, and nonhalogenated polycyclic aromatic hydrocarbons such as 3-methylcholanthrene, show inhibition of UROD activity and uroporphyria. Studies with the *Cyp1a2(-/-)* knockout mouse line[29] indicate that CYP1A2 is necessary but not sufficient in causing sporadic type I environmentally caused uroporphyria. The level of CYP1A2, therefore, pushes the porphyrin pathway in favor of excess uroporphyrins, and CYP1A2 inhibits UROD and coproporphyrinogen formation by a mechanism that is not yet understood (Figure 4.2).

Acquired type I PCT has been shown to be correlated with a particular DNA sequence variant in intron 1 of the *CYP1A2* gene.[30] In workers exposed to occupationally hazardous chemicals, or environmental chemicals that induce CYP1A2 levels (such as cigarette smoke), the AH receptor-mediated control of CYP1A2 (described in the previous section) represents another important polymorphism that may affect individual risk of PCT. Thus, whether or not PCT develops appears to depend on basal and inducible levels of CYP1A2, in addition to the amount of exposure to environmental toxicants and smoking history.

G. *Cadmium toxicity and CDM*

Cadmium (Cd^{++}) is a trace metal that exists at high concentrations in cigarette smoke, contaminated fish, and food, water, and soil in certain contaminated regions around the world. Cd^{++} has been designated a Group I human carcinogen, and inhaled Cd^{++} has been linked with respiratory tumors. Epidemiological evidence suggests that Cd^{++} exposure in humans might lead to testicular tumors, renal and pulmonary toxicity, and possibly osteoporosis. Striking interindividual variations have been found among people from the same area, same age group, and presumed to be exposed to the same amounts of Cd^{++}.[31] These findings suggest allelic differences probably exist in one or more human genes involved in relative sensitivity or resistance to heavy metal toxicity.

Inbred mouse strains exhibit striking differences in sensitivity or resistance to cadmium-induced testicular damage, and the resistance phenotype is inherited as an autosomal recessive trait. By classical genetics more than 25 years ago, the *Cdm* gene was localized to a 24-centiMorgan (cM) region of DNA.[32] Recently, with the use of genomics and semiquantitative histological phenotyping,[33] the region containing the *Cdm* gene was highly refined to about 0.64 cM, *viz.* 400–800 kb. The mouse gene responsible for resistance to Cd^{++} toxicity is thus expected to be identified and characterized imminently. Although clinical toxicity to numerous heavy metals is well known, virtually no molecular mechanisms have yet been uncovered in laboratory animals or humans. Knowledge about the mouse *Cdm* gene, and the human *CDM* gene in the near future, should therefore greatly facilitate our understanding of heavy metal toxicity by identifying and characterizing, for the first time, a major mammalian gene responsible for susceptibility to diseases caused by heavy metal toxicity.

IV. Variability in the human genome

Spontaneous mutations generally occur at rates between 1 in 10^6 and 1 in 10^8 bases of DNA. To a geneticist, any time an allele is found to exist in the population of any species at frequencies of greater than 0.001, there must be a good reason for this. For example, as described earlier, the frequency of the *NAT2* slow acetylator variant alleles is about 0.96 in Egyptians, of the *G6PD* mutant alleles is about 0.53 in Sephardic Jews, and of the *CYP2D6 PM* variant alleles is about 0.28 in Northern Europeans. Why do these mutant alleles persist at such high frequencies in human populations?

The founder effect, i.e., the overpropagation of a particular allele due to a genetic bottleneck (sparsity of breeding pairs due to geography, disease, or many other kinds of environmental factors), can explain increases in the frequency of a particular allele in local populations of recent origin. Neither spontaneous mutation nor the founder effect, however, can explain the striking interindividual differences, or the geographic and ethnic differences, in allelic variants that have been described in this chapter. Possible selective pressures are thus likely to play an important role. These include diet, climate, and geography as well as balanced polymorphisms.[9,34] The other possible explanation is that many, or all, genes are highly polymorphic.[1,16] A successful approach to answering this possibility has emerged only in the last couple of years due to the efficient and accurate high-throughput resequencing of alleles of the same gene (including all introns and 5′ and 3′ flanking regions) in dozens or hundreds of individuals.

There are several classes of DNA sequence variation: (1) single nucleotide polymorphisms (SNPs); (2) insertions or deletions of sometimes a single DNA base, but other times insertions/deletions of stretches of hundreds, or thousands, of bases (arising from unequal crossing over and other problems during DNA recombination); and (3) insertions or deletions of repetitive DNA (variable number of tandem repeats [VNTRs], microsatellites or simple tandem repeats [STRs], and *Alu* I segments).

The International SNP Map Working Group has recently compiled a map of more than 1.4 million SNPs.[35] It appears that about 120,000 coding SNPs will exist in the ~31,000 estimated genes of the human genome, and 40% of these 120,000 SNPs are expected to change an amino acid (i.e. nonsynonomouse mutation). Clearly, we can see high degrees of variability and large differences in DNA sequence variant frequencies, on a gene-by-gene basis, throughout the entire human genome.[1,16]

What fraction of all human DNA sequence variants is captured in the collection by the International SNP Map Working Group? It has been estimated that every site at which mutations are compatible with life has been mutated an average of 240 times — in just the most recent generation in human history — and similar rates of mutation of course have occurred in every previous generation of *Homo sapiens sapiens*.[36] Classic neutral theory of population genetics allows us to infer from this rate (and that two haploid genomes differ, on average, at one nucleotide per 1,331 base pairs[35]) that, worldwide, we will asymptotically approach 11 million SNPs having allelic frequencies of 0.01 or higher (so-called polymorphic variants).[36] For rare polymorphic variants (frequencies of <0.01), the number of SNPs will be even greater. This explosion of knowledge in allelic variants of all environmental susceptibility genes will no doubt translate in the near future to an exciting time for ecogenetics.

Acknowledgments

I thank my colleagues for valuable discussions and a careful reading of this manuscript. The artistic help of Marian Miller with the figures is greatly appreciated. This work was funded in part by NIH Grants P30 ES06096 and R01s ES06321, ES08147, and ES10416.

References

1. Nebert, D.W., Suggestions for the nomenclature of human alleles: relevance to ecogenetics, pharmacogenetics, and molecular epidemiology, *Pharmacogenetics*, 10, 279–290, 2000.

2. Harris, H., *Principles of Human Biochemical Genetics,* 3rd ed., Elsevier/North Holland Biomedical, New York, 1980, 331.

3. Dipple, K.M. and McCabe, E.R.B., Phenotypes of patients with "simple" Mendelian disorders are complex traits: thresholds, modifiers, and systems dynamics, *Am. J. Hum. Genet.*, 66, 1729–1735, 2000.

4. Nebert, D.W., Extreme discordant phenotype methodology: an intuitive approach to clinical pharmacogenetics, *Eur. J. Pharmacol.*, 410, 107–120, 2000.

5. Vesell, E.S., Pharmacogenetics, *N. Engl. J. Med.*, 287, 904–909, 1972.

6. Price-Evans, D.W., Pharmacogenetics in *Emery and Rimoin's Principles and Practice of Medical Genetics,* 3rd ed., Harcourt Publishers, England, 1994, 455-477.

7. Meyer, U.A., The molecular basis of genetic polymorphisms of drug metabolism, *J. Pharm. Pharmacol.*, 46, 409–415, 1994.

8. Kalow, W. and Bertilsson, L., Interethnic factors affecting drug response, *Advanc. Drug Res.*, 23, 1–53, 1994.

9. Nebert, D.W., Polymorphisms in drug-metabolizing enzymes: what is their clinical relevance and why do they exist?, *Am. J. Hum. Genet.*, 60, 265–271, 1997.

10. Nebert, D.W., Pharmacogenetics: 65 candles on the cake, *Pharmacogenetics*, 7, 435–440, 1997.
11. Nebert, D.W. and Carvan, M.J., III, Ecogenetics: from ecology to health, *Toxicol. Industr. Health*, 13, 163–192, 1997.
12. Weber, W.W., *Pharmacogenetics*, Oxford University Press, New York, 1997, 1–344.
13. Caraco, Y., Genetic determinants of drug responsiveness and drug interactions, *Ther. Drug Mon.*, 20, 517–524, 1998.
14. Kleyn, P.W. and Vesell, E.S., Genetic variation as a guide to drug development, *Science*, 281, 1820–1821, 1998.
15. Evans, W.E. and Relling, M.V., Pharmacogenomics: translating functional genomics into rational therapeutics, *Science*, 286, 487–491, 1999.
16. Nebert, D.W., Pharmacogenetics and pharmacogenomics: Why is this relevant to the clinical geneticist?, *Clin. Genet.*, 56, 247–258, 1999.
17. Weinshilboum, R.M., Otterness, D.M., and Szumlanski, C.L., Methylation pharmacogenetics: catechol *O*-methyltransferase, thiopurine methyltransferase, and histamine *N*-methyltransferase, *Annu. Rev. Pharmacol. Toxicol.*, 39, 19–52, 1999.
18. McLeod, H.L. and Evans, W.E., Pharmacogenomics: unlocking the human genome for better drug therapy, *Annu. Rev. Pharmacol. Toxicol.*, 41, 101–121, 2001.
19. Nebert, D.W. and Jorge-Nebert, L.F., Pharmacogenetics and pharmacogenomics, in *Emery & Rimoin's Principles and Practice of Medical Genetics*, 4th ed., Harcourt Brace, Edinburgh, 2002, 590–631.
20. Costa, L.G., The emerging field of ecogenetics, *Neurotoxicology*, 21, 85–89, 2000.
21. Price-Evans, D.A., Manley, K., and McKusick, V.A., Genetic control of isoniazid metabolism in man. *Brit. Med. J.*, 2, 485–498, 1960.
22. Hein, D.W., Grant, D.M., and Sim, E., Arylamine N-acetyltransferase (EC 2.3.1.5), Web site http://www.louisville.edu/medschool/pharmacology/NAT.html/, 2002.
23. Oscarson, M. et al., Human cytochrome *P450* (*CYP*) alleles, Web site http://www.imm.ki.se/CYPalleles/, 2002.
24. Furlong, C.E., *PON1* status and neurologic symptom complexes in Gulf War veterans, *Genome Res.*, 10, 153–155, 2000.
25. Nebert, D.W. et al., Role of the aromatic hydrocarbon receptor and [*Ah*] gene battery in the oxidative stress response, cell cycle control, and apoptosis, *Biochem. Pharmacol.*, 59, 65–85, 2000.
26. Nebert, D.W., The *Ah* locus: Genetic differences in toxicity, cancer, mutation and birth defects, *CRC Crit. Rev. Toxicol.*, 20, 153–174, 1989.
27. Maier, A. et al., Aromatic hydrocarbon receptor polymorphism: Development of new methods to correlate genotype with phenotype, *Environ. Health Perspect.*, 106, 421–426, 1998.
28. Smith, A.G. et al., Protection of the *Cyp1a2(-/-)* null mouse against uroporphyria and hepatic injury following dioxin exposure, *Toxicol. Appl. Pharmacol.*, 173, 89–98, 2001.
29. Sinclair, P.R. et al., Uroporphyria produced in mice by iron and 5-aminolevulinic acid does not occur in *Cyp1a2(-/-)* null mutant mice, *Biochem. J.*, 330, 149–153, 1998.
30. Christiansen, L. et al., Association between *CYP1A2* polymorphism and susceptibility to porphyria cutanea tarda, *Hum. Genet.*, 107, 612–614, 2000.

31. Elinder, C.G. et al., β_2-microglobulinuria among workers previously exposed to cadmium: follow-up and dose-response analyses, *Am. J. Industr. Med.*, 8, 553–564, 1985.

32. Taylor, B.A., Linkage of the cadmium resistance locus to loci on mouse chromosome 12, *J. Hered.*, 67, 389–390, 1976.

33. Dalton, T.P. et al., Refining the mouse chromosomal location of *Cdm*, the major gene associated with susceptibility to cadmium-induced testicular necrosis, *Pharmacogenetics*, 10, 141–151, 2000.

34. Nebert, D.W. and Dieter, M.Z., The evolution of drug metabolism, *Pharmacology*, 61, 124–135, 2000.

35. The International SNP Map Working Group. A map of human genome sequence variation containing 1.42 million single-nucleotide polymorphisms, *Nature*, 409, 928–933, 2001.

36. Kryglyak, L. and Nickerson, D.A., Variation is the spice of life, *Nature Genet.*, 27, 234–236, 2001.

chapter five

Toxicogenomics and pharmacogenomics: basic principles, potential applications, and issues

F. Peter Guengerich

Contents

Abstract Genomics and proteomics are terms used to describe the large-scale application of knowledge of genomes and proteins to problems in health issues, as well as other issues in science. Pharmacogenomics is the most developed field because of the applications with specific drugs and defined biological endpoints. Toxicogenomics involves the prediction and understanding of adverse biological effects produced by chemicals in experimental settings, and searches are underway to better define targets and intermediate responses. Environmental genomics is a more difficult area of study in the sense that issues of human health are dealt with in situations where exposures are difficult to control and the genes and pathways may be largely unknown. Nevertheless, the history of success with inherited diseases, the background of experimental animal research, and early successes with pharmacogenomics provide optimism that genomic approaches will be useful in environmental health sciences.

I. Principles and opportunities in genomics

The prospect of the availability of the sequence of the human genome has changed the approaches to many areas of science and application to practical problems. At the time of this writing, a nearly complete sequence was available. The traditional paradigm for biochemical medicine involved isolating enzymes or other proteins on the basis of a function (that can be assayed), determining part of the primary sequence of amino acids in the protein, using probes to isolate cDNAs, and then using probes to prepare genomic clones. Today, the opposite direction is possible and being exploited, i.e., using human sequence information to identify genes and putative coding sequences, to express proteins in heterologous systems, and then to characterize functions. Other work is being done on measuring changes in the transcription of genes under various conditions and the effects of sequence variations (polymorphism) on function (i.e., relationship of phenotype to genotype). This latter aspect is usually dealt with under the definition of genetics and in this sense would be considered a subset of genetics. Conversely, some aspects of what has been presented here under the guise of genomics can be considered a subset of genomics. The term proteomics is usually applied to other studies of changes in proteins, pre- and posttranslational.

Whatever the terms used, genomics has already changed the landscape in many areas of medicine and applied science. Pharmacogenomics is a term used for dealing with the discovery, pharmacological activity, and disposition of drugs. Toxicogenomics is a term applied to the study of the toxic actions of chemicals, whether drugs or other compounds. Environmental genomics is a term relevant to this discussion and is essentially toxicogenomics of physical and chemical agents to which humans are exposed in the environment (as opposed to pharmaceuticals). The term ecogenomics has also appeared but is somewhat confusing in the sense that ecology is a term usually applied to studies of ecosystems without reference to human health; thus, ecogenomics or ecogenetics is usually applied to genetic studies of wildlife, etc.

The technology used in genomic applications is changing rapidly and will not be discussed in any detail within this article. The reader is referred to articles on high-throughput screening and genomics.[1-9]

II. Pharmacogenomics: applications

Pharmacogenomics will be discussed because this area, although still in its infancy, is more advanced than toxicogenomics and environmental genomics. A number of useful applications have already been made.

One major development has been the availability of the genomic clones of pathological microorganisms.[10] Libraries have been prepared in which individual genes are deleted, and these can be screened in order to determine which are essential for life or virulence. Screening can then be done with large combinatorial or other chemical libraries to determine which inhibit the transcription of the essential genes or which inhibit the function of the product of the gene (protein). Chemicals that inhibit either can be developed as lead anti-infective agents.

Another use of the human genome sequence information base available today has been the identification of more genes with gene families. For instance, if a characteristic signature sequence exists, it is possible to identify relatives and characterize their function, whether it be related to pharmaco-dynamics or pharmacokinetics. An example is the use of database searches to find a new glutathione transferase.[11]

Another approach that will be used increasingly is the search for human orthologs of genes for which some function has been characterized in model organisms. This approach has already been utilized in finding human DNA repair genes[12-14] and a previously unrecognized glutathione transferase.[15] This strategy will find increasing use in the identification of human orthologs of cell cycle control and signal transduction genes discovered in lower organisms in the hope that some of these can be exploited as drug targets.

Another important aspect of pharmacogenomics is polymorphism. Polymorphism is defined as variability in a population, usually to the extent of > 1%. Polymorphisms in the human genome occur every few hundred bases, on average. Genetic variability is the result of gene polymorphisms that yield functional differences in the proteins they produce. The current collection of polymorphism data is still limited, in that most of the complete human genome sequence has been obtained from a small number of individuals. However, there is great interest in accumulating more information, more polymorphism data, particularly single nucleotide polymorphisms (SNPs). SNPs in drug targets can control the pharmacodynamics of how well different drugs work for different individuals.[16] SNPs in enzymes can strongly influence interindividual differences in the metabolism of and changes from drugs.[17] Appropriate exploitation of genomic differences has the potential to yield drugs that have better efficacy in individuals that fail to respond to some, as well as to minimize adverse drug-drug reactions.

Studies in pharmacogenomics have provided insight into what will be limitations and needs, and these points also apply to toxicogenomics and environmental genomics. The first issue is that we only know the functions of a relatively small fraction of genes. Even in the microbial models *Escherichia coli* and *Saccharomyces cerevisiae*, we only know the functions of less than one-half of the genes. More knowledge about the remainder of the genes' functions will be needed. The fraction of human genes with defined functions is even lower. A second general issue is that predicting enzyme function on the basis of sequence similarity is important. In several cases, similar protein folds are used and a chemical step such as proton abstraction may trigger any of several reactions. An example is the gene identified as glutathione transferase zeta (which conjugates haloacetic acids) and as maleylacetoacetic acid isomerase (a step in tyrosine degradation).[15,18]

A third problem involves polymorphisms and is related to the second problem. An SNP in the coding region of a gene may yield a functional difference in one activity but not another. The overall fraction of SNPs that have functional consequences is low, estimated to be as low as 10%.[19] As an example of the problem, the I359L polymorphism in cytochrome P450 2C9 affects its oxidation of the drugs warfarin[17] and losartan but not diclofenac. Thus, the effects of coding region SNPs must be considered for each protein ligand pair.

Other current issues involve how to find more SNPs, particularly SNPs that control function, and how to screen them faster. Technology has developed rapidly in these areas, particularly the latter.[20] One of the projected goals of pharmacogenomics is the tailoring of prescriptions to individuals based on their genomes.

III. Toxicogenomics: approaches

Currently, most of the interest in toxicogenomics is centered in two areas. One is the identification of SNPs associated with differences in toxic responses. This aspect has been considered with adverse effects of pharmaceuticals, and patients in clinical trials can be stratified on this basis as well as other pharmacokinetic and pharmacodynamic parameters.

The other major effort in toxicogenomics is on the utilization of gene expression patterns to predict toxicity. Most of the studies have been using cultured human cells and rodents (*in vivo*) to establish patterns of responses in the expression of different genes to chemicals with known toxicities. The basis for the approach is that most toxicities are believed to involve either changes in gene expression as a step in producing the toxic response or involve changes in gene expression that follow a critical toxic response. The major assay involves measurement of levels of individual mRNAs following an exposure using chip technology with oligonucleotides or cDNAs attached to a chip for hybridization analysis. Three approaches have been used, in terms of the number of genes used per chip: A relatively small number (~300) of probes to genes for which a literature basis for involvement has been proposed (e.g., the Phase One Molecular Toxicology Company); a larger number (2400) of genes with some suggestion of a potential role in toxicity (e.g., NIEHS Chip Version 1.0)[21]; and a large number of genes (10^4) many of unknown function (e.g., Incyte). The chips with the larger numbers of probes provide more information and pose greater challenges in bioinformatics. They also offer the potential of establishing new networks in the event of toxicity, an area that has been difficult to understand.

Measurement of mRNA levels in humans will be largely restricted to accessible tissues. Studies with skin and blood can readily be done, but validation will be needed when surrogates are used for other tissues such as liver and kidney. In the pharmaceutical industry, safety assessment has become a limiting step and the use of toxicogenomic assays with rodent and cell culture models will grow considerably.

IV. Environmental genomics: applications and issues

The basic techniques used in environmental genomics have already been mentioned under pharmacogenomics. In the case of environmental genomics, the studies are more complex for several reasons. The association of a single chemical and a disease response may not be well-established. Even if it is, the genes involved have probably not been defined well. The latter problem is manifested in cancer, a multifactorial disease often of poorly established etiology.

The general approach may be summarized as follows. A set of genes is identified for study, either on the basis of literature precedent for involvement in a disease or studies on toxicogenomic responses to a candidate chemical (physical agent) or mixture suspected to be associated with a disease. All polymorphisms in these genes are identified. Two arms of the approach now follow. One involves assays of the functions of the allelic variants in an appropriate system (this is essentially biochemistry). The other arm involves efforts to associate disease incidence in people directly to genomic differences (epidemiology). Both approaches have advantages and disadvantages.

The complexity of environmental genomics is much greater than pharmacogenomics, in which large numbers of subjects (thousands) can be administered defined doses of a single pure chemical and a relatively small number of parameters can be measured for association with SNP patterns.

The difficulty of establishing relationships between genetics and environmental health problems is far more challenging than doing work on pharmacogenomics and toxicogenomics with individual drugs and relatively well-defined outcomes, for a number of reasons.[22] *In vitro* work is complex because of the question of whether the assay is predictive, particularly regarding target organs. If the variability in expression of a gene has a nongenetic influence, the question arises as to what the level of expression was at the onset of disease (e.g., initiation of a tumor). Many environmental diseases are linked to complex mixtures (e.g., cigarette smoke, smog), and the precise etiology is uncertain. This issue of mixtures is particularly problematic for designing experimental exposures, whether *in vitro* or *in vivo*. Most pathways of metabolism of xenobiotic chemicals involve several enzymes, activating and detoxicating, so that multifactoral approaches need to be considered. In terms of epidemiology, the sample sizes are usually relatively small, particularly with some exposures to particular agents. Establishing the actual doses is also an issue in epidemiology, particularly with complex mixtures. Racial influences can also affect the sampling of genes or combinations of genes in terms of the sample size. Diet is also a complicating issue in the epidemiology.

Nevertheless, there is a good reason to pursue environmental genomics and that some important answers to health issues will result.

First of all, there is a long history of progress on the genetic basis of inherited diseases in terms of severe maladies.[23]

Second, numerous examples of interactions of genetic and environmental factors exist. Examples include the resistance of individuals with sickle cell hemoglobin trait to malaria, the resistance of individuals with a lack of *CCR5* gene function to human immunodeficiency virus, and the genetic influence (genes not well characterized) on development of lung cancer from smoking. Another example presented by Yokoyama et al.[24] deals with aldehyde dehydrogenase (ALDH2), ethanol, and the risk of head and neck cancer. Homozygous poor metabolizer (PM) individuals do not consume ethanol because of the immediate reaction and have a much lower risk than homozygous extensive metabolizer (EM) individuals. Heterozygous individuals can consume ethanol but do not metabolize acetaldehyde well; they have a 6-fold to 12-fold greater risk of cancer than homozygous EM individuals.

Third, many experimental animal models for the interaction of genetic and environmental factors are known. These include naturally discovered genetic variants[25] and mice with gene knockouts.[26-29] In many different cases, the deficient animals have rather normal physiology in the absence of environmental stresses but differ markedly in terms of their risk of toxicity or cancer when exposed to chemicals or physical agents.

Finally, in the area of genes coding for proteins involved in the metabolism of xenobiotics, we already have ample evidence that polymorphisms exist and can have major effects with pharmaceuticals. The differences in pharmacokinetics can have striking differences on the pharmacological responses.[17,30]

V. Conclusions

The use of pharmacogenomics to help develop better drugs and use them more effectively has already begun. However, the field has just begun and we will see many changes in the next few years. Toxicogenomics, the study of the relationships between genes and toxic effects of chemicals, has begun to develop and is largely limited to work with pure drugs and other chemicals and experimental model systems. Research in environmental genomics is based on principles of pharmacogenomics and toxicogenomics. The area has the potential to yield a better understanding of genetic influences on disease. However, the situation is necessarily more complex than that encountered with drugs, and genomics experience, innovative new approaches, and intelligent interpretation will be needed in this field.

References

1. Jones, D.A. and Fitzpatrick, F.A., Genomics and the discovery of new drug targets, *Curr. Opin. Chem. Biol.*, 3, 71–76, 1999.
2. Cheung, V.G. et al., Making and reading microarrays, *Nature Genet.*, 21, 15–19, 1999.
3. Service, R.F., Coming soon: the pocket DNA sequencer, *Science*, 282, 399–401, 1998.
4. Sittampalam, G.G., Kahl, S.D., and Jansen, W.P., High-throughput screening: advances in assay technologies, *Curr. Opin. Chem. Biol.*, 1, 384–391, 1997.
5. Veber, D.F., Drake, F.H., and Gowen, M., The new partnership of genomics and chemistry for accelerated drug development, *Curr. Opin. Chem. Biol.*, 1, 151–156, 1997.
6. Gerhold, D., Rushmore, T., and Caskey, C.T., DNA chips: promising toys have become powerful tools, *Trends Biochem. Sci.*, 24, 168–173, 1999.
7. Braxton, S. and Bedilion, T., The integration of microarray information in the drug development process, *Curr. Opin. Biotechnol.*, 8, 643–649, 1998.
8. Tarbit, M.H. and Berman, J., High-throughput approaches for evaluating absorption, distribution, metabolism, and excretion properties of lead compounds, *Curr. Opin. Chem. Biol.*, 2, 411–416, 1998.
9. Hajduk, P.J. et al., High-throughput nuclear magnetic resonance-based screening, *J. Med. Chem.*, 42, 2315–2317, 1999.
10. Allsop, A.E., Bacterial genome sequencing and drug discovery, *Curr. Opin. Biotechnol.*, 8, 637–642, 1998.
11. Liu, S., Stoesz, S.P., and Pickett, C.B., Identification of a novel human glutathione *S*-transferase using bioinformatics, *Arch. Biochem. Biophys.*, 352, 306–313, 1998.
12. Modrich, P., Mismatch repair, genetic stability, and cancer, *Science*, 266, 1959–1960, 1994.
13. Jiricny, J., Colon cancer and DNA repair: have mismatches met their match?, *Trends Genet.*, 10, 164–168, 1994.
14. Jiricny, J., Mismatch repair and cancer, *Cancer Surv.*, 28, 47–68, 1996.
15. Board, P.G. et al., Zeta, a novel class of glutathione transferases in a range of species from plants to humans, *Biochem. J.*, 328, 929–935, 1997.
16. Liggett, S.B., Pharmacogenetics of relevant targets in asthma, *Clin. Exp. Allergy*, Suppl.1, 77–79, 1998.
17. Steward, D.J. et al., Genetic association between sensitivity to warfarin and expression of *CYP2C9*3*, *Pharmacogenetics*, 7, 361–367, 1997.
18. Tong, Z., Board, P.G., and Anders, M.W., Glutathione transferase zeta-catalyzed biotransformation of dichloroacetic acid and other α-haloacids, *Chem. Res. Toxicol.*, 11, 1332–1338, 1998.
19. Shen, M.R., Jones, I.M., and Mohrenweiser, H., Nonconservative amino acid substitution variants exist at polymorphic frequency in DNA repair genes in healthy humans, *Cancer Res.*, 58, 604–608, 1998.
20. Picoult-Newberg, L. et al., Mining SNPs from EST databases, *Genome Res.*, 9, 167–174, 1999.
21. Nuwaysir, E.F. et al., Microarrays and toxicology: The advent of toxicogenomics, *Mol. Carcinogen.*, 24, 153–159, 1999.
22. Guengerich, F.P., The environmental genome project: Functional analysis of polymorphisms, *Environ. Health Perspect.*, 106, 365–368, 1998.

23. Scriver, C.R. et al., Eds., *The Metabolic and Inherited Bases of Inherited Disease*, McGraw-Hill, New York, 1995.
24. Yokoyama, A. et al., Multiple primary esophageal and concurrent upper aerodigestive tract cancer and the aldehyde dehydrogenase-2 genotype of Japanese alcoholics, *Cancer*, 77, 1896–1890, 1996.
25. Nebert, D.W., The *Ah* locus: Genetic differences in toxicity, cancer, mutation, and birth defects, *Crit. Rev. Toxicol.*, 20, 153–174, 1989.
26. Lee, S.S.T. et al., Role of CYP2E1 in the hepatotoxicity of acetaminophen., *J. Biol. Chem.*, 271, 12063–12067, 1996.
27. Valentine, J.L. et al., Reduction of benzene metabolism and toxicity in mice that lack CYP2E1 expression, *Toxicol. Appl. Pharmacol.*, 141, 205–213, 1996.
28. Radjendirane, V. et al., Disruption of the DT diaphorase (NQ01) gene in mice leads to increased menadione toxicity, *J. Biol. Chem.*, 273, 7382–7389, 1998.
29. Liang, H.C.L. et al., *Cyp1a2*(-/-) null mutant mice develop normally but show deficient drug metabolism, *Proc. Natl. Acad. Sci. USA*, 93, 1671–1676, 1996.
30. Kivistö, K.T., Neuvonen, P.J., and Klotz, U., Inhibition of terfenadine metabolism: Pharmacokinetic and pharmacodynamic consequences, *Clin. Pharmacokinet.*, 27, 1–5, 1994.

chapter six

Individual susceptibility to exposures: a role for genetic variation in DNA repair genes

Harvey W. Mohrenweiser

Contents

Abstract   Cells are continuously exposed to endogenous and exogenous agents capable of inflicting damage to DNA. To minimize the consequences of this damage, the lesions must be efficiently and accurately repaired prior to cell division, as unrepaired lesions transmitted to daughter cells can result in mutations capable of generating precancerous cells. A small number of individuals with different genetic diseases have only limited ability to repair one or more classes of DNA lesions and are at high risk of cancer. Cumulative data suggest that as many as 10% of the individuals in the population have a marginally reduced capacity to repair at least one class of DNA damage; these individuals are at modestly elevated risk of cancer. Over 50 of the more than 100 currently recognized DNA repair and repair related genes have been systematically screened for sequence variation in efforts from several laboratories. Most of the data are from screening a set of 50–100 unrelated individuals selected to be a sampling of the U.S. population. Over 170 different amino acid substitution variants have been identified in these screening efforts. Initial reports have associated several variants with variation in repair capacity, biomarkers of exposure and cancer risk.

I. DNA *damage*

DNA is routinely assaulted by intracellular and environmental agents that cause a wide range of damage.[1-3] The major intracellular agents are reactive oxygen species (e.g., superoxide and hydroxyl radicals) generated during cellular metabolism. These radicals attack DNA, producing modified bases, oxidized apurinic sites, and strand breaks. In the absence of exogenous mutagen exposure, as many as 750 thymidine glycol lesions and 1000 8-oxo-deoxyguanosine lesions are generated per cell per day.[2] The classes of damage caused by these endogenous agents are also observed in cells exposed to free radical generating agents such as anticancer drugs and radiomimetic antibiotics.[3,4] DNA strand breaks arise spontaneously during the course of DNA replication and are another class of potentially harmful damage.[5] Ionizing radiation exposure induces strand breaks. Ionizing radiation exposure also results in generation of oxygen radicals and elevated oxidative damage. One sievert of ionizing radiation is estimated to induce approximately 1000 single strand breaks, a similar number of damaged bases, and 40 Double Strand Breaks (DSBs) per cell.[6,7] Similarly, a range of covalently attached DNA adducts can be detected in cells exposed to different chemical mutagens from environmental or lifestyle exposures and pharmaceutical agents.[8-10] Many of these same DNA adducts can be identified in normal tissues from nominally unexposed individuals.[11,12] Thus, even in the absence of obvious exposures, cells must constantly deal with the generation of a wide range of premutagenic lesions in their DNA.

The level of DNA adducts are directly related to endogenous plus exogenous exposure, but the quantitative relationship is complex as large differences in the levels of adducts are noted among individuals with similar exposure histories.[13-16] Chromosomal translocations and gene mutations such as at the *HPRT* locus are monitors of damage to the genome and more indirect measures of exposure. These biomarkers of exposure have the advantage of integrating the accumulation of DNA damage over time. Again large differences in response are noted among individuals with similar exposures.[17-21] Although these differences could be associated with errors in the estimates of exposure, it has been documented that some of the variation is associated with biological differences among individuals. For example, genetic differences in the ability of cells to activate and/or detoxify mutagenic agents metabolically explain some of the variation observed among individuals with similar exposures.[15,22-27]

II. Repair of DNA damage

To counter the impact of this extensive damage to the DNA that is constantly occurring in each cell, organisms have developed several, generally nonredundant, pathways for repairing different DNA lesions. Nucleotide excision repair (NER) removes photoproducts induced in DNA by exposure to UV radiation and also the bulky DNA adducts formed following exposure to a multitude of chemicals.[28-32] The NER pathway involves more than 30 proteins, at least 13 of which form a single multiprotein complex. Two DSB repair pathways repair damage produced directly by exposure to ionizing radiation or indirectly by incomplete repair of single strand damage.[33,34] Homologous recombination repair (HR) relies on extensive nucleotide sequence complementarity between the intact homologous partner chromosome or sister chromatid and the damaged structure for strand exchange.[35] Non-homologous end joining repair (NHEJ) requires little or no sequence homology and is mediated by direct end-joining.[36] The more than 20 genes of the HR repair pathway are most active in repairing DSBs in cells in S and G2 phase. At least 15 genes are employed, primarily by noncycling cells, to repair DSBs via the more error prone NHEJ mechanism. Base excision repair (BER) involves at least 25 genes encoding proteins with functions in the direct processing of damaged DNA via two largely nonredundant BER pathways. The BER pathways repair damaged bases (oxidized bases are the most common class of DNA damage) and sites of base loss that arise spontaneously or result from the attack of DNA by free radicals generated by exposure to ionizing radiation or reactive chemicals as well as endogenous metabolism.[37-39] The choice of the "short patch" or "long patch" pathway appears to be initiated by different types of DNA base damage.[39] A fourth pathway, mismatch repair (MMR), repairs replication errors such as base mismatches and small loop-outs that arise during replication by misincorporation of nucleotides or slippage on the template strand and has only a limited role in repair of induced DNA damage.[40,41]

It is critical that DNA lesions be repaired prior to replication and cell division to prevent transfer of genetic damage and mutations to daughter cells. Thus, the DNA damage recognition checkpoint (DRC) genes (more than 30 currently identified) have important, although more indirect, roles in the repair of DNA damage by directing cell responses to limit the consequences of induced DNA damage.[42] Genes with roles in the DRC pathway include the well-characterized cancer genes, *ATM, p53, BRCA1,* and *BRCA2.*[43-46]

Over 120 genes encode proteins with roles in one or more of these pathways.[47,48] Roles for additional genes in the repair of damaged DNA continue to be identified. These pathways involve the interaction and sequential activity of multiple proteins. All of the proteins in these pathways must function efficiently to repair DNA damage accurately prior to replication in order to ameliorate the consequences of exposure. [43,46,49-53]

III. Variation in DNA repair capacity

DNA repair systems are responsible for maintaining the integrity of the genome by minimizing replication errors, removing DNA damage and minimizing deleterious rearrangements arising via aberrant recombination. The important role of DNA repair in maintenance of genomic integrity is most obvious in cancer families, where the presence of highly penetrant variant alleles of several genes are associated with a high risk of cancer. A classic example is xeroderma pigmentosum (XP), a prototypic cancer gene syndrome associated with an extreme risk of developing UV-induced skin cancers as the result of the loss of function of one of the genes of NER.[49,51] Ataxia telangiectasia, Bloom's Syndrome, Fanconi's anemia, and Nijmegen Breakage Syndromes are other examples of human genetic diseases associated with reduced DNA repair and increased cancer proneness.[1] Similarly, the high cancer risks associated with inherited loss-of-function variants of *p53, BRCA1,* and *BRCA2*[43,46] emphasize the important role of recognition of DNA damage and cell cycle delay in cancer risk. These rare familial cancers have been important models for expanding our understanding of carcinogenic processes although the disease associated alleles at these loci account for no more than 5% of the cancer cases in the population.

In noncancer families, the existence of first degree relatives with cancer constitutes a significant risk factor for an individual.[54-57] This is taken as evidence for a role for genetic variation in individual susceptibility in the 95% of the cancer cases that are sporadic. The role of genetic variation in cancer risk has also been directly demonstrated in molecular epidemiology studies, especially in studies of genes responsible for the metabolism and detoxification of chemical carcinogens [58-62.] The most significant associations with cancer incidence were noted in studies where the level of exposure to carcinogen was modest and alleles at more than a single locus in a pathway or process were included as potential risk factors.[59,60] Recent discussions have focussed on the relative contributions of genetics and exposure to cancer incidence. Some authors have suggested that "genetic factors make minor contributions to susceptibility,"[63-66] while others suggest that the same data "are still consistent with genes contributing high attributable risk."[67]

Direct evidence for a genetic basis for more subtle variation in capacity to repair damaged DNA has accumulated in the last several years. Cell-based assays of capacity to repair DNA damage induced by different agents have revealed considerable interindividual variation. DNA repair capacity that is only 60–80% of the population mean is observed in at least 10% of the individuals in the population.[68] Repair capacity in most studies has been determined by counting the number of chromatid breaks that remain in metaphase chromosomes of lymphocytes exposed *in vitro* to DNA damaging agents.[69,70] A micronucleous-based assay has also been employed to monitor residual chromosomal damage.[69] Another assay, the Host Reactivation Assay, measures the ability of lymphocytes to remove either UV-induced dimers or bulky DNA adducts from plasmid DNA transfected into cells.[71,73] More recently, electrophoretic assays have been used to monitor the ability of lymphocytes to rejoin DNA breaks induced by exposure of cells to DNA damaging agents.[74–76]

The reduced repair capacity phenotypes for damage induced by gamma radiation, bleomycin (a radiomimetic agent), and benzo[a]pyrene-diol epoxide (BPDE) behave as independent traits.[77] This is consistent with the expectation that damage induced by bleomycin and ionizing radiation is primarily strand breaks and oxidized bases, repaired by genes in the NHEJ, HR, and BER pathways, while BPDE induces bulky adducts (like UV-induced pyrimidine dimers) that are repaired by the NER pathway. Scott and colleagues have found decreased repair capacity for ionizing radiation induced damage in G0 cells from breast cancer cases using a micronucleus based lymphocyte assay[69,78,79] as well as reduced repair capacity in G2 cells from other individuals.[69,79] The independence of the G0 and G2 repair capacities was interpreted as indicative of the two assays measuring independent mechanisms or pathways, e.g., G0 cells primarily using genes in BER and NHEJ pathways, G2 cells using HR and DRC pathways.[69,80,81] The genetic contribution to the inter-individual differences in repair capacity for bleomycin, ionizing radiation and BPDE induced damage are estimated to range from 0.65–0.80.[82–84] Repair capacity is reduced threefold in G2 cells from AT patients and twofold in AT heterozygotes compared to controls with respect to repair of ionizing radiation induced chromatid breaks.[85] Similar reduction in repair capacity is observed in cells with defective BRCA1, emphasizing the role of DRCs.[86]

IV. Repair capacity and cancer risk

Epidemiology studies have demonstrated that a reduced repair capacity phenotype is associated with an increased risk of developing a range of different tumors, including breast, lung, skin, liver or head/neck (see Berwick and Vineis[87] and references cited therein). The elevated risk is associated with odds ratios generally ranging from 2–10 but tending to cluster around 4 and 5. Another pertinent finding is the association of the reduced repair capacity phenotype with the risk of second tumors.[88,89] Given that each of the reduced capacity phenotypes exists at an incidence of at least 10%, it would be expected that at least 1% of the individuals in the population will exhibit a reduced capacity in two repair pathways. Such individuals, with reduced capacity to repair both bleomycin and BPDE induced damage, are observed. These individuals exhibit a higher risk (OR > 30) of developing lung cancer[77] and hepatocellular carcinoma[90] than individuals with a reduced capacity in only one pathway (OR 5–7). Thus, the evidence supports the conclusion that DNA repair capacity is a highly heritable trait that is related to individual cancer risk.

V. Identification of sequence variation in human DNA repair genes

A. General strategies

The availability of the sequence for much of the human genome has made it feasible to initiate large-scale systemic screens to identify common genetic variants in the human population. Recent efforts have focused on identification of single nucleotide polymorphisms (SNPs) or substitutions as they are the most common class of variant. Two strategies have been utilized in screening for SNPs. The "whole-genome scan" strategy for identification of SNPs involves resequencing of random DNA fragments in a relatively small sample of individuals.[91] The SNP Consortium has identified over 1.4 million SNPs in such a genomewide effort.[92] This resource is useful for building high-density genetic maps. The vast majority of these SNPs are located in noncoding regions of the genome; thus, only a limited number of these variants are likely to have an impact on protein expression or function. A second strategy utilizes direct resequencing of DNA to identify SNPs but focuses on screening the coding and regulatory regions of genes selected because of their potential relevance to common diseases.[93–97] The DNA samples screened for variation in both of these strategies are usually obtained from less than 100 unrelated and normally healthy individuals, These individuals are selected to be representative of the general population as the major interest in these screens is identification of common polymorphisms. These population-based screening strategies are quite different from the efforts to identify the variant(s) segregating in a "disease family," where the focus is screening candidate genes in previously localized chromosomal region and the study of related individuals.

B. Variation in DNA repair genes

In our laboratory at Lawrence Livermore National Laboratory, we have initiated a systematic screen for common SNPs in genes of DNA repair and repair-related pathways.[98,99] We have emphasized amino acid substitution variants as they have higher potential than random or silent nucleotide

Table 6.1   Summary of Amino Acid Substitution Variants Identified in Genes in Different DNA Repair Pathways

Pathway	# of Genes in Pathway[a]	# of Genes Screened	# of Variants Identified	# of Alleles >2% Frequency[a]
Base Excision Repair	25	19	68	15
Nucleotide Excision Repair	33	16	51	17
Double Strand Break Repair	~30	11	31	8
Mismatch Repair	7	6	41	13
Damage Recognition and Cell Cycle Checkpoint	~30	6	13	4
Total	>110	51	177	50

[a] Genes with roles in more than one pathway are counted in all relevant pathways. The total is unique genes and variants, not the sum of the column.

substitutions to be associated with reduced protein function. This screening effort provides a catalog of variants that are reagents for subsequent bio-chemical and molecular epidemiology studies to address questions of the relationship of genetic variation and cancer risk. This has been described as a genotype to phenotype approach for molecular epidemiology studies.

The strategy for identification of variants has been to sequence the PCR fragments directly following amplification of the exons and adjacent intron or untranslated regions of a gene from genomic DNA. Most of the 35 genes screened thusfar have been screened in 92 samples from the "DNA Poly-morphism Discovery Resource" available from the Coriell Institute for Med-ical Research. The samples in this NIH-developed resource[100] are from U.S. residents selected to represent the major ethnic groupings of the population although the ethnic origin of specific individuals is unknown. The individ-uals in this sample set are from population groups as follows: 23 European-Americans; 23 African-Americans; 11 Mexican-Americans; 11 Native-Amer-icans; and 23 Asian-Americans. Data regarding the variants identified are available at http://greengenes.llnl.gov/dpublic/secure/reseq/reseq_home_page.html. A similar effort initiated by Dr. Maynard Olsen, in the laboratory at the University of Washington, has screened some 25 repair and repair related genes for variation, including 15 genes not screened by the Lawrence Livermore group. The results from their effort are available at http://www.genome.utah.edu/genesnps. Others have reported results from more limited population-based screens of these and additional repair genes.[101–105] The amino acid substitution data obtained in these screens are summarized by repair pathway in Table 6.1. This summary does not include data that can be gleamed from searches of the EST database,[106–108] the Gen-Bank genome sequence database[109,110], the different disease gene specific databases. It is difficult to ascertain the completeness of the variant screens and the degree to which the individuals screened are representative of the general population in many of these latter databases.

C. Extent of variation

The average of 3.5 different amino acid substitution variants per repair gene is higher than might have been expected for a set of genes that are highly conserved during evolution. It is also higher than the number of different variants observed (1.1–2.8/gene) in screening of other sets of candidate disease susceptibility genes (e.g., genes where functionally relevant is expected to influence the risk of cardiovascular disease, hypertension, or arthritis) for variation.[93–97] The average variant allele frequency for the repair genes is approximately 4%. Many variants were detected in only a single chromosome, while only five variants have estimated frequencies of over 40%. Only 30% of the repair gene variants exist at allele frequencies of 2% or more, while less than 10% of the variant alleles exist at >10% and only 5% have variant alleles frequencies of >20% in the current data set. Thus, low frequency variants contribute significantly to the variation among individuals in the population for these loci. It should be remembered that the estimated allele frequencies are based on data from only small numbers of individuals. Also, it is not possible to account for potential differences in allele frequencies among ethnic groups when screening the samples from the DNA Polymorphism Discovery Resource. Some of these variants could exist at high frequencies within specific subpopulations.

Homozygous variant individuals and individuals with multiple amino acid substitutions in a gene are observed, as are individuals with variants in multiple genes in a repair pathway. Extrapolating from the current estimate of the number of different variants per gene (3.5) and the average variant allele frequency (4%), the average individual should have approximately five variants among the 25 genes of the BER pathways and 6 variants among the genes of the DSBR and DRC pathways. Thus, the genotypes for each repair pathway are very complex. Most individuals in the population are expected to have variant subunits for several proteins that are components of multimeric complexes or sequential steps of each repair pathway. The potential exists for additive and even multiplicative effects among variants to contribute significantly to the disease risk in the population.

In addition, because of the linearity of the steps in the repair process or pathway, different variants may have a similar impact on repair capacity. That is, different variants in a gene or variants in different genes in a pathway could be considered to be equivalent in terms of impact on individual susceptibility and cancer risk. The amount of variation and the complexity of the genotypes when considering a process or pathway rather than individual variants or genes present significant challenges for molecular epidemiology studies.

VI. Impact of common DNA repair gene variants on function and cancer risk

A. Characteristics of variants

Over 70% of the substitutions observed are exchanges of amino acid residues with dissimilar physical and/or chemical properties. Similarly, approximately 70% of the substitutions occur at residues where the common allele is identical in human and mouse. Approximately 50% of the variants have both characteristics. These are characteristics expected of amino acid substitution variants that would be expected to have an impact on protein structure and function. Studies have now begun to address the question of the potential relevance of the polymorphic repair gene variants identified in the general (healthy) population.

B. DNA repair capacity

Several variants are associated with altered DNA repair capacity. Spitzet al.,[111] reported that variant alleles at amino acid residues 312 and 751 of *ERCC2 (XPD)* were associated with reduced repair capacity in lymphocytes from individuals in a lung cancer cohort. Being homozygous for a variant allele in either *XPC* or *XPD* was associated with reduced capacity to repair UV-induced DNA damage as assayed by the host reactivation assay in a cohort of healthy subjects (Q. Wei, pers. comm.). Not surprisingly, variation in the BER gene *XRCC1* was not associated with UV repair capacity. The *ERCC2* 751Gln variant was associated with a reduced capacity for repairing ionizing radiation using a cytogenetic-based assay[112] and for removing UV-induced DNA damage.[113] Hu et al.,[114] report that the *APE1* 148Glu allele was associated with prolonged mitotic delay in lymphocytes exposed to ionizing radiation in breast cancer patients. In addition, lymphocytes from women with at least three variant alleles of *APE1* and *XRCC1* are at increased likelihood of exhibiting ionizing radiation sensitivity.[114] At the cellular and gene specific level, the rate of repair of single strand breaks has been associated with variants of *XRCC1*. This reduced repair is related to the reduced ability of the variant protein to interact with Ligase 3,[115] another protein of the BER pathway. Although not associated with specific variants, reduced levels of Cdk1 and Cyclin B activity have been associated with increased chromosomal breaks after radiation exposure of cells in G2,[116] implicating DRC genes in this repair capacity phenotype.

C. Biomarkers of Exposure

The *XRCC1* 399Gln variant has been associated with several phenotypic measures of exposure, including increased aflatoxin adducts[117] and increased polyphenol adducts, glycophorin A mutations, and sister chromatid exchanges in smokers.[117,118] These reports indicate that the approximately

40% of the population with the variant allele of *XRCC1* incurred 20–50% more DNA damage than observed in individuals with the more common allele, following similar exposures. The *XRCC3* 241Met variant was associated with increased levels of bulky DNA adducts in smokers.[119] This study did not find an association of either the *XRCC1* 399Gln or the *ERCC2* 751Gln variants with adduct levels. The Gln allele at codon 399 of XRCC1 is associated with increased levels of DNA adducts following *in vitro* exposure of lymphocytes to a tobacco-specific nitrosamine.[120]

Variation in biomarkers of exposure to chemical carcinogens has also been associated with genetic variation in the carcinogen metabolizing enzymes.[22–27] The increase in adduct level associated with the *XRCC3* 241Met genotype was more marked in smokers with the NAT2-slow phenotype,[119] emphasizing the interaction between metabolism or damage induction and repair capacity. The sum of these studies emphasizes the complexity of the role of genetic variation in the individual response to an exposure. These studies documenting a role for individuals differences in response to common exposures begins to provide plausible explanations for the observation that the level of cytogenetic damage in lymphocytes is a better predictor of individual cancer risk than it is an estimator of historical exposure.[121, 122] This is consistent with the accumulating data indicating that the health consequences of exposure to environmental agents result from the interaction of dose and the genetic constitution of the individual.

D. Repair gene variants and cancer risk.

Nine amino acid substitution variants identified in the population based screening efforts of seven repair genes have been incorporated into molecular epidemiology studies of different cancer cohorts. Most of these molecular epidemiology studies have genotyped only a couple of variants, and in most instances only one of the 15–30 genes in a repair pathway. They have generally studied relatively small cohorts. The impact of the different variants on cancer risk is not obvious at this time as different studies have obtained different results. For example, four of ten studies of the variation at codon 399 of *XRCC1* on risk found that the Gln (variant) allele was associated with risk,[123–126] while three studies reported that the Arg (wildtype) allele was the risk allele[127–129] and three studies did not find an association with either allele.[119,130,131] Among seven studies of the variation at codon 194 of *XRCC1*, one reported that the variant (Trp) allele was associated with elevated risk,[124] four studies reported that the wildtype allele (Arg) was associated with elevated risk,[123,126,127,131] and four studies found no association.[119,128–130] In two studies,[123,126] being variant at codon 399 and wildtype at codon 194 increased risk above that observed for either risk allele though this combined association was not observed in other studies.

The data regarding the risk associated with the substitution of Gln for Lys at position 751 of ERCC2 (XPD), a protein of the NER complex, are equally discordant among seven studies to date. The variant allele at codon

312 (Asn) is the risk genotype in two studies[132,133] while the wildtype allele (Asp) is the risk allele in two other studies.[134,135] The same studies that observed the variant allele at codon 312 to be the risk allele also found the variant allele at codon 751 (Gln) to be associated with elevated risk.[132,133] Two other studies found the wildtype allele at codon 751 (Lys) to be associated with elevated risk.[136,137] Being variant at both codons was also associated with elevated risk.[135] Interestingly, a synonymous substitution in exon 6 (codon 156) of *ERCC2* is associated with elevated risk in four studies.[133,134,136,137] This variant may be in linkage disequilibrium with the three common amino acid substitution variants in this gene.[134,137]

A number of other variants have only been genotyped in single studies. One study has reported an association of the 241Met variant of *XRCC3*, a gene of the DSB pathway, with melanoma.[138] Similarly, an insertion/deletion polymorphism in intron 9 of *XPC* has been associated with risk of squamous cell carcinoma of the head and neck.[139] This polymorphism is in disequilibrium with an amino acid substitution polymorphism at codon 939 (Lys/Gln). Two molecular epidemiology studies have genotyped SNPs that do not result in amino acid substitutions. Variation in the 3′ untranslated regions of *ERCC1* (NER gene) has been associated with risk of glioma.[140] A nucleotide substitution in the 5′UTR of *RAD51*, a gene where only a single amino acid substitution has been identified (allele frequency <0.005) in screening over 250 unrelated individuals, is associated with cancer risk in *BRCA2* but not *BRCA1* carriers.[141] It is assumed that these substitutions in the UTRs are located in binding sites for factors that regulate gene expression, although the associations could reflect disequilibrium with variation in another gene.

At least two possibilities exist for explaining the discordance results obtained in different studies. First, the differences could be real and be reflective of differences in the interaction of the repair pathway with different exposures. The apparent inconsistencies could also be associated with tissue differences in metabolism or repair as the studies were studying potential relationships to a number different tumors. Second, although the associations are statistically significant, they could reflect chance associations as are often observed in molecular epidemiology studies with limited sample size.

The odds ratios for elevated risk for the individual variants are generally in the range of 1.3–2.0 and, as indicated, not all of the studies are consistent with regard to risk association.

The relatively low impact of one variant allele on cancer risk, as compared to the risk associated with the reduced capacity phenotype, is not unexpected. The repair capacity assays integrate the impact of all of the variants of one or more pathway, while the current molecular epidemiology studies are genotyping for only one or two of the four to eight variant alleles expected to be observed in each pathway in most individuals. The complexity of the problem to be addressed becomes apparent when it is realized that these four to eight variants anticipated to be observed in the average individuals are from among the 50–75 different repair gene variants expected to exist in each repair pathway. Although a single susceptibility allele may reduce activity of a protein only marginally, individuals with variants at multiple genes in a pathway may have more significantly reduced pathway function. In the end, it is the integrated function of the genes of the relevant pathways that is key to determining susceptibility to health risk from an exposure, not the variation at any one gene. The challenge is to develop strategies to utilize the rapidly growing catalogs of the extensive genetic variation among individuals and the resulting complex genotypes to refine the estimates of the health risks from exposure to relevant levels of environmental agents.

VII. Summary of genetic variation and cancer risk.

Genetic variation as exemplified by cancer susceptibility alleles is a key element in determining an individual's risk of cancer even in the absence of the highly penetrant variant alleles observed in cancer families. "Cancer susceptibility" alleles, although increasing risk only a few fold, exist at polymorphic frequency and have the potential to have a major impact on the population incidence of cancer.[72] This impact is in contrast to the generally rare cancer alleles, which have a major impact on risk for the affected individual but only limted impact of the population incidence. Note the use of the term alleles rather than genes. This reflects the growing realization that different variants, with different degrees of negative impact on normal protein expression or function, exist in these genes. That is, cancer alleles and susceptibility alleles are likely to exist at the same locus. Note also that the normal function of these "cancer" genes is to prevent cancer and that it is the variants of these genes with reduced or aberrant function that are associated with elevated cancer risk.

   The role of genetics in determining cancer susceptibility does not exclude a role for environmental exposure and lifestyle factors, e.g., smoking, in determining an individual's risk of cancer. Figure 6.1 illustrates the relationship between risk and exposure at different levels of genetic susceptibility.

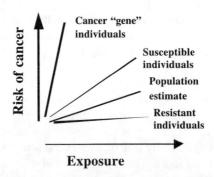

Figure 6.1 Risk of cancer from exposure varies among individuals within the population.

Exposure is nearly irrelevant for affected individuals in cancer families. Conversely, at high exposures, carcinogen exposure will be the dominant factor. However, it seems apparent that for most of the population, cancer risk is the result of the interaction of the altered function/activity expected for the gene products expressed by susceptibility alleles with moderate or even low level exposure to carcinogenic agents. As depicted, the cancer incidence in susceptible individuals would be higher than the population risk following an exposure. Obviously, the distribution of individual risk is not distributed into four distinct groups. Rather, cancer risk is a continuum reflecting the large number of genes with roles in the biological processes from exposure to tumor and the extensive variation in these genes. Knowledge of the extent and frequency of genetic variation for genes with the potential for conferring differences in susceptibility in human populations will provide resources for increased power in epidemiological studies. The challenge is to utilize these data to provide better estimates of risk and ultimately to reduce the incidence of disease.

Acknowledgments

This work was performed under the auspices of the U.S. Department of Energy by the University of California, Lawrence Livermore National Laboratory under contract No. W-7405-Eng-48, with additional support by an Interagency Agreement with NIEHS (Y1-ES-8054–05).

References

1. Friedberg, E.C., Walker, G.C., and Siede, W., *DNA Repair and Mutagenesis.* ASM Press, Washington, D.C., 1995.
2. Lindahl, T., Instability and decay of the primary structure of DNA, *Nature*, 362, 709–715, 1993.
3. Povirk, L.F., DNA damage and mutagenesis by radiomimetic DNA-cleaving agents: bleomycin, neocarzinostatin and other enediynes, *Mutat. Res.*, 355, 71–89, 1996.
4. von Sonntag, C., *The Chemical Basis of Radiation Biology*, Taylor and Francis, London, 1987.
5. Roth, R.B. and Wilson, J.H., Illegitimate recombination in mammalian cells, in *Genetic Recombination*, Kucherlapati, R. and Smith, G., Eds., American Society for Microbiology, Washington, D.C., 1988, pp. 621–633.
6. Ward, J.F., DNA damage produced by ionizing radiation in mammalian cells: identities, mechanisms of formation, and reparability, *Prog. Nucleic Acid Res. Mol. Biol.*, 35, 95–125, 1988.
7. Ward, J.F., Radiation mutagenesis: The initial DNA lesions responsible, *Radiat. Res.*, 142, 362–368, 1995.
8. Phillips, D.H. et al., Methods of DNA adduct determination and their application to testing compounds for genotoxicity, *Environ. Mol. Mutagen.*, 35, 222–233, 2000.

9. Garner, R.C., The role of DNA adducts in chemical carcinogenesis, *Mutat. Res.*, 402, 67–75, 1998.

10. Airoldi, L. et al., Carcinogen-DNA adducts as tools in risk assessment, *Adv. Exp. Med. Biol.*, 472, 231–240, 1999.

11. Povey, A.C., DNA adducts: endogenous and induced, *Toxicol. Pathol.*, 28, 405–414, 2000.

12. Gupta, R.C. and Lutz, W.K., Background DNA damage for endogenous and unavoidable exogenous carcinogens: A basis for spontaneous cancer incidence?, *Mutat. Res.*, 424, 1–8, 1999.

13. Hemminki, K., Koskinen, M., and Zhao, C., DNA adducts as a marker for cancer risk?, *Int. J. Cancer*, 92, 923–926, 2001.

14. Vineis, P. and Perera, F., DNA adducts as markers of exposure to carcinogens and risk of cancer, *Int. J. Cancer*, 88, 325–328, 2000.

15. Lutz, W.K., Dose-response relationships in chemical carcinogenesis reflect differences in individual susceptibility: Consequences for cancer risk assessment, extrapolation, and prevention, *Hum. Exp. Toxicol.*,18, 707–712, 1999.

16. Otteneder, M. and Lutz, W.K., Correlation of DNA adduct levels with tumor incidence: carcinogenic potency of DNA adducts, *Mutat. Res.*, 424, 237–247, 1999.

17. Thomas, C.B. et al., Elevated frequencies of hypoxanthine phosphoribosyl-transferase lymphocyte mutants are detected in Russian liquidators 6 to 10 years after exposure to radiation from the Chernobyl nuclear power plant accident *Mutat. Res.*, 439, 105–119, 1999.

18. Bigbee, W.L. et al., Human *in vivo* somatic mutation measured at two loci: individuals with stably elevated background erythrocyte glycophorin A (GPA) variant frequencies exhibit normal T-lymphocyte hprt mutant frequencies, *Mutat. Res.*, 397, 119–136, 1998.

19. Moore II, D.H. et al., A study of the effects of exposure on cleanup workers at the Chernobyl nuclear reactor accident using multiple end points, *Radiat. Res.*, 148, 463–475, 1997.

20. Jones, I.M. et al., Factors affecting HPRT mutant frequency in T-lymphocytes of smokers and nonsmokers, *Cancer Epidemiol. Biomarkers Prev.*, 2, 249–260, 1993.

21. MacGregor, J.T. et al., 'Spontaneous' genetic damage in man: Evaluation of interindividual variability, relationship among markers of damage, and influence of nutritional status, *Mutat. Res.*, 377, 125–135, 1997.

22. Au, W.W. et al., Biomarker monitoring for health risk based on sensitivity to environmental mutagens, *Rev. Environ. Health*, 16, 41–64, 2001.

23. Whyatt, R.M. et al., Biomarkers of polycyclic aromatic hydrocarbon-DNA damage and cigarette smoke exposures in paired maternal and newborn blood samples as a measure of differential susceptibility, *Cancer Epidemiol. Biomarkers Prev.*, 10, 581–588, 2001.

24. Hou, S.M. et al., Differential interactions between GSTM1 and NAT2 genotypes on aromatic DNA adduct level and HPRT mutant frequency in lung cancer patients and population controls, *Cancer Epidemiol. Biomarkers Prev.*, 10, 133–140, 2001.

25. Zhao, C. et al., DNA adducts of 1,3-butadiene in humans: relationships to exposure, GST genotypes, single-strand breaks, and cytogenetic end points, *Environ. Mol. Mutagen.*, 37, 226–230, 2001.

26. Butkiewicz, D. et al., Polymorphisms of the GSTP1 and GSTM1 genes and PAH-DNA adducts in human mononuclear white blood cells, *Environ. Mol. Mutagen.*, 35, 99–105, 2000.

27. Whyatt, R.M. et al., Association between polycyclic aromatic hydrocarbon-DNA adduct levels in maternal and newborn white blood cells and glutathione S-transferase P1 and CYP1A1 polymorphisms, *Cancer Epidemiol. Biomarkers Prev.*, 9, 207–212, 2000.

28. de Laat, W.L., Jaspers, N.G.J., and Hoeijmakers, J.H., Molecular mechanisms of nucleotide excision repair, *Genes Develop.*, 13, 768–785, 1999.

29. Cleaver, J.E. et al., Nucleotide excision repair, "a legacy of creativity," *Mutat. Res.*, 485, 23–36, 2001.

30. Balajee, A.S. and Bohr, V.A., Genomic heterogeneity of nucleotide excision repair, *Gene*, 250, 15–30, 2000.

31. Petit, C. and Sancar, A., Nucleotide excision repair: from E. coli to man, *Biochimie*, 81, 15–25, 1999.

32. Batty, D.P. and Wood, R.D., Damage recognition in nucleotide excision repair of DNA, *Gene*, 241, 193–204, 2000.

33. Thompson, L.H. and Schild, D., The contribution of homologous recombination in preserving genome integrity in mammalian cells, *Biochimie*, 81, 87–105, 1999.

34. Thompson, L.H. and Schild, D., Homologous recombinational repair of DNA ensures mammalian chromosome stability, *Mutat. Res.*, 477, 131–153, 2001.

35. Lieber, M.R. et al., Tying loose ends: roles of Ku and DNA-dependent protein kinase in the repair of double-strand breaks, *Current Opin. Genet. Dev.*, 7, 99–104, 1997.

36. Pfeiffer, P., Goedecke, W., and Obe, G., Mechanisms of DNA double-strand break repair and their potential to induce chromosomal aberrations, *Mutagenesis*, 15, 289–302, 2000.

37. Wilson III, D.M. and Barsky, D., The major human abasic endonucleolase: formation, consequences and repair of abasic lesions in DNA, *Mutat. Res.*, 485, 283–307, 2001.

38. Wilson III, D.M. and Thompson, L.H., Life without DNA repair, *Proc. Natl. Acad. Sci. USA*, 94, 12754–12757, 1997.

39. Lindahl, T., Suppression of spontaneous mutagenesis in human cells by DNA base excision-repair, *Mutat. Res.*, 462, 129–135, 2000.

40. Kolodner, R.D. and Marsischky, G.T., Eukaryotic DNA mismatch repair, *Curr. Opin. Genet. Dev.*, 9, 89–96, 1999.

41. Hsieh, P., Molecular mechanisms of DNA mismatch repair, *Mutat. Res.*, 486, 71–87, 2001.

42. Weinert, T., A DNA damage checkpoint meets the cell cycle engine, *Science*, 277, 1450–1505, 1997.

43. Fearon, E.R. and Dang, C.V., Cancer genetics: tumor suppressor meets oncogene, *Current Biology*, 9, R62–R65, 1999.

44. Khanna, K.K. and Jackson, S.P., DNA double-strand breaks: signaling, repair and the cancer connection, *Nature Genet.*, 27, 247–254, 2001.

45. Orr-Weaver, T.L., and Weinberg, R.A., A checkpoint on the road to cancer, *Nature*, 392, 223–224, 1998.

46. Welsch, P.S., Owens, K.N., and King, M.C., Insights into the functions of BRCA1 and BRCA2, *Current Opin. Genet. Develop.*, 7, 46–51, 2000.

47. Wood, R.D. et al., Human DNA repair genes, *Science*, 291, 1284–1289, 2001.

48. Romen, A. and Glickman, B.W., Human DNA repair genes, *Environ. Mol. Mutag.*, 37, 241–283, 2001.
49. Cleaver, J.E., Xeroderma pigmentosum: The first of the cellular caretakers, *Trends Biochem. Sci.*, 26, 398–401, 2001.
50. Cleaver, J.E., Common pathways for ultraviolet skin carcinogenesis in the repair and replication defective groups of xeroderma pigmentosum, *J. Dermatol. Sci.*, 23, 1–11, 2000.
51. Cleaver, J.E. et al., A summary of mutations in the UV-sensitive disorders: xeroderma pigmentosum, Cockayne syndrome, and trichothiodystrophy, *Hum. Mutat.*, 14, 9–22, 1999.
52. Ishikawa, T. et al., Importance of DNA repair in carcinogenesis: evidence from transgenic and gene targeting studies, *Mutat. Res.*, 477, 41–49, 2001.
53. Hoeijmakers, J.H., Genome maintenance mechanisms for preventing cancer, *Nature*, 411, 366–374, 2001.
54. Lynch, L.T. et al., Genetic epidemiology of breast cancer, in *Genetic Epidemiology of Cancer*, Lynch, H.T. and Hirayama, T., Eds., CRC Press, Boca Raton, FL, 1989, pp. 289–332.
55. Peto, J. et al., Cancer mortality in relatives of women with breast cancer: the OPCS Study, *Int. J. Cancer*, 65, 275–283, 1996.
56. Sellers, T.A. et al., Lung cancer detection and prevention: Evidence for interaction between smoking and genetic predisposition, *Cancer Res. Suppl.*, 52, 2694–2697, 1992.
57. Spitz, M.R. and Bondy, M.L., Genetic susceptibility to cancer, *Cancer*, 72, 991–995, 1993.
58. Perera, F.P., Molecular epidemiology: on the path to prevention?, *J. Natl. Cancer Inst.*, 92, 602–612, 2000.
59. Rothman, N. et al., The use of common genetic polymorphisms to enhance the epidemiologic study of environmental carcinogens, *Biochim. Biophys. Acta*, 1471, C1–C10, 2001.
60. Perera, F.P., Molecular epidemiology of environmental carcinogenesis, *Recent Results Cancer Res.*, 154, 39–46, 1998.
61. Shields, P.G. and Harris, C.C., Cancer risk and low-penetrance susceptibility genes in gene environment interactions, *J. Clin. Oncol.* 18, 2309–2315, 2000.
62. Au, W.W. et al., Usefulness of genetic susceptibility and biomarkers for evaluation of environmental health risk, *Environ. Mol. Mutagen.*, 37, 215–225, 2001.
63. Verkasalo, P.K. et al., Genetic predisposition, environment and cancer incidence: a nationwide twin study in Finland, 1976–1995, *Int. J. Cancer*, 83, 743–749, 1999.
64. Hemminki, K. and Mutanen, P., Genetic epidemiology of multistage carcinogenesis, *Mutat. Res.*, 473, 11–21, 2001.
65. Ahlbom, A. et al., Cancer in twins: Genetic and nongenetic familial risk factors, *J. Natl. Cancer Inst.*, 89, 287–93, 1997.
66. Lichtenstein, P. et al., Environmental and heritable factors in the causation of cancer: Analyses of cohorts of twins from Sweden, Denmark, and Finland, *N. Engl. J. Med.*, 343, 78–85, 2000.
67. Risch, N., The genetic epidemiology of cancer: interpreting family and twin studies and their implications for molecular genetic approaches, *Cancer Epidemiol. Biomarkers Prev.*, 10, 733–741, 2001.

68. Grossman, L. et al., DNA repair as a susceptibility factor in chronic diseases in human populations, in *Advances in DNA Damage and Repair*, Dizdaroglu, M. and Karakaya, A.E., Eds., Kluwer Academic/Plenum Publishers, New York, 1999, 149–167.

69. Scott, D. et al., Increased chromosomal radiosensitivity in breast cancer patients: a comparison of two assays, *Int. J. Radiat. Biol.*, 75, 1–10, 1999.

70. Wu, X. et al., Benzo[a]pyrene diol epoxide and bleomycin sensitivity and susceptibility to cancer of upper aerodigestive tract, *J. Natl. Cancer Inst.*, 90, 1393–1399, 1998.

71. Wei, Q. et al., DNA repair and aging in basal cell carcinoma: a molecular epidemiology study, *Proc. Natl. Acad. Sci. USA*, 90, 1614–1618, 1993.

72. Wei, Q. et al., Repair of tobacco carcinogen-induced DNA adducts and lung cancer risk: a molecular epidemiology study, *J. Natl. Cancer Inst.*, 92, 1764–1772, 2000.

73. Wei, Q. et al., DNA repair capacity correlates with mutagen sensitivity in lymphoblastoid cell lines, *Cancer Epidemiol. Biomarkers Prev.*, 5, 199–204, 1996.

74. Malcolmson, G.G. et al., Determination of radiation-induced damage in lymphocytes using the micronucleus and microgel electrophoresis "Comet" assay, *Eur. J. Cancer*, 31A, 2320–2323, 1995.

75. Olive, P.L., DNA damage and repair in individual cells: applications of the comet assay in radiobiology, *Int. J. Radiat. Biol.*, 75, 395–405, 1999.

76. Alapetite, C. et al., Analysis by alkaline comet assay of cancer patients with severe reactions to radiotherapy: defective rejoining of radioinduced DNA strand breaks in lymphocytes of breast cancer patients, *Int. J. Cancer*, 83, 83–90, 1999.

77. Wu, X. et al., A parallel study of *in vitro* sensitivity to benzo[a]pyrene diol epoxide and bleomycin in lung carcinoma cases and control, *Cancer*, 83, 1118–1127, 1998.

78. Burrill, W. et al., Heritability of chromosomal radiosensitivity in breast cancer patients: A pilot study with the lymphocyte micronucleus assay, *Int. J. Radiat. Biol.*, 76, 1617–1619, 2000.

79. Scot, D. et al., Genetic predisposition in breast cancer, *Lancet*, 344, 1444, 1994.

80. Boyle, J.M. et al., The relationship between radiation-induced G(1)arrest and chromosome aberrations in Li-Fraumeni fibroblasts with or without germline TP53 mutations, *Br. J. Cancer*, 85, 293–296, 2001.

81. Papworth, R. et al., Sensitivity to radiation-induced chromosome damage may be a marker of genetic predisposition in young head and neck cancer patients, *Br. J. Cancer*, 84, 776–782, 2001.

82. Cloos, J. et al., Inherited susceptibility to bleomycin-induced chromatid breaks in cultured peripheral blood lymphocytes, *J. Natl. Cancer Instit.*, 91, 1125–1130, 1999.

83. Roberts, S.A. et al., Heritability of cellular radiosensitivity: A marker of low-penetrance predisposition genes in breast cancer, *Am. J. Hum. Genet.*, 65, 784–794, 1999.

84. Wu, X. et al., Genetic influence on mutagen sensitivity: a twin study, *Proc. Amer. Assoc. Cancer Res.*, 41, 437, 2000.

85. Tchirkov, A. et al., Detection of heterozygous carriers of ataxia-telangiectasia (ATM) gene by G2 phase chromosomal radiosensitivity of peripheral blood lymphocytes, *Hum. Genet.*, 101, 312–316, 1997.

86. Speit, G. et al., Mutagen sensitivity of human lymphoblastoid cells with a BRCA1 mutation in comparison to ataxia telangiectasia heterozygote cells, *Cytogenet. Cell. Genet.*, 91, 261–266, 2000.

87. Berwick, M., and Vineis, P., Markers of DNA repair and susceptibility to cancer in humans: an epidemiologic review, *J. Natl. Cancer Inst.*, 92, 874–897, 2000.

88. Cloos, J. et al., Mutagen sensitivity as a biomarker for second primary tumors after head and neck squamous cell carcinoma, *Cancer Epidemiol. Biomarkers Prev.*, 9, 713–717, 2000.

89. Leprat, F. et al., Impaired DNA repair as assessed by the "comet" assay in patients with thyroid tumors after a history of radiation therapy: A preliminary study, *Int. J. Radiat. Oncol. Biol. Phys.*, 40, 1019–1026, 1998.

90. Wu, X. et al., Mutagen sensitivity as a susceptibility marker for human hepatocellular carcinoma, *Cancer Epidemiol. Biomarkers Prev.*, 7, 567–570, 1998.

91. Wang, D.G. et al., Large-scale identification, mapping, and genotyping of single-nucleotide polymorphisms in the human genome, *Science*, 280, 1077–1082, 1998.

92. Sachidanandam, R. et al., A map of human genome sequence variation containing 1.42 million single nucleotide polymorphisms, *Nature*, 409, 928–933, 2001.

93. Cambien, F. et al., Sequence diversity in 36 candidate genes for cardiovascular disorders, *Am. J. Hum. Genet.*, 65, 183–191, 1999.

94. Halushka, M.K. et al., Patterns of single-nucleotide polymorphisms in candidate genes for blood-pressure homeostasis, *Nature Genet.*, 22, 239–247, 1999.

95. Ohnishi, Y. et al., Identification of 187 single nucleotide polymorphisms (SNPs) among 41 candidate genes for ischemic heart disease in the Japanese population, *Hum. Genet.*, 106, 288–292, 2000.

96. Cargill, M. et al., Characterization of single-nucleotide polymorphisms in coding regions of human genes, *Nature Genet.*, 22, 231–238, 1999.

97. Yamada, R. et al., Identification of 142 single nucleotide polymorphisms in 41 candidate genes for rheumatoid arthritis in the Japanese population, *Hum. Genet.*, 106, 293–297, 2000.

98. Shen, M.R., Jones, I.M., and Mohrenweiser, H., Nonconservative amino acid substitution variants exist at polymorphic frequency in DNA repair genes in healthy humans, *Cancer Res.*, 58, 604–608, 1998.

99. Mohrenweiser, H.W. and Jones, I.M., Variation in DNA repair is a factor in cancer susceptibility: A paradigm for the promises and perils of individual and population risk estimation?, *Mutat. Res.*, 400, 15–24, 1998.

100. Collins, F.S., Brooks, L.D., and Chakravarti, A., A DNA polymorphism discovery resource for research on human genetic variation, *Genome Res.*, 8, 1229–1231, 1998.

101. Broughton, B.C., Steingrimsdottir, H., and Lehmann, A.R., Five polymorphisms in the coding sequence of the xeroderma pigmentosum group D gene, *Mutat. Res.*, 362, 209–211, 1996.

102. Fan, F. et al., Polymorphisms in the human DNA repair gene XPF, *Mutat. Res.*, 406, 115–120, 1999.

103. Bell, D.W. et al., Common nonsense mutations in RAD52, *Cancer Res.*, 59, 3883–3888, 1999.

104. Kato, M. et al., Identification of RAD51 alteration in patients with bilateral breast cancer, *J. Hum. Genet.*, 45, 133–137, 2000.

105. Ma, X. et al., Single nucleotide polymorphism analyses of the human proliferating cell nuclear antigen (PCNA) and flap endonuclease (FEN1) genes, *Int.
J. Cancer*, 88, 938–942, 2000.
106. Buetow, K.H., Edmonson, M.N., and Cassidy, A.B., Reliable identification of
large numbers of candidate SNPs from public EST data, *Nature Genet.*, 21,
323–325, 1999.
107. Clifford, R. et al., Expression-based genetic/physical maps of single-nucleotide polymorphisms identified by the cancer genome anatomy project, *Genome Res.*, 10, 1259–1265, 2000.
108. Picoult-Newberg, L. et al., Mining SNPs from EST databases, *Genome Res.* 9,
167–174, 1999.
109. Taillon-Miller, P. et al., Overlapping genomic sequences: a treasure trove of
single-nucleotide polymorphisms, *Genome Res.*, 8, 748–754, 1998.
110. Gu, Z., Hillier, L., and Kwok, P.Y., Single nucleotide polymorphism hunting
in cyberspace, *Hum. Mutat.*, 12, 221–225, 1998.
111. Spitz, M.R. et al., Modulation of nucleotide excision repair capacity by XPD
polymorphisms in lung cancer patients, *Cancer Res.*, 61, 1354–1357, 2001.
112. Lunn, R.M. et al., XPD polymorphisms: effects on DNA repair proficiency,
Carcinogenesis, 21, 551–555, 2000.
113. Moller, P. et al., Psoriasis patients with basal cell carcinoma have more repair-
mediated DNA strand-breaks after UVC damage in lymphocytes than psoriasis patients without basal cell carcinoma, *Cancer Lett.*, 151, 187–192, 2000.
114. Hu, J. et al., Amino acid variants of APE1 amd XRCC1 genes associated with
ionizing radiation sensitivity, *Carcinogenesis*, 22, 917–22, 2001.
115. Moore, D.J. et al., Mutation of a BRCT domain selectively disrupts DNA
single-strand break repair in noncycling Chinese hamster ovary cells, *Proc.
Natl. Acad. Sci. USA*, 97, 13649–13654, 2000.
116. Terzoudi, G.I. et al., Increased G2 chromosomal radiosensitivity in cancer
patients: the role of cdk1/cyclin-B activity level in the mechanisms involved,
Int. J. Radiat. Biol., 76, 607–615, 2000.
117. Lunn, R.M. et al., XRCC1 polymorphisms: effects on aflatoxin B1-DNA adducts and glycophorin A variant frequency, *Cancer Res.*, 59, 2557–2561, 1999.
118. Duell, E.J. et al., Polymorphisms in the DNA repair genes XRCC1 and ERCC2
and biomarkers of DNA damage in human blood mononuclear cells, *Carcinogenesis*, 21, 965–971, 2000.
119. Matullo, G. et al., DNA repair gene polymorphisms, bulky DNA adducts in
white blood cells and bladder cancer in a case-control study, *Int. J. Cancer*, 92,
562–567, 2001.
120. Abdel-Rahman, S.Z. and El-Zein, R.A., The 399Gln polymorphism in the
DNA repair gene XRCC1 modulates the genotoxic response induced in human lymphocytes by the tobacco-specific nitrosamine NNK, *Cancer Lett.*, 159,
63–71, 2000.
121. Bonassi, S. et al., Chromosomal aberrations in lymphocytes predict human
cancer independently of exposure to carcinogens: European Study Group on
Cytogenetic Biomarkers and Health, *Cancer Res.*, 60, 1619–1625, 2000.
122. Hagmar, L. et al., Chromosomal aberrations in lymphocytes predict human
cancer: A report from the European Study Group on Cytogenetic Biomarkers
and Health, *Cancer Res.*, 58, 4117–4721, 1998.
123. Sturgis, E.M. et al., Polymorphisms of DNA repair gene XRCC1 in squamous
cell carcinoma of the head and neck, *Carcinogenesis*, 20, 2125–2129, 1999.

124. Abdel-Rahman, S.Z. et al., Inheritance of the 194Trp and the 399Gln variant alleles of the DNA repair gene XRCC1 are associated with increased risk of early-onset colorectal carcinoma in Egypt, *Cancer Lett.*, 159, 79–86, 2000.

125. Divine, K.K. et al., The XRCC1 399 glutamine allele is a risk factor for adenocarcinoma of the lung, *Mutat. Res.*, 461, 273–278, 2001.

126. Shen, H. et al., Polymorphisms of the DNA repair gene XRCC1 and risk of gastric cancer in a Chinese population, *Int. J. Cancer*, 88, 601–606, 2000.

127. Stern, M.C. et al., DNA repair gene XRCC1 polymorphisms, smoking, and bladder cancer risk, *Cancer Epidemiol. Biomarkers Prev.*, 10, 125–131, 2001.

128. Duell, E.J. et al., Polymorphisms in the DNA repair gene XRCC1 and breast cancer, *Cancer Epidemiol. Biomarkers Prev.*, 10, 217–222, 2001.

129. Lee, J.M. et al., Genetic polymorphisms of XRCC1 and risk of the esophageal cancer, *Int. J. Cancer*, 95, 240–246, 2001.

130. Butkiewicz, D. et al., Genetic polymorphisms in DNA repair genes and risk of lung cancer, *Carcinogenesis*, 22, 593–597, 2001.

131. Ratnasinghe, D. et al., Polymorphisms of the DNA repair gene XRCC1 and lung cancer risk, *Cancer Epidemiol. Biomarkers Prev.*, 10, 119–123, 2001.

132. Sturgis, E.N. et al., XPD/ERCC2 polymorphisms and risk of head and neck cancer: a case-control analysis, *Carcinogenesis*, 21, 2219–2223, 2000.

133. Tomescu, D. et al., Nucleotide excision repair gene XPD polymorphisms and genetic predisposition to melanoma, *Carcinogenesis*, 22, 403–408, 2001.

134. Vogel, U. et al., Polymorphisms of the DNA repair gene XPD: correlations with risk of basal cell carcinoma revisited, *Carcinogenesis*, 22, 899–904, 2001.

135. Hemminki, K. et al., XPD exon 10 and 23 polymorphisms and DNA repair in human skin *in situ*, *Carcinogenesis*, 22, 1185–1188, 2001.

136. Dybdahl, M. et al., Polymorphisms in the DNA repair gene XPD: correlations with risk and age at onset of basal cell carcinoma, *Cancer Epidemiol. Biomarkers Prev.*, 8, 77–81, 1999.

137. Caggana, M. et al., Associations between ercc2 polymorphisms and gliomas, *Cancer Epidemiol. Biomarkers Prev.*, 10, 355–360, 2001.

138. Winsey, S.L. et al., A variant within the DNA repair gene XRCC3 is associated with the development of melanoma skin cancer, *Cancer Res.*, 60, 5612–5616, 2000.

139. Shen, H. et al., An intronic poly (AT) polymorphism of the DNA repair gene XPC and risk of squamous cell carcinoma of the head and neck: a case-control study, *Cancer Res.*, 61, 3321–3325, 2001.

140. Chen, P. et al., Association of an ERCC1 polymorphism with adult-onset glioma, *Cancer Epidemiol. Biomarkers Prev.*, 9, 843–847, 2000.

141. Levy-Lahad, E. et al., A single nucleotide polymorphism in the RAD51 gene modifies cancer risk in BRCA2 but not BRCA1 carriers, *Proc. Natl. Acad. Sci. USA*, 98, 3232–3236, 2001.

chapter seven

Studies of DNA repair and human cancer: an update

Marianne Berwick, Giuseppe Matullo, and Paolo Vineis

Contents

Abstract Deficiencies in DNA repair systems may lead to the development of cancer. The epidemiology of DNA repair capacity and its effect on cancer susceptibility in humans are important to critically investigate. We have summarized all the published epidemiologic studies on DNA repair in human cancer between 1999 and the first half of 2001 (n = 43) that addressed the association of cancer susceptibility with a putative defect in DNA repair capacity, either phenotypic or genotypic. We have reviewed study design, epidemiologic and laboratory methods, evaluation of study results, causality, bias, and confounding and statistical considerations. Assays have been evaluated as phenotypic assays and genotypic assays, focusing on the association of polymorphic variants with cancer. Future directions include the need for standardized nomenclature and the development of reproducible phenotypic methods that can be used in epidemiologic studies to assess function.

I. Introduction

The number of DNA repair genes so far identified in humans is 130, and it is anticipated that additional repair genes will be discovered. Enzymes encoded by these genes constantly monitor the genome and repair damaged nucleotides resulting from exposure to environmental and endogenous DNA damaging agents.[1] If a DNA repair pathway is compromised or inactivated, cells can become hypermutable and more susceptible to cancer.[2] Thus, DNA repair capacity can be considered a biomarker of cancer susceptibility and may potentially be a biomarker of susceptibility to other human diseases as well. However, DNA repair capacity is difficult to assess using an epidemiological approach because of redundancy in DNA repair pathways (i.e., DNA repair enzymes perform overlapping functions).

This chapter presents an update to our 2000 review of markers of DNA repair and susceptibility to cancer in humans[3] focusing on studies published after 1998 and through the first half of 2001. This compilation is not meant to be exhaustive. Mismatch repair genes, potentially important biomarkers for hereditary nonpolyposis colorectal cancer and other human cancers, have been excluded. Not all the results are shown in the tables, since several studies considered multiple genes. More information is given in the text.

There has been an understandable shift from phenotypic assays of DNA repair capacity to genotyping for alterations in DNA repair genes. This review documents that change. Sixteen studies focus on phenotypic assays[4–19] and 28 studies focus on genotypic assays;[17,20–45] only one study published after 1998 combined phenotype and genotype in relationship to human cancer.[17] Study design and results are discussed, and epidemiologic factors that affect the integrity and the generalizability of these studies are considered.

II. Phenotypic studies — methods

A.   Epidemiologic methods

1.   Study design

Phenotypic studies of DNA repair are cell-based assays that measure the response to *in vitro* damaging agents. The design of these studies is presented in Table 7.1. Most phenotypic studies are case control studies (n = 12)[4–6,8–9,11–13,16–20] while one assessed families,[7] one examined a cohort of subjects and cases who developed second primary cancers.[10] Two were cohort

Table 7.1 Study Design of Studies Based on Phenotyping

Study	Method, Ethnic Group	Design	Cancer Site	No. Cases/Controls
Dybdahl[4] (Denmark)	HCRA Not stated	H-B Case control	Basal cell carcinoma (BCC)	40 cases (20 with psoriasis) 40 controls (20 with psoriasis)
D'Errico[5] (Italy)	HCRA Not stated	H-B Case control	BCC	49/68
El-Zein[6] (USA)	Chromosomal instability Caucasian	H-B Case control	Gliomas	25/28
Ankathil[7] (India)	Chromosomal aberrations Asian (Indian)	H-B Family study	Colorectal cancer	26 familial cases 30 sporadic cases 60 unaffected family members 30 healthy controls
Yu[8] (USA)	Mutagen sensitivity White and nonwhite	H-B Case control	Head and neck cancer	170/175
Cheng[9] (USA)	Gene expression Non-Hispanic white African-American Other	H-B Case control	Lung cancer	75/95
Cloos[10] (The Netherlands)	Mutagen sensitivity Note stated	H-B Second primary	Head and neck cancer	19 cases with second primaries 218 cases with first primaries
Zhang[11] (USA)	Mutagen sensitivity White and nonwhite	H-B Case control	Head and neck cancer	173/176 ETS exposure

Table 7.1 Study Design of Studies Based on Phenotyping (*Continued*)

Study	Method, Ethnic Group	Design	Cancer Site	No. Cases/Controls
Wei[12] (USA)	HCRA Non-Hispanic Caucasians	H-B Case control	Lung cancer	316/316
Schmezer[13] (Germany)	Comet Not stated	H-B Case control	Lung cancer	100/110
Bonassi[14] Nordic countries (Denmark, Finland, Norway, Sweden)	Chromosomal aberrations Not stated	Nested case control	All cancers	150/590
Herman[15] (Israel)	UDS Not stated	Cohort kidney transplant	All cancers	
Wu[16]	Mutagen sensitivity Caucasian Hispanic-American African-American	H-B Case control	Lung cancer	183/227
Spitz[17] (USA)	HCRA Caucasian	H-B Case control	Lung cancer	341/360
Rajaee-Behbahani[18] (Germany)	Comet Not stated	H-B Case control	Nonsmall cell lung cancer	160/180
Aukely[19] (USA)	DNA-PK activity	H-B Case control 1/3 Hispanic, 2/3 Caucasian, other	Lung cancer	41/41

HB = hospital-based; BCC = basal cell carcinoma; HCRA = host cell reactivation assay; PB = population-based; UDS = unscheduled DNA synthesis; DNA-PK = DNA protein kinase.

studies, one an international cohort[14] and the other a cohort of kidney transplant patients.[15] The response rate is an indicator of whether the sample is representative of the population being studied; however, the response rate is rarely reported.

2.   Controls
The source of controls is not usually described in detail. Controls are often obtained in a similar manner to the cases, either as a subset of a larger study or as a convenience sample. Sometimes, they are matched by age and sex or other characteristics that might be important to cancer susceptibiltiy, such as smoking status.

B.   Laboratory methods

1.   Host cell reactivation assay (HCRA)
The host cell reactivation assay compares the ability of lymphocytes to repair a damaged, transfected plasmid. The transfection step and the use of frozen lymphocytes are additional sources of variation in this assay. Cheng et al.[46] recently compared cryopreserved blood with cryopreserved lymphocytes in this assay. They reported acceptable correlations (0.77, $P < 0.001$) and similar mean values between sets. An advantage of this assay is that the lymphocytes are not subject to DNA damage as in most other functional assays.

2.   Comet assay
The comet assay has historically been used to measure DNA damage within a single cell. Schmezer et al.,[13] have developed methods for determining DNA repair capacity. The DNA damage assay can measure either single strand breaks or double strand breaks depending on the conditions, alkaline or neutral, respectively. There are several variations on the standard comet assay.[47] Many of these assays measure the distance migrated by putatively damaged DNA relative to the distance migrated by undamaged DNA or the core of the nucleus.

   In epidemiologic analyses, lymphocytes are commonly used. However, many factors influence their response in DNA repair assays, including traits of the donor such as age, cigarette smoking, and exercise. Cell cycle status is potentially even more important. Because the comet assay is widely used in many laboratories in a large number of different applications, the consistency of results obtained with this assay is limited.

3.   Mutagen sensitivity
The mutagen sensitivity assay was developed by T.C. Hsu in 1989[48] and has been used to measure repair of double strand breaks and single strand breaks induced by bleomycin, benzo[a]pyrene diol epoxide (BPDE), ultraviolet (UV) radiation, ionizing radiation, and 4-nitroquinoline-1-oxide (4-NQO), all of which are different aspects of DNA repair.[49] This assay consists of a short-

term growth of blood cells in culture, during which cells are exposed to a damaging agent, allowed to repair DNA damage, and then analyzed by chromosome spreading techniques during metaphase. Chromosomal spreads are visualized and inspected for gaps and breaks. This assay has been criticized for potential observer variability; however, Berwick and Wei (unpublished data) found approximately 75% correlation between laboratories in this assay. Berwick[50] has found a 70% intra-individual correlation and a 70% interobserver correlation for this assay. These values could possibly be improved. The problem of assay variability over time can be solved by freezing lymphocytes or whole blood and then assaying in batch, as Cheng et al.[46] showed a 61% correlation between assays using frozen blood with assays using fresh lymphocytes. Furthermore, they did not find any relationship in rank order of individuals. The problem of differential death in cell populations may be critical to the results of this assay.

4.   Chromosomal aberrations assay

Chromosomal aberrations, as measured by Hagmar et al.[51] and Bonassi et al.[52] are chromosome breaks, not including chromatid gaps, in unstimulated metaphase spreads. These measurements are classic cytogenetic techniques that are laborious and time consuming and not suitable for large epidemiologic evaluation. Hagmar and Bonassi have taken advantage of the pre-existence of these measures on subjects in the assembled cohorts.

5.   Unscheduled DNA synthesis (UDS)

UDS is one of the earliest assays for DNA repair.[53] In this assay, autoradiographic methods are used to detect the incorporation of [3H]thymidine (dT) into the DNA of cultured cells during repair of ultraviolet light damage. In irradiated cells, [3H]dT is incorporated into DNA at all stages of the cell cycle whereas normally [3H]dT is incorporated into S-phase cells during semiconservative DNA replication. Low levels of excision repair can be detected using appropriate methods. Some controversy exists as to the utility of scintillation counting versus counting by eye for UDS because it may be difficult to separate the high background levels from the cells performing unscheduled synthesis (Setlow, pers. comm.). Thus, the most stringent UDS assays are extremely laborious and unsuited for epidemiologic studies.

III. Genotyping studies — methods

A.   Epidemiologic methods

The designs of epidemiological studies for genotyping DNA repair are described in Table 7.2. Fourteen studies were hospital-based,[17,20–22,25,27,30,33,37,39,40–41,45] seven were population-based,[23–24,32,35–36,42] one was a cohort study (nested case-control),[38] one was a case-case study[31] and five had a different design.[26,29,34,43,44] Concerning the choice of controls, it is noteworthy that Shen et al.[34] compared

Table 7.2 Study Design of Studies Based on Genotyping

Study	Gene (Method)	Design Ethnic Group	Cancer Site	No. Cases/Controls + Response Rates (%)	Exposures
Sturgis[20] (USA)	XRCC1 (PCR-RFLP)	H-B Case-control, non-Hispanic white, African-American, Hispanic	Head and neck cancer	203/424 (?)	Smoking, alcohol
Dybdahl[21] (Denmark)	XPD (PCR-RFLP)	H-B Case-control, Caucasian	BCC, psoriasis/BCC	40/40 (?)	Psoriasis genotoxic treatment
Gu[22] (USA)	PADPRP Pseudogene (PCR-LP)	H-B Case-control Mexican-American African-American Caucasian	Lung cancer	288/292 (?)	Smoking
Healey[23] (UK)	BRCA2 (TaqMan)	P-B Case-control, Caucasians	Breast carcinoma	3459/3013 (?)	None
Infante-Rivard[24] (Canada)	XRCC1 (PCR-RFLP)	P-B Case-control	Childhood leukemia	491/491 (?)	Radiation
Wikman[25] (Germany)	hOGG1 (LightCycler, +PCR-RFLP)	H-B Case-control Caucasian	Lung cancer	105/105 (?)	Smoking
Abdel-Rahman[26] (Egypt)	XRCC1 (PCR-RFLP)	Case-control (1), Arabic	Colorectal cancer	48/48 (?)	Occupation, diet, reproductive history
Kaur[27] (USA)	AGT (PCR-RFLP)	H-B Case-control Causasians, African-American, or Asians	Lung cancer	139/139 (93/?)	Smoking

Table 7.2 Study Design of Studies Based on Genotyping (Continued)

Study	Gene (Method)	Design Ethnic Group	Cancer Site	No. Cases/Controls + Response Rates (%)	Exposures
Chen[28] (USA)	ERCC1 (PCR-SSCP)	P-B Case-control, white	Glioma	122/159 (?)	None
Winsey[29] (UK)	XRCC1 XRCC3 XPF, XPD, ERCC1 (PCR-SSP)	Case-control (2), Caucasians	Melanoma	125/211 (?)	None
Sturgis[30] (USA)	XPD/ERCC2 (PCR-RFLP)	H-B Case-control, non-Hispanic white, African-American, Hispanic	Head and neck cancer	189/496 (?)	Smoking, alcohol
Stanulla[31] (Germany)	NBS1 657del5 (PCR-RFLP)	Case/case Caucasian	Non-Hodgkin's lymphoma (age <18)	109 with mutation 984 without (?)	None
Shen[32] (China)	XRCC1 (PCR-RFLP)	P-B Case-control Chinese	Gastric adeno-carcinoma	243/243 (79/71)	Diet, smoking, occupation, H. pylori
Lee[33] (Taiwan)	XRCC1 (PCR-RFLP)	H-B Case-control, Chinese	Esophageal cancer	105/264 (86/92)	Smoking, drinking, diet
Shen[34] (USA and China)	XPC (PCR-RFLP)	Mixed design, case-control (3), Caucasian	Head and neck cancer	287/311 (?)	Smoking, alcohol
Spitz[17] (USA)	XPD (PCR-RFLP) +HCRA	H-B Case-control, white	Lung cancer	341/360 (?)	Smoking, alcohol
Duell[35] (USA)	XRCC1 (PCR-RFLP)	P-B Case-control, white, African-American	Breast cancer	862/790 (74/53)	Risk factors for breast cancer

Study	Genes (method)	Study design, population	Cancer type	Cases/controls	Confounders
Butkiewicz[36] (Poland)	XRCC1, XRCC3, XPD (PCR-RFLP)	P-B Case-control, white	Lung cancer	96/96 (?)	Smoking, occupation
Vogel[37] (USA)	XPD (PCR-RFLP)	H-B Case-control, Caucasians	BCC	70/117 (?) +52 members of 4 families	Sunburns, skin type
Ratnasinghe[38] (China)	XRCC1 (TaqMan)	Nested case-control in a cohort of tin miners, Chinese	Lung cancer	108/216 (50% of the cohort provided blood)	Smoking, alcohol, radon and arsenic
Stern[39] (USA)	XRCC1 (PCR-RFLP)	H-B Case-control, white or black	Bladder cancer	235/213 (?)	Smoking
Matullo[40] (Italy)	XRCC1, XRCC3, XPD (PCR-RFLP)	H-B Case-control, Caucasian	Bladder cancer	124/85 (DNA not available for 38/19)	Smoking, diet, drugs, occupation
Tomescu[41] (Scotland)	XPD, ERCC1 (PCR-RFLP)	H-B Case-control, Caucasian	Melanoma	28/28 (?)	None
Caggana[42] (USA)	XPD (PCR-RFLP)	P-B Case-control, Caucasian, other	Glioma	187/169 (?)	None
Divine[43] (USA)	XRCC1 (PCR-RFLP)	P-B cases, H-B controls, non-H whites, Hispanics	Adenocarcionoma of the lung	172/143 (?)	Smoking
Levy-Lahad[44] (?)	RAD51 (PCR-RFLP)	H-B family study, BRCA1/2, mutation carriers and healthy Ashkenazi Jewish	Breast and/or ovarian cancer	164/93/73 (?)	None
Xing[45] (China)	hOGG1 (SSCP-sequencing)	H-B Controls, Chinese	Esophageal cancer	196/201 (100/?)	Smoking

P-B = population-based; H-B = hospital-based; (1) = Controls were friends of the cases; (2) = Controls were cadaveric transplant donors; (3) = Controls were blood donors.

Chinese cases living in the United States with Chinese controls living in China. One study[29] used cadaveric transplant donors as controls, and one used friends of the cases.[26] Several studies match for exposure (typically, studies on lung cancer match on smoking). The response rate was usually not reported although the study by Duell et al.[25] reported an important difference in response rates between cases and controls.

B.   Laboratory methods

Three studies used a relatively high-throughput genetic technology (Taqman or LightCycler).[23,25,38] PCR-RFLP was the most frequently used method although one study used PCR/SSCP with sequencing.[45] Genotype-phenotype correlation and cancer status were sought in one study,[17] with a comparison between the HCRA assay and genotyping for XPD in lung cancer patients. Cases and controls with the wild genotype had the most proficient DNA repair capacity as measured by the host cell reactivation assay. One study compared PCR-RFLP and DHPLC,[40] others compared RFLP or SSCP with sequencing in a subgroup.

III. Evaluation of study results

Results of the phenotypic studies are shown in Table 7.3 and the genotyping studies in Table 7.4. Interactions were often assessed in the genotyping studies but only formally in two phenotypic studies,[8,11] and these are shown in Table 7.5. The study by Spitz et al.[17] examined genotype-phenotype and cancer status correlation and is in all tables.

A.   Assessment of causality

1.   Strength of association

Phenotypic studies tend to have relatively high odds ratios (ORs). As the sample sizes are quite small, the variabilities of the assays are often not reported, the reliability most often not measured, and the populations generally subsets of larger and hospital-based studies; the large estimates of effect (i.e.,, odds ratios) are common. Larger studies are needed to determine whether phenotypes of low DNA repair capacity do actually have such large effects. The cohort study should demonstrate the most reliable results as measurement of phenotype would be made prior to the development of disease. It should be noted that ORs vary widely with the referent group used. Phenotyping is generally expensive and time consuming. Therefore, studies that attempt to correlate phenotype and genotype in such a way that genotyping, which is far less time consuming and expensive, would be valuable. Only one study[17] to date has correlated phenotype and genotype among cancer cases and controls. Among those studies that have used genotyping, ORs tend to be moderately low (on the order of two or less), with

Table 7.3   Results of Studies Based on Phenotyping

Study	Assay and Cancer Site	Assay Estimates	Odds Ratio (95% CI) (or Comparisons)
Dybdahl[4]	HCRA, BCC		Psoriasis patients, lower HCRA (P = 0.015, one-sided) Low repair psoriasis patients OR = 6.4, (95% CI 1.4–28.5) No data for BCC without psoriasis and controls
D-Errico[5]	HCRA, BCC	Significant decline of HCRA with age in controls not cases	Young BCC cases, lower HCRA (NS); older BCC cases, higher HCRA (P < 0.001)
El-Zein[6]	Chromosomal instability, glioma	Cases 2.4 breaks/1000 cells Controls 1.4 breaks/1000 cells	OR = 8.5 (95% CI 2.1–34.9) adj OR = 15.3 (95% CI 2.7–87.8)
Ankathil[7]	Chromosomal abberations, colorectal cancer	Familial cases, mean 1.64 b/c Sporadic cases, mean 1.08 b/c Unaffected relatives, mean 0.62 b/c Controls, mean 0.52 b/c	P < 0.001
Yu[8]	Mutagen sensitivity, head and neck cancer		Cases with family history and mutagen sensitive, OR = 7.9 (95% CI 2.5–25.3)
Cheng[9]	Gene expression, lung	XPG, CSB 12.2, 12.5% decrease in baseline expression levels in cases cf. controls	XPG OR = 2.3 (95% CI 1.2–4.4) CSB OR = 2.5 (95% CI 1.3–4.8) Dose-response between reduced expression levels and increased lung cancer risk (P < 0.01)
Cloos[10]	Mutagen sensitivity, head and neck cancer	ND between 2nd primary and 1st primary in MS If 2nd primary ≥3 yrs from 1st primary, P = 0.005	RR = 7.8 (95% CI 0.9–61.7) for 2nd primary

Table 7.3   Results of Studies Based on Phenotyping (*Continued*)

Study	Assay and Cancer Site	Assay Estimates	Odds Ratio (95% CI) (or Comparisons)
Zhang[11]	Mutagen sensitivity, head and neck cancer		> multiplicative interaction between MS and ETS exposure OR = 17.5 (95% CI 1.9–162)
Wei[12]	HCRA, lung cancer	Lower HRCA in cases than controls ($P < 0.001$)	reduced HCRA OR = 4.3 (95% CI 2.6–7.2) for 4th quartile vs. 1st
Schmezer[13]	Comet, lung cancer		Cases more sensitive to mutagens than controls ($P < 0.001$)
Bonassi[14]	Chromosomal abberations, all cancers	Occupational exposure and cigarette smoking had no effect	Nordic countries: Adj OR = 2.4 (95% CI 1.3–4.2); Italy: OR = 2.7 (95% CI 1.3–5.6)
Herman[15]	UDS, all cancers	Triple therapy group had lower HCRA (679 cpm) than control group (1049 cpm), $P < 0.02$ Double therapy group similar to controls	5 tumors in triple therapy group; 1 tumor in the double therapy group; Cyclosporin A reduced DNA repair significantly
Wu[16]	Mutagen sensitivity, lung cancer	Mean levels of IGF-I and IGF-1/IGFBP-3 higher in advanced cases	Mutagen sensitivity-bleomycin (R = 2.5 (95% CI 1.5–4.2), BPDE OR = 2.9 (95% CI 1.7–5.1)
Spitz[17]	HCRA, lung cancer	HCRA lower in cases (7.8%) than controls (9.5%; $P = 0.001$)	HCRA OR = 2.0 (95% CI 1.5–2.7)
Rajaee-Behbahani[18]	Comet, lung	Cases: 67% repair (15 min) Controls: 79.3%	OR = 2.1 (95% CI 1.1–4.0) reduced repair, increased sensitivity OR = 4.0 (95% CI 2.2–7.4)
Auckley[19]	DNA-PK activity, lung	Interindividual variability in cases and controls.	Enzyme activity in bronchial cells strongly associated with that in lymphocytes, r = 0.83; $P = 0.003$)

HCRA = host cell reactivation assay; BCC = basal cell carcinoma; OR = odds ratio; CI = confidence interval; B/c = breaks per cells; UDS = Unscheduled DNA synthesis; cpm = counts per minute; DNA-PK = DNA protein kinased.

Table 7.4 Results of Studies Based on Genotyping*

Study	Polymorphisms[§] and Cancer Site	Genotype, Frequency[a]	Odds Ratio (95% CI)
Sturgis[20]	XRCC1, Codons 194, 399 Head and neck cancer	Cod. 194, TT 0% Cod. 399, AA 15.8%	Cod. 194, CC 1.34 (0.8–2.25) Cod. 399, GG+GA 1.6 (0.97–2.6)
Dybdahl[21]	XPD, Codons 156, 751 BCC, psoriatics	Cod. 156, AA: psoriatics 15%, non-psoriatics 25% Cod. 751, AA: psoriatics 40%, non-psoriatics 45%	Cod. 156, psoriatics only: AA 5.3 (0.8–36.3) Cod. 751, all subjects: AA 4.3 (0.8–23.6)
Gu[22]	PADPRP 193bp deletion Lung	Caucasians BB 3.9% H-A BB 5.8% A-A BB 41.3%	BB+AB *Caucasians 0.5 (0.1–1.9) H-A 2.3 (1.2–4.4) *A-A 30.3 (1.7-547.5)
Healey[23]	BRCA2 Codon 372 (Asn→His) Breast	His/His 7.5%	(all) 1.31 (1.09–1.58) British samples 1.46 (1.13–1.89)
Infante-Rivard[24]	XRCC1, Codon 194 Childhood leukemia		Not given, except for Interaction (Table 7.5)
Wikman[25]	hOGG1 Codon 326 (Ser→Cys) Lung	Cys/Cys 1.9%	Cys/Cys 2.2 (0.4–11.8) Ser/Cys+Cys/Cys 0.7 (0.4–1.3)
Abdel-Rahman[26]	XRCC1, Codons 194, 399 Colorectal	Cod. 194,TT 0% Cod. 399, AA 4.1%	Cod. 194, CT 2.6 (0.7–9.4) Cod. 399, AA+AG 3.9 (1.5–10.6)
Kaur[27]	AGT Codons 143, 160 Lung	African-Americans: Cod. 143 Ile/Val 6% Cod. 160 Gly/Arg 4% Caucasians: Cod. 143 Ile/Val 15% Cod. 160 Gly/Arg 2%	African-Americans: Cod. 143 Ile/Val 2.3 (0.7–8.3) Cod. 160 Gly/Arg ND Caucasians: Cod. 143 Ile/Val 2.0 (0.8–5.7) Cod. 160 Gly/Arg ND

Table 7.4   Results of Studies Based on Genotyping* (*Continued*)

Study	Polymorphisms and Cancer Site	Genotype, Frequency[a]	Odds Ratio (95% CI)
Chen[28]	ERCC1, 8092 3′-UTR Polymorphism Glioma	AA 5%	all cases 1.4 (0.9–2.3) oligoastrocytoma 4.6 (1.6–13.2)
Winsey[29]	XRCC3, Codon 241	Cod. 241[1] TT 11%	XRCC3, Cod 241 crude OR
	XRCC1, Codons 194, 399	XRCC1, Cod. 194, CC 0%	TT+CT vs. CC 2.36 (P = 0.004)
	XPD, Codons 156, 312, 751	XRCC1, Cod. 399, AA 9%	
	XPF, Codon 824, 5′UTR	XPD, Cod. 156, AA 13%	
	ERCC1, Codon 118	XPD, Cod. 312, AA 13%	
	Melanoma	XPD, Cod. 751, CC 15%	
		XPF, Cod. 824, CC 12%	
		XPF, 5′UTR, AA 11%	
		ERCC1, Cod. 118, GG 18%	
		XRCC3, Cod. 241, TT 11%	
		XRCC3, 5′, AA 4%	
Sturgis[30]	XPD/ERCC2 Codons 156, 751 Head and neck	Cod. 156, AA 20.4% Cod. 751, CC 11.5%	Cod. 156, AA 0.9 (0.52–1.56) Cod. 751, CC 1.65 (0.98–2.77)
Stanulla[31]	NBS1 657del5 Non-Hodgkin's lymphoma	—	0.0 (no cases)
Shen[32]	XRCC1, Codons 194, 399 Stomach	Cod. 194 TT 11.4% Cod. 399, AA 7.8%	Cod. 194, CC 1.6 (0.8–3.5) Cod. 399, GA+AA 1.4 (0.9–2.4)
Lee[33]	XRCC1 Codons 194, 280, 399 Esophagus	Cod. 194, TT 7.2% Cod. 280, AA 0.8% Cod. 399, GG 9.1%	—[2]

Shen[34]	XPC intron 9 Poly (AT) Head and neck	+/+ 11.9% (12% in non-Hispanic white, Chinese and H-A; 9% in A-A)	+/+ 1.85 (1.12–3.05)
Spitz[17]	XPD Codons 751, 312 + HCRA Lung	Cod. 751, CC 10.8% Cod. 312, AA 7.0% HCRA suboptimal 2.0 (1.4–2.7)	Cod. 751, CC 1.4 (0.8–2.2) Cod. 312, AA 1.5 (0.8–3.0)
Duell[35]	XRCC1 Codon 194 and 399 Breast	African-Americans: Cod. 194, TT 0%, Cod. 399, AA 2% White: Cod. 194, TT 0.9%, Cod. 399, AA 16%	African-Americans: Cod. 194, TT+CT 0.7 (0.3–1.5) Cod. 399, AA+AG 1.7 (1.1–2.4) White: Cod. 194, TT+CT 0.7 (0.4–1.3), Cod. 399, AA +AG 1.0 (0.8–1.4)
Butkiewicz[36]	XRCC1 Codons 194, 280, 399 XRCC3 Codon 241, XPD Codons 312, 751 Lung	XRCC1: Cod. 194, ND XRCC1: Cod. 280, ND XRCC1; Cod. 399, ND XRCC3: Cod. 241, ND XPD: Cod. 751, ND XPD: Cod. 312, AA 18%	Cod. 194, ND Cod. 280, ND Cod. 399, ND Cod. 241, ND Cod. 751, ND Cod. 312: AA 1.86 (1.02–3.40)
Vogel[37]	XPD Codon 156, 312 and 751 BCC	Cod. 156, AA 20% Cod. 312, AA 19% Cod. 751, CC 10%	Cod. 156, AA 1.7 (0.7–4.0) Cod. 156, AC 2.0 (0.9–4.1) Cod. 312 AA 1.11 (0.49–2.54) Cod. 751 CC 1.8 (0.7–4.7)
Ratnasinghe[38]	XRCC1, Codons 194, 280, 399 Lung	Cod. 194, TT 10% Cod. 280, AA 0% Cod. 399, AA 5%	Cod. 194, TT 0.7 (0.3–1.8) Cod. 280, AA+AG 1.8 (1.0–3.4) Cod. 399, AA 1.3 (0.5–3.5)
Stern[39]	XRCC1, Codons 194, 280, 399 Bladder	Cod. 194, TT 0%[3] Cod. 280, AA 1% Cod. 399, AA 13%	Cod. 194, CT 0.6 (0.3–1.0)[4] Cod. 280, GA 1.2 (1.0–3.4)[4] Cod. 399, AA 0.7 (0.4–1.3)[4]

Biomarkers of environmentally associated disease

Table 7.4 Results of Studies Based on Genotyping* *(Continued)*

Study	Polymorphisms and Cancer Site	Genotype, Frequency[a]	Odds Ratio (95% CI)
Matullo[40]	XRCC1 Codon 399 XRCC3 Codon 241 XPD Codon 751 Bladder	XRCC1, Cod. 399, AA: urological ctrls 16%, non-urological 14% XRCC3, Cod. 241, TT: urological ctrls 13%, non-urological 23% XPD, Cod. 751, CC: urological ctrls 8%, non-urological 17%	XRCC1, Cod. 399, AA+AG: urolog. ctrls 0.6 (0.2–1.7), non-urol 0.80 (0.28–2.25) XRCC3, Cod. 241, TT+CT: urolog. ctrls 2.8 (1.3–6.0), non-urol 2.7 (1.4–5.4) XPD, Cod. 751, CC+AC: urolog ctrls 0.94 (0.42–2.1), non-urol 0.71 (0.32–1.58)
Tomescu[41]	XPD Codons 156, 711, 751 ERCC1 Codon 118 Melanoma	ND	XPD Cod. 156, A: 2.0 (0.9–4.5) XPD Cod. 711, C: 2.6 (1.1–6.7) XPD Cod. 751, A: 2.8 (1.2–7.0) ERCC1 Cod. 118, A: 1.6 (0.7–3.7)
Caggana[42]	XPD Codons 156, 312, 711, 751 Glioma	Cod. 156: AA 17% Cod. 312: AA 12% Cod. 711: TT 11% Cod. 751: CC 16%	Cod. 156: 2.3 (1.3–4.2) Cod. 312: 0.7 (0.5–1.2) Cod. 711: 0.8 (0.5–1.3) Cod. 751: 0.7 (0.4–1.1)
Divine[43]	XRCC1 Codon 399 Adenocarcinoma of the lung	Caucasian AA 8.2% Hispanic AA 11.4%	AA (all) 2.5 (1.1–5.8) Caucasian 3.3 (1.2–10.7) Hispanic 1.4 (0.3–5.9)
Levy-Lahad[44]	RAD51 135C/G 5'-UNT	Healthy women CG 6.1%	—
Xing[45]	hOGG1 Codon 326 (Ser→Cys)	Cys/Cys 13.4%%	1.95 (1.3–2.6)

Abbreviations: A-A = African-Americans; H-A = Hispanic-Americans.

* When not specified otherwise, ORs are age-adjusted and refer to the "high risk" homozygotes vs. "low-risk" homozygotes.

§ To avoid ambiguity, for almost all the polymorphisms is reported the codon and the nucleotide change or, in some cases, the three letter code amino acid change.

[1] In controls; only the rarest homozygous genotype is considered;[2] OR calculated from allelic distribution in cases/controls;[3] Given the nature of the controls, we omit reporting the frequencies of all the genotypes considered;[4] The results are presented only stratified by drinking habits (see Table 7.3).[5] Figures are for whites; in blacks they are all 0%;[6] Whites and blacks combined.

Table 7.5 Study Results: Interactions[1]

Study	Polymorphism and Cancer Site	Interactions: OR (95% CI)
Sturgis[20]	XRCC1 Head and neck	Both risk genotypes 2.02 (1.00–4.05) Cod. 194, oral cavity/pharynx 2.5 (1.22–4.97) Cod. 399, oral cavity/pharynx 1.41 (0.80–2.48) Cod. 399, current smokers 3.18 (1.28–7.94)
Gu[22]	PADPRP pseudogene 193bp deletion Lung	H-A adenocarcinoma 3.21 (1.14–9.08) H-A large cell carcinoma 10.8 (1.2–100.2)
Infante-Rivard[24]	XRCC1, Cod. 194	≥2 X rays in girls 6.66 (0.78–56.6)
Abdel-Rahman[26]	XRCC1 Colorectal	Cod. 399, age <40 11.90 (2.30–51.50) Cod. 399, urban residence 9.97 (1.98–44)
Sturgis[30]	XPD/ERCC2 Head and neck	>65 years Cod. 751, 2.22 (1.03–4.80) current drinkers Cod. 751, 2.59 (1.25–5.34)
Shen[32]	XRCC1 Stomach	"Other genotypes" 1.73 (1.12–2.69) Cardias, "other genotypes" 2.18 (1.21–3.94) Cod. 194, *H. pylori* negative 2.01 (1.04–3.89) Cod. 399, non-drinkers 1.89 (1.05–3.41)
Lee[33]	XRCC1 Esophagus	Drinkers Cod. 399, GG vs AA+AG 2.78 (1.15–6.67) Nondrinkers 0.76 (0.29–2.0)
Shen[34]	XPC intron 9 Head and neck	age ≥ 66 5.55 (2.19–14.03) Nondrinkers 3.3 (1.06–10.50)
Spitz[17]	XPD +HCRA Lung	XPD Cod. 312+751, 1.84 (1.11–3.04) XPD Cod. 312+CAT suboptimal 3.50 (1.06–11.59) (in cases)
Duell[35]	XRCC1 Breast	OR for former smokers (African-Americans): XRCC1 Cod. 399, GG 2.2 (1.3–3.6) AA+AG 1.4 (0.6–3.2) OR for duration of smoking >20 (African-Americans): XRCC1 Cod. 399, GG 2.9 (1.6–5.2) AA+AG 0.7 (0.3–1.5)
Butkiewicz[36]	XPD Codon 156 Lung cancer	Smokers ≤ 34.5 pack-years 5.32 (1.35–21.02) >34.5 pack-years, no difference

Table 7.5 Study Results: Interactions[1] *(Continued)*

Study	Polymorphism and Cancer Site	Interactions: OR (95% CI)

Vogel[37]	XPD
	BCC
		No family history Cod. 156 AC 3.42 (1.34–8.75)
		    AA 3.09 (1.05–9.10)
		Family history Cod. 312 AG 4.50 (0.67–30.20)
		    AA 6.0 (0.93–39)
		Family history Cod 751 AC 3.25 (0.66–16)
		    CC 6.0 (0.48–75)

Ratnasinghe[38]	XRCC1 Lung cancer	Drinkers Cod 194 TT+CT 0.4 (0.2–0.9) Drinkers Cod 280 AA+AG 3.7 (1.5–9.5)
Stern[39]	XRCC1, Codons 194, 399	Tests of interaction between genotype and smoking duration not significant
Matullo[40]	XRCC1-399, XRCC3-241, XPD-751 Bladder	Ex-smokers, XRCC3-241 5.2 (1.2–21.5) Nonsmokers, XRCC3-241 4.8 (1.1–21.2)
Caggana[42]	XPD Codons 156, 312, 711, 751 Glioma	Oligoastrocytoma Cod. 156, 3.2 (1.1–9.5)
Levy-Lahad[44]	RAD51 Breast/ovarian	BRCA1 mutation carrier (Breast + ovarian), HR[2] 1.18 BRCA2 mutation carrier (Breast + ovarian), HR[2] 4.0 (1.6–9.8) BRCA2 mutation carrier (only breast), HR[2] 3.46 (1.3–9.2)
Xing[45]	hOGG1-326 Esophagus	Only smoking 4.9 (2.8–8.6) Cys/Cys + smoking 4.8 (2.0–11.1)

[1] When not specified otherwise, ORs are age-adjusted and refer to the "high risk" homozygotes vs. "low-risk" homozygotes.

[2] Hazard ratio (95% CI).

the exception of oligoastrocytoma and *ERCC1*,[28] with an OR of 4.6. Low ORs are to be expected for low-penetrance genetic variants.

2. Consistency

a. Phenotypic studies. Consistency of results is impossible to evaluate among these phenotypic studies of DNA repair. Eight different methods have been applied (Table 7.1) in 16 studies. As all results are positive and significant, there may be important associations between reduced DNA repair capacity and the development of cancer. Vogel et al.[54] assessed the association of gene expression of a series of repair genes in relation to a functional assay, HCRA. They found that mRNA levels for ERCC1 and XPD approximated the DNA repair capacity of the functional assay. Few subjects were evaluated, but the paradigm is useful and should be developed further.

   b. Genotyping studies.   Twelve studies considered *XRCC1* polymorphisms in ten different cancer sites: head and neck, colo-rectum, melanoma, gastric cancer, breast, lung (three studies); bladder (two studies); and stomach, esophagus, leukemia (Table 7.2). For bladder cancer, findings are inconsistent and based on relatively small numbers.[39-40] Therefore, it is presently impossible to draw inferences on consistency of the findings.

   c. General considerations.   Particular attention must be given to comparing and interpreting results from different studies. For instance, contrasting results have been described by Lunn et al.[55] for *XPD*-Lys751Gln, in which the Lys/Lys *XPD*-751 genotype was found associated with reduced repair of x-ray-induced cytogenetic damage measured by chromosome breaks and gaps (OR 7.2, 95%CI 1.01-87.7) although cancer status was not measured. Another study by Dybdahl et al.[21] reported that individuals with the common allele (Lys751) had an elevated risk of basal cell carcinoma, whereas a recent study found no effect of *XPD* on polyphenol DNA adducts.[56] Moreover, Spitz et al.[17] found that subjects homozygous for the variant genotype *XPD*-751Gln/Gln had a deficient DRC compared with Lys/Lys wild-type homozygotes.
   A possible explanation for these results could be that not only amino acid variants in different domains of XPD may affect different protein interactions, resulting in the expression of different phenotypes, but also the same *XPD*-Lys751Gln polymorphism could have divergent effects in different DNA repair pathways and on different types of DNA damage. Moreover, this conservative substitution could be in linkage with another responsible XPD variant; in this case, it is possible that different populations have different alleles in linkage disequilibrium with the responsible *XPD* variant. Or, finally, selection bias could account for discrepancies.

3.   Internal coherence

One study compared two control groups with similar results.[40] In general, large variation was found among ethnic groups (Table 7.3), but this can be a genuine finding rather than an artifact. Also variation with age was noted (Table 7.5). Results from Auckley et al.[19] comparing DNA-PK activity in peripheral lymphocytes with DNA-PK activity in lung tissue show a strong correlation between the measurements. Whenever possible, a comparison between the proxy tissue, usually lymphocytes, with the target tissue should be made.

4.   Time sequence

   a. Phenotypic studies.   Most studies using phenotyping assays use cases that have been diagnosed and possibly treated for cancer. This treatment and the presence of cancer, if untreated, may have influences on the measures used. Two studies used prospectively collected data.[10,14] Both studies found that a relatively lengthy time was necessary to see results; Bonassi et al. followed subjects for 25 years; Cloos et al. found a significant difference only among those cases who developed a second primary three years or more after the diagnosis of the first primary.

    b. Genotyping studies. The criterion of a time sequence applies only if we suspect that the genotype influences survival. In this case, the recruitment of prevalent (not newly diagnosed) cases would lead to a bias in favor of the genotypes associated with longer survival. Apparently no study included prevalent cases except for Sturgis et al. 1999.[20] Genotype does not change with time, so that case-control studies are not affected by the usual forms of bias that characterize the retrospective collection of information on exposures.

5. Experimental evidence

    a. Phenotypic studies. Although there is ample evidence from animal studies and those in lower eukaryotes and prokaryotes, little of this is cited and none is shown.

    b. Genotyping studies. No comparison with relevant animal data is shown. Evidence of genotype–phenotype correspondence is given by Spitz et al.[17] who found that cases and controls with the wildtype genotype had the most proficient DNA repair capacity.

B. Bias

1. Confounding

Most studies have considered at least some potential confounders (e.g., age). The concept of confounding is related to the existence of a variable that is a risk factor for the disease and is associated with the genotype. If the prevalence of a certain genotype changes with age because of a survival effect, age is a confounder since it is strongly predictive of cancer onset. Often, which relevant confounders could exist in genotyping studies is difficult to hypothesize. As the genotype is fixed, confounding related to enzyme induction is irrelevant. On the other hand, in the phenotypic studies, potentially many potential confounders may affect results, e.g., cancer status, diet, exercise. These make the results of phenotypic studies all the more difficult to interpret.

2. Publication bias

Publication bias occurs when positive studies are preferentially published by scientific journals compared to negative studies and particularly to small negative studies. In the genotyping studies we are examining, all results except Duell et al.[35] were small or moderately small. Curiously, the lowest ORs come from Duell et al.[35] and from another relatively larger investigation, Stern et al.[39] This is only a weak suggestion in favor of potential publication bias. With the increasing recognition of this problem, more negative studies are being published.

3.   Other bias

One limitation of several studies is that the control group is poorly defined or clearly inadequate. For example, Shen et al.[32] compared Chinese cases living in the United States with Chinese controls living in China. One study[29] used cadaveric transplant donors as controls, and one used friends of the cases.[26]

C.   Statistical considerations (sample size, power, multiple comparisons)

1.   Phenotypic studies

Phenotypic studies are generally small to moderate in size and often consist of a highly selected sample. The differences reported between cases and controls are difficult to evaluate. A large issue is that of separating the results from cases who have cancer and the effects that cancer may exert on the phenotype from a susceptibility to develop cancer. As phenotyping to date is labor intensive, this issue is unlikely to be resolved soon. The cohort studies represented by Bonassi et al.[14] give the best insight into the likely true nature of the effect of lower DNA repair capacity on cancer, an approximately twofold increase for those with a higher level of aberrations. In phenotypic studies the statistical power is much stronger than genotyping studies because continuous variables can be used. Multiple comparisons can become a problem whenever the group is split into subgroups, as in any study.

2.   Genotyping studies

The sample size is always a problem in this type of investigations. However, relatively common genotypes were considered, so that statistical power is usually moderate. An important methodological problem refers to multiple comparisons (Table 7.5): Splitting the data into several subgroups increases the probability of finding statistically significant results. For this reason, the reported interactions in Table 7.5 have to be considered cautiously.

D. Interactions

Table 7.5 shows potential interactions between genotypes or phenotypes and exposures or other characteristics of the cases and controls. Most of the reported interactions cannot be evaluated because of multiple comparisons and lack of a formal statistical analysis for interaction. Interestingly, a few studies find an association with the relevant genotypes only among nondrinkers,[32,34,38] one finds the association only in *H. pylori*-negative subjects with stomach cancer,[32] and one finds that the genotype influences the basal cell carcinoma risk among those with a family history of basal cell carcinoma.[37] This is consistent with functional studies that show a relationship between DNA repair capacity and cancer in those with a family history.[8] None of the studies (except probably[35]) had a sufficient power to detect weak interactions.

V. Future directions

A. Standardized nomenclature

Many symbols are used to name human genes and the number of symbols and names will increase tremendously. It is important to establish rules for systematic naming of all new genes, and it is equally important to standardize nomenclature for naming allelic variants of genes. New genes and allelic variants are being discovered rapidly because of high-throughput sequencing and resequencing efforts and DNA chip technologies, and this rapid expansion of new genomic data will undoubtedly overwhelm the scientific journals that publish these data. Therefore, it will be appropriate and necessary to place the data on Web sites. For examples, see *CYP450, UGT, NAT2, NAT1, ALDH* genes, etc.; these Web sites have numerous links between sites, and frequently update information, so that colleagues in all fields of medical and genetic research can remain consistent in terminology.

It is also important to use clear terminology for genotypes; symbols should be described accurately to avoid misunderstandings (e.g., for *XRCC1*-399 polymorphism, A/G variation could be both a nucleotide or an amino acid variation). In the field of molecular epidemiology, it is extremely important to attempt to correlate an informative DNA repair genotype with an unequivocal phenotype in a specific ethnic group.

B. Function and mechanistic interactions

A great deal of progress has been made in developing methods for population-based study of DNA repair in the last several years; however, there is room for much more progress in the future. Researchers are genotyping large numbers of subjects for newly identified repair gene polymorphisms and association studies are being carried out. Information on the function and mechanistic interactions of these genotypic alterations is urgently needed. New technologies are being developed rapidly. It may soon be possible to determine whether a small decrement in DNA repair capacity predisposes to cancer. Ultimately, measures of DNA repair capacity may be developed as viable biomarkers for human susceptibility to cancer and possibly other human disease.

References

1. Wood, R.D. et al., Human DNA repair genes, *Science*, 291, 1284–1289, 2001.
2. Jasin, M., Chromosome breaks and genomic instability, *Cancer Invest.*, 18, 78–86, 2000.
3. Berwick, M. and Vineis, P., Markers of DNA repair and susceptibility to cancer in humans: a review, *J. Natl. Cancer Inst.*, 92, 874–897, 2000.
4. Dybdahl, M. et al., Low DNA repair is a risk factor in skin carcinogenesis: a study of basal cell carcinoma in psoriasis patients, *Mutat. Res.*, 433, 15–22, 1999a.
5. D'Errico, M. et al., Factors that influence the DNA repair capacity of normal and skin cancer-affected individuals, *Cancer Epidemiol. Biomarkers Prev.*, 8, 553–559, 1999.
6. El-Zein, R. et al., Increased chromosomal instability in peripheral lymphocytes and risk of human gliomas, *Carcinogenesis*, 20, 811–815, 1999.
7. Ankathil, R. et al., Deficient DNA repair capacity: a predisposing factor and high risk predictive marker in familial colorectal cancer, *J. Exp. Clin. Cancer Res.*, 18, 33–37, 1999.
8. Yu, G.P. et al., Family history of cancer, mutagen sensitivity, and increased risk of head and neck cancer, *Cancer Lett.*, 146, 93–101, 1999.
9. Cheng, L. et al., Reduced expression levels of nucleotide excision repair genes in lung cancer: a case-control analysis, *Carcinogenesis*, 21, 1527–1530, 2000.
10. Cloos, J. et al., Mutagen sensitivity as a biomarker for second primary tumors after head and neck squamous cell carcinoma, *Cancer Epidemiol. Biomarkers Prev.*, 9, 713–717, 2000.
11. Zhang, Z.F. et al., Environmental tobacco smoking, mutagen sensitivity, and head and neck squamous cell carcinoma, *Cancer Epidemiol. Biomarkers Prev.*, 9, 1043–1049, 2000.
12. Wei, Q. et al., Repair of tobacco carcinogen-induced DNA adducts and lung cancer risk: A molecular epidemiologic study, *J. Natl. Cancer Inst.*, 92, 1764–1772, 2000.
13. Schmezer, P. et al., Rapid screening assay for mutagen sensitivity and DNA repair capacity in human peripheral blood lymphocytes, *Mutagenesis*, 16, 24–30, 2001.
14. Bonassi, S. et al., for the European Study Group on Cytogenetic Biomarkers and Health, Chromosomal aberrations in lymphocytes predict human cancer independently of exposure to carcinogens, *Cancer Res.*, 60, 1619–1625, 2000.
15. Herman, M. et al., Effect of cyclosporin A on DNA repair and cancer incidence in kidney transplant recipients, *J. Lab. Clin. Med.*, 137, 14–20, 2001.
16. Wu, X. et al., Joint effect of insulin-like growth factors and mutagen sensitivity in lung cancer risk, *J. Natl. Cancer Inst.*, 92, 737–743, 2000.
17. Spitz, M.R. et al., Modulation of nucleotide excision repair capacity by XPD polymorphisms in lung cancer patients, *Cancer Res.*, 61, 1354–1357, 2001.
18. Rajaee-Behbahani, N. et al., Altered DNA repair capacity and bleomycin sensitivity as risk markers for non-small cell lung cancer, *Int. J. Cancer*, 94, 86–91, 2001.
19. Auckley, D.H. et al., Reduced DNA-dependent protein kinase activity is associated with lung cancer, *Carcinogenesis*, 22, 723–727, 2001.
20. Sturgis, E.M. et al., Polymorphisms of DNA repair gene XRCC1 in squamous cell carcinoma of the head and neck, *Carcinogenesis*, 20, 2125–2129, 1999.

21. Dybdahl, M. et al., Polymorphisms in the DNA repair gene XPD: correlations with risk and age at onset of basal cell carcinoma, *Cancer Epidemiol. Biomarkers Prev.*, 8, 77–81, 1999b.

22. Gu, J. et al., Ethnic differences in poly(ADP-ribose) polymerase pseudogene genotype distribution and association with lung cancer risk, *Carcinogenesis*, 20, 1465–1469, 1999.

23. Healey, C.S. et al., A common variant in BRCA2 is associated with both breast cancer risk and prenatal viability, *Nat. Genet.*, 26, 362–364, 2000.

24. Infante-Rivard, C., Mathonnet, G., and Sinnett, D., Risk of childhood leukemia associated with diagnostic irradiation and polymorphisms in DNA repair genes, *Environ. Health Perspect.*, 108, 495–498, 2000.

25. Wikman, H. et al., hOGG1 polymorphism and loss of heterozygosity (LOH): significance for lung cancer susceptibility in a Caucasian population, *Int. J. Cancer*, 88, 932–937, 2000.

26. Abdel-Rahman, S.Z. et al., Inheritance of the 194Trp and the 399Gln variant alleles of the DNA repair gene XRCC1 are associated with increased risk of early-onset colorectal carcinoma in Egypt, *Cancer Lett.*, 159, 79–86, 2000.

27. Kaur, T.B. et al., Role of polymorphisms in codons 143 and 160 of the O^6–alkylguanine DNA alkyltransferase gene in lung cancer risk, *Cancer Epidem. Biomarkers Prev.*, 9, 339–342, 2000.

28. Chen, P. et al., Association of an ERCC1 Polymorphism with adult-onset glioma, *Cancer Epidemiol. Biomarkers Prev.*, 9, 843–847, 2000.

29. Winsey, S.L. et al., A variant within the DNA repair gene XRCC3 is associated with the development of melanoma skin cancer, *Cancer Res.*, 60, 612–616, 2000.

30. Sturgis, E.M. et al., Molecular epidemiology and cancer prevention: XPD/ERCC2 polymorphism and risk of head and neck cancer: A case-control analysis, *Carcinogenesis*, 21, 2219–2223, 2000.

31. Stanulla, A. et al., No evidence for a major role of heterozygous deletion 657del5 within the NBS1 gene in the pathogenesis of non-Hodgkin's lymphoma of childhood and adolescence, *Br. J. Haematol.*, 109, 117–120, 2000.

32. Shen, H. et al., Polymorphisms of the DNA repair gene XRCC1 and risk of gastric cancer in a Chinese population, *Int. J. Cancer*, 88, 601–606, 2000.

33. Lee, J.-M. et al., Genetic polymorphisms of *XRCC1* and risk of esophageal cancer, *Int. J. Cancer*, 95, 240–246, 2001.

34. Shen, H. et al., An intronic poly (AT) polymorphisms of the DNA repair gene XPC and risk of squamous cell carcinoma of the head and neck: a case control study, *Cancer Res.*, 61, 3321–3325, 2001.

35. Duell, E.J. et al., Polymorphisms in the DNA repair gene XRCC1 and breast cancer, *Cancer Epidemiol. Biomarkers Prev.*, 10, 217–222, 2001.

36. Butkiewicz, D. et al., Molecular epidemiology and cancer prevention — genetic polymorphisms in DNA repair genes and risk of lung cancer, *Carcinogenesis*, 22, 593–597, 2001.

37. Vogel, U. et al., Molecular epidemiology and cancer prevention – Polymorphisms of the DNA repair gene XPD: correlations with risk of basal cell carcinoma revisited, *Carcinogenesis*, 22, 899–904, 2001.

38. Ratnasinghe, D. et al., Polymorphisms of the DNA repair gene XRCC1 and lung cancer risk, *Cancer Epidem. Biomarkers and Prev.*, 10, 119–123, 2001.

39. Stern, M.C. et al., DNA repair gene XRCC1 polymorphisms, smoking, and bladder cancer risk, *Cancer Epidemiol. Biomarkers Prev.*, 10, 125–131, 2001.

40. Matullo, G. et al., DNA repair gene polymorphisms, bulky DNA adducts in white blood cells and bladder cancer in a case-control study, *Int. J. Cancer*, 92, 562–567, 2001.

41. Tomescu, D. et al., Nucleotide excision repair gene XPD polymorphisms and genetic predisposition to melanoma, *Carcinogenesis*, 3, 403–408, 2001.

42. Caggana, M. et al., Associations between ERCC2 polymorphisms and gliomas, *Cancer Epidemiol. Biomarkers Prev.*, 10, 355–360, 2001.

43. Divine, K.K. et al., The XRCC1 399 glutamine allele is a risk factor for adenocarcinoma of the lung, *Mutat. Res.*, 461, 273–278, 2001.

44. Levy-Lahad, E. et al., A single nucleotide polymorphism in the RAD51 gene modifies cancer risk in BRCA2 but not BRCA1 carriers, *Proc. Natl. Acad. Sci. USA*, 98, 3232–3236, 2001.

45. Xing, D.Y. et al., Ser326Cys polymorphism in hOGG1 gene and risk of esophageal cancer in a Chinese population, *Int. J. Cancer*, 95, 140–143, 2001.

46. Cheng, L. et al., Cryopreserving whole blood for functional assays using viable lymphocytes in molecular epidemiology studies, *Cancer Lett.*, 166, 155–163, 2001.

47. McGlynn, A.P. et al., The bromodeoxyuridine comet assay: Detection of maturation of recently replicated DNA in individual cells, *Cancer Res.*, 59 5912–5916, 1999.

48. Hus, T.C. et al., Sensitivity to genotoxic effects of bleomycin in humans: Possible relationship to environmental carcinogenesis, *Int. J. Cancer*, 43, 403–409, 1989.

49. Gu, J. et al., Three measures of mutagen sensitivity in a cancer-free population, *Cancer Genet. Cytogenet.*, 110, 65–69, 1999.

50. Berwick, M. et al., Mutagen sensitivity as an indicator of soft tissue sarcoma risk, *Environ. Molec. Mut.*, in press.

51. Hagmar, L. et al., Cancer risk in humans predicted by increased levels of chromosome aberrations in lymphocytes: Nordic Study Group on the Health Risk of Chromosome Damage, *Cancer Res.*, 54, 2919–2922, 1994.

52. Bonassi, S. et al., Are chromosome aberrations in circulating lymphocytes predictive of a future cancer onset in humans? Preliminary results of an Italian cohort study, *Cancer Genet. Cytogenet.*, 79, 133–135, 1995.

53. Djordjevic, B. and Tolmach, L.J., Responses of synchronous populations of HeLa cells to ultraviolet irradiation at selected stages of the generation cycle, *Radiat. Res.*, 32, 327–346, 1967.

54. Vogel, U. et al., DNA repair capacity: Inconsistency between effect of overexpression of five NER genes and the correlation to mRNA levels in primary lymphocytes, *Mutat. Res.*, 461, 197–210, 2000.

55. Lunn, R.M. et al., XPD polymorphisms: Effects on DNA repair proficiency, *Carcinogenesis*, 21, 551–555, 2000.

56. Duell, E. et al., Polymorphisms in the DNA repair genes XRCC1 and ERCC2 and biomarkers of DNA damage in human blood mononuclear cells, *Carcinogenesis* 21, 965–971, 2000.

chapter eight

DNA adducts as biomarkers of DNA damage in lung cancer

Suryanarayana V. Vulimiri, and John DiGiovanni

Contents

Abstract Lung cancer, which is primarily caused by smoking, provides a useful model for studying the interactions between genetic predisposition and environmental exposure. Cigarette smoke contains several chemicals, of

1-56670-596-7/02/$0.00+$1.50
© 2002 by CRC Press LLC

which two major classes, i.e., PAHs (e.g., B[a]P) and *N*-nitrosamines (e.g., NNK) have unequivocally been shown to be associated with lung carcinogenesis. Phase I (e.g., CYP enzymes) and phase II (e.g., GST) enzymes play a major role in the metabolic activation and detoxification of cigarette smoke carcinogens and their metabolites, respectively. Reactive intermediates and/or free radicals generated during the metabolic activation of cigarette smoke carcinogens bind or modify DNA bases covalently resulting in DNA adduct formation. DNA adducts, which represent the biologically effective dose of the carcinogens can be categorized as bulky DNA adducts (e.g., BPDE-dG), alkylation products (e.g., O^6-mG), and oxidative lesions (e.g., 8-oxodG) and are formed in target and surrogate tissues of cigarette smokers. Animal studies clearly show that DNA adducts are markers of exposure and risk, although the data are less definitive in human beings. Nonetheless, DNA adducts have been widely used as biomarkers of exposure to cigarette smoke carcinogens in several human biomonitoring studies. Through these studies, it has been determined that the levels of DNA adducts in cigarette smokers are modulated by several factors, such as the extent and duration of exposure, type of smoking (e.g., active or passive), smoking status (never, former, and current smokers), ethnicity, gender, allelic polymorphisms of enzymes that are involved in metabolic activation and detoxification, DNA repair, and mutagen sensitivity. In spite of the complexity of factors affecting the formation and persistence of DNA adducts, numerous studies support that DNA adduct levels in target and in surrogate tissues are appropriate biomarkers of DNA damage induced by tobacco carcinogen exposure. Much further work is needed to determine, however, the extent to which DNA adducts may be used as biomarkers of risk or susceptibility to lung cancer.

I. Lung cancer — incidence, predisposition, and environment

Lung cancer is the paradigm of an environmentally induced disease. It is estimated that there will be 169,400 new cases and an estimated 154,900 deaths accounting for 28% of cancer-related deaths in the United States in 2002.[1] The prognosis for lung cancer remains poor, with a 13% overall 5-year relative survival rate. About 87% of the cancer cases are attributed to cigarette smoking and the relative risk for lung cancer in current smokers compared with those who have never smoked is up to 20-fold higher.[2] However, fewer than 20% of the cigarette smokers develop lung cancer. In genetic epidemiological studies, which utilize twin comparisons, the effects of genetic and environmental factors on current smoking, smoking initiation, and smoking persistence have been estimated. One of the conclusions that may be drawn from these studies is that, for most individuals, the liability to become and remain a smoker is explained by additive genetic factors.[3] It has been reported that smoking and nonsmoking relatives of lung cancer patients have an increased risk of developing lung cancer.[4] This increased risk appears to be independent of age, sex, occupation, and smoking[5] and has been confirmed by segregation analyses as compatible with Mendelian co-dominant inheritance of a rare major autosomal gene for lung cancer pre-disposition.[6] However, it is important to note that genetically susceptible individuals will have higher risk only if they are exposed to relevant carcin-ogens.[7] Thus, lung cancer, primarily caused by smoking, serves as a useful model for the study of the interactions between genetic predisposition and environmental exposure.[8]

II. Cigarette smoke carcinogens

Tobacco smoke and its condensate contain a complex mixture of over 3800 compounds; of these ~50 have been identified as being either animal or human carcinogens.[9] Of the several carcinogens present in cigarette smoke, polycyclic aromatic hydrocarbons (PAHs) (e.g., benzo[a]pyrene or B[a]P) and the tobacco-specific nitrosamines (e.g., 4-(methylnitrosamino)-1-(3-pyridyl)-1-butanone or NNK) appear to play a major role in inducing lung cancer.[10] Other classes of chemicals such as aza-arenes, aromatic amines (e.g., 4-ami-nobiphenyl or 4-ABP), heterocyclic aromatic amines, aldehydes, miscella-neous organic compounds (e.g., 1,3-butadiene), and inorganic compounds (e.g., Ni, Cr) are also known to be carcinogenic.[9,10] In addition, tobacco smoke contains volatile compounds (e.g., benzene) and radioelements (e.g., polo-nium-210) that may also play a role in its carcinogenicity.[10] Cigarette smoke also contains free radicals capable of inducing oxidative DNA damage. For example, the tar phase contains stable free radicals, such as catechols, semi-quinones, quinones, hydroquinones, and carbon-centered radicals capable of redox-cycling, which generates reactive oxygen species (ROS) such as hydro-gen peroxide (H_2O_2) and hydroxyl radical ($\cdot OH$). The gas phase of cigarette smoke contains the highly reactive oxygen- and carbon-centered radicals, nitric oxide and peroxynitrile.[11,12]

III. Metabolic activation of cigarette smoke carcinogens and DNA adduct formation

Several of the chemical carcinogens present in cigarette smoke are inactive per se and require metabolic activation (e.g., PAH). Biotransformation of these chemical carcinogens results in the formation of reactive electrophilic intermediates, which covalently bind to nucleophilic sites in DNA forming carcinogen DNA adducts.[13] In a human lung, several of the PAH carcinogens are predominantly activated by cytochrome P4501A1 (CYP1A1) enzyme to their ultimate carcinogenic metabolites, such as diol epoxides which bind covalently to DNA forming bulky aromatic DNA adducts.[14] For example, the major metabolic pathway of B[a]P involves its conversion to benzo[a]pyrene 7,8-diol-9,10-epoxide (BPDE) via B[a]P-7,8-diol by CYP1A1, CYP3A4, and epoxide hydrolase in a two-step reaction. BPDE is an ultimate carcinogenic metabolite of B[a]P and binds covalently to deoxyguanosine (dG) bases forming the major BPDE-dG adduct in lung DNA from cigarette smokers.[15]

Activation of NNK (a nitrosamine present in cigarette smoke) and its main metabolite, 4-(methylnitrosamino)-1-(3-pyridyl)-1-butanol (NNAL) is carried out through α-hydroxylation (α-hydroxylation of the carbons adjacent to the N-nitroso group), predominantly by CYP1A2 enzymes in human lung although CYP2A6, CYP2E1, and CYP2D6 contribute to a lesser extent.[10,16] Alpha-hydroxylation of NNK at the methyl carbon produces α-hydroxymethyl-NNK, which, upon spontaneous loss of formaldehyde, converts to a pyridyloxobutyldiazohydroxide. The latter metabolite reacts with DNA (predominantly dG residues) yielding bulky pyridyloxobutylation adducts. On the other hand, α-hydroxylation of NNK at the methylene carbon produces α-methylenehydroxy-NNK, which spontaneously produces methanediazohydroxide and a keto aldehyde. The former metabolite is converted to methanediazonium ion, which methylates dG or thymidine (T) in DNA forming methylating adducts, such as 7-methylguanine (7-mG), O^6-methylguanine (O^6-mG) and O^4-methylthymidine (O^4-mT).[16]

Similarly, several reactive metabolites generated through metabolic activation of PAHs, N-nitrosamines and aromatic amines are capable of forming bulky DNA adducts.[15] In addition to the alkylating products induced by NNK, several other carcinogens and free radicals generated in the gas and particulate phases of cigarette smoke are known to hydroxylate DNA bases causing simple base modifications or oxidative DNA lesions, e.g., 8-oxodeoxyguanosine (8-oxodG).[17] Some of the aldehydes present in cigarette smoke such as acrolein and crotonaldehyde are known to induce lipid peroxidation generating lipid peroxides capable of forming exocyclic DNA adducts (e.g., etheno adducts, propanodeoxyguanosine adducts) in target tissues of cigarette smokers.[18]

IV. Role of smoking-related DNA adducts in lung cancer

DNA adducts represent the biologically effective dose, defined as the amount of carcinogen bound to DNA in either target tissue or a surrogate tissue taking into account the individual differences in absorption and metabolism of a carcinogen to its DNA reactive intermediate(s), detoxification of the reactive intermediates, and repair of DNA damage. DNA adduct formation is generally accepted as one of the key events in tumor initiation during chemical carcinogenesis.[19] The biological potential of a given DNA adduct depends on its mutagenic potential, ability to be repaired, location within a target gene, and the nature of the target gene.[20] For example, formation of BPDE-dG adducts in mutational hot spots of the *p*53 tumor suppressor gene in lung cancer patients substantiates the association between tobacco carcinogen-induced DNA damage and lung cancer.[21] Total DNA adduct levels, and in some cases specific adducts, have been correlated with *in vitro* mutations,[22,23] chromosomal aberrations,[19,24,25] and generally with carcinogenicity.[26,27] Thus, quantitation of DNA adducts in human cells provides a useful parameter of risk assessment in terms of monitoring human carcinogen exposure, the determination of a biologically effective dose, and individual cell type-specific DNA repair capacity. Detection and quantitation of DNA adducts in human tissues also confirms epidemiological associations of cancer and risk factors and provides information on the identification of carcinogenic hazard and quantitative risk assessments of accumulative genetic damage. Several other biomarkers, such as carcinogen-protein adducts, chromosomal aberrations, polymorphisms in drug metabolizing enzymes, and host DNA repair capacity are also used to estimate the interindividual variation in response to carcinogen exposure and thus assessment of cancer risk.[28,29]

V. Analysis of smoking-related DNA adducts

With recent advances in analytical biochemistry techniques, DNA adducts have been detected and quantified in target and nontarget tissues or surrogate tissues of smokers using highly sensitive methods, such as immunoassays, the ^{32}P-postlabeling assay, and fluorescence assays, with detection limits of one adduct per 10^6–10^{10} normal nucleotides.[30,31] Immunological methods such as competitive enzyme-linked immunosorbent assay (ELISA) require the development of antibodies to the specific adduct of interest. Several polyclonal and monoclonal antibodies are available which can be used in large population studies.[30,31] However, some of these antibodies frequently cross react with structurally related DNA adducts formed by other carcinogens. For example, the antibody raised against BPDE-dG cross reacts with diol epoxide adducts of other PAH carcinogens. Recently, immunoslotblot assays are becoming increasingly popular owing to their high sensitivity (detects one adduct in 10^8 nucleotides) and relatively small amounts of DNA that are required to perform the assay.[30] Monoclonal antibodies raised against specific DNA adducts have also been used in detecting DNA damage *in situ* in tissue sections by immunohistochemical techniques.[32] Scanning synchronous fluorescence spectroscopy (SFS) takes advantage of the fluorescent properties of PAH carcinogens, carcinogen metabolites, and their adducts by simultaneously scanning excitation and emission wavelengths at a fixed wavelength difference; the sensitivity of this method is three to ten adducts per 10^8 nucleotides.[33]^{32}P-postlabeling is another method developed to detect DNA damage caused by structurally diverse chemical classes.[34,35] This method has been widely used in human biomonitoring studies owing to its high sensitivity of detection (detects one adduct in 10^{10} normal nucleotides).[36] Often the cigarette smoke-induced aromatic DNA adducts analyzed by the ^{32}P-postlabeling assay, which uses urea-based solvents, display one or two diagonal radioactive zones (DRZs) representing a mixture of DNA adducts of diverse polarity.[36] Several other physical methods, such as gas chromatography/mass spectrometry (GC/MS), liquid chromatography/mass spectrometry (LC/MS), and high-performance liquid chromatography (HPLC) have been utilized for structural elucidation or fractionation of unknown DNA adducts.[37] In addition, methods to detect oxidative DNA lesions (e.g., 8-oxodG) by HPLC with electrochemical detection (HPLC-ECD) have also been developed.[17,38]

VI. DNA adducts as intermediate biomarkers of lung cancer

A.   Tobacco carcinogen-induced bulky DNA adducts

1.   Target and nontarget tissues

The presence of smoking-related DNA adducts was first established in human placenta,[39] and the lung and larynx of smokers.[40–42] The principal target organs of tobacco smoke-induced respiratory tract carcinogenesis (lung, bronchus/bronchial epithelium, and larynx) consistently exhibit high adduct levels in smokers, while adduct levels in these tissues are absent or low in nonsmokers.[15,36] Smoking-related DNA adducts have also been detected in other target and nontarget tissues, including kidney, bladder, esophagus, pancreas, ascending aorta, liver, and cervix.[42–44] BPDE-dG adducts have also been detected by immunohistochemistry in human sperm and in embryos from smoking couples.[45]

2.   Surrogate tissues

One limitation of DNA adduct measurement in humans is the difficulty of accessing target tissues by noninvasive method. Hence, a number of studies have investigated cigarette smoking related DNA adducts in a surrogate tissue such as peripheral blood cells. These studies have utilized either total white blood cells (WBCs) or fractionated subpopulations of specific blood cell types. However, results from these studies have not been consistent. Of a total of six studies in which the presence of PAH-DNA adducts in total WBC was analyzed by either immunoassay or ^{32}P-postlabeling, in only two were the investigators able to detect a significant increase in adduct levels in smokers compared to nonsmokers.[44, 46–50] Studies in specific white blood cell populations have also produced conflicting results. One study involving analysis of a cell population reportedly containing >90% lymphocytes by ^{32}P-postlabeling demonstrated no difference in adduct levels when compared with smoking exposure.[51] Three other studies using mononuclear cells as surrogates found 1.4-fold to 4-fold higher levels of adducts in smokers than in nonsmokers.[51–53] The ability to detect smoking exposure more easily when using specific white blood cell types may be due to the variable life span and the metabolic capacity of the specific cells. Total white blood cells are approximately 70–75% short-lived granulocytes (half-life of 7 h to 1 day) and 25–30% mononuclear cells.[54] Mononuclear cells isolated by Ficoll are comprised of about 15% monocytes and 85% lymphocytes. Lymphocytes have a reported average life span of 1–5 years[54] while monocytes have a much shorter half-life of 8 h.[55] Therefore, in those studies in which the WBC samples consisted mainly of short-lived granulocytes, the inability to detect differences in adduct levels between smokers and nonsmokers may have been because there was insufficient time to accumulate stable adducts in these cells. Furthermore, it has been reported that smoking-related DNA adduct levels in blood mononuclear cells correlate with smoking-related DNA adduct levels in lung tissue although the levels of adducts in the surrogate tissue are five to six times lower than the levels found in the lung.[56] In addition, it has been shown *in vitro*, that monocytes have a higher metabolic capacity for activating B[a]P than other white blood cell type.[57] Recently, studies from our laboratory have shown a significant increase in lymphocyte aromatic DNA adduct levels of cancer cases compared to age-, gender-, and ethnicity-matched controls among current smokers from a case-control study of lung cancer involving minority populations.[58] Several nonblood cells, such as buccal mucosal cells, nasal mucosal cells, sputum cells, cells from bronchoalveolar lavage, cervical epithelial cells, exfoliated urothelial cells, and spermatocytes have also been used successfully to correlate DNA adduct levels with smoking exposure in human biomonitoring studies employing ^{32}P-postlabeling or immunological methods.[15,59,60]

B. Oxidative DNA damage in smokers

Cigarette smoke, which is a complex mixture of chemical carcinogens, also contains high levels of oxidants. These ROS generated from cigarette smoke can induce DNA single strand breaks *in vitro*[61] as well as cause oxidative DNA damage in cultured human cells[62,63] and are known to be involved in chemical carcinogenesis.[64] Among the various forms of oxidative lesions, 7,8-dihydro-8-oxo-2'-deoxyguanosine (8-oxodG) is the most abundant product of oxidative DNA damage and is a sensitive marker of free radical-mediated DNA damage.[17] Formation of 8-oxodG in DNA has been shown to be associated with mutagenesis[65,66] and carcinogenesis,[67,68] and in several animal studies this oxidative lesion has been shown to be associated with the incidence of cancer.[69–71] Urinary excretion of 8-oxodG has been used extensively as a noninvasive biomarker of oxidative DNA damage in humans *in vivo*.[72] Cigarette smoking has also been shown to generate high levels of 8-oxodG, which is detected in urine.[73] However, Lindahl[74] pointed out that 8-oxodG could also be produced from the oxidation of DNA breakdown products in the kidney and in which case it may not represent generalized oxidative damage of the individual. Alternatively, Kiyosawa et al.,[75] suggested that human peripheral blood lymphocyte DNA could be used as a surrogate tissue to evaluate oxidative DNA damage. Subsequently, cigarette smoke has been shown to induce 8-oxodG formation in human leukocyte DNA.[76,77] In a recent study, we have shown that levels of 8-oxodG in the lymphocyte DNA of lung cancer cases who were current smokers were significantly higher compared to the levels in the corresponding current smokers without lung cancer.[58]

VII. Factors affecting DNA adduct levels in smokers

A. Exposure

Aromatic DNA adduct levels in surrogate or target tissues may also reflect the extent and amount of cigarettes smoked by a smoker. Nontumor bronchial and larynx tissue samples from autopsies or biopsies of smokers have been reported to have high levels of aromatic DNA adducts that were proportional to the number of cigarettes smoked per day and the number of packyears. In contrast, tissues from nonsmokers from these studies had almost undetectable levels of DNA adducts,[15,36,42,78,79] suggesting that DNA adducts formed in the individuals who smoked were exposure-related. Another group found that aromatic DNA adduct levels reached a plateau in lymphocytes and alveolar macrophages at higher exposures to cigarette smoke (> 20 cigarettes/day) suggesting less efficient adduct formation.[80] In a recent study, we have also shown that the levels of aromatic DNA adducts present in peripheral blood lymphocytes peaked at approximately one pack per day and showed a decline beyond this dose in lung cancer cases who were current smokers.[58] There have been reports indicating that African-American or Mexican-American smokers may have an increased risk of developing lung cancer at lower exposures.[81] It has also been reported that individuals who smoked nonfiltered cigarettes had higher levels of DNA adducts compared to those smoking filtered brands.[82]

B. Smoking status

Several studies have reported higher levels of aromatic and/or hydrophobic DNA adducts in target or surrogate tissues of current smokers compared to nonsmokers.[15,36,83] In general, current smokers reportedly have higher levels of DNA damage due to continual exposure to cigarette smoke carcinogens. In contrast, former smokers tend to have less DNA damage compared to current smokers. Although persistent aromatic DNA adducts have been detected in major target tissues of ex-smokers, their levels are often below the levels of the adducts present in current smokers.[36,82] However, two studies reported higher levels of aromatic DNA adducts in the leukocyte or lymphocyte DNA of ex-smokers in comparison to nonsmokers.[58,84] In a different study, the levels of aromatic DNA adducts in former smokers showed a rapid decline initially followed by a slower later phase after smoking cessation.[53] The levels of DNA damage in former smokers is time-dependent; the longer the interval following smoking cessation, the lower the DNA adduct levels. For example, it has been reported that in the lung biopsies of former smokers who quit smoking only 3 months prior to surgery, adduct levels were comparable to current smokers, while those ex-smokers who had quit smoking for at least 5 years had low levels of DNA adducts similar to the levels seen in nonsmokers.[42,79] In a recent study in which all the subjects were Caucasian, we presented preliminary data showing lung cancer cases that were former smokers (quit smoking six months prior to the study) had aromatic DNA adduct levels similar to current smokers.[85] However, one of the problems with such analyses could be the uncertainty regarding the accuracy of self-reporting of smoking behavior by the individuals in the study population.

C. Ethnicity

It has been shown that, because of allelic differences, the ability to metabolize carcinogens may vary with race and hence ethnic background may be one of the factors contributing to DNA damage and hence to lung cancer risk.[86] Lung cancer risk associated with cigarette smoking is higher for African-Americans than Caucasians.[1, 87, 88] However, the subjects in most of the studies that have analyzed aromatic DNA adducts have been Caucasian. To date, there has not been a comprehensive study that compared cigarette smoke-induced DNA damage among different ethnic groups. In a recent study, we examined aromatic DNA adduct levels as related to smoking exposure in a cohort consisting of primarily Mexican-Americans and some African-Americans.[58] We have shown that lung cancer cases of Mexican-Americans had higher levels of DNA damage compared to African-American subjects. However, the relatively small number of subjects of African-American descent in our study may have precluded accurate evaluation of aromatic DNA adduct levels in this ethnic group.[58]

D. Gender

Epidemiological evidence indicates a greater risk of lung cancer for women than men.[89] Gender differences in susceptibility to smoking-related DNA damage have also been reported.[8, 90, 91] Ryberg et al., reported higher DNA adduct levels in lung tissue of female smokers compared to male smokers after adjusting for total tobacco exposure.[90] Also, in another lung cancer study, higher levels of hydrophobic DNA adducts were detected in normal adjacent tissue and a higher frequency of G to T transversions was observed in the p53 tumor suppressor gene in lung tumors in females than in males.[92] In both these studies female smokers had a lower level of exposure to cigarette smoke carcinogens compared to male smokers.[90,91] The differences in susceptibility to cigarette smoke-induced DNA damage could also be explained by the gender differences in the expression of polymorphic drug metabolizing enzymes. For example, in lung cancer patients, the expression levels of lung CYP1A1 are higher in females compared to males and that the CYP1A1 levels of both sexes combined correlated with PAH-DNA adduct levels.[93] The increased risk of developing lung cancer in female smokers compared to male smokers has also been attributed to a greater *GST M1* null (0/0) genotype in females.[91] In contrast, in a recent study we did not find significant differences in aromatic DNA adduct levels in lymphocyte DNA between male and female smokers from a case-control study involving minority populations.[58] Another group has shown higher adduct levels in men than in women, however, their study did not find a significant difference in aromatic DNA adduct levels between lung cancer cases and controls who were current smokers.[94]

E.   Metabolic polymorphisms

Inherited polymorphisms in phase I (CYP enzymes) and phase II (predominantly GSTM1) drug metabolizing enzymes have been suggested to contribute to DNA damage and cancer susceptibility of an individual and may contribute to interindividual differences following exposures to carcinogens or mutagens.[83,95,96] Aromatic DNA adduct levels in lung tissue of cigarette smokers have been related to the presence of particular *CYP1A1*, *GST M1*, and *GST P1* polymorphisms in smokers.[8, 97] In the *CYP1A1* gene (*CYP1A1*1* or *wt*), which encodes for the P4501A1 enzyme involved in the metabolic activation of PAH carcinogens, an *Msp*1 polymorphism in the 3' noncoding region (*CYP1A1*2*) and an Ile-Val polymorphism in exon 7 (*CYP1A1*3* or *m2*) — both of which are found predominantly in Japanese populations and rarely if at all in Caucasians — as well as a *CYP1A1*4* polymorphism specific to African-Americans, have been shown to be associated with increased risk of developing lung cancer in several studies.[98] Higher levels of DNA adducts in WBCs of smokers with the *CYP1A1*3* polymorphism and in cord blood and placenta of newborns with *CYP1A1*2* polymorphisms compared to those without these variants have been reported.[97] *GST M1*, the gene coding for the phase II enzyme *GST M1* involved in detoxification of diol epoxides of PAHs, ethylene oxide and styrene, is absent in 40–50% of the U.S. population.[99] Individuals with a *GSTM1* null (0/0) genotype are shown to have higher aromatic DNA adduct levels in their lung tissue compared to those with a wildtype *GSTM1* and a highly significant correlation has been reported between the absence of GSTM1 activity and adenocarcinoma of the lung.[100] Another subclass of GST, *GST* P1, which has a polymorphic site at codon 104 (A to G substitutions replace isoleucine with valine) was reported to be associated with higher PAH-DNA adducts in WBC of *GST P1 ile/val* and *ile/ile* newborns compared with the adduct levels in *GSTP1 val/val* newborns.[101] Gene-gene interactions also play a critical role in lung cancer susceptibility.[8,97] Although *GST M1 0/0* is a moderately strong susceptibility factor for lung cancer it could become a dominant risk factor when combined with a particular polymorphic form of *CYP1A1*. For example, in the Japanese population, it has been shown that individuals that possessed both *CYP1A1 m2/m2* and *GST M1 0/0* genotypes had an eightfold increased frequency of p53 mutations than those with neither genotype.[97, 102] Also, among lung cancer cases with at least one G allele of the *GST P1* gene there is an increase in the frequency of the *GST M1 0/0* genotype relative to the frequency observed in the normal population, suggesting that the combined genotype (*GST M1* and *GST P1*) may play an important role in lung cancer susceptibility.[103]

F.   DNA repair

Not only is the balance between metabolic activation and detoxification of xenobiotic compounds critical in determining individual susceptibility to lung cancer, but the DNA repair capacity of an individual can modulate susceptibility as well.[29,104,105] Diminished DNA repair capacity of an individual may result in increased accumulation of DNA adducts in target and nontarget tissues, leading to mutations in specific genes and eventually leading to carcinogenesis.[29] It has been shown that under *in vitro* culture conditions, lymphocytes from lung cancer cases tend to form higher levels of DNA adducts compared to the lymphocytes from cancer free controls upon BPDE treatment.[106] These observations parallel with the frequency of BPDE-induced chromosomal aberrations, which were significantly higher in lung cancer cases than in controls.[107] A reduction in the DNA repair of specific DNA lesions, such as O^6-methylguanine has also been reported in fibroblast cultures of lung cancer patients.[108]

Epidemiological studies have indicated that increased DNA adduct levels and reduced DNA repair capacity are associated with increased risk of lung cancer.[104] Nucleotide excision repair (NER) is one major pathway capable of removing a variety of structurally unrelated bulky DNA adducts including those induced by tobacco carcinogens.[109] A reduction in the expression level of NER genes could also be associated with a higher risk of lung cancer. Recently, Cheng et al.[110] in an age-, gender-, ethnicity-, and tobacco use-matched case-control study of lung cancer have shown that lung cancer patients are more likely than the controls to have a reduced expression level of XPG/ERCC5 (XPG, xeroderma pigmentosum complementary group B; ERCC, excision-repair cross-complementing) and CSB/ERCC6 (CSB, Cockayne's syndrome complementary group B).

G.   Mutagen sensitivity

Inherited predisposition is an important component of risk characterization for a given cancer. Cytogenetic assays have been used to examine the exposure and response of human beings to carcinogens based on the hypothesis that the extent of genetic damage in lymphocytes may reflect critical events in carcinogenesis in the target tissues. To examine the interindividual differences in susceptibility to carcinogens, Hsu et al.[111] developed the mutagen sensitivity assay in which the frequency of *in vitro* bleomycin-induced chromatid breaks (BICS) in short-term peripheral lymphocyte cultures is quantified as a measure of cancer susceptibility. Mutagen sensitivity (defined as equaling one break/cell) has been shown to be an independent risk factor for upper aerodigestive tract cancers after adjustment for tobacco and alcohol consumption[112,113] and lung cancer.[111] In a recent case-control study of lung cancer involving minority populations, we determined that individuals who were mutagen sensitive had higher levels of aromatic DNA adducts compared to those who were not sensitive, and that adduct levels of cases but not controls (without lung cancer) correlate with BICS in lymphocyte cultures.[58] In a separate epidemiological study that examined DNA repair capacity in relation to susceptibility, approximately 53.8% of 132 African-American and Mexican-American lung cancer cases were mutagen sensitive compared with 22.4% of the 232 age-, sex-, and ethnicity-matched controls.[29] Furthermore, it has been shown that mutagen sensitivity correlates with DNA repair capacity, suggesting that individuals with reduced DNA repair capacity tend to be at a greater risk of developing lung cancer.[114] It is worth noting, however, that because of the complexity of the factors involved in conferring individual susceptibility and risk, it is likely that a subset of individuals who are not mutagen sensitive may still develop cancer due to reasons other than DNA repair.

References

1. American Cancer Society, *Cancer Facts and Figures 2000*, American Cancer Society, Atlanta, GA, 2002.
2. Shopland, D.R., Eyre, H.J., and Pechacek, T.F., Smoking-attributable cancer mortality in 1991: is lung cancer now the leading cause of death among smokers in the United States? *J. Natl. Cancer Inst.*, 83, 1142–1148, 1991.
3. Bergen, A.W. and Caporaso, N., Cigarette smoking, *J. Natl. Cancer Inst.*, 91, 1365–1375, 1999.
4. Ooi, W.L. et al., Increased familial risk for lung cancer, *J. Natl. Cancer Inst.*, 76, 217–222, 1986.
5. Sellers, T.A. et al., Familial risk of cancer among randomly selected cancer probands, *Genet. Epidemiol.*, 5, 381–391, 1988.
6. Sellers, T.A. et al., Evidence for Mendelian inheritance in the pathogenesis of lung cancer, *J. Natl. Cancer Inst.*, 82, 1272–1279, 1990.
7. Sellers, T.A. et al., Lung cancer detection and prevention: evidence for an interaction between smoking and genetic predisposition, *Cancer Res.*, 52, 2694s–2697s, 1992.
8. Haugen, A. et al., Gene-environment interactions in human lung cancer, *Toxicol. Lett.*, 112–113, 233–237, 2000.
9. IARC, IARC monographs on the evaluation of carcinogenic risk of chemicals to human beings: tobacco Smoking, *IARC Monogr. Eval. Carcinog. Risk Chem. Hum.*, IARC, Lyon, France, 1986.
10. Hecht, S.S., Tobacco smoke carcinogens and lung cancer, *J. Natl. Cancer Inst.*, 91, 1194–1210, 1999.
11. Church, D.F. and Pryor, W.A., Free-radical chemistry of cigarette smoke and its toxicological implications, *Environ. Health Perspect.*, 64, 111–126, 1985.
12. Pryor, W.A., Prier, D.G., and Church, D.F., Electron-spin resonance study of mainstream and sidestream cigarette smoke: nature of the free radicals in gas-phase smoke and in cigarette tar, *Environ. Health Perspect.*, 64, 345–355, 1983.
13. Guengerich, F.P., Metabolism of chemical carcinogens, *Carcinogenesis*, 21, 345–351, 2000.
14. Baird, W.M. and Ralston, S.L., in *Chemical Carcinogens and Anticarcinogens* Bowden, G.T. and Fischer, S.M., Eds., Elsevier Science, New York, 1997, pp. 171–200.

15. Phillips, D.H., DNA adducts in human tissues: biomarkers of exposure to carcinogens in tobacco smoke, *Environ. Health Perspect.*, 104 (Suppl. 3), 453–458, 1996.
16. Hecht, S.S., DNA adduct formation from tobacco-specific *N*-nitrosamines, *Mutat. Res.*, 424, 127–142, 1999.
17. Kasai, H., Analysis of a form of oxidative DNA damage, 8-hydroxy-2'-deoxyguanosine, as a marker of cellular oxidative stress during carcinogenesis, *Mutat. Res.*, 387, 147–163, 1997.
18. Nath, R.G. et al., 1,N2-propanodeoxyguanosine adducts: potential new biomarkers of smoking-induced DNA damage in human oral tissue, *Cancer Res.*, 58, 581–584, 1998.
19. Bartsch, H., DNA adducts in human carcinogenesis: etiological relevance and structure-activity relationship, *Mutat. Res.*, 340, 67–79, 1996.
20. Bohr, V.A., Gene-specific damage and repair of DNA adducts and cross-links, *IARC Sci. Publ.*, 125, 361–369, 1994.
21. Denissenko, M.F. et al., Preferential formation of benzo[a]pyrene adducts at lung cancer mutational hotspots in *p53*, *Science*, 274, 430–432, 1996.
22. McCormick, J.J. and Maher, V.M., Cytotoxic and mutagenic effects of specific carcinogen-DNA adducts in diploid human fibroblasts, *Environ. Health Perspect.*, 62, 145–155, 1985.
23. Beland, F.A. et al., Arylamine-DNA adducts *in vitro* and *in vivo*: their role in bacterial mutagenesis and urinary bladder carcinogenesis, *Environ. Health Perspect.*, 49, 125–134, 1983.
24. Morris, S.M. et al., Relationships between specific DNA adducts, mutation, cell survival, and SCE formation, *Basic Life Sci.*, 29, 353–360, 1984.
25. Talaska, G. et al., The correlation between DNA adducts and chromosomal aberrations in the target organ of benzidine exposed, partially-hepatectomized mice, *Carcinogenesis*, 8, 1899–1905, 1987.
26. Poirier, M.C. and Beland, F.A., DNA adduct measurements and tumor incidence during chronic carcinogen exposure in animal models: implications for DNA adduct-based human cancer risk assessment, *Chem. Res. Toxicol.*, 5, 749–755, 1992.
27. Otteneder, M. and Lutz, W.K., Correlation of DNA adduct levels with tumor incidence: carcinogenic potency of DNA adducts, *Mutat. Res.*, 424, 237–247, 1999.
28. Bartsch, H., Studies on biomarkers in cancer etiology and prevention: a summary and challenge of 20 years of interdisciplinary research, *Mutat. Res.*, 462, 255–279, 2000.
29. Wei, Q. and Spitz, M.R., The role of DNA repair capacity in susceptibility to lung cancer: a review. *Cancer Metastasis Rev.*, 16, 295–307, 1997.
30. Farmer, P.B. and Shuker, D.E., What is the significance of increases in background levels of carcinogen-derived protein and DNA adducts? Some considerations for incremental risk assessment, *Mutat. Res.*, 424, 275–286, 1999.
31. Poirier, M.C., Santella, R.M., and Weston, A., Carcinogen macromolecular adducts and their measurement, *Carcinogenesis*, 21, 353–359, 2000.
32. Santella, R.M., Immunological methods for detection of carcinogen-DNA damage in humans, *Cancer Epidemiol. Biomarkers Prev.*, 8, 733–739, 1999.
33. Vahakangas, K., Haugen, A., and Harris, C.C., An applied synchronous fluorescence spectrophotometric assay to study benzo[a]pyrene-diolepoxide-DNA adducts, *Carcinogenesis*, 6, 1109–1115, 1985.

34. Reddy, M. and Randerath, K., Nuclease P1-mediated enhancement of sensitivity of ^{32}P-postlabeling test for structurally diverse DNA adducts, *Carcinogenesis*, 7, 1543–1551, 1986.

35. Gupta, R. and Early, K., ^{32}P-adduct assay: comparative recoveries of structurally diverse DNA adducts in the various enhancement procedures, *Carcinogenesis*, 9, 1687–1693, 1988.

36. Beach, A. and Gupta, R., Human biomonitoring and the ^{32}P-postlabeling assay, *Carcinogenesis*, 13, 1053–1074, 1992.

37. Weston, A., Physical methods for the detection of carcinogen-DNA adducts in humans, *Mutat. Res.*, 288, 19–29, 1993.

38. Floyd, R.A. et al., Hydroxyl free radical adduct of deoxyguanosine: sensitive detection and mechanism of formation, *Free Rad. Res. Commun.*, 1, 163–172, 1986.

39. Everson, R.B. et al., Detection of smoking-related covalent DNA adducts in human placenta, *Science*, 231, 54–57, 1986.

40. Randerath, E. et al., Comparative ^{32}P-analysis of cigarette smoke-induced DNA damage in human tissues and mouse skin, *Cancer Res.*, 46, 5869–5877, 1986.

41. Randerath, E. and Randerath, K., Monitoring tobacco smoke-induced DNA damage by ^{32}P-postlabelling, *IARC Sci. Publ.*, 124, 305–314, 1993.

42. Randerath, E. et al., Covalent DNA damage in tissues of cigarette smokers as determined by ^{32}P-postlabeling assay, *J. Natl. Cancer Inst.*, 81, 341–347, 1989.

43. Cuzick, J. et al., DNA adducts in different tissues of smokers and non-smokers, *Int.J. Cancer*, 45, 673–678, 1990.

44. Jones, N.J., McGregor, A.D., and Waters, R., Detection of DNA adducts in human oral tissue: correlation of adduct levels with tobacco smoking and differential enhancement of adducts using the butanol extraction and nuclease P1 versions of ^{32}P-postlabeling, *Cancer Res.*, 53, 1522–1528, 1993.

45. Zenzes, M.T., Smoking and reproduction: gene damage to human gametes and embryos, *Hum. Reprod. Update*, 6, 122–131, 2000.

46. Perera, F.P. et al., DNA adducts, protein adducts, and sister chromatid exchange in cigarette smokers and nonsmokers, *J. Natl. Cancer Inst.*, 79, 449–456, 1987.

47. Perera, F.P. et al., Comparison of DNA adducts and sister chromatid exchange in lung cancer cases and controls, *Cancer Res.*, 49, 4446–4451, 1989.

48. Savela, K. et al., Interlaboratory comparison of the ^{32}P-postlabelling assay for aromatic DNA adducts in white blood cells of iron foundry workers, *Mutat. Res.*, 224, 485–492, 1989.

49. Reddy, M.V., Hemminki, K., and Randerath, K., Postlabeling analysis of polycyclic aromatic hydrocarbon-DNA adducts in white blood cells of foundry workers, *J. Toxicol. Environ. Health*, 34, 177–185, 1991.

50. Liou, S.H. et al., Biological monitoring of fire fighters: sister chromatid exchange and polycyclic aromatic hydrocarbon-DNA adducts in peripheral blood cells, *Cancer Res.*, 49, 4929–4935, 1989.

51. Savela, K. and Hemminki, K., DNA adducts in lymphocytes and granulocytes of smokers and nonsmokers detected by the ^{32}P-postlabelling assay, *Carcinogenesis*, 12, 503–508, 1991.

52. Jahnke, G.D. et al., Multiple DNA adducts in lymphocytes of smokers and nonsmokers determined by ^{32}P-postlabeling analysis, *Carcinogenesis*, 11, 205–211, 1990.

53. Schoket, B., Kostic, S., and Vincze, I., Determination of smoking-related DNA adducts in lung-cancer and noncancer patients, *IARC Sci. Publ.*, 124, 315–319, 1993.

54. Buckton, K.E., Brown, W.M., and Smith, P.G., Lymphocyte survival in men treated with x-rays for ankylosing spondylitis, *Nature*, 214, 470–473, 1967.

55. Meuret, G., Human monocytopoiesis, *Exp. Hematol.*, 2, 238–249, 1974.

56. Wiencke, J.K. et al., Correlation of DNA adducts in blood mononuclear cells with tobacco carcinogen-induced damage in human lung, *Cancer Res.*, 55, 4910–4914, 1995.

57. Holz, O., Krause, T., and Rudiger, H.W., Differences in DNA adduct formation between monocytes and lymphocytes after *in vivo* incubation with benzo[a]pyrene, *Carcinogenesis*, 12, 2181–2183, 1991.

58. Vulimiri, S.V. et al., Analysis of aromatic DNA adducts and 7,8-dihydro-8-oxo-2'-deoxyguanosine in lymphocyte DNA from a case-control study of lung cancer involving minority populations, *Mol. Carcinog.*, 27, 34–46, 2000.

59. Salama, S.A., Serrana, M., and Au, W.W., Biomonitoring using accessible human cells for exposure and health risk assessment, *Mutat. Res.*, 436, 99–112, 1999.

60. Santella, R.M., DNA damage as an intermediate biomarker in intervention studies, *Proc. Soc. Exp. Biol. Med.*, 216, 166–171, 1997.

61. Borish, E.T. et al., Cigarette tar causes single-strand breaks in DNA, *Biochem. Biophys. Res. Comm.*, 133, 780–786, 1985.

62. Nakayama, T., et al., Cigarette smoke induces DNA single-strand breaks in human cells, *Nature*, 314, 462–464, 1985.

63. Leanderson, P. and Tagesson, C., Cigarette smoke-induced DNA damage in cultured human lung cells: role of hydroxyl radicals and endonuclease activation, *Chem.-Biol. Interact.*, 81, 197–208, 1992.

64. Pryor, W.A., Cigarette smoke and the involvement of free radical reactions in chemical carcinogenesis, *Br. J. Cancer*, 55, 19–23, 1987.

65. Moriya, M. et al., Site-specific mutagenesis using a gapped duplex vector: A study of translesion synthesis past 8-oxodeoxyguanosine in *E. coli*, *Mutat. Res.*, 254, 281–288, 1991.

66. Cheng, K.C. et al., 8-hydroxydeoxyguanosine, an abundant form of oxidative DNA damage causes G to T and A to C substitutions, *J. Biol. Chem.*, 267, 166–172, 1992.

67. Kamiya, H. et al., c-Ha-*ras* containing 8-hydroxyguanine at codon 12 induces point mutations at the modified and adjacent positions, *Cancer Res.*, 52, 3483–3485, 1992.

68. Floyd, R.A., The role of 8-hydroxydeoxyguanine in carcinogenesis, *Carcinogenesis*, 11, 1447–1450, 1990.

69. Kasai, H. et al., Oral administration of the renal carcinogen, potassium bromate, specifically produces 8-hydroxydeoxyguanosine in rat target organ DNA, *Carcinogenesis*, 8, 1959–1961, 1987.

70. Fiala, E.S., Conaway, C.C., and Mathis, J.E., Oxidative DNA and RNA damage in the livers of Sprague-Dawley rats treated with the hepatocarcinogen 2-nitropropane, *Cancer Res.*, 49, 2603–2605, 1989.

71. Umemura, U. et al., Oxidative DNA damage and cell proliferation in kidneys of male and female rats during 13-weeks exposure to potassium bromate (KBrO3), *Arch. Toxicol.*, 72, 264–269, 1998.

72. Loft, S. et al., 8-hydroxydeoxyguanosine as a urinary biomarker of oxidative DNA damage, *J. Toxicol. Environ. Health*, 40, 391–404, 1993.
73. Loft, S. et al., Oxidative DNA damage estimated by 8-hydroxydeoxyguanosine excretion in humans: influence of smoking, gender and body mass index, *Carcinogenesis*, 13, 2241–2247, 1992.
74. Lindahl, T., Instability and decay of the primary structure of DNA, *Nature*, 362, 709–715, 1993.
75. Kiyosawa, H. et al., in *Medical, Biochemical and Chemical Aspects of Free Radicals*, Hayashi, O., Niki, E., Kondo, M. and Yoshikawa, T., Eds., Elsevier Science Publishers, New York, 1991, pp. 1511–1512.
76. Kiyosawa, H. et al., Cigarette smoking induces formation of 8-hydroxydeoxyguanosine, one of the oxidative DNA damages in human peripheral leukocytes, *Free Rad. Res. Commun.*, 11, 23–27, 1990.
77. Asami, S. et al., Increase of a type of oxidative DNA damage, 8-hydroxyguanine, and its repair activity in human leukocytes by cigarette smoking, *Cancer Res.*, 56, 2546–2549, 1996.
78. Kriek, E. et al., DNA adducts as a measure of lung cancer risk in humans exposed to polycyclic aromatic hydrocarbons, *Environ. Health Perspect.*, 99, 71–75, 1993.
79. Phillips, D.H. et al., Correlation of DNA adduct levels in human lung with cigarette smoking, *Nature*, 336, 790–792, 1988.
80. van Schooten, F.J. et al., ^{32}P-postlabelling of aromatic DNA adducts in white blood cells and alveolar macrophages of smokers: saturation at high exposures, *Mutat. Res.*, 378, 65–75, 1997.
81. Ishibe, N. et al., Susceptibility to lung cancer in light smokers associated with CYP1A1 polymorphisms in Mexican- and African-Americans, *Cancer Epidemiol. Biomarkers Prev.*, 6, 1075–1080, 1997.
82. van Schooten, F.J. et al., Polycyclic aromatic hydrocarbon-DNA adducts in lung tissue from lung cancer patients, *Carcinogenesis*, 11, 1677–1681, 1990.
83. Spivack, S.D. et al., Molecular epidemiology of lung cancer, *Crit. Rev. Toxicol.*, 27, 319–365, 1997.
84. Tang, D. et al., A molecular epidemiological case-control study of lung cancer, *Cancer Epidemiol. Biomarkers Prev.*, 4, 341–346, 1995.
85. Vulimiri, S.V. et al., Analysis of aromatic DNA adducts and oxidative DNA damage in lymphocyte DNA of Caucasians from a lung cancer case-control study, Nieburgs, N.E., Ed., 5th International Symposium on Predictive Oncology and Therapy, International Society for Preventive Oncology, Geneva, Switzerland, 2000.
86. Perera, F.P., Molecular epidemiology: insights into cancer susceptibility, risk assessment, and prevention, *J. Natl. Cancer Inst.*, 88, 496–509, 1996.
87. Harris, R.E. et al., Race and sex differences in lung cancer risk associated with cigarette smoking, *Int. J. Epidemiol.*, 22, 592–599, 1993.
88. Caraballo, R.S. et al., Racial and ethnic differences in serum cotinine levels of cigarette smokers: Third National Health and Nutrition Examination Survey, 1988–1991, *J. Amer. Med. Assn.*, 280, 135–139, 1998.
89. Risch, H.A. et al., Are female smokers at higher risk for lung cancer than male smokers? A case-control analysis by histologic type, *Am. J. Epidemiol.*, 138, 281–293, 1993.
90. Ryberg, D. et al., Different susceptibility to smoking-induced DNA damage among male and female lung cancer patients, *Cancer Res.*, 54, 5801–5803, 1994.

91. Tang, D.L. et al., Associations between both genetic and environmental bio-markers and lung cancer: evidence of a greater risk of lung cancer in women smokers, *Carcinogenesis*, 19, 1949–1953, 1995.

92. Kure, E.H. et al., p53 mutations in lung tumours: relationship to gender and lung DNA adduct levels, *Carcinogenesis*, 17, 2201–2205, 1996.

93. Mollerup, S. et al., Sex differences in lung CYP1A1 expression and DNA adduct levels among lung cancer patients, *Cancer Res.*, 59, 3317–3320, 1999.

94. Hou, S.M., Lambert, B. and Hemminki, K., Relationship between hprt mutant frequency, aromatic DNA adducts and genotypes for GSTM1 and NAT2 in bus maintenance workers, *Carcinogenesis*, 16, 1913–1917, 1995.

95. Wormhoudt, L.W., Commandeur, J.N., and Vermeulen, N.P., Genetic poly-morphisms of human N-acetyltransferase, cytochrome P450, glutathione-S-transferase, and epoxide hydrolase enzymes: relevance to xenobiotic metab-olism and toxicity, *Crit. Rev. Toxicol.*, 29, 59–124, 1999.

96. Perera, F.P., Environment and cancer: who are susceptible? *Science*, 278, 1068–1073, 1997.

97. Bartsch, H. et al., Genetic polymorphism of CYP genes, alone or in combina-tion, as a risk modifier of tobacco-related cancers, *Cancer Epidemiol. Biomarkers Prev.*, 9, 3–28, 2000.

98. IARC, *Metabolic polymorphisms and susceptibility to cancer*, Vineis, P. et al., Eds., *IARC Sci. Publ.*, Lyon, France, 1999.

99. Strange, R.C. and Fryer, A.A., Chapter 19. The glutathione S-transferases: Influence of polymorphism on cancer susceptibility, *IARC Sci. Publ.*, 148, 231–249, 1999.

100. Ketterer, B. et al., The human glutathione S-transferase supergene family, its polymorphism, and its effects on susceptibility to lung cancer, *Environ. Health Perspect.*, 98, 87–94, 1992.

101. Whyatt, R.M. et al., Association between polycyclic aromatic hydrocarbon-DNA adduct levels in maternal and newborn white blood cells and glu-tathione S-transferase P1 and CYP1A1 polymorphisms, *Cancer Epidemiol. Bio-markers Prev.*, 9, 207–212, 2000.

102. Kawajiri, K. et al., Association of CYP1A1 germ line polymorphisms with mutations of the p53 gene in lung cancer, *Cancer Res.*, 56, 72–76, 1996.

103. Ryberg, D. et al., Genotypes of glutathione transferase M1 and P1 and their significance for lung DNA adduct levels and cancer risk, *Carcinogenesis*, 18, 1285–1289, 1997.

104. Qingyi, W. et al., Repair of tobacco carcinogen-induced DNA adducts and lung cancer risk: a molecular epidemiologic study, *J. Natl Cancer Inst.*, 92, 1764–1772, 2000.

105. Spitz, M.R. et al., Genetic susceptibility to tobacco carcinogenesis, *Cancer Invest.*, 17, 645–659, 1999.

106. Li, D. et al., *In vitro* induction of benzo[a]pyrene diol epoxide-DNA adducts in peripheral lymphocytes as a susceptibility marker for human lung cancer, *Cancer Res.*, 56, 3638–3641, 1996.

107. Wei, Q. et al., Benzo[a]pyrene diol epoxide-induced chromosomal aberrations and risk of lung cancer, *Cancer Res.*, 56, 3975–3979, 1996.

108. Rudiger, H.W. et al., Reduced O^6-methylguanine repair in fibroblast cultures from patients with lung cancer, *Cancer Res.*, 49, 5623–5626, 1989.

109. Cleaver, J.E., Defective repair replication of DNA in xeroderma pigmentosum, *Nature*, 218, 652–656, 1968.

110. Cheng, L. et al., Reduced expression levels of nucleotide excision repair genes in lung cancer: a case-control analysis, *Carcinogenesis*, 21, 1527–1530, 2000.
111. Hsu, T.C. et al., Sensitivity to genotoxic events of bleomycin in humans: possible relationship to environmental carcinogens, *Int. J. Cancer*, 43, 403–409, 1989.
112. Spitz, M.R. et al., Chromosome sensitivity to bleomycin-induced mutagenesis, an independent risk factor for upper aerodigestive tract cancers, *Cancer Res.*, 49, 4626–4628, 1989.
113. Spitz, M.R. et al., Mutagen sensitivity as a predictor of tumor recurrence in patients with cancer of the upper aerodigestive tract, *J. Natl. Cancer Inst.*, 90, 243–245, 1998.
114. Wei, Q. et al., Reduced DNA repair capacity in lung cancer patients, *Cancer Res.*, 56, 4103–4107, 1996.1.

chapter nine

Tamoxifen–DNA adducts: biomarkers for drug-induced endometrial cancer

Shinya Shibutani, Naomi Suzuki, and Arthur P. Grollman

Contents

Abstract Effective utilization of the biomarker paradigm in molecular epidemiology requires accurate information regarding exposure and internal dose. These parameters are difficult to assess for environmental carcinogens though, for therapeutic agents, they are precisely known. Therefore, drugs provide useful models for biomarker research. The antiestrogen tamoxifen, used in the treatment of breast cancer, has recently been approved for chemoprophylaxis in women at high risk of developing this disease. We have shown that tamoxifen forms covalent DNA adducts in the endometrial tissue of a large fraction of women taking this drug and that tamoxifen–DNA adducts are strongly mutagenic in mammalian cells. Tamoxifen–DNA adducts fulfill an important criteria for biomarkers, namely, their presence reflects a genotoxic effect of a chemical agent, which in this case, may be used to predict a woman's risk of developing endometrial cancer. Our data illustrate the importance of individual variability and target tissue specificity and suggest that individuals who form tamoxifen–DNA adducts may have a predisposition to develop endometrial cancer. The methodology described can be used to test the hypothesis that the carcinogenic effects of tamoxifen on the endometrium can be separated from the chemoprotective effects of this drug.

I. Introduction

Molecular epidemiologic studies involving biomarkers are frequently designed to identify environmental carcinogens.[1] Effective utilization of the biomarker paradigm requires accurate information regarding background level, level of exposure, internal dose, biologically effective dose, early biological effect, altered structure–function, and importantly, genetically determined susceptibility factors.[2] However, even when information regarding these parameters is available, as in the case of aflatoxin, a human hepatocarcinogen,[3] it is difficult to predict risks to the individual exposed person.

For environmental carcinogens, the most difficult parameters to assess are exposure and internal dose. In contrast, these factors are precisely known for therapeutic agents. Detailed information regarding metabolic pathways and mechanism(s) of toxicity are also available for established drugs.

Paradoxically, cancer chemotherapeutic agents themselves may be carcinogenic. Several drugs in this class form covalent adducts with DNA. DNA adducts are reliable biomarkers for DNA damage, an initiating event in the development of cancer.[4]

In this paper, we review our investigations into the genetic toxicology of tamoxifen, a drug which serves as a model for biomarker research. This assessment of genotoxicity has practical importance for the large numbers of otherwise healthy women who consider taking tamoxifen to reduce their risk of developing breast cancer. Our data illustrate the importance of interindividual variability and target tissue specificity and suggest that individuals that form tamoxifen–DNA adducts may have a predisposition to develop endometrial cancer.

II. Tamoxifen is a carcinogen

The antiestrogen tamoxifen is widely used as first-line endocrine therapy for breast cancer patients; more than 500,000 women in the United States are currently being treated with this drug.[5] A randomized clinical trial designed for healthy women at high risk of developing this disease showed that therapeutic doses of tamoxifen reduced the risk of invasive breast cancer by approximately 50%.[6] Subsequently, tamoxifen was approved in 1998 for use as a chemopreventive agent. Unfortunately, the use of tamoxifen in breast cancer patients is associated with an increased risk of endometrial cancer.[7–12] A similar observation was made during the breast cancer chemo-prevention trial.[6]

Tamoxifen is listed as a human carcinogen by the IARC.[13] The cellular mechanism responsible for this carcinogenic effect has not been defined.[14–16] Since tamoxifen induces hepatocellular carcinomas in rats[17–19] and tamoxifen-DNA adducts have been detected in rat liver, [14,20–22] it is possible that tamoxifen or its metabolites may initiate endometrial cancer by a genotoxic mechanism. However, women treated with tamoxifen do not develop hepatocellular cancer. The failure to detect tamoxifen-DNA adducts[23, 24] in human endometrial tissue[25] or in the liver[26] has been cited in favor of an alternative hypothesis[16, 24] that estrogenic and/or other epigenetic effects of tamoxifen account for the carcinogenic properties of the drug.

A. Formation of tamoxifen-DNA adducts in rodents

Tamoxifen is metabolized in the liver of rodents and humans to α-hydroxytamoxifen (α-OHTAM), N-desmethyltamoxifen (N-desmethylTAM), tamoxifen N-oxide (TAM N-oxide), and 4-hydroxytamoxifen (4-OHTAM).[27–29] We recently found that tamoxifen α-sulfate reacts with the exocyclic amino group of guanine in DNA to form two *trans* (fr-1 & fr-2) and two *cis* (fr-3 & fr-4) diastereoisomers of α-(N^2-deoxyguanosyl)tamoxifen (dG-N^2-TAM) (Figure 9.1),[30] as had previously been observed with α-acetoxytamoxifen.[21,30,31]. *Trans* and *cis* forms of dG-N^2-TAM are produced via a short-lived carbocation intermediate.[32] α-OHTAM is a substrate for rat and human hydroxysteroid sulfotransferases; suggesting a metabolic pathway by which tamoxifen could be activated to react with DNA and thereby exert genotoxic effects in target tissues.[33–35]

α-(N^2-Deoxyguanosinyl)tamoxifen N-oxide (dG-N^2-TAM N-oxide) was formed when dG was reacted with α-acetoxytamoxifen N-oxide.[36]. Although dG-N^2-TAM adducts were formed primarily in the liver of mice treated with tamoxifen, dG-N^2-TAM N-oxide adduct also was detected; the *trans*- and *cis*-forms of dG-N^2-TAM N-oxide accounted for 7.2% and 0.7%, respectively, of the total tamoxifen-DNA adducts observed.[37] Formation of dG-N^2-TAM N-oxide adducts may be due to the N-oxidation activity by flavin-containing monooxygenase in the mouse liver.[38–41] Massspectroscopic analysis indicated that α-(N^2-deoxyguanosinyl)-N-desmethyltamoxifen (dG-N-desmethyl-TAM) is a major adduct, in addition to dG-N^2-TAM, in the liver of rats treated with tamoxifen.[42] α-Acetoxy-N-desmethylTAM was recently prepared as a model for activated forms of N-desmethylTAM.[43,44] This derivative was highly reactive, resulting in a mixture of two *trans*-diastereoisomers or two *cis*-diastereoisomers of dG-N^2-N-desmethylTAM. This compound could also be used as a standard to identify tamoxifen adducts in the liver of rats treated with N-desmethylTAM.[45,46]

α-(N^2-Deoxyguanosinyl)-4-hydroxytamoxifen (dG-N^2-4-OHTAM) is formed when 4-OHTAM quinone methide,[47] produced by oxidation of 4-

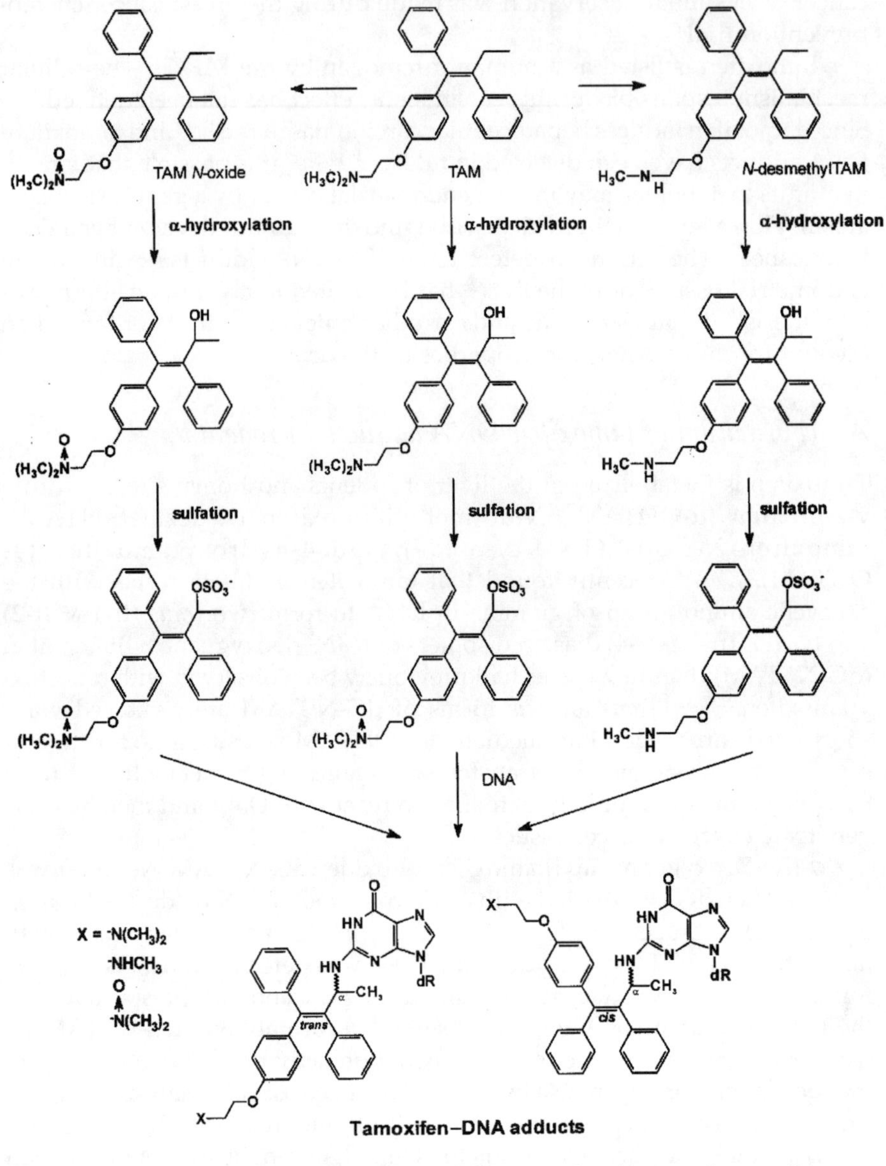

Figure 9.1 Formation of tamoxifen–DNA adducts via O-sulfation of tamoxifen metabolites.

OHTAM, reacts with dG *in vitro*.[48] However, this adduct was not found in the liver of rats treated with tamoxifen, α-OHTAM, or 4-OHTAM.[49,50] 4-OHTAM may not be involved in the formation of tamoxifen–DNA adducts. Thus, tamoxifen–DNA adducts are formed primarily via sulfation or acetylation of α-hydroxylation of tamoxifen and its metabolites, N-desmethylTAM and TAM N-oxide.

Table 9.1 Mutagenic Potential of dG-N2-TAM in Mammalian Cells

Mutation	dG-N^2-TAM			
	trans-1	*trans*-2	*cis*-1	*cis*-2
G → T	1.1%	9.6%	10.9%	12.3%
G → A	1.5	2.8	1.7	1.7
G → C	0.7	0	0.8	0
Δ^1	0	0	0	0
Total	3.3	12.4	13.4	14.0

B.   Mutagenic potential of dG-N^2-tamoxifen DNA adducts

Among several tamoxifen–DNA adducts described above, the mutagenic potential of dG-N^2-TAM were established using site-specific mutagenesis.[51] dG-N^2-TAM adducts promoted primarily G→T transversions, along with small numbers of G→A transitions (Table 9.1). Except for a *trans*-diastereoisomer (fr-1) of dG-N^2-TAM, mutational frequencies were 12.4–14.0%, slightly higher than that observed with dG-C8-AAF, a model chemical carcinogen.[52] The mutagenic specificities were similar to those observed in primer extension reactions catalyzed by mammalian DNA polymerases on dG-N^2-TAM-modified DNA templates[53] and in the liver DNA of lambda/*lacI* transgenic rats.[54] Thus, dG-N^2-TAMs are mutagenic lesions in mammalian cells.

C.   Detection of tamoxifen–DNA adducts in human endometrium

[32]P-postlabeling combined with chromatography has been used to detect tamoxifen–DNA adducts in endometrial tissue.[23,25,55] However, conflicting evidence has been published regarding the detection of tamoxifen–DNA adducts in human tissues. Using a ^{32}P-postlabelling-TLC technique, Carmichael and his colleagues failed to detect tamoxifen adducts in the endometrium of tamoxifen-treated patients.[23] Applying a ^{32}P-postlabeling HPLC analysis, Hemminki et al. detected a putative tamoxifen-induced adduct in endometrial tissues obtained from breast cancer patients:[25] the level of tamoxifen adducts reported was 0.29–0.82 adducts/10^8 bases; standard markers were not used. Also using a ^{32}P-postlabeling-HPLC analysis, Carmichael et al.[24] reported they were unable to reproduce the results of the study by Hemminki et al.[25]

Using a butanol extraction procedure and ^{32}P-postlabeling-TLC[55] we have unequivocally identified tamoxifen-DNA adducts in endometrial tissue; the level of tamoxifen adducts were 1.5–13.1 adducts/10^8 bases. This analytical method does not resolve the two *trans*-diastereoisomers of dG-N^2-TAM or separate them from other tamoxifen-DNA adducts. Therefore, the analytical procedure was improved by coupling high resolution ^{32}P-postlabeling HPLC[49] with partial purification of DNA adducts.[56] Using this method, we found that *cis*- and *trans*-tamoxifen-DNA adducts are present in significant amounts in endometrial tissue in eight of 16 women treated

Table 9.2   Level of Tamoxifen–DNA Adducts
in Human Endometrium

Sample	Age	Duration of Therapy (Months)	dG-N^2-TAM (adducts/10^8 dNs) trans-form	cis-form	Total
T1	46	4	7.4	8.3	15.7
T4	45	37	3.5	2.8	6.3
T7	49	40	4.7	N.D.	4.7
T8	75	34	N.D.	4.0	4.0
T10	38	7	2.1	N.D.	2.1
T11	73	72	11.6	6.4	18.0
T12	76	15	0.20	N.D.	0.20
T14	52	60	N.D.	1.6	1.6

Source: Data taken from Shibutani, S., Suzuki, N., and Grollmon, A.P., *Biochemistry*, 37, 12034–12041, 1998.

with tamoxifen (Table 9.2). The level of tamoxifen adducts were 0.2–18.0 adducts/10^8 bases, reproducing the results described in our previous study with ^{32}P-postlabeling-TLC.[55] We attribute the failure to detect tamoxifen adducts[23,24] to the relative lack of sensitivity of methods used by other investigators. Tamoxifen–DNA adduct levels in target organs correlate with tumor incidence in experimental animal[57]; comparable data do not exist for human subjects. Tamoxifen–DNA adducts are miscoding lesions[53] and have been shown to be mutagenic in mammalian cells.[51] This fact, coupled with their demonstrated presence in the endometrium, suggests that a genotoxic mechanism is likely to be responsible for tamoxifen carcinogenic effect on this tissue.

A marked interindividual variation were observed in the level of tamoxifen–DNA adducts formed in the endometrium of women treated with this drug[56]; this variability may be due to differences in the activity of enzymes involved in the α-hydroxylation of tamoxifen and its metabolites and/or cellular sulfotransferases that converts α-OHTAM to an activated form that reacts with DNA. Adduct levels may also depend on the ability of nucleotide excision repair to excise tamoxifen adducts from DNA.[58]

III. Other antiestrogen drugs

Tamoxifen is a hepatocarcinogen in rats while toremifene, a chlorinated tamoxifen analog (Figure 9.2), is not.[19,59] Although toremifene, like tamoxifen, has estrogenic effects on the human endometrium,[60] the formation of toremifene-DNA adducts in the liver of rats was two-orders of magnitudes less than that of tamoxifen.[19] Therefore, genotoxic effects of tamoxifen are thought to be involved in development of rat hepatocarcinoma. Toremifene has been used for breast cancer chemotherapy in the United States since 1998.

Raloxifene (Figure 9.2), a selective estrogen response modifier, reduced the incidence of breast cancer in women at high risk of developing this disease.[61] Unlike tamoxifen, raloxifene is unlikely to react with DNA due to

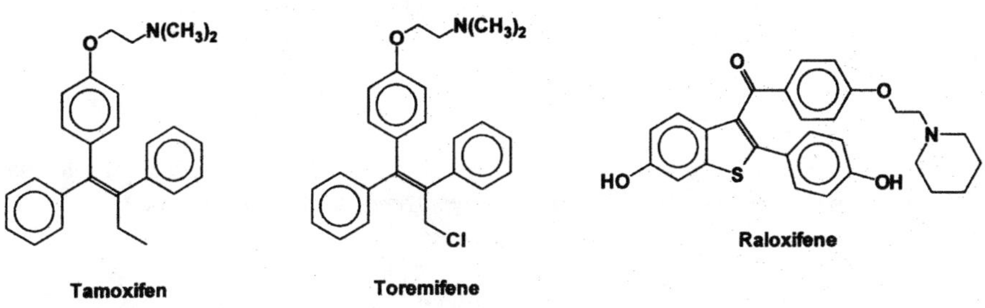

Figure 9.2 Structures of several antiestrogen drugs.

the absence of the ethyl moiety and does not demonstrate proliferative effects on the uterus of postmenopausal women.[62] The incidence of endometrial cancer was not increased in women enrolled in the raloxifene chemopreventive trial.[61]

In view of their reduced genotoxicity, there is reason for physicians to consider raloxifene and toremifene in recommending drugs for the chemoprevention of breast cancer in women at high risk of developing this disease.

III. Tamoxifen–DNA adducts as biomarkers

Tamoxifen-DNA adducts fulfill an important criteria for biomarkers, namely, their presence reflects a genotoxic effect of this drug, which in principle, can be used to predict individual's risk of developing endometrial cancer. Our findings also illustrate an axiom of biomarker research — that gene-environment interactions are at the core of most chronic human diseases. Methodology developed for this research could be used in a randomized clinical trial to test the hypothesis that the carcinogenic effects of tamoxifen on the endometrium can be separated from the chemoprotective effects of this drug in women at high risk of developing breast cancer.[56]

Acknowledgments

We thank Drs. L. Dasaradhi and A. Ravindernath for synthesizing tamoxifen metabolites, Dr. I. Terashima for exploring mutagenic properties of tamoxifen–DNA adducts, and Drs. M. Pearl and S. Sugarman for collecting endometrial samples for our studies. This research was supported in part by a Grant ES09418 from the National Institute of Environmental Health Sciences.

References

1. Wogan, G.N., Molecular epidemiology in cancer risk assessment and prevention: recent progress and avenues for future research, *Environ. Health Perspect.*, 98, 167–178, 1992.

2. Groopman, J.D. and Kensler, T.W., The light at the end of the tunnel for chemical-specific biomarkers: daylight or headlight? *Carcinogenesis*, 20, 1–11, 1999.
3. Eaton, D.L. and Groopman, J.D., Eds., *The Toxicology of Aflatoxins: Human Health, Veterinary and Agricultural Significance*, Academic Press, San Diego, CA, 1994.
4. Poirier, M.C., Santella, R.M., and Weston, A., Carcinogen macromolecular adducts and their measurement, *Carcinogenesis*, 21, 353–359, 2000.
5. Osborne, C.K., Tamoxifen in the treatment of breast cancer, *New Eng. J. Med.*, 339, 1609–1617, 1998.
6. Fischer, B. et al., Tamoxifen for prevention of breast cancer: report of the National Surgical Adjuvant Breast and Bowel Project P-1 Study, *J. Natl. Cancer Inst.*, 90, 1371–1388, 1998.
7. Killackey, M., Hakes, T.B., and Pierce, V.K., Endometrial adenocarcinoma in breast cancer patients receiving antiestrogens, *Cancer Treat. Rep.*, 69, 237–238, 1985.
8. Seoud, M.A.F., Johnson, J., and Weed, J.C., Gynecologic tumors in tamoxifen-treated women with breast cancer, *Obstet. Gynecol.*, 82, 165–169, 1993.
9. Fischer, B. et al., Endometrial cancer in tamoxifen-treated breast cancer patients: Findings from the National Surgical Adjuvant Breast and Bowel Project (NSABP) B-14, *J. Natl. Cancer Inst.*, 86, 527–537, 1994.
10. van Leeuwen, F.E. et al., Risk of endometrial cancer after tamoxifen treatment of breast cancer, *Lancet*, 343, 448–452, 1994.
11. Clarke, M. et al., Tamoxifen for early breast cancer: an overview of the randomized trials, *Lancet*, 351, 1451–1467, 1998.
12. Bernstein, L. et al., Tamoxifen therapy for breast cancer and endometrial cancer risk, *J. Natl. Cancer Inst.*, 91, 1654–1662, 1999.
13. IARC Monographs on the evaluation of carcinogenic risks to humans, *Some Pharmaceutical Drugs*, Vol. 66, IARC, Lyon, France, 1996.
14. Tannenbaum, S.R., Comparative metabolism of tamoxifen and DNA adduct formation and *in vitro* studies on genotoxicity, *Semin. Oncol.*, 24, 81–86, 1997.
15. Wogan, G.N., Review of the toxicology of tamoxifen, *Semin. Oncol.*, 24 (Suppl. 1), 87–97, 1997.
16. Stearns, V. and Gelman, E.P., Does tamoxifen cause cancer in humans? *J. Clin. Oncol.*, 16, 779–792, 1998.
17. Williams, G.M. et al., The triphenylethylene drug tamoxifen is a strong liver carcinogen in the rat, *Carcinogenesis*, 14, 315–317, 1993.
18. Greaves, P. et al., Two-year carcinogenicity study of tamoxifen in Alderley Park Wister-derived rats, *Cancer Res.*, 53, 3919–3924, 1993.
19. Hard, G.C. et al., Major difference in the hepatocarcinogenicity and DNA adduct forming ability between toremifene and tamoxifen in female Crl:CD(BR) rats, *Cancer Res.*, 53, 4534–4541, 1993.
20. Han, X. and Liehr, J.G., Induction of covalent DNA adducts in rodents by tamoxifen, *Cancer Res.*, 52, 1360–1363, 1992.
21. Osborne, M.R. et al., Identification of the major tamoxifen-deoxyguanosine adduct formed in the liver DNA of rats treated with tamoxifen, *Cancer Res.*, 56, 66–71, 1996.
22. Divi, R.L. et al., Tamoxifen-DNA adduct formation in rat liver determined by immunoassay and ^{32}P-postlabeling, *Cancer Res.*, 59, 4829–4833, 1999.
23. Carmichael, P.L. et al., Lack of genotoxicity of tamoxifen in human endometrium, *Cancer Res.*, 56, 1475–1479, 1996.

24. Carmichael, P.L. et al., Lack of evidence from HPLC ^{32}P-post-labeling from tamoxifen–DNA adducts in the human endometrium, *Carcinogenesis*, 20, 339–342, 1999.
25. Hemminki, K. et al., Tamoxifen-induced DNA adducts in endometrial samples from breast cancer patients, *Cancer Res.*, 56, 4374–4377, 1996.
26. Martin, E.A. et al., ^{32}P-postlabelled DNA adducts in liver obtained from women treated with tamoxifen, *Carcinogenesis*, 16, 1651–1654, 1995.
27. Phillips, D.H. et al., α-Hydroxytamoxifen, a metabolite of tamoxifen with exceptionally high DNA-binding activity in rat hepatocytes. *Cancer Res.* 54, 5518–5522 (1994a).
28. Jarman, M. et al., The deuterium isotope effect for the α-hydroxylation of tamoxifen by rat liver microsomes accounts for the reduced genotoxicity of [D_5-ethyl]tamoxifen, *Carcinogenesis*, 16, 683–688, 1995.
29. Poon, G.K. et al., Identification of tamoxifen metabolites in human Hep G2 cell line, human liver homogenate, and patients on long-term therapy for breast cancer, *Drug Metab. Dispos.*, 23, 377–382, 1995.
30. Dasaradhi, L. and Shibutani, S., Identification of tamoxifen-DNA adducts formed by α-sulfate tamoxifen and α-acetoxytamoxifen, *Chem. Res. Toxicol.*, 10, 189–196, 1997.
31. Osborne, M.R., Hardcastle, I.R., and Phillips, D.H., Minor products of reaction of DNA with α-acetoxytamoxifen, *Carcinogenesis*, 18, 539–543, 1997.
32. Sanchez, C. et al., Lifetime and reactivity of an ultimate tamoxifen carcinogen: The tamoxifen carbocation, *J. Am. Chem. Soc.*, 120, 13513–13514, 1998.
33. Shibutani, S. et al., α-Hydroxytamoxifen is a substrate of hydroxysteroid (alcohol) sulfotransferase, resulting in tamoxifen DNA adducts, *Cancer Res.*, 58, 647–653, 1998.
34. Shibutani, S. et al., Sulfation of α-hydroxytamoxifen catalyzed by human hydroxysteroid sulfotransferase results in tamoxifen DNA adducts, *Carcinogenesis*, 19, 2007–2011, 1998.
35. Davis, W., Venitt, S., and Phillips, D.H., The metabolic activation of tamoxifen and α-hydroxytamoxifen to DNA-binding species in rat hepatocytes proceeds via sulphation, *Carcinogenesis*, 19, 861–866, 1998.
36. Umemoto, A. et al., Tamoxifen-DNA adducts formed by α-acetoxytamoxifen N-oxide, *Chem. Res. Toxicol.*, 12, 1083–1089, 1999.
37. Umemoto, A. et al., Identification of hepatic tamoxifen-DNA adducts in mice: α-(N^2-deoxyguanosinyl)tamoxifen and α-(N^2-deoxyguanosinyl)tamoxifen N-oxide, *Carcinogenesis*, 21, 1737–1744, 2000.
38. Mani, C. and Kupfer, D., Cytochrome P-450-mediated activation and irreversible binding of the antioestrogen tamoxifen to proteins in rats and human liver: possible involvement of flavin-containing mono-oxygenases in tamoxifen activation, *Cancer Res.*, 51, 6052–6058, 1991.
39. Mani, C., Hodgson, E., and Kupfer, E., Metabolism of the antimammary cancer antiestrogenic agent tamoxifen. II. Flavin-containing monooxygenase-mediated N-oxidation, *Drug Metab. Dispos.*, 21, 657–66, 1993.
40. Lim, C.K. et al., A comparative study of tamoxifen metabolism in female rat, mouse and human liver microsomes, *Carcinogenesis*, 15, 589–593, 1994.
41. Hengstler, J.G. et al., Interspecies differences in cancer susceptibility and toxicity, *Drug Metab. Rev.*, 31, 917–970, 1999.
42. Rajaniemi, H. et al., Identification of the major tamoxifen–DNA adducts in rats liver by mass spectroscopy, *Carcinogenesis*, 20, 305–309, 1999.

43. Gamboa da Costa, G. et al., Characterization of the major DNA adduct formed by α-hydroxy-N-desmethyltamoxifen *in vitro* and in vivo, *Chem. Res. Toxicol.*, 13, 200–207, 2000.

44. Kitagawa, M. et al., Identification of tamoxifen-DNA adducts induced by α-acetoxy-N-desmethyltamoxifen, *Chem. Res. Toxicol.*, 13, 761–769, 2000.

45. Phillips, D.H. et al., *N*-demethylation accompanies α-hydroxylation in the metabolite activation of tamoxifen in rat liver cells, *Carcinogenesis*, 20, 2003–2009, 1999.

46. Brown, K. et al., Further characterization of the DNA adducts formed in rat liver after the administration of tamoxifen, *N*-desmethyltamoxifen or *N,N*-didesmethyltamoxifen, *Carcinogenesis*, 20, 2011–2016, 1999.

47. Moorthy, B. et al., Tamoxifen metabolic activation: comparison of DNA adducts formed by microsomal and chemical activation of tamoxifen and 4-hydroxytamoxifen with DNA adducts formed *in vitro*, *Cancer Res.*, 56, 53–57, 1996.

48. Marques, M.M. and Beland, F.A., Identification of tamoxifen-DNA adducts formed by 4-hydroxytamoxifen quinone methide, *Carcinogenesis*, 18, 1949–1954, 1997.

49. Martin, E.A. et al., Evaluation of tamoxifen and α-hydroxytamoxifen ^{32}P-post-labelled DNA adducts by the development of a novel automated on-line solid-phase extraction HPLC method, *Carcinogenesis*, 19, 1061–1069, 1998.

50. Beland, F.A., McDaniel, L.P., and Marques, M.M., Comparison of the DNA adducts formed by tamoxifen and 4-hydroxytamoxifen *in vivo*, *Carcinogenesis*, 20, 471–477, 1999.

51. Terashima, I., Suzuki, N., and Shibutani, S., Mutagenic potential of α-(N^2-deoxyguanosinyl)tamoxifen lesions, the major DNA adducts detected in endometrial tissues of patients treated with tamoxifen, *Cancer Res.*, 59, 2091–2095, 1999.

52. Shibutani, S., Suzuki, N., and Grollman, A.P., Mutagenic specificity of (acetylamino)fluorene-derived DNA adducts in mammalian cells, *Biochemistry*, 37, 12034–12041, 1998.

53. Shibutani, S. and Dasaradhi, L., Miscoding potential of tamoxifen-derived DNA adducts: α-(N^2-deoxyguanosinyl)tamoxifen, *Biochemistry*, 36, 13010–13017, 1997.

54. Davies, R. et al., Tamoxifen causes gene mutations in the livers of lamda/lacI transgenic rats, *Cancer Res.*, 57, 1288–1293, 1997.

55. Shibutani, S. et al., Tamoxifen–DNA adducts detected in the endometrium of women treated with tamoxifen, *Chem. Res. Toxicol.*, 12, 646–653, 1999.

56. Shibutani, S. et al., Identification of tamoxifen–DNA adducts in the endometrium of women treated with tamoxifen, *Carcinogenesis*, 21, 1461–1467, 2000.

57. Ottender, M. and Lutz, S.K., Correlation of DNA adduct levels with tumor incidence: carcinogenic potency of DNA adducts, *Mutat. Res.*, 424, 237–247, 1999.

58. Shibutani, S. et al., Excision of tamoxifen–DNA adducts by the human nucleotide excision repair system, *Cancer Res.*, 60, 2607–2610, 2000.

59. White, I.N.H. et al., Genotoxic potential of tamoxifen and analogues in female Fischer F344/n rats, DBA/2 and C57B1/6 mice and in human MCL-5 cells, *Carcinogenesis*, 13, 2197–2203, 1992.

60. Maenpaa, J.U. and Ala-Fossi, S.L., Toremifene in postmenopausal breast cancer: efficacy, safety, and cost, *Drugs Aging*, 11, 261–270, 1997.
61. Commings, S.R. et al., The effect of raloxifene on risk of breast cancer in postmenopausal women: results from the more randomized trial, *JAMA*, 281, 2189–2197, 1999.
62. Paech, K. et al., Differential ligand activation of estrogen receptors ERα and ERβ at AP1 sites, *Science*, 277, 1508–1510, 1997.

chapter ten

Polycyclic aromatic hydrocarbon DNA adducts and human cancer

Mikhail F. Denissenko, Moon-shong Tang, and Gerd P. Pfeifer

Contents

Abstract Formation of specific DNA adducts in target organs is implicated as a critical initiating event in carcinogenesis. Levels of adducts in DNA are used as biological markers in molecular epidemiology studies of human tumors. Adducts of polycyclic aromatic hydrocarbons (PAHs) are suspected as causative agents of certain human cancers, in particular of tumors of the respiratory system in chronic smokers. Measurements of PAH–DNA adduct levels are performed to document human environmental exposure to these carcinogens, determine internal dosimetry of exposure, and assess individual

1-56670-596-7/02/$0.00+$1.50
© 2002 by CRC Press LLC

risk for cancer development. A variety of laboratory methods for detection and quantitation of carcinogen–DNA adducts are available. The most sensitive methods are used to determine adduct levels in global genomic DNA in exposed target tissues. Adducts at the positions of individual nucleotides of the human genome can be measured by the ligation-mediated polymerase chain reaction technique. Using this method, it was possible to correlate increased PAH–DNA adduct levels at certain nucleotides in the p53 gene with mutational hotspots for G to T transversions in lung tumors of smokers.

I. Introduction

A majority of chemical carcinogens, either man-made or natural, exert their biological effects through interaction with DNA. The covalent reaction between an electrophilic carcinogen and a nucleophilic site in DNA leads to formation of modified nucleotides or DNA adducts. In addition to DNA-reactive chemicals, ionizing radiation and the ultraviolet (UV) light component of a natural sunlight form specific DNA lesions. Exogenously induced DNA adducts are a threat to genome integrity and are thought to be critical initiating factors in tumorigenesis, at least for some cancers.[1-5] There has been growing realization that adducts formed from endogenous sources may possess similar if not higher mutagenic potential.[6] Complex cellular DNA repair systems have evolved to remove DNA adducts from the genome thus counteracting a constant challenge to DNA encountered by all living species. These repair processes result in a consistent reduction of the cellular levels of DNA adducts. However, if unrepaired damage is still present during DNA replication, it may either cause DNA polymerases to stop at the site of a lesion (resulting in premature termination of replication, cell growth arrest, cytotoxicity, or chromosomal aberrations), or the polymerase will bypass the altered base with the possibility of base misincorporation and mutagenesis.[7,8] DNA adducts may also undergo hydrolysis forming an abasic site and increasing the probability of strand scission and mutagenesis.[6] The resulting genetic alterations may lead to creation of a new phenotype and, if growth controlling genes are involved, to cellular transformation and the development of tumors. Protooncogenes and tumor suppressor genes are critical targets for carcinogens.[7,9] Identification of a link between DNA damage and mutations will strengthen the understanding to what extent elements of the environment are responsible for initiation of tumorigenesis in humans.

DNA adducts induced by different carcinogens may have significantly different mutational efficiencies. The steady-state levels of DNA adducts in human target tissues following chronic exposure to carcinogens are found to be dose-dependent and, in some cases, predictive of cancer incidence.[1,5,10,11] One of the goals of these studies has been the development of preventive intervention methods to lower the human health impact from carcinogen exposures of environmental (including dietary), occupational, or clinical origin.

Over the last two decades, several sufficiently sensitive technologies have been developed to measure protein and DNA adducts at levels consistent with environmental and occupational exposure to genotoxic agents. However, the progress in this area faces the difficulties of accurate interpretation of the adduct type, individual and, importantly, interlaboratory variabilities in quantitative measurements. A plethora of published results notwithstanding, more data have to be obtained in order to establish a better correlation between the levels of adducts and specific types of cancer. Further methodological advances are necessary to make analytical methods more amenable to molecular epidemiological studies.

II. Molecular epidemiology of human cancer and DNA adducts

The emerging field of molecular epidemiology is based on the detection of biological markers of human chronic disease with environmental components for the purpose of identifying its causes and outcomes.[9,12] Identification of human carcinogens is considered most accurate when supported by cancer epidemiology data.[13] The biomarkers used in the molecular epidemiology of human cancer include specific changes at the gene and chromosome level (polymorphisms, mutations and cytogenetic alterations), concentrations of potential carcinogens and/or metabolites in cells, tissues and biological fluids, and levels of covalent carcinogenic adducts to proteins and to DNA.[14] In certain cases changes found at the molecular level may give insights into possible origins of the disease.[9,15] Development of a biomarker normally involves thorough laboratory characterization in order to recognize its utility for molecular epidemiological studies. The ultimate level of validation of a biomarker is achieved after extensive studies involving human samples.[12]

DNA adducts have proven to be useful biomarkers in human population research.[13,14] Carcinogen–DNA adducts in target tissues are considered to be more valuable than carcinogen-protein adducts because the former reflect not only individual differences in absorption and distribution but also differences in metabolism and detoxification of the chemical and effects of DNA repair.[14] Whether DNA adducts are always carcinogenic is an unanswered question because of the limitations of carcinogenicity testing in animals.[5] Studies show that tissue DNA adduct levels do not always simply translate into cancer risk. It should be noted that DNA from target tissue is not always available and surrogate tissues (e.g., peripheral blood cells) are often used.[14]

Adducts may be generated in DNA by a variety of pathways either involving metabolic activation of a procarcinogen into a reactive species or through a direct reaction with DNA.[5,16] Examples of directly acting DNA damaging genotoxins include alkylating agents such as nitrosoureas, nitrosamines, and mustards. PAHs, aflatoxins, vinyl chloride, and many other carcinogens undergo metabolic activation to generate DNA-binding species. Adducts seem to form in a site-specific manner although this is technically difficult to prove in human samples exposed to low environmental doses of carcinogens.[13,17,18] As a result, in most molecular epidemiology studies, DNA adduct frequencies are usually measured in a nonsequence-specific manner.

Based on DNA adducts, human exposure was documented to aflatoxins,[19,20] 4-aminobiphenyl,[21] benzo[a]pyrene, other PAHs[11,22,23] (see below), styrene, cisplatin, 8-methoxypsoralen, nitrosamines, and other compounds.[10,24] The adducts of interest included the following: PAH, styrene, and diesel exhaust for occupational exposure; PAH, psoralen, procarbazine, and cisplatin for clinical exposure;and PAH, tobacco-specific nitrosamines, and 4-aminobiphenyl for lifestyle (smoking) and dietary factors.[25] Additionally, a variety of endogenous DNA adducts were identified.[13] In general, elevated mean levels of specific adducts were found in exposed subjects, with high background level variability identified among few apparently unexposed individuals.[10,25] The latter may result from the widespread environmental exposure of control group subjects to carcinogens in the food and/or air.

Another important factor may be the interindividual variations in efficiency of DNA repair. Obvious difficulties of this task include finding the appropriate control group versus occupational or environmental group, limited availability of DNA samples, especially from the target tissues, the nature of the DNA repair assays used, and significant variability of results obtained in different laboratories.[26] This direction of research is closely related to attempts to assess DNA adduct levels by correlating the measured DNA damage and the degree of exposure to carcinogens. Examples of such studies include measurements of the PAH adducts in white blood cells of iron foundry workers[27,28] and wildland firefighters.[29] These data have clearly demonstrated that lowering the exposure to carcinogenic PAH resulted in a decrease of the molecular dose of PAH–DNA adducts. In addition to PAH, aflatoxins are among the few environmental carcinogens for which quantitative dosimetry assessments have been attempted.[20,30] Experiments assessing the effect of dietary intake of aflatoxin B1 on DNA damage levels and urinary adduct excretion have shown an excellent molecular dosimetry correlation in two different areas of the world.[31-33] Molecular dose of DNA adducts constitutes a balance between adduct formation (metabolic activation and detoxification) and adduct repair. In addition, multiple adducts formed by many carcinogens may impede correct dose and risk assessments. All these parameters should be carefully looked at when performing quantitative studies on the accumulation and repair of specific adducts in humans.

DNA damage biomarkers have been utilized for determinations of individual risk for cancer development. Risk assessment studies require a type of design different than the one used for the adduct dosimetry. Prospective design with a nested case-control analysis using banked samples from patients who are followed for cancer development is crucial to ensure that the marker is not influenced by the disease itself.[25] Complex studies using this type of approach are still rare. The most successful ones have measured the relationship between exposure to aflatoxin B1 (including DNA adducts) and hepatocellular carcinoma also taking into consideration the markers for hepatitis B virus.[33-35] There was a strong interaction between markers of chronic hepatitis B infection and aflatoxin exposure in liver cancer risk. The importance of reducing the carcinogen exposure in prevention of liver cancer was clearly demonstrated. Therefore, DNA adducts are likely to be more closely associated with human tumor risk than other internal markers.[26]

Another aspect of utilizing DNA adduct measurements for human cancer risk assessment is the use of animal models.[36] DNA adducts may be measured in laboratory animals under conditions of chronic exposure to carcinogens and the tumor incidence data extrapolated to human populations. The fact that DNA adducts detected in humans are chemically identical to those found in animals further substantiates the usefulness of DNA adduct data and the growing need in reliable methods for such measurements. Estimates of risk from suspected carcinogens are generally obtained from tumor incidence models based on high animal exposures whereas probable human exposures may be orders of magnitude lower. Molecular dosimetry based on DNA damage may significantly extend the measurable range of data and permit more accurate risk determinations.[37]

III. Role of PAH–DNA adducts in human carcinogenesis

PAHs are products of incomplete pyrolysis of organic matter and are widespread in the environment.[38,39] Humans are largely exposed to complex mixtures of these compounds. Sizable amounts of PAHs are generated as a result of industrial coke and iron and other metal production. In addition, PAHs are found in combustion emissions of fuels, including diesel and biomass, in certain types of cooked food, and in tobacco smoke. It is believed that the largest concentrations of PAHs are inhaled by smokers with the mainstream smoke of cigarettes.[40,41] Smoking is considered to be the major cause of upper aerodigestive tract cancers including cancer of the lung.[42] Lung cancer is a leading cause of cancer death for American women and men[43] and is one of the most common types of cancer worldwide. Therefore, smoking-related PAHs are strongly implicated as agents responsible for initiation and development of lung cancer. Interestingly, recent work still does not ascribe a role of diesel engine exhaust exposure in lung cancer development.[44]

PAHs are metabolically activated by the cytochrome P450 enzymes to generate the ultimate carcinogenic derivatives capable of covalently binding to DNA. Most PAHs are activated through diol-epoxidation at the hydrocarbon bay region. Activated diol epoxides of PAHs form bulky covalent DNA adducts primarily at the exocyclic N2 position of guanine residues and the N6 position of adenines.[39] An ultimate reactive metabolite of the well studied PAH benzo[a]pyrene (BP) is (+/-) anti-7β,8α-dihydroxy-9α,10α-epoxy-7,8,9,10-tetrahydrobenzo[a]pyrene (BPDE). BPDE generates in DNA predominantly covalent (+) trans adducts at the N2 position of guanine.[45-47]

These covalent adducts have been shown to be the basis for the mutagenic and carcinogenic effects of BPDE.[48,49] The genomic targets for BPDE may include oncogenes and tumor suppressor genes. We have recently reported a strong correlation between the sites of BPDE adduct formation in the human p53 gene and lung cancer mutational hotspots.[17] These results provided an etiological link between a PAH present in cigarette smoke and human lung cancer mutations.

Because of their importance, biomarkers of exposure to PAHs have been a subject of numerous human biomonitoring, risk assessment, and intervention studies.[11,23,25] PAH–DNA adducts were measured mostly by ^{32}P-postlabeling, by fluorescence, and by immunoassays. There have been consistent increases in PAH–DNA damage in cases with occupational, clinical, and high environmental exposures. Elevated levels of damage were detected in target tissues of smokers vs. nonsmokers[50-58] and in white blood cells of lung cancer patients after adjustment for smoking.[52,59-61] Otherwise, the correlation between target lung and total surrogate white blood cell adducts was poor[60,62] PAH–DNA adducts were also found at higher frequencies in coke oven, iron foundry, and aluminum plant workers, and in roofers as compared with nonexposed controls,[27,28,63-67] as well as in psoriasis patients treated with coal tar[68] and in wildland firefighters.[29] High levels of air pollution in coal-based industrial regions appeared to be a source for increase in PAH–DNA damage in Silesia, Poland[65,69] and in Northern Bohemia, Czech Republic.[70] The PAH damage levels were higher in winter conditions. The control groups in these cases consisted of subjects from agricultural regions with no heavy industry. Sources of PAH–DNA adducts also included urban environment and vehicle exhaust[71,72] and dietary consumption of grilled or charbroiled meats[73-75] Notably, the responses to PAH ingestion were clearly dose dependent. In a feeding study, the adduct frequencies, once increased in responsive subjects, rapidly returned to initial levels after cessation of charbroiled beef consumption[73] Evidence of segregation of the subjects into separate response groups was observed.[75] Most studies have been performed on DNA from blood cells although lung,[52,53,55,56,60,62] placental[76] and other tissue DNA samples were also investigated.

Determined levels of PAH adducts were generally within the range from 1 to 10, and not higher than 40 adducts per 10^8 nucleotides.[23] Typical statistically significant differences between the exposed and control groups were from 1.5- to 3-fold.[23] All studies detected a considerable adduct level variability in subjects belonging to the same experimental group. This suggests a strong effect of individual genetic susceptibility factors. These factors include pathways of absorption, activation, distribution, and detoxification of carcinogens, DNA repair capacity, and various lifestyle and dietary effects. Therefore, a recent trend in research on molecular epidemiology of DNA adducts has been to correlate DNA damage in populations with genotype or phenotype of xenobiotic metabolism and DNA repair genes, e.g., *CYP1A1*, *GSTM1*, *NAT2*, *XP* genes, and *XRCC1*.

IV. Assays for PAH–DNA adducts

Several sensitive methods have been developed over the years for measurement of PAH–DNA adducts in human tissues or cells. These assays normally require between 1 and 100 μg DNA for adduct analysis. The most common laboratory techniques for damage detection and quantitation are described below.

A. ^{32}P-postlabeling

The ^{32}P-postlabeling assay is a versatile method developed in 1981[77] for a variety of DNA damage types.[78] This method is extremely sensitive with the limit of detection in the range of 1 adduct per 10^{9-10} or even 10^{11} nucleotides. In this method, DNA is hydrolyzed to a mixture of normal and adducted nucleotides and the hydrophobic adducted nucleotides are either additionally extracted with butanol for enrichment or the normal nucleotides removed from the assay by digestion with nuclease P1. The mixture is then phosphorylated using [γ^{32}P] ATP and polynucleotide kinase, separated by TLC or HPLC and the adduct levels calculated from radioactivity counts. The choice of enrichment procedure is determined by the type of target DNA lesion. ^{32}P-postlabeling is a relatively nonspecific assay and is generally applicable to unidentified lesions. Most adducts of very different origin may be monitored, e.g., adducts of PAHs, aromatic amines, small alkylating agents, mycotoxins, chemotherapeutic drugs, oxidative damage generators, and adducts from complex mixtures. Another advantage of this technique is the requirement for small amounts of sample DNA making it ideal for target tissue measurements. There are several limitations of ^{32}P-postlabeling. The adduct identification is frequently not possible because this method does not yield structural adduct data. However, with the further improvements of separation procedures for the ^{32}P-postlabeled adducts (capillary electrophoresis or capillary electrochromatography, CE or CEC), much better characterization is expected.[13]

B. Mass spectrometry and other methods

Gas chromatography-mass spectrometry (GC-MS) offers both high sensitivity and specificity of damage detection. Adducts of PAH, nitrosamines, malondialdehyde and oxidatively damaged bases have been measured in human samples by this technique. The high cost of necessary equipment has limited the use of GC-MS in human biomonitoring. Also, the thermal stability and volatility GC-MS requirements of the sample are not always met by carcinogen-adducted DNA.[13] These limitations may be overcome by combining liquid chromatography/electrospray ionization with mass spectrometry (LC/ESI-MS) or tandem MS.[79] LC/ESI-MS/MS is extremely useful for analysis of DNA adducts, which are not amenable to GC and derivatization owing to the presence of several adjacent polar functional groups.[79] Another methodological improvement may be the use of accelerator mass spectrometry (AMS) detecting ^{14}C with exceptional sensitivity.[13] Unfortunately, none of these methods is simple and inexpensive.

C. Immunoassays

Since the development in the 1970s of antisera against carcinogen-DNA adducts, immunological assays have been widely used for detection and quantitative assessment of DNA damage.[10,24-26] Immunoassays may generally be applied for measurement of any structural DNA alteration caused by carcinogens or mutagens, for which a specific antibody is available. The inception of antibodies recognizing BPDE-DNA adducts has opened new and growing avenues for monitoring human exposure to this ubiquitous carcinogen by directly assessing damage in DNA.[80] Furthermore, antibodies against many different types of DNA lesions have been elicited and utilized for measurements *in vivo*. Usually, two types of immunogens are used for antibody development, either adducted unhydrolyzed DNA electrostatically complexed to methylated carrier protein (bovine serum albumin) or monoadducts coupled to a carrier protein (keyhole limpet hemocyanin). Polyclonal and monoclonal antibodies have been used for biomonitoring studies. The advantage of polyclonal antibodies is the rapid generation and easy handling of antisera. Monoclonal antibodies are more difficult to develop and maintain but the major and hard-to-match advantage of their use is the unlimited supply of antibody achievable by propagating the immortalized hybrid clones. The specificity of antibodies is important, since cross-reactivity with undamaged DNA or structurally related adducts may complicate the results.[37] This feature may actually become advantageous if there is a need to detect a whole group of structurally similar damages, e.g., total aromatic or PAH DNA adducts. A major limitation of immunoassays is the need to develop and characterize antibodies for each DNA adduct of interest. However, the progress in the area of new anti-DNA adduct antibody development has been remarkable and more specific clones have become available. Once the antibody is fully characterized and, importantly, validated by independent techniques, it may be used in comparative experiments using human samples. Besides quantitative assays, antibodies have been employed in damage visualization techniques including immunohistochemistry and immuno-electron microscopy, and for immuno-affinity purification. Quantitative measurements were primarily accomplished by competitive radioimmunoasssay (RIA) and enzyme-linked immunosorbent assay (ELISA) and by noncompetitive immunoslot blot (ISB) analysis. Noncompetitive immuno-slot or -dot blot methods have been developed for adduct detection in DNA immobilized directly onto nitrocellulose membrane surfaces.[81] Incubation with primary antibody is followed by incubation with enzyme-conjugated secondary antibody, or an intermediate avidin-biotin binding step is performed.[82] The signal is produced with either colored[81,83] or chemiluminescent[13,84] substrates. Using a polyclonal antiserum recognizing BPDE–DNA adducts, we were able to reach the sensitivity of 2 adducts/10^7 nucleotides with a colored endpoint.[85]

D. Ligation-mediated PCR

The ligation-mediated PCR (LMPCR) technique was initially applied to the detection of DNA adducts in the early 1990s.[86] The method does not have the exquisite sensitivity of, for example, ^{32}P-postlabeling, but it can be used with adduct levels of one adduct in 10 to 20 kilobases of DNA. Thus, it has a similar sensitivity as the genomic Southern blot techniques used in gene-specific DNA repair assays.[87] In contrast to the other methods which measure adduct levels in total genomic DNA or within whole genes, the LMPCR method is a technique for the detection of DNA adducts at individual nucleotide positions in mammalian genes. Adduct-specific enzymes such as T4 endonuclease V, various base excision repair enzymes, UvrABC nuclease, and chemical cleavage techniques can be used to convert DNA adducts into DNA strand breaks. The positions of these breaks are then detected by LMPCR and displayed on DNA sequencing gels. This method has been used primarily to map the distribution of UV-induced DNA lesions[86,88–90] and for adducts of polycyclic aromatic hydrocarbons.[17,91–93]

The LMPCR technique itself is based on the ligation of an oligonucleotide linker onto the 5′ end of each DNA molecule that was created by strand cleavage and a gene-specific primer extension reaction (Figure 10.1). This ligation provides a common sequence for all 5′ ends allowing exponential PCR to be used for amplification of the intervening sequence. The PCR amplification of the linker-ligated fragments is done using the longer oligonucleotide of the linker (linker-primer) and a nested gene-specific primer (primer 2). After 18 to 20 PCR amplification cycles, the DNA fragments are separated on sequencing gels, electroblotted onto nylon membranes, and hybridized with a gene-specific probe to visualize the sequence distribution of lesions. Alternative methods to visualize the sequence include a primer extension reaction with a ^{32}P or infrared dye labeled oligonucleotide. The whole procedure has been partially automated.[94]

In principle, LMPCR can be used whenever it is possible to convert the DNA adduct into a strand break. The method has been used with base excision repair enzymes to detect oxidative lesions at pyrimidine and purine bases.[95] Chemical or heat treatment can be used for a variety of labile DNA lesions including adducts of DNA binding drugs[96] and aflatoxin B1.[97] A technique that does not depend on specific adduct cleavage is terminal transferase-dependent PCR.[98] This method has been used for mapping of sites of UV damage,[98] aflatoxin B1,[97] aristolochic acid DNA adducts,[99] and psoralen adducts.[100] The terminal transferase method has a high sensitivity but in most cases adducts cannot be mapped as precisely to a single nucleotide position as with LMPCR.

For mapping of PAH adducts at nucleotide resolution by ligation-mediated PCR, DNA or cells are treated with the DNA binding metabolites of

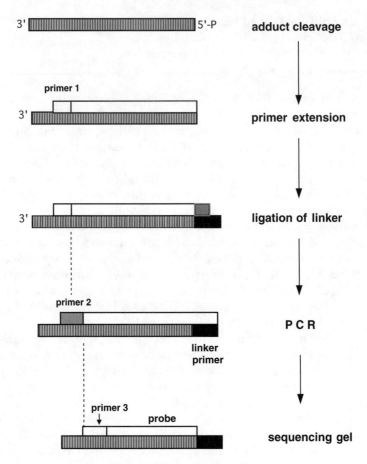

Figure 10.1 Outline of the ligation-mediated PCR procedure.

PAHs such as BPDE. This DNA is then cleaved at the sites of modified bases with the UvrABC nuclease complex from *Escherichia coli*. It has been shown that UvrABC nuclease incises 6 to 7 bases 5′ and 4 bases 3′ to a (+/−) anti-BPDE-modifed purine, and that under these reaction conditions, the cleavage at BPDE–DNA adducts by UvrABC nucleases is quantitative.[101] These results validate the UvrABC incision method for analysis of the sequence selectivity of BPDE binding. Since the UvrABC incision at the 3′ side of BPDE–DNA adducts is very specific (4 bases 3′ to the adduct), LMPCR can be used to determine the BPDE adduct distribution at nucleotide resolution.

V. PAH–DNA adducts, p53 mutations and lung cancer

Lung cancers are strongly associated with cigarette smoking.[42] The *p53* mutation spectra are different between smokers and nonsmokers and this difference is statistically highly significant — G to T transversions are 30% versus 10%; p<0.0001, $\chi2$ test.[102] A similar difference is seen between lung cancers

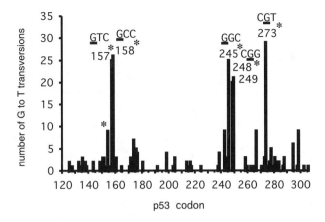

p53 codon

Figure 10.2 The *p53* mutational spectrum of G to T transversions in lung cancer. The data was obtained from the *p53* mutation database.[113] Data for nonsmokers and occupationally exposed individuals were excluded as specified.[102] Mutational hotspots of G to T transversions involving methylated CpG sequences are indicated by asterisks. The mutated guanines are underlined. These same guanines are sites of preferential formation of PAH–DNA adducts in human bronchial epithelial cells.[93] The spectrum of G to T transversions of nonsmokers is not drawn here since there were only seven data points available (one G to T mutation occurred at codon 148, one at codon 158, one at codon 242, and four at codon 249).

and nonlung cancers. In the *p53* gene of lung cancers from smokers, many G to T transversions are found at codons 157, 158, 245, 248, and 273 (see Figure 10.2).[102] Lung cancers from nonsmokers or from uranium miners do not contain these hotspot mutations. Codons 157 and 158 are not usually seen as mutational hotspots in other types of malignancies. The other three hotspot codons are common mutation sites in many different types of human tumors. It is important to note that G to T mutations are frequent at codons 245, 248, and 273 in lung tumors from smokers. However, this type of mutation is almost absent at the same codons in other types of tumors, including common malignancies such as colon, breast, and brain cancers.[102]

Important mutagens present in cigarette smoke are compounds belonging to the classes of nitrosamines and PAHs. The mutational spectrum of the *p53* gene in lung cancers does not contain a conspicuous fingerprint of mutations that may be derived from nitrosamine-induced DNA lesions such as O^6-methylguanine. These would be G to A transitions at non-CpG sites. The lung cancer *p53* spectrum is more consistent with the mutational patterns induced by certain PAHs. Benzo[a]pyrene is a prominent member of the PAH class, is present in cigarette smoke and is one of the strongest carcinogens known. This compound induces G to T transversions in many experimental systems.[103–105]

The distribution BPDE and other PAH diol epoxide adducts was mapped at nucleotide resolution along exons of the *p53* gene in PAH-treated normal human bronchial epithelial cells.[17,93] Frequent adduct formation occurred at guanine positions in codons 157, 158, 245, 248, and 273. These same positions are major mutational hotspots in human lung cancers (Figure 10.2).[93,102] The mechanistic basis for the selective occurrence of these adduct hotspots is an enhancement of adduct formation by 5-methylcytosine bases present at CpG sequences.[91,106–108] All CpG sequences in the *p53* coding exons five through nine are completely methylated in all tissues examined.[109] Methylation at CpG sites probably increases carcinogen binding at the intercalation step. The precise mechanism of how cytosine methylation at CpG sites enhances carcinogen binding is unclear at present. In more recent work, we have observed that the preferential formation of BPDE adducts at methylated CpG sites is reflected in strongly enhanced mutagenesis at CpG sequences in three different CpG-methylated mutational reporter genes.[110]

Our results strongly suggest that targeted adduct formation in addition to phenotypic selection appears to be responsible for shaping the *p53* mutational spectrum in lung tumors. The vast majority (>85%) of G to T transversions in lung cancers are targeted to guanines on the nontranscribed DNA strand implying a strand-specific DNA repair phenomenon. DNA repair experiments analyzing BPDE adducts in the *p53* gene have shown that the nontranscribed strand is in fact repaired more slowly than the transcribed strand.[92] These findings suggest that both the initial DNA adduct levels and a strand bias in repair may contribute strongly to the mutational spectrum of the human *p53* gene in lung cancer.

VI. Other DNA adducts, other types of cancer?

Monitoring of the sequence-specific formation of PAH–DNA adducts along the *p53* gene has provided important clues as to the potential origin of G to T transversion mutations in human lung cancers. A similar case can be made for the involvement of sunlight-induced cyclobutane pyrimidine dimers and *p53* mutations in human nonmelanoma skin tumors.[89,111] Interestingly, the DNA base 5-methylcytosine also plays an important role in enhancing formation of DNA damage by sunlight although the mechanism is, of course, completely different from that of enhanced PAH adduct formation at methylated CpG sites.[112] One other obvious example, where an exogenous agent has been clearly implicated in human carcinogenesis, is the connection between aflatoxin B1 and human liver cancer. Hepatocellular carcinomas from geographic areas of the world with high suspected food contamination by aflatoxin B1 carry a unique G to T transversion mutation signature at codon 249 of the *p53* gene. LMPCR and terminal transferase-dependent PCR were used to map aflatoxin B1 adducts in carcinogen-exposed human cells along the *p53* gene.[97] Although codon 249 was a prominent site of adduct formation, many other guanines along different exons of the *p53* gene formed equal or even higher levels of aflatoxin B1 adducts. Thus, additional and yet undiscovered processes of mutagenic selectivity or phenotypic selection need to be invoked to explain the specificity of the codon 249 mutation in hepatocellular carcinoma. It remains to be seen if adduct mapping using

LMPCR technology will help to identify suspected causative agents for other human tumors, for which there is much less evidence at present to implicate specific agents in the carcinogenic process.

Acknowledgments

The original work of the authors was supported by grants from the National Cancer Institute (CA84469 to G.P.P.) and the National Institute of Environmental Health Sciences (ES06070 to G.P.P. and ES08389 to M.-s.T.).

References

1. Lutz, W.K., *In vivo* covalent binding of organic chemicals to DNA as a quantitative indicator in the process of chemical carcinogenesis, *Mutation Res.*, 65, 289–356, 1979.
2. Weinstein, I.B., The origins of human cancer: molecular mechanisms of carcinogenesis and their implications for cancer prevention and treatment (27th G.H.A. Clowes Memorial Award Lecture), *Cancer Res.*, 48, 4135–4143, 1988.
3. Harris, C.C., Chemical and physical carcinogenesis: advances and perspectives for the 1990s, *Cancer Res.*, 51, 5023s–5044s, 1991.
4. Ames, B.N., Shigenaga, M.K., and Gold, L.S., DNA lesions, inducible DNA repair, and cell division: three key factors in mutagenesis and carcinogenesis, *Environ. Health Perspect.*, 101, 35–44, 1993.
5. Hemminki, K., DNA adducts, mutations and cancer, *Carcinogenesis.*, 14, 2007–2012, 1993.
6. Marnett, L.J. and Burcham, P.C., Endogenous DNA adducts: potential and paradox, *Chem. Res. Toxicol.*, 6, 771–785, 1993.
7. Strauss, B.S., The origin of point mutations in human tumor cells, *Cancer Res.*, 52, 249–253, 1992.
8. Echols, H. and Goodman, M.F., Fidelity mechanisms in DNA replication, *Annu. Rev. Biochem.*, 60, 477–511, 1991.
9. Hussain, S.P. and Harris, C.C., Molecular epidemiology of human cancer: contribution of mutation spectra studies of tumor suppressor genes, *Cancer Res.*, 58, 4023–4037, 1998.
10. Santella, R.M., DNA adduct in humans as biomarkers of exposure to environmental and occupational carcinogens, *Environ. Carcinog. Rev.*, C9, 57–81, 1991.
11. Phillips, D.H., DNA adducts in human tissues: Biomarkers of exposure to carcinogens in tobacco smoke, *Environ. Health Perspect.*, 104, 453–458, 1996.
12. Albertini, R.J., Biomarker responses in human populations: towards a worldwide map. *Mutation Res.* 428, 217–226, 1999.
13. Farmer, P.B., Studies using specific biomarkers for human exposure assessment to exogenous and endogenous chemical agents, *Mutation Res.*, 428, 69–81, 1999.
14. Vainio, H., Use of biomarkers: new frontiers in occupational toxicology and epidemiology, *Toxicol. Lett.*, 102–103, 581–589, 1998.
15. Pfeifer, G.P. and Denissenko, M.F., Formation and repair of DNA lesions in the p53 gene: relation to cancer mutations? *Environ. Mol. Mutagen.*, 31, 197–205, 1998.

16. Dipple, A., DNA adducts of chemical carcinogens, *Carcinogenesis*, 16, 437–441, 1995.
17. Denissenko, M.F. et al., Preferential formation of benzo[a]pyrene adducts at lung cancer mutational hotspots in *p53*, *Science*, 274, 430–432, 1996.
18. Denissenko, M.F. et al., Quantitation and mapping of aflatoxin B1-induced DNA damage in genomic DNA using aflatoxin B1–8,9-epoxide and microsomal activation systems, *Mutation Res.*, 425, 205–211, 1999.
19. Choy, W.N., A review of the dose-response induction of DNA adducts by aflatoxin B1 and its implications to quantitative cancer-risk assessment, *Mutation Res.*, 296, 181–198, 1993.
20. Groopman, J.D., Wang, J.S., and Scholl, P., Molecular biomarkers for aflatoxins: from adducts to gene mutations to human liver cancer, *Can.J. Physiol. Pharmacol.*, 74, 203–209, 1996.
21. Lin, D. et al., Analysis of 4-aminobiphenyl-DNA adducts in human urinary bladder and lung by alkaline hydrolysis and negative ion gas chromatography-mass spectrometry, *Environ. Health Perspect.*, 102 (Suppl. 6), 11–16, 1994.
22. Weston, A. et al., Fluorescence and mass spectral evidence for the formation of benzo[a]pyrene anti-diol-epoxide-DNA and -hemoglobin adducts in humans. *Carcinogenesis*, 10, 251–257, 1989.
23. Schoket, B., DNA damage in humans exposed to environmental and dietary polycyclic aromatic hydrocarbons, *Mutation Res.*, 424, 143–153, 1999.
24. Poirier, M.C., DNA adducts as exposure biomarkers and indicators of cancer risk, *Environ. Health Perspect.*, 105, 907–912, 1997.
25. Santella, R.M., DNA damage as an intermediate biomarker in intervention studies, *Proc. Soc. Exp. Biol. Med.*, 216, 166–171, 1997.
26. Poirier, M.C., Human exposure monitoring, dosimetry, and cancer risk assessment: the use of antisera specific for carcinogen-DNA adducts and carcinogen-modified DNA, *Drug Metab. Rev.*, 26, 87–109, 1994.
27. Perera, F.P. et al., Detection of polycyclic aromatic hydrocarbon-DNA adducts in white blood cells of foundry workers, *Cancer Res.*, 48, 2288–2291, 1988.
28. Santella, R.M. et al., Polycyclic aromatic hydrocarbon-DNA adducts in white blood cells and urinary 1-hydroxypyrene in foundry workers, *Cancer Epidemiol. Biomarkers Prev.*, 2, 59–62, 1993.
29. Rothman, N. et al., Contribution of occupation and diet to white blood cell polycyclic aromatic hydrocarbon-DNA adducts in wildland firefighters, *Cancer Epidemiol. Biomarkers Prev.*, 2, 341–347, 1993.
30. Groopman, J.D. et al., Molecular biomarkers for aflatoxins and their application to human cancer prevention, *Cancer Res.*, 54, 1907s–1911s, 1994.
31. Groopman, J.D. et al., Molecular dosimetry of aflatoxin-N7-guanine in human urine obtained in the Gambia, West Africa, *Cancer Epidemiol. Biomarkers Prev.*, 1, 221–227, 1992a.
32. Groopman, J.D. et al., Molecular dosimetry of urinary aflatoxin-DNA adducts in people living in Guangxi Autonomous Region, People's Republic of China *Cancer Res.*, 52, 45–52, 1992b.
33. Qian, G.S. et al., A follow-up study of urinary markers of aflatoxin exposure and liver cancer risk in Shanghai, People's Republic of China, *Cancer Epidemiol. Biomarkers Prev.*, 3, 3–10, 1994.
34. Ross, R.K. et al., Urinary aflatoxin biomarkers and risk of hepatocellular carcinoma, *Lancet*, 339, 943–946, 1992.

35. Wang, L.Y. et al., Aflatoxin exposure and risk of hepatocellular carcinoma in Taiwan, *Int. J. Cancer*, 67, 620–625, 1996.

36. Beland, F.A. and Poirier, M.C., Significance of DNA adduct studies in animal models for cancer molecular dosimetry and risk assessment, *Environ. Health Perspect.*, 99, 5–10, 1993.

37. La, D.K. and Swenberg, J.A., DNA adducts: biological markers of exposure and potential applications to risk assessment, *Mutation Res.*, 365, 129–146, 1996.

38. Gelboin, H.V. and Ts'o, P.O.P., *Polycyclic Aromatic Hydrocarbons and Cancer*, Vol.1, Academic Press, New York. 1978.

39. Harvey, R.G., *Polycyclic Aromatic Hydrocarbons: Chemistry and Carcinogenicity*, Cambridge University Press, Cambridge, U.K., 1991.

40. Wynder, E.L. and Hoffmann, D.A., A study of tobacco carcinogenesis. VII. The role of higher polycyclic hydrocarbons, *Cancer* 12, 1079–1086, 1959.

41. Hoffmann, D., Hoffmann, I., and El-Bayoumi, K., The less harmful cigarette: a controversial issue (A tribute to Ernst L. Wynder), *Chem. Res. Toxicol.*, 14, 767–790, 2001.

42. Hecht, S.S., Tobacco smoke carcinogens and lung cancer, *J. Natl. Cancer Inst.*, 91, 1194–1210, 1999.

43. Wingo, P.A. et al., Annual report to the nation on the status of cancer, 1973–1996, with a special section on lung cancer and tobacco smoking, *J. Natl. Cancer Inst.*, 91, 675–690, 1999.

44. Muscat, J.E. and Wynder, E.L., Diesel engine exhaust and lung cancer: An unproven association, *Environ. Health Perspect.*, 103, 812–818, 1995.

45. Jeffrey, A.M. et al., Letter: Benzo[a]pyrene-nucleic acid derivative found in vivo: structure of a benzo[a]pyrenetetrahydrodiol epoxide-guanosine adduct, *J. Am. Chem. Soc.*, 98, 5714–5715, 1976.

46. Meehan, T. and Straub, K., Double-stranded DNA steroselectively binds benzo(a)pyrene diol epoxides, *Nature*, 277, 410–412, 1979.

47. Cheng, S.C. et al., DNA adducts from carcinogenic and noncarcinogenic enantiomers of benzo[a]pyrene dihydrodiol epoxide, *Chem. Res. Toxicol.*, 2, 334–340, 1989.

48. Levin, W. et al., (+,-)-trans-7,8-dihydroxy-7,8-dihydrobenzo (a)pyrene: a potent skin carcinogen when applied topically to mice, *Proc. Natl. Acad. Sci. USA*, 73, 3867–3871, 1976.

49. Slaga, T.J. et al., Marked differences in the skin tumor-initiating activities of the optical enantiomers of the diastereomeric benzo(a)pyrene 7,8-diol-9,10-epoxides, *Cancer Res.*, 39, 67–71, 1979.

50. Phillips, D.H. et al., Correlation of DNA adduct levels in human lung with cigarette smoking, *Nature*, 336, 790–792, 1988.

51. Randerath, E. et al., Covalent DNA damage in tissues of cigarette smokers as determined by ^{32}P-postlabeling assay, *J. Natl. Cancer Inst.*, 81, 341–347, 1989.

52. van Schooten, F.J. et al., Polycyclic aromatic hydrocarbon-DNA adducts in lung tissue from lung cancer patients, *Carcinogenesis*, 11, 1677–1681, 1990.

53. Alexandrov, K. et al., An improved fluorometric assay for dosimetry of benzo(a)pyrene diol- epoxide-DNA adducts in smokers' lung: comparisons with total bulky adducts and aryl hydrocarbon hydroxylase activity, *Cancer Res.*, 52, 6248–6253, 1992.

54. Corley, J. et al., Solid matrix, room temperature phosphorescence identification and quantitation of the tetrahydrotetrols derived from the acid hydrolysis of benzo[a]pyrene-DNA adducts from human lung, *Carcinogenesis*, 16, 423–426, 1995.

55. Kato, S. et al., Human lung carcinogen-DNA adduct levels mediated by genetic polymorphisms *in vivo*, *J. Natl. Cancer Inst.*, 87, 902–907, 1995.

56. Andreassen, A. et al., Comparative synchronous fluorescence spectrophotometry and [32]P- postlabeling analysis of PAH–DNA adducts in human lung and the relationship to TP53 mutations, *Mutation Res.*, 368, 275–282, 1996.

57. Kriek, E. et al., Polycyclic aromatic hydrocarbon-DNA adducts in humans: relevance as biomarkers for exposure and cancer risk, *Mutation Res.*, 400, 215–231, 1998.

58. Besarati-Nia, A. et al., Immunoperoxidase detection of polycyclic aromatic hydrocarbon-DNA adducts in mouth floor and buccal mucosa cells of smokers and nonsmokers, *Environ. Mol. Mutagen.*, 36, 127–133, 2000.

59. Perera, F.P. et al., Comparison of DNA adducts and sister chromatid exchange in lung cancer cases and controls, *Cancer Res.*, 49, 4446–4451, 1989.

60. Tang, D.L. et al., A molecular epidemiological case-control study of lung cancer, *Cancer Epidemiol. Biomarkers Prev.*, 4, 341–346, 1995.

61. Tang, D.L. et al., Associations between both genetic and environmental biomarkers and lung cancer: evidence of a greater risk of lung cancer in women smokers, *Carcinogenesis*, 19, 1949–1953, 1998.

62. van Schooten, F.J. et al., Polycyclic aromatic hydrocarbon-DNA adducts in white blood cells from lung cancer patients: no correlation with adduct levels in lung, *Carcinogenesis*, 13, 987–993, 1992.

63. Harris, C.C. et al., Detection of benzo[a]pyrene diol epoxide-DNA adducts in peripheral blood lymphocytes and antibodies to the adducts in serum from coke oven workers, *Proc. Natl. Acad. Sci. USA*, 82, 6672–6676, 1985.

64. Shamsuddin, A.K. et al., Detection of benzo(a)pyrene: DNA adducts in human white blood cells, *Cancer Res.*, 45, 66–68, 1985.

65. Hemminki, K. et al., DNA adducts in humans environmentally exposed to aromatic compounds in an industrial area of Poland, *Carcinogenesis*, 11, 1229–1231, 1990.

66. van Schooten, F.J. et al., Determination of benzo[a]pyrene diol epoxide-DNA adducts in white blood cell DNA from coke-oven workers: the impact of smoking, *J. Natl. Cancer Inst.*, 82, 927–933, 1990.

67. Ovrebo, S. et al., Detection of polycyclic aromatic hydrocarbon-DNA adducts in white blood cells from coke oven workers: correlation with job categories, *Cancer Res.*, 52, 1510–1514, 1992.

68. Santella, R.M. et al., Polycyclic aromatic hydrocarbon-DNA and protein adducts in coal tar treated patients and controls and their relationship to glutathione S-transferase genotype *Mutation Res.*, 334, 117–124, 1995.

69. Perera, F.P. et al., Molecular and genetic damage in humans from environmental pollution in Poland, *Nature*, 360, 256–258, 1992.

70. Topinka, J. et al., Influence of GSTM1 and NAT2 genotypes on placental DNA adducts in an environmentally exposed population, *Environ. Mol. Mutagen.*, 30, 184–195, 1997.

71. Nielsen, P.S. et al., Environmental air pollution and DNA adducts in Copenhagen bus drivers: effect of GSTM1 and NAT2 genotypes on adduct levels, *Carcinogenesis*, 17, 1021–1027, 1996.

72. Autrup, H. et al., Biomarkers for exposure to ambient air pollution-comparison of carcinogen-DNA adduct levels with other exposure markers and markers for oxidative stress, *Environ. Health Perspect.*, 107, 233–238, 1999.

73. Rothman, N. et al., Formation of polycyclic aromatic hydrocarbon-DNA adducts in peripheral white blood cells during consumption of charcoal-broiled beef, *Carcinogenesis*, 11, 1241–1243, 1990.

74. van Maanen, J.M. et al., Formation of aromatic DNA adducts in white blood cells in relation to urinary excretion of 1-hydroxypyrene during consumption of grilled meat, *Carcinogenesis*, 15, 2263–2268, 1994.

75. Kang, D.H. et al., Interindividual differences in the concentration of 1-hydroxypyrene-glucuronide in urine and polycyclic aromatic hydrocarbon-DNA adducts in peripheral white blood cells after charbroiled beef consumption, *Carcinogenesis*, 16, 1079–1085, 1995.

76. Everson, R.B. et al., Detection of smoking-related covalent DNA adducts in human placenta, *Science*, 231, 54–57, 1986.

77. Randerath, K., Reddy, M.V., and Gupta, R.C., ^{32}P-labeling test for DNA damage, *Proc. Natl. Acad. Sci. USA*, 78, 6126–6129, 1981.

78. Beach, A.C. and Gupta, R.C., Human biomonitoring and the ^{32}P-postlabeling assay, *Carcinogenesis*, 13, 1053–1074, 1992.

79. Tretyakova, N.Y. et al., Quantitative analysis of 1,3-butadiene-induced DNA adducts *in vivo* and *in vitro* using liquid chromatography electrospray ionization tandem mass spectrometry, *J. Mass. Spectrom.*, 33, 363–376, 1997.

80. Poirier, M.C. et al., Quantitation of benzo(a)pyrene-deoxyguanosine adducts by radioimmunoassay, *Cancer Res.*, 40, 412–416, 1980.

81. Nehls, P., Adamkiewicz, J., and Rajewsky, M.F., Immuno-slot-blot: a highly sensitive immunoassay for the quantitation of carcinogen-modified nucleosides in DNA, *J. Cancer Res. Clin. Oncol.*, 108, 23–29, 1984.

82. Wani, A.A. and D'Ambrosio, S.M., Immunological quantitation of O[4]-ethylthymidine in alkylated DNA: repair of minor miscoding base in human cells, *Carcinogenesis*, 8, 1137–1144, 1987.

83. Wani, A.A., D'Ambrosio, S.M., and Alvi, N.K., Quantitation of pyrimidine dimers by immunoslot blot following sublethal UV-irradiation of human cells, *Photochem. Photobiol.*, 46, 477–482, 1987.

84. Leuratti, C. et al., Determination of malondialdehyde-induced DNA damage in human tissues using an immunoslot blot assay, *Carcinogenesis*, 19, 1919–1924, 1998.

85. Venkatachalam, S., Denissenko, M.F., and Wani, A.A., DNA repair in human cells: quantitative assessment of bulky anti-BPDE-DNA adducts by non-competitive immunoassays, *Carcinogenesis*, 16, 2029–2036, 1995.

86. Pfeifer, G.P. et al., *In vivo* mapping of a DNA adduct at nucleotide resolution: Detection of pyrimidine (6–4) pyrimidone photoproducts by ligation-mediated polymerase chain reaction, *Proc. Natl. Acad. Sci. USA*, 88, 1374–1378, 1991.

87. Bohr, V.A. et al., DNA repair in an active gene: removal of pyrimidine dimers from the DHFR gene of CHO cells is much more efficient than in the genome overall, *Cell*, 40, 359–369, 1985.

88. Pfeifer, G.P. et al., Binding of transcription factors creates hot spots for UV photoproducts *in vivo*, *Mol. Cell. Biol.*, 12, 1798–1804, 1992.

89. Tommasi, S., Denissenko, M.F., and Pfeifer, G.P., Sunlight induces pyrimidine dimers preferentially at 5-methylcytosine bases, *Cancer Res.*, 57, 4727–4730, 1997.

90. Yoon, J.H. et al., The DNA damage spectrum produced by simulated sunlight, *J. Mol. Biol.*, 299, 681–693, 2000.
91. Denissenko, M.F. et al., Cytosine methylation determines hot spots of DNA damage in the human *p53* gene, *Proc. Natl. Acad. Sci.USA*, 94, 3893–3898, 1997.
92. Denissenko, M.F. et al., Slow repair of bulky DNA adducts along the non-transcribed strand of the human *p53* gene may explain the strand bias of transversion mutations in cancers, *Oncogene*, 16, 1241–1247, 1998.
93. Smith, L.E. et al., Targeting of lung cancer mutational hotspots by polycyclic aromatic hydrocarbons, *J. Natl. Cancer Inst.*, 92, 803–811, 2000.
94. Dai, S.M. et al., Ligation-mediated PCR for quantitative *in vivo* footprinting, *Nature Biotechnol.*, 18, 1108–1111, 2000.
95. Rodriguez, H. et al., Mapping of copper/hydrogen peroxide-induced DNA damage at nucleotide resolution in human genomic DNA by ligation-mediated polymerase chain reaction, *J. Biol. Chem.*, 270, 17633–17640, 1995.
96. Lee, C.S., Pfeifer, G.P., and Gibson, N.W., Mapping of DNA alkylation sites induced by adozelesin and bizelesin in human cells by ligation-mediated polymerase chain reaction, *Biochemistry*, 33, 6024–6030, 1994.
97. Denissenko, M.F. et al., The *p53* codon 249 mutational hotspot in hepatocellular carcinoma is not related to selective formation or persistence of aflatoxin B1 adducts, *Oncogene.*, 17, 3007–3014, 1998.
98. Komura, J.I. and Riggs, A.D., Terminal transferase dependent PCR: A versatile and sensitive method for *in vivo* footprinting and detection of DNA adducts, *Nucleic Acids Res.*, 26, 1807–1811, 1998.
99. Arlt, V.M., Schmeiser, H.H., and Pfeifer, G.P., Sequence-specific detection of aristolochic acid-DNA adducts in the human *p53* gene by terminal transferase-dependent PCR, *Carcinogenesis*, 22, 133–140, 2001.
100. Komura, J.I. et al., Mapping psoralen cross-links at the nucleotide level in mammalian cells: suppression of cross-linking at transcription factor- or nucleosome-binding sites, *Biochemistry*, 40, 4096–4105, 2001.
101. Tang, M.S. et al., Use of UvrABC nuclease to quantify benzo[a]pyrene diol epoxide-DNA adduct formation at methylated versus unmethylated CpG sites in the *p53* gene, *Carcinogenesis*, 20, 1085–1089, 1999.
102. Hainaut, P. and Pfeifer, G.P., Patterns of *p53* G to T transversions in lung cancers reflect the primary mutagenic signature of DNA-damage by tobacco smoke, *Carcinogenesis*, 22, 367–374, 2001.
103. Eisenstadt, E. et al., Carcinogenic epoxides of benzo(a)pyrene and cyclopenta(cd)pyrene induce base substitutions via specific transversions, *Proc. Natl. Acad. Sci. USA*, 79, 1945–1949, 1982.
104. Mazur, M. and Glickman, B., Sequence specificity of mutations induced by benzo(a)pyrene-7,8-diol-9,10-epoxide at endogenous APRT gene in CHO cells, *Somat. Cell Mol. Genet.*, 14, 393–400, 1988.
105. Chen, R.H., Maher, V.M., and McCormick, J.C., Effect of excision repair by diploid human fibroblasts on the kinds and locations of mutations induced by (±)-7β,8α-dihydroxy-9α,10α-epoxy-7,8,9,10-tetrahydrobenzo[a]pyrene in the coding region of the *HPRT* gene, *Proc. Natl. Acad. Sci. USA*, 87, 8680–8684, 1990.
106. Chen, J.X. et al., Carcinogens preferentially bind at methylated CpG in the *p53* mutational hotspots, *Cancer Res.*, 58, 2070–2075, 1998.

107. Weisenberger, D.J. and Romano, L.J., Cytosine methylation in a CpG sequence leads to enhanced reactivity with benzo[a]pyrene diol epoxide that correlates with a conformational change, *J. Biol. Chem.*, 274, 23948–23955, 1999.

108. Das, A. et al., Reactivity of guanine at m5CpG steps in DNA: evidence for electronic effects transmitted through the base pairs, *Chem. Biol.*, 6, 461–471, 1999.

109. Tornaletti, S. and Pfeifer, G.P., Complete and tissue-independent methylation of CpG sites in the *p53* gene: implications for mutations in human cancers, *Oncogene.*, 10, 1493–1499, 1995.

110. Yoon, J.-H. et al., Methylated CpG dinucleotides are the preferential targets for G to T transversion mutations induced by benzo[a]pyrene diol epoxide in mammalian cells: similarities with the p53 mutation spectrum in smoking-associated lung cancers, *Cancer Res.*, 61, 7110–7117, 2001.

111. You, Y.-H. and Pfeifer, G.P., Similarities in sunlight-induced mutational spectra of CpG-methylated transgenes and the *p53* gene in skin cancer point to an important role of 5-methylcytosines in solar UV mutagenesis, *J. Mol. Biol.*, 305, 389–399, 2001.

112. You, Y.-H., Li, C., and Pfeifer, G.P., Involvement of 5-methylcytosine in sunlight-induced mutagenesis, *J. Mol. Biol.*, 293, 493–503, 1999.

113. Hernandez-Boussard, T. et al., IARC *p53* mutation database: a relational database to compile and analyze *p53* mutations in human tumors and cell lines, *Human Mutation.*, 14, 1–8, 1999.

chapter eleven

Quantitative PCR: a sensitive biomarker of gene-specific DNA damage and repair

Bennett Van Houten, Yiming Chen, Janine H. Santos, and Bhaskar S. Mandavilli

Contents

Abstract DNA damage underlies the genesis of many important human diseases such as cancer, neurodegenerative diseases, and aging. Over the last three decades several biomarkers of DNA injury, including chromosome aberrations or mutation assay have been used in small populations studies. While these approaches have been informative they do not allow a direct assessment of DNA damage and repair. Several methods have been developed that measure the formation and removal of DNA lesions. However, these approaches are not often amenable to large population studies. We have developed a robust DNA damage and repair assay that is based on the quantitative polymerase chain reaction (QPCR) that can be easily set up in most laboratories and has the high throughput necessary for a large number of samples.

I. Introduction

Our genetic material is under constant threat of damage by a wide variety of chemical and physical agents found in our environment. In addition, endogenous sources of DNA damage, primarily reactive oxygen, can be generated during normal energy production or at the site of inflammation. DNA damage if not repaired can lead to mutations, and/or cell death. At the organismal level, DNA damage can lead to cancer, neurodegenerative diseases, birth defects, and aging. During the last three decades, a series of methods have been developed that allow detection and quantitation of DNA damage. After surveying a number of methods used to assess the formation and repair of DNA adducts, this chapter describes a novel method based on the polymerase chain reaction (PCR) for the analysis of DNA damage.

A.   DNA biomarkers

Several assays have been developed that have been used as biomarkers of DNA damage, these include micronuclei assay, sister chromatic exchanges, chromosome aberrations, mutations in the hypoxanthine phosphoribosyl transferase (HPRT) gene, and DNA adducts. However, all but the last approach gives an indirect measure of damage processing. A number of rigorous analytical techniques have been used to analyze the formation and repair of DNA adducts in human samples, primarily human lymphocytes (see Table 11.1). These approaches have been nicely reviewed by Poirer, Santella, and Weston.[1] Most of the techniques outlined in Table 11.1 have high sensitivity of 1 adduct in 10^7–10^9.

B.   Limitations of current techniques

Great strides have been made at developing sensitive assays for the detection of DNA damage. However, many laboratories do not have access to the sophisticated analytical instruments, such as mass spectrometry, that are necessary for these assays. Consequently the applications of these tools have not been fully realized. Furthermore the large amounts of DNA required by several of the assays makes sample size a concern. The ^{32}P-postlabeling assay has a high sensitivity but requires high levels of radioactivity. An ideal biomarker of DNA damage should have the following attributes to allow widespread utility. First, the biomarker should be robust with the ability to

Table 11.1   Techniques for measuring DNA damage

Technique	Adduct	Sensitivity	DNA	Problems/Comments
Comet Assay	Wide range	$1/10^{10}$	per cell	Must score at least 100 cells
Electrochemical Detection	8-oxodG	$1/10^6$–10^7	50 µg	Notoriously high background
Fluorescence Spectroscopy	AFB, PAH	$1/10^5$–10^6	100 µg	Relatively low sensitivity
Immunoassays (ELISA, IH, IB)	AFB, PAH, 8-oxodG, UV	$1/10^7$–10^9	50–100 µg	Need large amounts of DNA
Mass Spectrometry (MS)	Wide range oxidative	$1/10^5$–10^6	100 µg	Very sensitive, but equipment is
GC-IDMS	nitrosamines	$1/10^{12}$		expensive and technically challenging
Accelerator MS				
^{32}P-Post-labeling	PAH, oxidative lesions	$1/10^9$–10^{10}	5–15 µg	Must use high levels of ^{32}P
QPCR	Wide range	$1/10^5$	15 ng	Gene-specific

detect a large variety of lesions. Second, the assay should only require small amounts of DNA, keeping tissue sample size to a minimum. Third, the method should have high sensitivity so to detect biological relevant levels of damage. Fourth, the assay should have the ability to examine adduct frequencies in defined regions of DNA, such as mitochondrial genome or specific nuclear genes. Fifth, the methd should use common routine procedures without the need for expensive equipment or radioactivity making it widely accessible to research laboratories. Finally, the assay should have high throughput so that hundreds of samples could be processed in a day. Over the last decade, my laboratory has been working on an assay that is based on the QPCR. We recently have succeeded in measuring the amounts of PCR products using a 96-well fluorescence plate reader, circumventing the need for analysis by agarose gel electrophoresis gel or the use of radionucleotides. Thus, the QPCR assay fulfills all six of the criteria outlined above.

II. Use of quantitative PCR in the assessment of DNA damage and repair

A.   Technique

The QPCR gene-specific assay is based on the principle that lesions present in the DNA block the progression of any thermostable polymerase on the template, resulting in the decrease of DNA amplification in the damaged template when compared to the undamaged DNA.[2,3] The PCR actually measures the fraction of undamaged template molecules, which decreases with increasing amount of lesions. Gene-specific damage is measured as loss of template amplification and DNA repair activity as the restoration of the amplification signal.[2,3] Analysis of long PCR products allowed increased sensitivity of the assay.[4,5] We have successfully used this approach to follow damage and repair of a wide variety of DNA lesions in DNA from yeast, rodent, and human cells and tissues.[6,7] The assay is outlined in Figure 11.1. The QPCR technique is a sensitive and reliable assay for the detection of gene-specific damage and repair. Current detection limits are on the order of 1 lesion/10^5 nucleotides from as little as 5 ng of total genomic DNA. The assay can be performed on cultured cells or tissue (step 1) and can detect insults from a number of genotoxicants such as environmental carcinogens and anticancer. These include the following: alkylating agents, asbestos, benzo[a]pyrene diol epoxide, butadiene diepoxide, cisplatin, reactive oxygen species, and UV light. The DNA is isolated using a series of commercially available kits from QIAGEN, Inc., (Chatsworth, CA) (step 2) and the amount of DNA is accurately determined using Picogreen fluorescence (Molecular Probes) using a fluorescence plate reader (step 3). We routinely isolate DNA from about one million cells or about 25 mg of frozen tissue, with typical yields on the order of 1–3 µg. The DNA is subjected to QPCR using specially designed oligonucleotide primers. We have developed a number of PCR

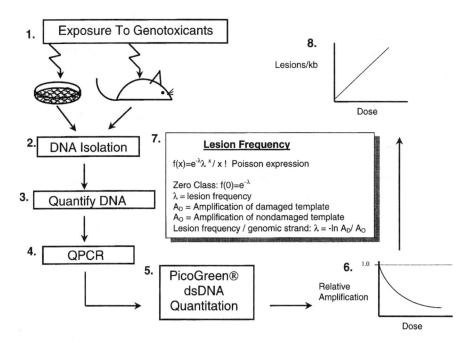

1. Exposure To Genotoxicants

8. Lesions/kb — Dose

2. DNA Isolation

7. **Lesion Frequency**

$f(x)=e^{-\lambda}\lambda^x / x\,!$ Poisson expression

Zero Class: $f(0)=e^{-\lambda}$
λ = lesion frequency
A_D = Amplification of damaged template
A_O = Amplification of nondamaged template
Lesion frequency / genomic strand: $\lambda = -\ln A_D/A_O$

3. Quantify DNA

4. QPCR

5. PicoGreen® dsDNA Quantitation

6. Relative Amplification — Dose

Figure 11.1 Schematic of the QPCR gene-specific assay. See text for details. (From Ayala-Torres, S. et al., in *Methods: A Companion to Methods in Enzymology*, Academic Press, New York, 2000. With permission.)

products using gene specific primers[6,7] in several species including yeast, mouse, rat, and humans. The level of amplification of the PCR products is determined using Picogreen fluorescence. The amount of PCR product is inversely proportional to the amount of damage; thus, a decrease in the PCR product is indicative of DNA damage (step 6). DNA lesion frequency is calculated using the Poisson expression such that lesion frequency per fragment at a particular dose, $D = -\ln A_D/A_C$, where A_D is the amount of amplification after dose, D, and A_C is the amount of amplification of the nondamaged control sample (steps 6 and 7).

B.   Assessment of DNA repair capacity in human cells

This year, over 800,000 people in the United States will be diagnosed with skin cancer, the most common form of cancer. There are large amounts of epidemiological data that indicate that sun exposure causes genetic changes in skin cells leading to basal cell carcinoma or squamous cell carcinoma. UV light-induced photoproducts are repaired by NER in which a complex series of over 30 proteins work together to remove the DNA lesion. Several photosensitive human disorders have been associated with defects in NER. Xeroderma pigmentosum is characterized by extreme sun sensitivity, hyper- and hypopigmentation, and a 3000-fold increased incidence of skin cancer. Due to the large number of gene products, it has been suggested that DNA repair capacity in humans may vary as much as tenfold. While there have been some measurements of repair capacity in human populations, a direct and simple assay which measures the rate of UV-induced photoproduct removal has not been well developed. To this end, we have recently validated the QPCR assay for the detection and removal of biological relevant doses of ultraviolet radiation.[8] We found that we could detect DNA damage following as little as 2.5 joules/m^2 of UVC. This dose of UV light produces about the same of amount of DNA damage as a 15-min walk on a sunny beach in June. Furthermore, this assay could detect differences in the rates of repair of an expressed and nonexpressed gene in human cells[8] (Figure 11.2). It is interesting to note that the rate of repair was more rapid in the EB-transformed B-cell lymphoblasts (Figure 11.2b) as compared to a SV40-transformed fibroblast cell line (Figure 11.2a). Cellular transformation with viruses might alter repair or it is possible that these differences in repair reflect differences in cell type and/or the human donor's repair capacity.

C.   Mitochondrial DNA damage as a biomarker of oxidative stress

One of the most ubiquitous DNA damaging agents is oxidative stress, an endogenous DNA damaging agent generated by normal cellular respiration. During the formation of ATP, electrons flow down the electron chain complexes, which generate a proton-gradient that is harvested by complex V, ATPase synthase. The final acceptor of electrons is molecular oxygen which is reduced in a four-electron addition to water. However reactive oxygen products of one- and two-electron additions, superoxide and hydrogen peroxide, respectively, are generated in the mitochondria. It has been estimated that as much as 1–2% of all the oxygen we consume is released as reactive oxygen species.

Interestingly, we found that hydrogen peroxide induced three times more damage to mitochondrial DNA (mtDNA) than nuclear targets in human fibroblasts.[9] This damage was repaired within 1.5 h after a 15-min treatment with 200 μM hydrogen peroxide. After 3 h of repair, amplification of both genomes increased over the control levels. This unexpected result probably represents an induction of DNA repair enzymes by the oxidative stress such that their action reduces DNA damage to levels lower than the basal levels.[10,11] Strangely, a 60-min treatment with 200 μM hydrogen peroxide lead to persistent mitochondrial DNA damage even though the nuclear damage was repaired within 1.5 h (Figure 11.3). What is the basis of this persistent mtDNA damage? For a long time, it was believed that mitochondria lack efficient repair, but work in many laboratories[12] during the last decade have clearly demonstrated that mitochondria have some but not all the repair pathways associated with nucleus. For example, mitochondria are deficient in the removal of UV-induced pyrimidine dimers or cisplatin adducts, lesions which are repaired by NER (mentioned above). However, base excision repair enzymes, which are primarily responsible for the removal of oxidative lesions, have been isolated from mitochondria.[12] Glu-

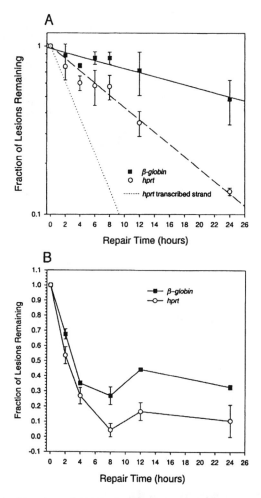

Figure 11.2 **Repair kinetics of UV-induced damage in human cells.** SV-40 transformed fibroblasts (Panel A) or human lymphoblastoid cells (Panel B) were treated with 10 J/m² of UVC and allowed various periods of time for repair. The DNA was extracted and subjected to quantitative PCR. Repair was examined in two gene regions, nonexpressed beta-globin region (closed symbols) and the expressed *HPRT* gene (open symbols). (From Van Houten, B., Cheng, S., and Chen, Y., *Mut. Res.*, 460, 81–94, 2000. With permission.)

cose oxidase (GO), a steady generator of more physiological concentrations of hydrogen peroxide, resulted in large amounts of mtDNA damage with little or no nuclear damage.[13] Human fibroblasts treated with 6 milliunits GO/ml of media generated about 10 μM hydrogen peroxide during a 15-min treatment and produced approximately 1.0 lesion/mitochondrial genome. This mtDNA damage was repaired within 4 h after the treatment.[13] This finding of increased mtDNA damage after an oxidant has been repeated in a large number of cell types from both human and rodent cells.[14–16] It is

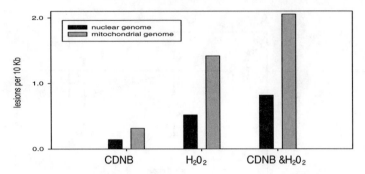

Figure 11.3 **Oxidative damage to mitochondrial and nuclear DNA.** Yeast wild type strain ale 1000 (1×10^7 cells/mL) were inoculated into YPD broth. Eight hours after inoculation, cells were harvested and resuspended in PBS. The cells were treated for 30 min with either 1mM of 1-chloro-2,4dinitrobenzene (CDNB) or 20 mM of hydrogen peroxide alone, or a combined exposure in which CDNB was administered first followed by 30 min to H_2O_2. Lesions (per 10 Kb), observed in both nuclear and mitochondrial genomes, are plotted as a function of treatments (data of mitochondria normalized for mitochondria copy number).

interesting to note that the antiapoptotic protein, Bcl-2, while not capable of blocking ROS-induced mitochondrial damage, lead to increased rates of mtDNA repair.[15]

Even yeast cells suffer significantly more mtDNA damage than nuclear damage following oxidant stress.[7] Oxidative stress was induced in yeast by treatment with hydrogen peroxide or of 1-chloro-2,4dinitrobenzene (CNDB), a glutathione reactive compound. CNDB leads to the loss of glutathione, an important co-factor of glutathione peroxidase, the primary hydrogen peroxide detoxifying enzyme in mitochondria. It is interesting to note that depletion of glutathione leads to an increase damage in mtDNA as compared to a nuclear gene. Hydrogen peroxide produced approximately three times more damage in the mitochondrial DNA than a nuclear gene. Furthermore combination of CNDB and hydrogen peroxide leads to synergistic amounts of damage in both genomes.

These observations have lead to two working hypotheses: increased divalent metal ions in mitochondria produce more mtDNA damage and damaged mitochondria can generate ROS and lead to a vicious cycle of damage, which even in the presence of an active DNA repair system results in persistent mtDNA damage. These hypotheses provide a paradigm for the molecular aspects of many chronic diseases associated with aging, such as Alzheimer's disease and Parkinson's disease discussed in more detail below.

D.   Validation studies in animals: insights into Parkinson's disease and aging

Prior to using the QPCR in human populations to assess rates of repair and endogenous DNA damage we felt it was important to validate the assay in

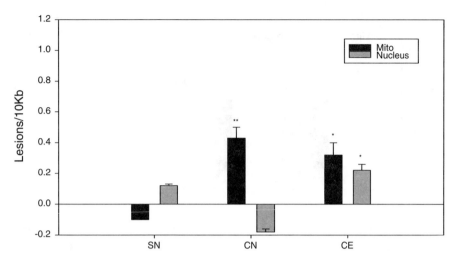

Figure 11.4 **Age-related differences in DNA damage in brain regions of mice.** DNA from specific brain regions were isolated from 22-day-old or 1-year-old mice (n = 6) and subjected to QPCR for mitochondrial DNA or clusterin, a nuclear gene. Lesions/10kb for the one year old animals as compared to the 22-day animals, are plotted for substantia nigra (SN), caudate nucleus (CN), and cerebellum (CE) in black for mitochondrial DNA and gray for nuclear DNA (±S.E.M.). "Negative lesions" would indicate that the 22-day animals had more damage than the older animals. Student's T-test was used to determine the significant differences, where P < 0.05 = *, P < 0.01 = ** (From Mandavilli, B.S., Ali, S.F., and Van Hauten, B., *Brain Research*, 885, 45–52, 2000. With permission.)

animals. The free radical theory of aging suggests that during the life of an organism, free radicals lead to damage in macromolecules causing a loss of homeostasis and lower enzyme efficiency. This damage ultimately leads to symptoms associated with aging. Since mitochondria are a primary source of reactive oxygen intermediates, then they might play an important role in the aging process. One such idea is that as mitochondria accumulate damage to their macromolecules they become less efficient and generate more reactive oxygen species.[17] This ROS generation leads to a vicious cycle of damage and ultimately to a catastrophic demise of mitochondrial function resulting in the death of the cell by apoptosis. We, therefore, wanted to test the idea that mitochondrial DNA will accumulate damage with age. The QPCR assay was able to show that specific regions of the brain of older mice do accumulate both nuclear and mitochondrial DNA damage (Figure 11.4).

During the late 1970s it was discovered that a byproduct of underground Demerol synthesis, 1-methyl-4-phenyl-1,2,3,6-tetrahydropyridine (MPTP) leads to marked depletion of dopaminergic neurons in rodent and human brains leading to a rapid and irreversible onset of Parkinsonian-like symptoms. MPTP is converted to MPP+ by monoamine oxidase in glial cells and is taken up by dopaminergic neurons. MPP+ is believed to inhibit complex I, NADH dehydrogenase, of the electron transport machinery, leading to the

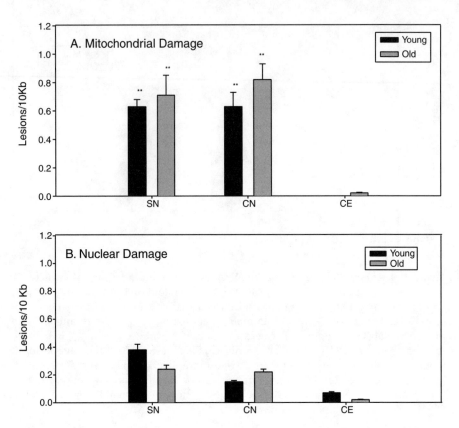

Figure 11.5 **MPTP-induced DNA damage in young and old mice.** DNA from specific brain regions were isolated from either control or mice treated with MPTP (4 x 10 mg/Kg, i.p.) and subjected to QPCR for mitochondrial DNA or clusterin, a nuclear gene. Both 22-day and 1-year-old animals were treated (n = 6). MPTP-induced lesions/10kb (±S.E.M.) is plotted for substantia nigra (SN), caudate nucleus (CN), and cerebellum (CE) in black for the 22-day (young) and gray for the 1-year-old mice (old). Student's T-test was used to determine the significant differences, where P < 0.05 = *, P<0.01 = **. (From Mandavilli, B.S., Ali, S.F., and Van Houten, B., *Brain Research,* 885, 45–52, 2000. With permission.)

generation of ROS, and loss of mitochondrial function. Using the QPCR assay we are able to show that within 72 h of treatment with MPTP, mice developed two to three times more mtDNA than nuclear damage in the substantia nigra and the caudate nucleus, two regions with high numbers of dopaminergic neurons.[18] Whereas the cerebellum, which is unaffected by MPTP showed no damage to either genome (Figure 11.5). These data strongly support the idea that MPTP leads to ROS generation in the mitochondria leading to subsequent damage to macromolecules like DNA. Since mtDNA encodes 13 polypeptides which are involved in electron transport and ATP generation, damaged mtDNA would lead to a decrease in mitochondrial function due to the loss of these proteins. This partial loss of mitochondrial activity could lead to a vicious cycle of ROS generation and a catastrophic demise of mitochondrial function.[16]

III. Future directions

The QPCR assay holds great promise for measuring DNA repair capacity and levels of endogenous DNA damage in human populations. However a number of questions remain about the use of the QPCR assay as well as others listed in Table 11.1 for the measurement of DNA damage in human populations. If endogenous DNA damage is 100-fold higher than xenobiotic damage, as has been suggested, then it is unclear whether the more infrequent lesions from exogenous chemicals pose a real threat to biological systems.[19] Peripheral lymphocytes are often used as surrogate markers for DNA, but does this cell population serve as a good marker for DNA damage throughout the body? More studies must be done in animals to compare levels of adducts in target tissue versus peripheral blood cells. Since cancer can take as long as 20 years to develop in humans following the initial injury, any study of persistent DNA adducts must take into account the biological half-life of these DNA adducts. Finally, molecular epidemiology holds great promise in defining the basis of interindividual susceptibility to disease. We live in an age where a correlation of specific single nucleotide polymorphisms in target genes involved in DNA repair to slow rates of repair and accumulation of DNA damage is now possible. However, screening large populations for these end points is not feasible due to the slow throughput and high costs associated with these technologies. Clearly, advances in biomarkers of DNA damage will certainly allow a better understanding of role of endogenous and exogenous factors in causing genetic instability leading to human disease.

References

1. Poirier, M.C., Santella, R.M., and Weston, A., Carcinogen macromolecular adducts and their measurement, *Carcinogenesis*, 21, 353–359, 2000.
2. Kalinowski, D., Illenye, S., and Van Houten, B., Analysis of DNA damage and repair in murine leukemia L1210 cells using a quantitative polymerase chain reaction assay, *Nucleic Acids Res.*, 20, 3485–3494, 1992.
3. Chandrasekhar, D. and Van Houten, B., High resolution mapping of UV-induced photoproducts in the *E. coli lacI* gene: inefficient repair in the non-transcribed strand correlates with high mutation frequency, *J. Mol. Biol.*, 238, 319–332, 1994.
4. Yakes, F.M., Chen, Y., and Van Houten, B., PCR-based assays for the detection and quantitation of DNA damage and repair, in *Technologies for Detection of DNA Damage and Mutations*, Pfeifer, G.P., Ed., Plenum Press, New York, 1996, pp. 171–184.
5. Cheng, S. et al., Template integrity is essential for PCR amplification of 20- to 30kb sequences from genomic DNA, *PCR Meth. and Appl.*, 4, 294–298, 1995.

6. Ayala-Torres, S. et al., Analysis of gene-specific DNA damage and repair using quantitative PCR, in *Methods. A Companion to Methods in Enzymology,* Doetsch, Paul W. (Ed). Academic Press, New York, 2000, 22, 135-147.

7. Santos, J., Mandavilli, B., and Van Houten, B., Measurement of oxidative damage and repair in mitochondria using quantitative PCR, in *Methods in Molecular Biology,* Copeland, W., Ed., 2001, in press.

8. Van Houten, B., Cheng, S., and Chen, Y., Measuring DNA damage and repair in human genes using quantitative amplification of long targets from nanogram quantities of DNA, *Mut. Res.,* 460, 81–94, 2000.

9. Yakes, F.M. and Van Houten, B., Mitochondrial DNA damage is more extensive and persists longer than nuclear DNA damage in human cells following oxidative stress, *Proc. Natl. Acad. Sci.,* 94, 514–519, 1997.

10. Chen, K.H. et al., Up-regulation of base excision repair correlates with enhanced protection against a DNA damaging agent in mouse cell lines, *Nucleic Acids Res.,* 26, 2001–2007, 1998.

11. Fung, H. et al., Patterns of 8-hydroxydeoxyguanosine (8OHdG) Formation in DNA and indications of oxidative stress in rat and human pleural mesothelial cells after exposure to crocidolite asbestos, *Carcinogenesis,* 18, 825–832, 1997.

12. Sawyer, D.E. and Van Houten, B., Repair of DNA damage in mitochondria, *Mut. Res.,* 434, 161–176, 1999.

13. Salazar, J.J. and Van Houten, B., Preferential mitochondrial DNA injury caused by glucose oxidase as a steady generator for hydrogen peroxide in human fibroblasts, *Mut. Res.,* 385, 139–149, 1997.

14. Ballinger, S.W. et al., Hydrogen peroxide causes significant mitochondrial DNA damage in human RPE cells, *Exp Eye Res.,* 68, 765–772, 1999.

15. Deng, G. et al., Bcl-2 facilitates recovery from DNA damage after oxidative stress, *Exp. Neurol.,* 159, 309–18, 1999.

16. Ballinger, S.W. et al., Hydrogen peroxide- and peroxynitrite-induced mitochondrial DNA damage and dysfunction in vascular endothelial and smooth muscle cells, *Circ. Res.,* 86, 960–966, 2000.

17. Shigenaga, M.K., Hagen, T.M., and Ames, B.N., Oxidative damage and mitochondrial decay in aging, *Proc. Natl. Acad. Sci. USA,.* 91, 10771–10778, 1994.

18. Mandavilli, B.S., Ali, S.F., and Van Houten, B., DNA damage in brain mitochondria caused by aging and MPTP treatment, *Brain Research,* 885, 45–52, 2000.

19. Ames, B.N., DNA damage from micronutrient deficiencies is likely to be a major cause of cancer, *Mut. Res..,* 475, 7–20, 2001.

chapter twelve

Fidelity of DNA synthesis as a molecular biomarker

Kandace Williams

Contents

Abstract The primer-template extension reaction was initially developed for precise measurements of polymerase kinetic activity during DNA synthesis. Additional insights into individual polymerase mechanistic abilities have been derived by altering different components of this *in vitro* assay. Although the diversity and complexity of information derived from this assay has broadened in scope, several fields of research have begun to focus on the phenomenon of mutational hot spots within genomic DNA. For instance, we have used this assay to define a specific sequence context of hypervariable regions within the HIV envelope gene — oligoadenylyl tracts — that also corresponds to a unique heritable mutation in the APC tumor suppressor gene. We have further employed this assay to define more precisely mammalian mismatch repair versus mismatch extension activity at the well-known *H-RAS* codon 12 hot spot of mutation. Here we have found that synthetic activities of DNA polymerases α and β are less inhibited by the insertion of a mismatched base at codon 12 that will produce a G:C→A:T transition mutation than by insertion of a mismatch at this same location that will produce a G:C→T:A transversion mutation. This report contains descriptions of these and other experiments that are part of an ongoing study to improve our understanding of potential biological consequences of spe-

1-56670-596-7/02/$0.00+$1.50
© 2002 by CRC Press LLC

cific sequences during cellular replication or repair synthesis under adverse conditions. Sequence-specific molecular biomarkers altered by different events contributing to cellular stress will improve predictions of subsequent pathological processes triggered by such things as inheritable conditions, biological exposures, or environmental insults.

I. Introduction

We are entering the fourth decade of genetic, biochemical, and molecular studies targeted towards elucidating the contribution of individual DNA polymerases to mutational events. It has long been recognized that error discrimination during DNA synthesis is important for maintaining genomic integrity. Initial genetic manipulations of the bacteriophage T4 polymerase have revealed large variations in error frequencies, encompassing mutator and antimutator phenotypes, as compared to wildtype T4.[1] Subsequent biochemical and molecular studies have provided a significant number of details to the initial concept of how DNA polymerases contribute to the accuracy of DNA synthesis. Recently, this field has expanded tremendously with the identification of several different procaryotic and eucaryotic polymerases exhibiting vastly different fidelities of replication and repair synthesis on DNA templates.[2]

The primer-template extension reaction has been developed as an elegantly simple *in vitro* method for precise biochemical and molecular studies of polymerase replication activities.[3–10] This assay was the first to document 3′→5′ exonuclease proofreading activity now associated with the majority of replicative DNA polymerases within the cell.[11] It is now clearly defined that exonuclease-to-polymerase activity is highest for polymerases having highest overall replication fidelity although there is sequence-dependence associated with fidelity of this proofreading activity.[12] In contrast, DNA polymerases that lack this proofreading function have significantly lower replication fidelity.[13] Additional insights derived from primer-template extension reactions have been highly informative in regard to specific biochemical characteristics of different polymerases. These include effects such as dNTP pool concentration, local sequence context, and next-nucleotide and nearest neighbor effects, in addition to other aspects of DNA stability. Each conditional influence has been determined by measuring mutational spectra and error rates, including primer-template slippage producing frameshift errors.[1,14,15] It is also possible to measure individual polymerization and proofreading steps in exquisite detail by the primer extension method on a pre-steady state time scale.[6] By the use of a steady state time scale, several additional characteristics of individually purified polymerases have been defined. These include measurements of progressive or distributive activity of individual polymerases, kinetics of incorrect or base analogue insertion and subsequent incorporation, extension kinetics from an incorrect primer-3′-termini, including a mismatched or damaged primer-3′-termini, and extension kinetics over a specific DNA template lesion resulting in lesion bypass.[8–10,16–23]

Beginning in the 1980s, molecular studies within such diverse fields as microbiology, virology, carcinogenesis, genetic diseases, developmental biology, and aging research began to focus on the recognition of mutational hot spots — frequent mutational events targeted to specific sites within DNA — that often result in distinct cellular changes in function. It is not yet understood if specific mutational hot spots represent decreased fidelity of replication, are targets for DNA damaging events, or perhaps suffer reduced fidelity or incapacity of specific repair pathways because of physiologic or structural hindrances. It is now generally accepted that sequence context is an important influence upon DNA replication and repair events. With the emergence of DNA damage and replication/repair studies targeted to hot spots of mutation within specific DNA sequences, primer extension assays have been further developed to investigate mechanisms of mutational activity targeted to sensitive template sites. These experiments progress beyond initial protocols that have primarily focused upon individual polymerase activity, by examining the interaction between individual polymerases and unique template sequences.[14] Information obtained using templates containing well-defined hot spots of mutation can be useful as molecular biomarkers defining specific factors, endogenous and exogenous, predisposing towards enhanced mutation frequency at these critical locations within the genome.[2,3,24]

We have used primer extension assays, in combination with other experimental systems, to determine individual characteristics of polymerases traversing specifically designed DNA templates. These experiments were initially conceived to interpret the effect of template nucleotide sequence on the processivity of individual polymerases. Ultimately, we wish to better understand potential biological consequences of specific sequences during polymerase replication or repair synthesis within normal cells and environmentally stressed cells. The identification of sequence-specific molecular biomarkers within the human genome that contribute to different pathological cellular events has tremendous potential to define and, therefore, to improve predictions of molecular mechanisms of disease processes.

II. Hypervariable regions as molecular biomarkers

The ability of HIV to escape host immune response chronically appears to be intimately related to the high mutation frequency of the viral genome, most notably within the envelope gene. The calculated mutation rate during replication of the HIV genome is as much as 10^6-fold higher than for other genomes.[25] Furthermore, HIV reverse transcriptase (RT) has been found to be singularly error prone even among other RTs also lacking $3' \rightarrow 5'$ exonuclease activity.[26,27] In addition, errors produced by HIV RT appear nonrandom, with several mutational hotspots located within homo-oligomeric (dA) sequences.[25] Therefore, to study the ability of HIV RT to traverse specific homo-oligomeric stretches, we used M13 DNA templates altered to contain oligo(purine) and oligo(prymidine) tracts. The progress of HIV RT along these templates was potently and uniquely inhibited from further progres-

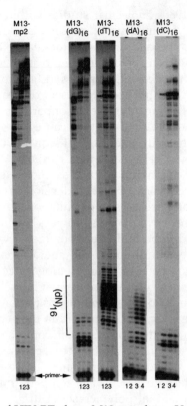

Figure 12.1 Pause sites of HIV RT along M13 templates. Unmodified M13 and four recombinant M13 DNA templates were annealed to a 5'-^{32}P-labeled 17-mer universal primer. Each primer template was extended at 37°C for 5 min. by purified HIV RT at a molar excess to template of 1)12:1, 2)37:1, 3)74:1, 4)100:1. (From Williams, K.J., Loeb, L.A., and Fry, M., *J. Biol. Chem.*, 265, 180682–180689, 2001. With permission.)

sion only at a d(A)$_{16}$ insert, even at 100:1 molar excess of HIV RT to template[4] (Figure 12.1).

Because our primer extension assays demonstrated exceedingly strong pausing of HIV RT opposite dA stretches, and because dA sequences have been associated with increased pausing and errors by HIV RT,[25] we analyzed the variability of different repetitive sequences existing within the HIV genome[24] (Table 12.1). Locations of dA tracts of five or more nucleotides were compared amongst aligned sequences of all available sequenced HIV genomes derived from human infections at the time of publication.[24]

We discovered that the fraction of oligoadenylyl tracts located within the *env* gene is disproportionately high (0.41) as compared to the size of the *env* gene in relation to the rest of the genome (0.27). Most importantly, the extent of degeneracy of nucleotide sequence homology within oligoadenylyl tracts in the *env* gene is 1.5 to 2.2-fold higher (only 34% homology) than the variance of these or any of the other homo-digomeric or random sequences within the HIV genome[24] (Table 12.1). These combined observations suggest a specifi-

Table 12.1   Analysis of Variant HIV Sequence Homology[1]

Nucleotide	No. of Other Analyzed[2]	Other[3] (%)	No. of *env* Analyzed[2]	*env*[3] (%)
≥ AAAAA	85 (114)	56	21 (47)	34
≥ TTTTT	13 (14)	70	5 (6)	51
≥ CCCCC	12 (20)	40	2 (4)	75
≥ GGGGG	15 (23)	63	3 (8)	70
rndm[4]	40	69	16	46.5

[1] Locations of homo-oligomeric sequences of five or more nucleotides were identified by a computer search in all aligned HIV genomes. The average number of aligned genomes analyzed within the envelope gene and within the rest of the genome was 26.6 and 18.5 respectively

[2] Only those homo-oligomeric sequences which were found in at least two of all aligned genomes at each location were analyzed for percent homology. Other indicates all genomic locations outside of envelope sequences. *Env* is only locations within envelope sequences. Number in parenthesis is actual total number of locations of repetitive sequences identified, including unique sequences in alignment not analyzed for percent homology.

[3] Locations of all sequences identified outside the envelope gene (Other), or within the envelope gene (*env*) were compared for percent homology with all other sequences in alignment, except for unique sequences as noted above.

[4] Two random sequences were generated to determine average percent homology among all aligned isolates: GACAT and GCTGC. The percent homology of all consensus sequences at all locations of each random sequence was determined individually. Results were not significantly different between the two random sequence locations, therefore the combined average is present.

Source: Williams, K.J. and Loeb, L.A., in *Genetic Diversity of RNA Viruses*, Vol. 176, Springer Verlag, Berlin, 1992. With permission.

cally high frequency of mutational alteration of oligoadenylyl tracts within the HIV *env* gene. Thus, the possibility of cooperative interaction between HIV RT and specific genetic sequences within the *env* gene provides a mechanism for increased hypervariability at precisely defined locations. Therefore, this unique polymerase-template interaction has potential as a molecular biomarker of HIV genetic variability over time, directly related to immune evasion. Experiments to determine interactions of HIV RT and its own genome, particularly the *env* genetic region, should provide more information regarding template-directed infidelity of replication within the HIV genome. As an interesting analogy to these findings, Vogelstein and colleagues have recently reported an inherited mutation in the adenomatous polyposis coli (APC) tumor suppressor gene associated with familial adenomatous polyposis (FAP) that appears to indicate a similar function of hypervariable oligoadenylyl tracts in human cancer.[28] The hereditary mutation of T→A at APC nucleotide 3920, found in 28% of Ashkenazim with a family history of colorectal cancer, results in an eight-nucleotide string of A sequences which subsequently appear to be highly susceptible to frameshift mutation. Therefore, this mutation contains the potential of silencing the APC tumor suppressor by the creation of a small, hypermutable region of the gene. This indirect cancer predisposition consisting of an oligoadenylyl mutation is a unique biomarker for a human heritable disease process, but perhaps is not so unique in living organisms overall, as a mechanism for mutation induction.

III. Human H-RAS hotspot of mutation as a molecular biomarker

It has been well documented that the *ras* oncogene family has a high incidence of single-base substitution at selected codon locations (codons 12, 13, 61), subsequently producing an activated gene product in a large percentage of human cancers.[29] Initially, we analyzed the effects of sequence context within the human *H-RAS* exon 1, containing codons 12 and 13, on the progression of synthesis of purified DNA polymerase α primase (DNA pol α) a major eucaryotic replication enzyme (lacking 3′→5′ exonuclease activity)[3] (Figure 12.2). As demonstrated within Figure 12.2, using the wildtype sequence of human *H-RAS* gene as a template, major pause sites were observed at codons 12 and 13, as well as two other pause sites upstream (intronic region) and downstream (codons 23 and 24). By the use of the activated human *H-RAS* gene as the template, containing a G→T middle base substitution in codon 12, the pause site at codons 12 and 13 is completely removed at all concentrations of DNA pol α used in these experiments. Further analysis of complete primer extension products from the wildtype template sequence experiments (copied past the strong pause site at codons 12 and 13) did not substantiate a high frequency of mutation at codons 12 or 13. Thus, successful synthesis by DNA pol α does not appear to result in increased mutations at these *H-RAS* hot spots. Unfortunately, the 3′-termini of the unextended chains at codons 12 and 13 were not analyzed for incorrect nucleotide insertion events that may have inhibited further polymerase progression without the benefit of 3′→5′ exonuclease activity.

However, subsequent experiments within our lab using mammalian cells in culture have demonstrated that mismatch repair activity at the middle base position of codon 12 can be severely hampered, depending not only on the type of mismatch but whether the cells are proliferating or are in G1 arrest[30-33] (Figure 12.3). Our studies have further demonstrated that mismatch-specific repair at the codon 12 site is inefficient rather than inaccurate in proliferating mammalian cells, resulting in DNA replication without mismatch repair. This lack of mismatch repair consistently produces a mixed colony genotype, rather than incorrect mismatch repair that would result in a pure mutated colony genotype. As an interesting aside, these studies have repeatedly demonstrated that this site is not a hot spot of either inadequate or incorrect mismatch repair for *E. coli*, regardless of the nature of the mismatch located at this site.

Because competent mismatch repair within the cell depends strongly on the specific mismatch present at codon 12, we then asked if efficiency of primer

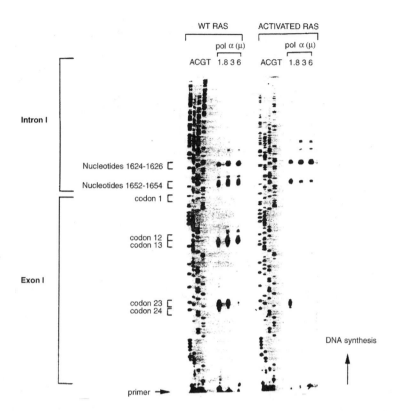

Figure 12.2 Pause sites of DNA polymerase α along wildtype and activated *ras* gene. One hundred ng of 5'-³²P-labeled single-stranded M13mp19/wt *ras* recombinant DNA or M13mp19/activated *ras* recombinant DNA were extended in presence of 500 μM dNTP at 37°C for 1 h by the indicated amounts of DNA pol α. (From Hoffman, J.-S. et al., *Cancer Res.*, 53, 2895–2900, 1993. With permission.)

extension, from a codon 12 middle base mismatch at the 3′ terminus, would depend upon the type of mismatch. We, therefore, compared the progression of DNA synthesis catalyzed by DNA polymerase α or β from specific 3′ mispaired bases at the codon 12 middle nucleotide position[34] (Figure 12.4).

These primer extension data indicate that both DNA polymerases catalyze extension from a 3′ A(primer):C(transcribed template) mismatch more efficiently than from any other mismatch tested at this sensitive location. In contrast, significantly decreased extension product by both polymerases was observed from 3′ T(primer):C(transcribed template). These results agree with our cell culture experiments[31] (Figure 12.3) in that T:C mismatches are repaired significantly more often than A:C mismatches in proliferating cells; thus, the polymerase is more reluctant to extend the T:C mismatch (Figures 12.3 and 12.4). Surprisingly, extension opposite the coding template was more easily achieved by both polymerases from a 3′ T(primer):G(coding template) than from a 3′ A(primer):G(coding template). These results appear in contrast to our cell culture experiments, as G:T was the most efficiently repaired

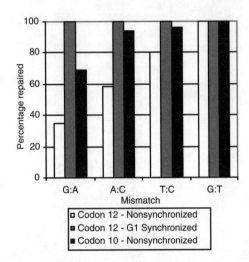

Figure 12.3 Overall comparison of mismatch repair rates of *H-RAS* codons 10 & 12 in nonsynchronized and G$_1$-synchronized cells. (From Mattor, N., Simonetti, J., and Williams, K.J., *Carcinogenesis*, 20, 8, 1417–1424, 1999. With permission.)

mismatch within the proliferating cell, while G:A was repaired at a significantly lower frequency than all other mismatches tested[31] (Figures 12.3 and 12.4). However, both of these 3' mismatched primers were extended with less than 50% efficiency from the coding template by both polymerases, as compared to the 3' termini matched base pair of G:C. Perhaps this is because G:T within the cell does have alternate base excision repair pathways available to correct this mismatch back to G:C, such as thymine DNA glycosylase.

opposite the transcriptional template:

T or

5'-^{32}P- GTG GTG GTG GGC GCC G^A - - - - >

. . . TTC GAC CAC CAC CAC CCG CGG **CCG** CCA CAC . .

codon=   6   7   8   9   10   11   12   13   14

Polymerase α extension		
G:C	A:C	T:C
(100%)	79%	28%

(row label: % Extended relative to G:C)

Polymerase β extension		
G:C	A:C	T:C
(100%)	81%	28%

opposite the coding template:

T or

5'-^{32}P- ACT CTT AAG CAC ACC G^A - - - - >

. . . GTC GCG TGA GAA TTC GTG TGG **CGG** CCG CGG . .

codon=   18   17   16   15   14   13   12   11   10

Polymerase α extension		
G:C	G:T	G:A
(100%)	46%	42%

(row label: % Extended relative to G:C)

Polymerase β extension		
G:C	G:T	G:A
(100%)	36%	25%

Figure 12.4 Primer extension from *H-RAS* codon 12, middle mismatched base.

Overall, these primer extension data are consistent in that the synthetic activities of DNA polymerases α and β are less inhibited by the insertion of a mismatched base at codon 12 that will produce a G:C→A:T transition mutation (A opposite transcriptional template C, or T opposite coding template G) than by insertion of a mismatch at this same location that will produce a G:C→T:A transversion mutation (T opposite transcriptional template C, or A opposite coding template G) at this site. These results also agree with others who have previously determined that the majority of animal model tumors contain an activated *H-RAS* because of a codon 12 G→A transition.[35,36] Most human tumors that have an *H-RAS* codon 12 mutation do contain a G→T transversion, however,[29] which may indicate additional accessory proteins (or lack thereof) that play a role during human *H-RAS* mutagenesis *in vivo*.

Perhaps surprisingly, these primer extension experiments do not demonstrate significant differences between DNA pol α or DNA pol β for individual mismatch extension efficiencies at this location. However, our primer extension results using the same 3' mismatches closely resemble mismatch extension kinetics of Perrino and Loeb[8] using a φX174 *am*3 DNA template, and of Mendelman et al.[9] using a M13 bacteriophage template. In addition, the purified pol α and pol β used in these studies did not contain exonuclease activity, nor is either polymerase highly processive. Finally, it has been observed that pol β can have an equal or even lower mutation frequency than pol α, depending upon the template sequence.[37]

The above primer extension experiments reveal high variability of polymerase activities that appear primarily dependent upon whether the template is either a proto-oncogene, an activated oncogene, or on the specific 3' termini mismatched base. More detailed primer extension kinetic analysis using other mammalian polymerases, such as DNA polymerase δ, and other base analogues at the *H-RAS* codon 12 location (in comparison to nonhot spot sites) will be required before more detailed mechanistic information can be obtained in regard to polymerase kinetic activity at this sensitive site.

IV. Conclusion

The primer extension assay is a wonderful tool because of the investigator's ability to explicitly control experimental conditions. Although this assay was initially designed to determine individual polymerase kinetics on a few standard templates, it has since been expanded to explore the effects of different templates on polymerase kinetic activity. We have used this assay to ask diverse experimental questions, and have found fascinating answers that correlate closely with cell culture and animal model systems, as well as sequence results from human epidemiological surveys. The obvious next step towards developing the versatility of this precise analytical tool as a valuable molecular biomarker assay will be to compare polymerase extension results of specific primer-template sequences in the presence of environmental quantities of known genotoxic chemicals that are of such growing concern within our world. Information elicited from these future experiments will likely be of highest value when compared with results of other types of experimental design, such as cell culture experiments, animal model studies, and/or epidemiological data.

References

1. Goodman, M.F. and Fygenson, D.K., DNA polymerase fidelity: from genetics toward a biochemical understanding, *Genetics*, 148, 1475–1482, 1998.
2. Friedberg, E.C., Feaver, W.J., and Gerlach, V.L., The many faces of DNA polymerases: strategies for mutagenesis and for mutational avoidance, *Proc. Natl. Acad. Sci. USA*, 97, 11, 5681–5683, 2000.
3. Hoffmann, J.-S. et al., Codons 12 and 13 of *H-RAS* proto-oncogene interrupt the progression of DNA syntheiss catalyzed by DNA polymerase α, *Cancer Res.*, 53, 2895–2900, 1993.
4. Williams, K.J., Loeb, L.A., and Fry, M., Synthesis of DNA by human immunodeficiency virus reverse transcriptase is preferentially blocked at template oligo(deoxyadenosine) tracts, *J. Biol. Chem.*, 265, 180682–180689, 1990.
5. Patel, P.H. et al., A single highly mutable catalytic site amino acid is critical for DNA polymerase fidelity, *J. Biol. Chem.*, 276, 5044–5051, 2001.
6. Einolf, H.J. and Guengerich, F.P., Kinetic analysis of nucleotide incorporation by mammalian DNA polymerase δ, *J. Biol. Chem.*, 275, 21, 16316–16322, 2000.
7. Perrino, F.W. et al., Extension of mismatched 3′ termini of DNA is a major determinant of the infidelity of human immunodeficiency virus type 1 reverse transcriptase, *Proc. Natl. Acad. Sci. USA*, 86, 8343–8347, 1989.
8. Perrino, F.W. and Loeb, L.A., Differential extension of 3′ mispairs is a major contribution to the high fidelity of calf thymus DNA polymerase-α, *J. Biol. Chem.*, 264, 5, 2898–2905, 1989.
9. Mendelman, L.V., Petruska, J., and Goodman, M.F., Base mispair extension kinetics comparison of DNA polymerase α and reverse transcriptase, *J. Biol. Chem.*, 265, 4, 2338–2346, 1990.
10. Boosalis, M.S., Petruska, J., and Goodman, M.F., DNA polymerase insertion fidelity gel assay for site-specific kinetics, *J. Biol. Chem.*, 262, 30, 14689–14696, 1987.
11. Brutlag, D. and Kornberg, A., Enzymatic synthesis of deoxyribonucleic acid, XXXVI: A proofreading function for the 3′ — >5′ exonuclease activity in deoxyribonucleic acid polymerases, *J. Biol. Chem.*, 247, 241–248, 1972.
12. Bloom, L.B. et al., Pre-steady-state kinetic analysis of sequence-dependent nucleotide excision by the 3′-exonuclease activity of bacteriophage T4 DNA polymerase, *Biochemistry*, 33, 7576–7586, 1994.
13. Kunkel, T.A. and Bebenek, K., DNA replication fidelity, *Ann. Rev. of Biochem.*, 69: p. 497–529, 2000.
14. Singer, B. and Hang, B., Nucleic acid sequence and repair: role of adduct, neighbor bases and enzyme specificity, *Carcinogenesis*, 21, 6, 1071–1078, 2000.
15. Mendelman, L.V. et al., Nearest neigbor influences on DNA polymerase insertion fidelity, *J. Biol. Chem.*, 264, 24, 14415–14423, 1989.
16. Viguera, E., Canceill, D., and Ehrlich, S.D., Replication slippage involves DNA polymerase pausing and dissociation, *EMBO J.*, 20, 10, 2587–2595, 2001.
17. Fry, M. and Loeb, L.A., A DNA polymerase α pause site is a hot spot for nucleotide misinsertion, *Proc. Natl. Acad. Sci. USA*, 89, 763–767, 1992.

18. Miller, H. et al., 8-OxodGTP incorporation by DNA polymerase β is modified by active-site residue Asn279, *Biochemistry*, 39, 1029–1033, 2000.

19. Kosa, J.L. and Sweasy, J.B., The E249K mutator mutant of DNA polymerase β extends mispaired termini. *J. Biol. Chem.*, 274, 50, 35866–35872, 1999.

20. Kamiya, H. and Kasai, H., Formation of 2-hydroxydeoxyadenosine triphosphate, an oxidatively damaged nucleotide, and its incorporation by DNA polymerases, *J. Biol. Chem.*, 270, 33, 19446–19450, 1995.

21. Kamath-Loeb, A.S. et al., Incorporation of the guanosine triphosphate analogs 8-oxodGTP and 8-NH$_2$-dGTP by reverse transcriptases and mammalian DNA polymerases, *J. Biol. Chem.*, 272, 9, 5892–5898, 1997.

22. Voigt, J.M. and Topal, M.D., O^6-methylguanine-induced replication blocks, *Carcinogenesis*, 16, 8, 1775–1782, 1995.

23. Daube, S.S., Arad, G., and Livneh, Z., Translesion replication by DNA polymerase β is modulated by sequence context and stimulated by fork-like flap structures in DNA, *Biochemistry*, 39, 397–405, 2000.

24. Williams, K.J. and Loeb, L.A., Retroviral reverse transcriptases: error frequencies and mutagenesis, in *Genetic Diversity of RNA Viruses*, Vol. 176, Holland, J.J., Ed., Springer-Verlag, Berlin, 1992, pp. 165–180.

25. Bebenek, K., et al., Specificity and mechanism of error-prone replication by human immunodeficiency virus-1 reverse transcriptase, *J. Biol. Chem.*, 264, 28, 16948–16956, 1989.

26. Preston, B.D., Poiesz, B.J., and Loeb, L.A., Fidelity of HIV-1 reverse transcriptase, *Science*, 242, 1168–1171, 1988.

27. Roberts, J.D., Bebenek, K., and Kunkel, T.A., The accuracy of reverse transcriptase from HIV-1, *Science*, 242, 1171–1173, 1988.

28. Laken, S.J. et al., Familial colorectal cancer in Ashkenazim due to a hypermutable tract in APC, *Nature Genetics*, 17, 1, 79, 1997.

29. Bos, J., *H-RAS* oncogenes in human cancer: a review, *Cancer Res.*, 49, 4682–4689, 1989.

30. Arcangeli, L. and Williams, K.J., Mammalian assay for site-specific DNA damage processing using the human *H-RAS* proto-oncogene, *Nucleic Acids Res.*, 23, 2269–2276, 1995.

31. Arcangeli, L. et al., Site- and strand-specific mismatch repair of human *H-RAS* genomic DNA in a mammalian cell line, *Carcinogenesis*, 18, 1311–1318, 1997.

32. Matton, N., Simonetti, J., and Williams, K.J., Inefficient *in vivo* repair of mismatches at an oncogenic hotspot correlated with lack of binding by mismatch repair proteins and with phase of the cell cycle, *Carcinogenesis*, 20, 8, 1417–1424, 1999.

33. Matton, N., Simonetti, J., and Williams, K.J., Identification of mismatch repair protein complexes in HeLa nuclear extracts and their interaction with heteroduplex DNA, *J. Biol. Chem.*, 275, 17808–17813, 2000.

34. Clark, D. et al., Differential nucleotide extension from mismatched base pairs at human *H-RAS* codon 12 by human DNA polymerase α and β, *Proc. Am. Assoc. Cancer Res.*, 38, 64, 1997.

35. Barbacid, M., *H-RAS* Genes, *Ann. Rev. Biochem.*, 56, 779–827, 1987.

36. Guerrero, I. and Pellicer, A., Mutational activation of oncogenes in animal model systems of carcinogenesis, *Mutat. Res.*, 185, 293–308, 1987.

37. Kunkel, T.A. and Alexander, P.S., The base substitution fidelity of eucaryotic DNA polymerases, *J. Biol. Chem.*, 261, 1, 160–166, 1986.

chapter thirteen

Mutational specificity of environmental carcinogens

Kathleen Dixon and Mario Medvedovic

Contents

Abstract Many environmental carcinogens are genotoxic and mutagenic. The mutagenic specificity of genotoxic environmental agents was characterized to elucidate mechanisms of action. Mutation spectra were generated for a wide variety of genotoxic agents, particularly in the pZ189-SupF assay system. Cluster analysis was applied to these spectra to generate groupings based on similarities in the distribution and frequency of mutations within the SupF coding sequence. In general, these groupings support existing evidence for the mutagenic mechanism of different agents. This analysis provides a basis for deducing mutational mechanisms for unknown agents. Furthermore, it is formally possible that one could back-extrapolate from mutational patterns in exposed individuals to determine the agent to which that individual was exposed.

I. Introduction

The link between exposure to certain genotoxic environmental agents and the development of human cancer has been established. In most cases, this link has been deduced from epidemiological studies of cancer incidence in workers exposed to specific chemicals in an industrial setting. For example, workers exposed to chromate in the chromate production industry and in the manufacture of pigments exhibit an elevated risk of lung cancer.[1,2] In a few cases, the exposure to specific agents in the environment has been shown to cause an increase in cancer risk in the general population. For example, groups exposed to elevated levels of arsenic in drinking water, exhibit an increased incidence of skin, bladder, and liver cancer.[3,4] The carcinogenic potential of many chemicals can be confirmed by animal bioassays and by any number of *in vitro* assays of mutagenic activity (e.g., the Ames assay).[5,6] Predictions of the carcinogenic potential of a much larger number of environmental agents for which epidemiological data are lacking have been made by extrapolation from animal bioassays and mutagenesis data.[7] However, despite the availability of large amounts of such data, it is usually not possible to identify the specific environmental exposures responsible for the majority of human cancers. This is partly because of the small number of confirmed human carcinogens[7] and partly due to the large variety of agents to which any one individual is exposed. In order to provide the missing link between exposure and cancer, investigators have sought to identify a signature of specific environmental agents that can be identified in the exposed population. One potential signature is the mutational specificity of genotoxic environmental agents.

Mutational specificity depends in part on the nature of the initial damage to the DNA and on cellular mutation avoidance pathways (Figure 13.1). Genotoxic environmental agents may cause a wide variety of damage to cellular DNA. For example, sunlight induces the formation of covalent bonds between adjacent pyrimidines to form pyrimidine cyclobutane dimers and 6-4 photoproducts; it also induces a range of oxidative DNA damage.[8] Benzo[a]pyrene (found in cigarette smoke and incomplete combustion of fossil fuels) is metabolized by the cell to reactive intermediates that bind covalently to DNA bases, primarily guanine and adenine, to form DNA adducts.[9] Other agents induce DNA strand breaks or DNA crosslinks. This initial damage must be processed by the cell either during DNA replication or DNA repair to be fixed in the DNA as a permanent genetic mutation.[10] Depending in part on the nature of the initial damage, the resulting mutations can be point mutations such as base substitution mutations or they can be more extensive changes such as deletions or DNA rearrangements. Not all of the initial damage is fixed as mutations because the cell has a variety of protective mechanisms or mutation avoidance pathways. Nucleotide exci-

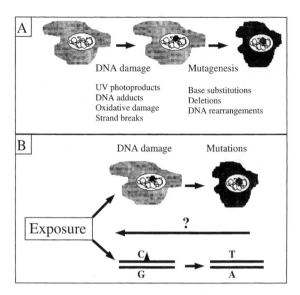

Figure 13.1 Do mutation spectra provide signatures of environmental genotoxicants for back-extrapolation? Panel A: Environmental genotoxic agents induce DNA damage (e.g., UV photoproducts, DNA adducts, oxidized bases and sugars, DNA strand breaks) characteristic of the particular agent. The cell processes this damaged DNA by DNA repair mechanisms to restore the normal DNA structure or faulty DNA replication or repair can lead to mutations. The spectrum of mutations reflects the type of initial damage and cellular repair and replication processes. Panel B: By characterizing the mutation spectra generated by unknown agent or agents that function by unknown mechanisms it may be possible to identify the mutagenic agent and mutational mechanism.

sion repair (NER) pathways can often remove the initial damage from the DNA and restore the normal DNA sequence.[11] There are also recombination pathways for repairing DNA strand breaks.[12] In addition, certain auxiliary DNA polymerases may be recruited to avoid errors during replication of damaged DNA templates.[13] However, not all of the cellular parameters that determine mutational specificity have been identified.

In recent years, a number of different assay systems have been developed that allow the characterization of the mutational specificity of genotoxic environmental agents.[14–19] In general, these systems provide an easily select-able mutational target, and mutational specificity is determined by DNA sequence analysis of collections of mutants resulting from treatment with the particular agent of interest. The most widely used systems for studying mutational specificity in human cells are the hypoxanthine phosphoribosyl transferase (HPRT) system[14] and the SupF system.[15] In the HPRT system, the endogenous human gene is used as the mutagenesis target and mutations in this gene can be analyzed either in lymphoblasts from exposed individuals or in cells treated in culture with the agent of interest. In the SupF system, the bacterial SupF gene is carried on a plasmid that can replicate in human cells. The plasmid can be treated with the agent of interest and then introduced into human cells where mutations are fixed, or cells containing the plasmid can treated directly. In either case, the mutated plasmid is recovered from the cells, retransformed into bacteria and screened for mutations. Both systems have been used to analyze the mutational specificity of many environmental agents.[20,21] We will address data from only the SupF system here.

The SupF system has been used to analyze the mutational specificity of over 40 different environmental agents,[20] including UV, benzo[a]pyrene-diolepoxide, aflatoxin, and a variety of others. In each case, the mutational specificity is presented as a "mutation spectrum," that is the position and type of mutation in the *SupF* gene sequence in each of a collection of independent mutants. In some cases, the mutation spectrum has been compared to the distribution of initial damage in the *SupF* gene.[22,23] In other cases, the spectrum for the same agent has been compared in the presence or absence of specific DNA repair pathways.[24,25] In general, these comparisons, have been made in a pairwise fashion, using either the statistical method described by Adams and Skopek[26] or without statistical analysis.

Here we describe the analysis of mutational spectra in the SupF system by clustering procedures that provide for an overall comparison of the spectra and group the spectra in terms of overall similarity. The purpose of this analysis is to determine whether the existing data are sufficient to generate groupings that are meaningful chemically and biologically and to determine whether this analysis might be adapted to back-extrapolate from mutational spectrum to type of mutagen. The capacity to back-extrapolate is a necessary requirement for using mutational specificity as a "signature" for carcinogen exposure.

II. Materials and methods

A.   Cells

Typically, monolayer cultures of monkey or human cells are used. The CV-1 African green monkey kidney cell line (TC-7 clone)[27] or the human cells lines, GM0637 (SV40-transformed fibroblasts), or Ad293 cells (adenovirus p16-transformed fibroblasts) are used most frequently. These cells are usually grown in Dulbecco's modified Eagle medium (Gibco BRL, Gaithersburg, MD) supplemented with antibiotics and 5-10% fetal calf serum in a 5% CO_2 humidified atmosphere at 37°C.

B.   Plasmids

The shuttle vector pZ189[15] has the SV40 virus origin and T-antigen gene, allowing its replication in human and monkey cells (Figure 13.2). The pBR327 origin enables the vector to grow in bacterial cells. The SupF tyrosine tRNA suppressor gene serves as the mutagenesis target; its activity is assayed by the ability to suppress an amber mutation in the *lacZ* gene in the competent

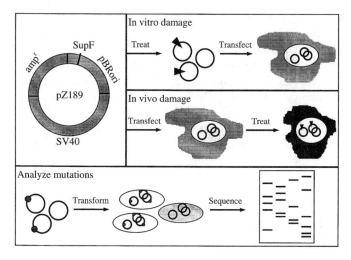

Figure 13.2 Use of the SupF mutation assay system. Top left: The structure of the pZ189 vector carrying the SupF target gene. Top right: The vector DNA can be treated *in vitro* with a DNA damaging agent and then transfected into human or primate cells where it replicates and mutagenesis occurs. Middle right: Undamaged vector DNA can be transfected into cells and then the cells can be treated with the test agent during replication of the vector DNA to cause mutagenesis. Bottom panel: The vector DNA is recovered from the mammalian cells and transformed into bacterial tester cells; blue colonies contain vector with wildtype SupF and white (or light blue) colonies contain vector with mutant SupF. The mutant vector DNA is then isolated and sequenced.

bacterial host *Escherichia coli* MBM7070. Almost any change in this small tRNA gene will cause the loss of suppressor function and give rise to a white or light blue phenotype, instead of a dark blue wild phenotype on agar plates containing isopropyl-β-D-thiogalactoside (IPTG, an inducer of the lac operon), 5-bromo-4-chloro-3-indoyl-β-D-galactoside (X-gal, an artificial substrate for β-galactosidase), and ampicillin.[28] Modifications of this vector have been made over the years; the most widely used modified form is the pS189 which carries a site into which randomized oligonucleotides can be inserted to allow confirmation of independence of resulting mutants.[29] The pS189 vector also carries a modified SupF promotor, increasing the sensitivity of the vector for detecting partial loss of function mutants.

C. Shuttle vector system

The steps in the pZ189 mutagenesis assay are diagrammed in Figure 13.2. Rapidly growing, subconfluent cell cultures were transfected with pZ189 vector by one of several different methods: DEAE-dextran,[30] calcium phosphate,[31] or more recently, lipofectin.[32] Once inside the cell, the viral T-antigen is expressed and the plasmid DNA is replicated by the cellular DNA replication apparatus. About 48 to 72 h after transfection, the vector DNA is

harvested from the cells by the Hirt extraction method[33] After further purification, the vector DNA is treated with the ^{me}A-dependent restriction endonuclease, *DpnI* to digest nonreplicated fully methylated input plasmid, leaving intact the unmethylated or hemimethylated replicated plasmid
molecules.[34] The replicated plasmid DNA was then transformed into competent bacterial cells (MBM7070). Plasmid mutants are scored as white or
light blue colonies after plating on nutrient agar plates containing IPTG, X-
Gal, and ampicillin. Transformants that carry the wildtype *SupF* gene form
dark blue colonies on these plates. To generate a mutation spectrum, the
SupF genes of a collection of independently arising mutants are sequenced.

D.   Statistical analysis

Statistical procedure for identifying groups of "similar" spectra that we used
is based on the Finite Mixture Model approach to clustering.[35] In this
approach, we assume that similar observed spectra that cluster together were
generated by the same underlying probability distribution (underlying
mutational spectra). Suppose that $(n_1,...n_{85})$ represents an observed mutational spectrum with n_i representing the number of base substitution
observed at the ith position. The probability distribution generating this
spectrum is characterized by relative probabilities (mutational intensities),
p_i i=1,..,85, that a base substitution occurring in a mutant will occur on the
i^{th} position. Given $(p_1,...,p_{85})$, and the total number of mutations observed,
the probability of observing a mutational spectrum $(n_1,...,n_{85})$ is defined by
the multinomial probability density function:

$$f_{mult}(n_1,...,n_{85}) = \frac{(n_1+...+n_{85})!}{n_1!...n_{85}!} p_1^{n_1}...p_{85}^{n_{85}}$$

Assuming that each of a set of T mutational spectra is generated by one
of G multinomial distribution functions, a clustering procedure is performed
to identify groups of spectra generated by the same distribution functions.
This is done by maximizing classification likelihood using Gibbs sampler
and classification expectation maximization algorithm as describe by Medvedovic et al.[36]

III.   Results

A.   Analysis of UV-induced and spontaneous spectra

The earliest studies with the SupF system focused on the characterization of
mutations induced by UV light in monkey and human cells.[24,37] The striking
similarity of UV-induced spectra determined in cells of the two species and
the differences between the UV-induced spectra and the spontaneous
spectra[38,39] was immediately apparent. The data from these studies, repro-

duced here in Figures 13.3 and 13.4, were compared mostly on the basis of hot spot locations. Although in many studies hot spots were defined simply as the four or five sites in the sequence at which the most mutations occurred, Tarone[40] has established a more rigorous definition. The number of mutations needed for a position to be called a hot spot was n, such that the probability of n mutations occurring at any position (assuming a random distribution of mutations) was less than 0.05. According to this definition, the hot spots in these spectra are as shown in Table 13.1. Despite the overall similarity of the human (GM) and monkey (CV-1) UV spectra, only two hot spots were common to both. In addition to the base substitution mutations displayed in the mutation spectra, deletion mutations were also recorded. The frequency of deletion mutations also differed between the UV-induced spectra (about 10%) and the spontaneous spectra (about 50%).

These spectra can also be compared using the method developed by Adams and Skopek,[26] which provides an overall comparison of two mutation spectra. In this method, deviations of frequencies of mutations at each position from the frequency expected (under the assumption that the two spectra are equal) were combined in order to assess the overall difference between the spectra. This method tells us whether the two spectra are similar or not; however, it is not designed to evaluate groups of spectra. In Table 13.2 the pair-wise comparisons of the UV and spontaneous spectra in the two species are presented. This analysis shows the similarity between the two UV spectra and the two spontaneous spectra. By this test, the monkey (CV-1) UV spectrum differed significantly from either spontaneous spectrum, but the human (GM) UV spectrum did not differ significantly from either spontaneous spectrum. Mutation frequencies at individual positions can be compared using the multiple comparison procedure described by Westfall and Young.[41] In this method, frequencies of mutations were individually compared at each position using Fisher's Exact Test, adjusted for the multiple comparisons being made. The method is not valid for the comparison of a few preselected sites within the spectrum, especially if those sites were selected on the basis of observed differences.

In order to provide a method that allows a more global evaluation of the similarities among spectra we have adapted techniques of cluster analysis to the comparison of mutation spectra.[36] In this approach, we assume that similar observed spectra that cluster together were generated by the same underlying probability distribution (underlying mutational spectra). A clustering procedure is performed to identify groups of spectra generated by the same distribution functions. This is shown schematically in Figure 13.5. Thus, the mutation spectra are grouped on the basis of similarity. As the specified number of groups is increased, the distinctions among groups become more subtle.

When this technique is applied to a collection of UV-induced and spontaneous spectra from the literature, the two major groupings (UV and spontaneous) are immediately apparent (Table 13.3). Interestingly, the single UVB spectrum (UV-313) groups with the spontaneous mutations. This grouping

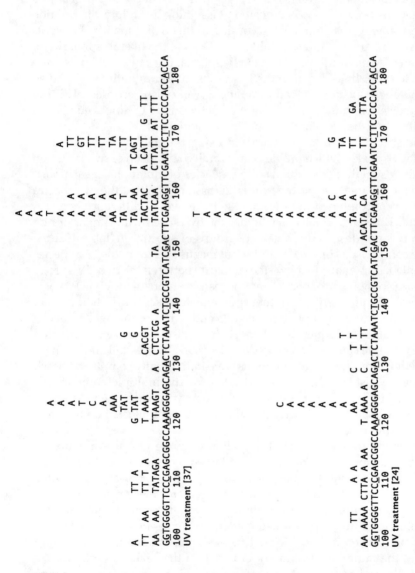

Figure 13.3 Mutations induced by UVC radiation (254 nm) in the SupF gene. The coding sequence of the SupF gene is shown along with the mutations is the spectra. The specific base changes in the collection of mutants are indicated by letter above the site of the mutated base. The numbers below the sequence indicate the DNA sequence position within the pZ189 vector. (From Bredberg, A., Kraemer, K.H., and Seidmen, M.M., *Proc. Natl. Acad. Sci. USA*, 83, 8273–8277, 1986; Protic-Sablejic, M. and Kraemer, K., *Proc. Natl. Sci. USA*, 82, 6622–6626, 1985. With permission.)

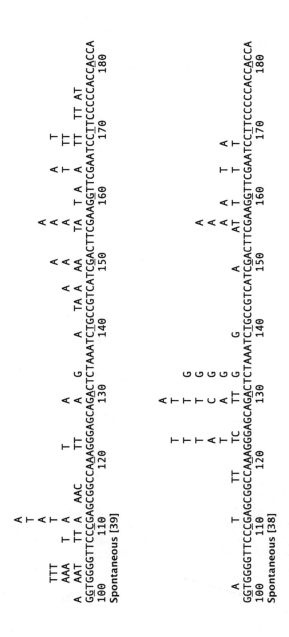

Figure 13.4 Spontaneous mutations induced during passage of the undamaged vector in mammalian cells. The coding sequence of the SupF gene is shown along with the mutations is the spectra. The specific base changes in the collection of mutants are indicated by letter above the site of the mutated base. The numbers below the sequence indicate the DNA sequence position within the pZ189 vector. (From Seidmen, M.M. et al., *Proc. Natl. Acad. Sci. USA*, 84, 4944–4948, 1987; Hauser, J., Levine, A.S., and Dixon, K., *EMBO J.*, 6, 63–67, 1987. With permission.)

Table 13.1 Mutational Hotspots

Spectrum	Hotspot Sites	Total No.[a]	Needed No.[b]
UV CV-1[37]	123, 156, 159, 168, 169	138	8
UV GM[24]	123, 156	80	6
Spont CV-1[39]	111	56	5
Spont GM[38]	123, 129, 133	37	5

[a] Total number of observed base substitutions in the spectrum.

[b] The sufficient number of base substitutions for a position to be a hot spot.

Table 13.2 Pairwise Comparisons of Spectra;[26]
p-values

		CV-1		GM	
		UV[37]	Spont[39]	UV[24]	Spont[38]
CV-1	UV[37]	—	0.0082	0.90	0.0013
	Spont[39]	—	—	0.16	0.90
GM	UV[24]	—	—	—	0.13
	Spont[38]	—	—	—	—

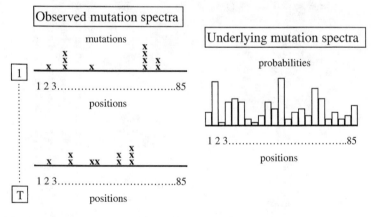

Figure 13.5 Cluster analysis of mutation spectra. Schematic representing the steps in the cluster analysis of mutation spectra.[36] The number of mutations at each site in the SupF coding sequence are entered for each mutation spectra (observed mutation spectra). These are used to generate an underlying mutation spectrum for each observed mutation spectrum. The number of groups to be generated is selected. The underlying mutation spectra are then grouped according to similarity and the underlying mutation spectra for the group are determined.

is consistent with the hypothesis that oxidative damage is responsible for spontaneous mutations and a subset of UVB-induced mutations. The single UV spectrum generated in pSP189 also does not consistently group with the other UV spectra; this observation will be discussed later. When three groups are chosen, all of the UV spectra, except the UVB spectrum, group together and the spontaneous spectra segregate according to species (i.e., monkey vs.

Table 13.3 Cluster Analysis of UV and Spontaneous Spectra

Mutagen	Plasmid	Cell Line	No. Mutations[a]	No. Clusters 2	3	4	Reference
UV	pZ189	XP-A	71	1	1	1	24
UV	pZ189	HeLa In Vitro	44	1	1	2	42
UV	pZ189 f1	HeLa In Vitro	28	1	1	1	42
UV	pZ189R2	HeLa In Vitro	30	1	1	1	42
UV	pYZ289	HL18	55	1	1	1	43
UV	pZ189	WI38VA13	84	1	1	1	43
UV	pZ189	XP-A	62	1	1	1	43
UV	pZ189	CV-1	138	1	1	1	37
UV	pZ189	XP-D	67	1	1	1	25
UV	pZ189	GM0637	80	1	1	1	24
UV	pSP189	XP-A	137	2	1	2	29
UV	pZ189	XP-F	73	1	1	1	43
UV	pZ189	CV-1	179	1	1	1	44
UV(313)	pZ189	CV-1	247	2	2	3	44
Spont	pZ189	CV-1	71	2	2	3	44
Spont	pZ189	CV-1	125	2	2	3	45
Spont	pZ189	CV-1	211	2	2	3	46
Spont	pZ189	CV-1	56	2	2	3	39
Spont	pSP19/pZ189	AD293	85	2	3	2	47
Spont	pSP189	Human Fibroblasts	44	2	3	4	48
Spont	pSP189	AD293	51	2	3	2	49
Spont	pZ189	GM606	77	2	3	4	50
Spont	pZ189	GM606	108	2	3	4	51
Spont	pZ189	GM606/GM0637	37	2	3	2	38

[a] Number of base substitutions in a spectrum.

human). With four groups, the groupings appear less rational, although most of the UV spectra remain grouped together. The underlying spectra for the three groups are shown in (Figure 13.6).

B. Analysis of a collection of published spectra

We have expanded the application of the clustering methods to include spectra generated with a wide range of environmental chemicals.[36] These are listed in Table 13.4 along with the results of the clustering analysis. We have presented the results for five, six, and seven groups and with and without inclusion of chromium spectra (see below). If we examine the assortment of spectra into the seven groups, we see that the UV spectra (except for UVB) remain grouped together. The spontaneous mutations split into groups according to species and vector (pZ189 vs. pSP189). Most of the diolepoxides of chrysene derivatives form their own cluster as do the diolepoxides of benzo[c]phenanthrene. Agents that cause oxidative damage tend to group with spontaneous mutations. Ionizing radiation spectra group

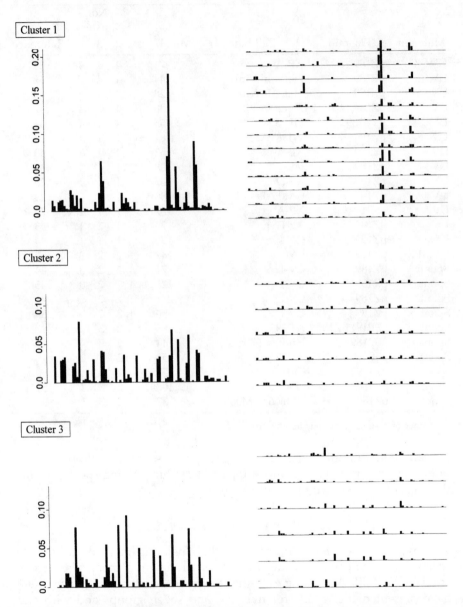

Figure 13.6 Results of cluster analysis of UV and spontaneous spectra. The underlying spectra of the three groups are shown on the left along with the individual underlying spectra on the right.

Table 13.4 Clustering all Mutational Spectra

Mutagen	Plasmid	Cell Line	No. Mutations[a]	No. Clusters						Reference
				5	5Cr[b]	6	6Cr	7	7Cr	
Anti-benzo[g]chrysene 11,12-dihydrodiol 13,14-epoxide	pSP189	Ad293	96	1	1	1	1	1	1	52
Syn-benzo[g]chrysene 11,12-dihydrodiol 13,14-epoxide	pSP189	Ad293	105	1	1	1	1	1	1	53
Anti-5,6-dimethylchrysene 1,2-dihydrodiol 3,4-epoxides	pSP189	Ad293	84	1	1	1	1	1	1	54
Syn-5,6-dimethylchrysene 1,2-dihydrodiol 3,4-epoxides	pSP189	Ad293	86	1	1	1	1	1	1	54
Syn-1,2-dihydrodiol 3,4-epoxide of 5-methylchrysene	pSP189	Ad293	97	1	1	1	1	1	1	55
7,8-dihydrxy-9-epoxy-7,8,9,10-tetrahydrobenzo[a]pyrene	pZ189	CV-1	39	2	2	2	2	2	2	56
2-Amino-3-methylimidao[4,5-f]quinoline	pSP189	Human Fibroblasts	62	2	1	2	1	2	2	57
2-Amino-1-methyl-6-phenylimidazo[4,5-b]pyridine	pSP189	Human Fibroblasts	73	2	3	2	4	2	2	57
Aflatoxin B1 –8,9-epoxide	pSP189	Ad293	127	2	3	2	3	2	2	58
Hydrogen peroxide — *in vitro*	pZ189	CV-1	75	2	5	2	4	2	2	46
Monofunctionally activated Mitomycin C	pSP189	Ad293	57	3	2	2	2	2	2	59
Peroxynitrite	pSP189	AD293	116	2	3	2	3	2	2	49
Spont	pSP189	AD293	51	3	2	2	2	2	2	49
Spont	pSP19/pZ189	AD293	85	3	2	2	2	2	2	47
Spont	pSP189	Human Fibroblasts	44	3	2	6	2	7	2	48
7-bromomethylbenz[a]anthracene	pS189	Ad293	93	2	3	3	3	3	3	60
7-bromomethyl-12-methylbenz[a]anthracene	pS189	Ad293	110	2	3	3	3	3	3	60

Table 13.4 Clustering all Mutational Spectra (*Continued*)

Mutagen	Plasmid	Cell Line	No. Mutations[a]	No. Clusters						Reference
				5	5Cr[b]	6	6Cr	7	7Cr	
Aflatoxin B1 –8,9-epoxide	pS189	Ad293	93	2	3	3	3	3	3	58
Anti-1,2-dihydrodiol 3,4-epoxide of 5-methylchrysene	pS189	Ad293	98	2	3	3	3	3	3	61
1,6-dinitropyrene	pS189	Ad293	85	2	3	3	3	3	3	62
Aflatoxin B1 –8,9-epoxide	pS189	Ad293	48	2	3	3	3	3	3	63
N-acetoxy-N-trifluoroacetyl-2-aminofluorene	pS189	Human Fibroblasts	49	2	3	3	3	3	3	64
7,8-dihydrxy-9-epoxy-7,8,9,10-tetrahydrobenzo[a]pyrene	pZ189	Ad293	68	2	3	3	3	3	3	65
1-Nitrosopyrene	pZ189	Ad293	54	2	3	3	3	3	3	66
anti 3,4-dihydrodiol 1,2-epoxide of 7-methylbenz[a]anthracene	pS189	Ad293	75	2	3	3	3	3	3	61
(–)-benzo[c]phenanthrene 3,4-dihydrodiol 1,2 epoxide-2	pS189	Ad293	146	1	4	4	5	4	4	67
(+)-benzo[c]phenanthrene 3,4-dihydrodiol 1,2 epoxide -2	pS189	Ad293	86	1	4	4	5	4	4	67
(–)-benzo[c]phenanthrene 3,4-dihydrodiol 1,2 epoxide 1	pS189	Ad293	140	1	4	4	5	4	4	67
(+)-benzo[c]phenanthrene 3,4-dihydrodiol 1,2 epoxide -1	pS189	Ad293	106	1	4	4	5	4	4	67
(–)-benzo[c]phenanthrene 3,4-dihydrodiol 1,2 epoxide -2	pZ189	Ad293	98	1	4	4	5	4	4	68
UV	pZ189	XP-A	71	4	5	5	6	5	5	24
UV	pZ189	HeLa — in vitro	44	4	5	5	6	5	5	42
UV	pZ189 f1	HeLa — in vitro	28	4	5	5	6	5	5	42
UV	pZ189R2	HeLa — in vitro	30	4	5	5	6	5	5	42
UV	PYZ289	HL18	55	4	5	4	6	5	5	43
UV	pZ189	WI38VA13	84	4	5	5	6	5	5	43

UV	pZ189	XP-A	62	4	5	5	6	5	5	43
UV	pZ189	Monkey	138	4	5	5	6	5	5	37
UV	pZ189	XP-D	67	4	5	5	6	5	5	25
UV	pZ189	GM0637	80	4	5	5	6	5	5	24
UV	pSP189	XP-A	137	4	5	5	6	5	5	29
UV	pZ189	XP-F	73	4	5	5	6	5	5	43
UV	pZ189	CV-1	179	4	5	5	6	5	5	44
Spont	pZ189	GM606/GM0637	37	3	2	2	2	5	6	37
UV(313)	pZ189	CV-1	247	5	5	5	4	5	6	44
Hydrogen peroxide — *in vivo*	pZ189	CV-1	123	5	5	5	4	5	6	45
Spont	pZ189	CV-1	71	5	5	5	4	5	6	44
Spont	pZ189	CV-1	125	5	5	5	4	5	6	45
Spont	pZ189	CV-1	211	5	5	5	4	5	6	46
Spont	pZ189	CV-1	56	5	5	5	4	5	6	39
Alpha	pZ189	GM606	61	3	2	6	2	6	7	51
X_ray_5_40	pZ189	GM606	246	3	2	6	2	6	7	50
X_ray_160	pZ189	GM606	131	3	2	6	2	6	7	50
Alpha	pZ189	GM606	58	3	2	6	2	6	7	51
Gamma	pZ189	GM606	126	3	2	6	2	6	7	69
Spont	pZ189	GM606	77	3	2	6	2	6	7	50
Spont	pZ189	GM606	108	3	2	6	2	6	7	51
Cr-Cys	pSP189	Human Fibroblasts	43	*	2	*	2	5	2	48
Cr-his	pSP189	Human Fibroblasts	44	*	5	*	4	5	2	48
Cr-gsh	pSP189	Human Fibroblasts	40	*	2	*	2	5	2	48
Cr(III)-KP	pSP19	AD293	56	*	2	*	2	5	2	47
Cr(III)-TE	pSP189/pZ189	AD293	61	*	2	*	2	5	2	47
Cr-CL3	pSP189	Human Fibroblasts.	41	*	2	*	2	5	2	48
Cr — *in vivo*	pZ189	CV-1	56	3	3	*	3	7	7	70
Cr — *in vitro*	pZ189	CV-1	99	*	2	*	2	7	7	71

[a] Number of base substitutions in a spectrum.

[b] Cr denotes analyses with Chromium spectra included.

together along with spontaneous spectra from the same cell type and vector. Likewise, hydrogen peroxide (*in vivo*, CV-1) groups with spontaneous mutations from that species. An exception is the spectrum from hydrogen peroxide *in vitro*; this spectrum groups among apparently dissimilar agents, although the group does include some spontaneous spectra. The underlying spectra for the six groups are presented in Figure 13.7.

There are several additional issues related to the clustering method itself. In this approach, the clustering that optimally fits the data is the one that maximizes the classification likelihood criterion. Finding such clustering is generally a difficult problem due to the likely existence of a number of locally optimal clusterings. Consequently, the clustering identified as optimal by a traditional "hill-climbing" algorithm is likely to depend on the initial clustering supplied to the algorithm.[35] In this analysis we used the Gibbs sampler-based algorithm that is capable of consistently finding the globally optimal solution regardless of the initial clustering.[36] For each analysis, the algorithm converged to the same solution after it was started from five different randomly selected initial clusterings.

Finally, the true number of underlying spectra that generated the data is unknown, but it is likely to be larger than in the models we used (maximum number of clusters was eight). Because of that, and because the influence that a spectrum has on the classification likelihood depends on the spectrum size, it is generally possible that differences between two underlying spectra within the same cluster are higher than between two spectra that are separated in two different clusters. Previously, we addressed this issue by imputing mutations to observed spectra with a smaller than the maximum number of mutations and showed that differences in spectra sizes did not affect results of the analysis.[36]

The generalizability of this type of analysis to other mutation detection systems depends on parameters of the target gene. The most important of these is the number of mutable sites detected in the system. For a large number of mutable sites, most sites will have either zero or one mutation in a normal size spectrum so different distributions are not easily distinguished. For this reason, large target genes like *HPRT* are not as amenable to analysis.

C.   *Grouping of chromium-induced spectra with others*

Recently, we began to investigate the mutagenic activity of chromate, Cr(VI), a known human carcinogen. Although chromate had been found to be mutagenic in a variety of test systems,[72,73] the mechanism of mutagenesis was less understood. There was evidence that chromate was reduced in the cell by glutathione and other cellular reducing agents, generating reactive intermediates that cause oxidative damage to DNA.[74,75] It had also been demonstrated that Cr(III), the stable end product of intracellular reduction of chromate, bound tightly to DNA.[48,76] We reasoned that the mutational mechanism could be revealed through an analysis of mutation spectra. For these studies we used both protocols outlined in Figure 13.2. First, we treated

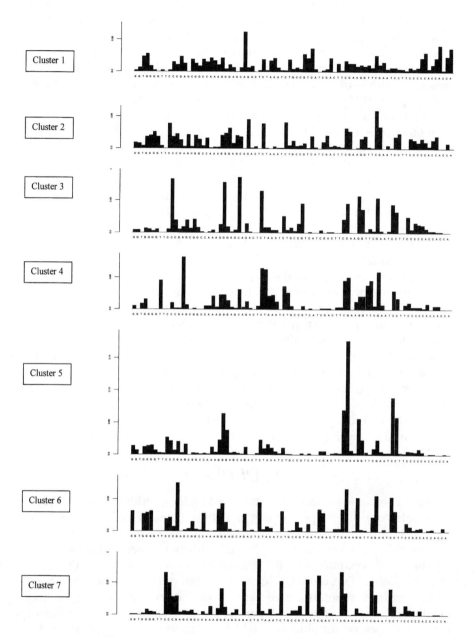

Figure 13.7 Results of cluster analysis of a collection of published spectra. The underlying spectra of seven groups are shown.

the vector DNA *in vitro* with chromate and glutathione, removed any bound chromium from the DNA and transfected it into mammalian cells (Cr-*in vitro*).[71] For comparison, we introduced the undamaged vector into the mammalian cells and then treated the cells with chromate (Cr-*in vivo*).[70] The results of these studies are reproduced here in Figure 13.8. We compared these results for the SupF gene (position 99-183) to similar studies with the oxidizing agent, H_2O_2 using the method of Adams and Skopek.[26] This analysis revealed that the chromium spectra were not significantly different from each other (p = 0.11) and the Cr-*in vitro* spectrum did not differ significantly from the H_2O_2-*in vitro* spectrum (p = 0.20). However, the H_2O_2-*in vivo* spectrum differed significantly from the Cr-*in vivo* spectrum (p < 0.001), the Cr-*in vitro* spectrum (p = 0.043) and the H_2O_2-*in vitro* spectrum (p < 0.001).

In order to compare these Cr-induced spectra to a broader range of mutation spectra, we have employed the clustering methods and we have expanded the collection of mutation spectra to include published spectra in which pZ189 was treated with Cr(III) in the presence of amino acids to form DNA/Cr/amino acid complexes[47,48] (Table 13.4, bottom). This analysis for seven groups revealed that the two chromium spectra (Cr-*in vivo* and Cr-*in vitro*) group with each other and together with ionizing radiation spectra (group 7). Interestingly, most of the spectra from chromium-induced crosslinks grouped together (group 2) with several other agents that form DNA adducts and also with H_2O_2-*in vitro*. These results suggest that the mutational specificity of the chromium-induced crosslinks differs from that of the chromium-induced oxidative damage. Furthermore, they suggest that mutational specificity of chromium *in vivo* may be more like chromium-induced oxidative damage than like chromium-induced crosslinks. However, some of these differences could also be related to the cell type and vector used in the experiments.

IV. Discussion

We have compared mutation spectra based on the distribution and frequency of mutations at the 85 potential sites within the SupF coding sequence. In general, the cluster analysis reveals that spectra generated by similar mutagenic agents are similar. The UV-induced spectra cluster strongly together and separate from the spontaneous spectra, whereas the spectra from oxidative DNA damage cluster in several different groups and are not as distinct from spontaneous mutations. The chromium-induced mutations clustered into two groups dependent on whether chromium-induced crosslinks or chromium-induced oxidative damage was being examined. All of these observations argue that the distribution of mutations within the mutational target gene provides a basis for characterizing mutagens.

So far, we have only included sites of base substitutions in these analyses. The types of base substitutions (transitions and transversions) also appear to be characteristic of specific mutagenic agents. For example, UV-induced mutations are predominantly G:C to A:T transitions, while mutations induced

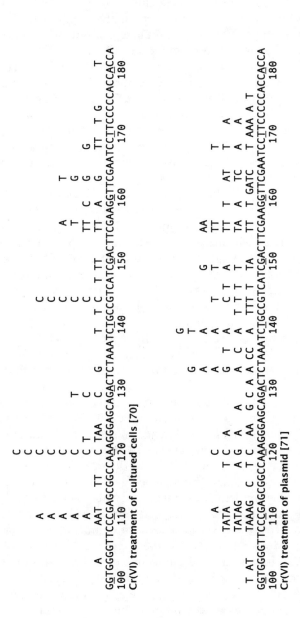

Figure 13.8 Mutation spectra for chromium. Top panel: CV-1 cells carrying pZ189 vector were treated with potassium dichromate. Bottom panel: pZ189 was treated in vitro with potassium dichromate in the presence of glutathione, bound chromium was removed from the DNA and the DNA was introduced into CV-1 cells. The coding sequence of the SupF gene is shown along with the mutations is the spectra. The specific base changes in the collection of mutants are indicated by letter above the site of the mutated base. The numbers below the sequence indicate the DNA sequence position within the pZ189 vector. (From Liu, S., Medvedovic, M., and Dixon, K., *Environ. Mol. Mutagen.*, 33, 313–319, 1999; Liu, S. and Dixon, K., *Environ. Mol. Mutagen.*, 28, 71–79, 1996. With permission.)

by benzo[a]pyrene diolepoxide exhibit similar frequencies of all three possible substitutions at G:C base pairs. These differences can be evaluated overall, however, most mutation spectra are not large enough to incorporate the site specificity of each type of base change into a cluster analysis. We have also not included deletion mutations into the present analysis although the frequency of deletion mutations differs among mutagens. Comparison of deletion mutations would likely require a different type of analysis.

As discussed in the introduction, mutagenic activity depends on metabolism of the compound, the type and level of initial DNA damage, the activity of DNA repair processes and the mechanism of mutation fixation. In theory, alterations in any of these processes could lead to changes in mutation spectrum. Therefore, another application of the determination and analysis of mutation spectra is the elucidation of role of these processes in mutagenesis and how genetic defects in these processes may alter mutagenesis. Most of the studies included in the analyses reported here were done in "normal" cell lines. A large study of the influence of DNA repair defects on mutation spectra was done in the yeast-based sup4-o system developed by Pierce and colleagues.[17] The sup4-o suppressor tRNA target gene is carried on a low copy-number plasmid and mutants are selected on the basis of suppression of nonsense mutations in specific yeast genes. Spontaneous mutations were characterized in a variety of DNA repair-deficient yeast lines and alterations in mutational spectra were observed. This type of analysis could also be carried out examining the influence of repair deficiencies on patterns of carcinogen-induced mutations.

It is clear that clustering analysis of mutation spectra generates groupings of mutation spectra that make chemical and biological sense. In general, agents that have similar modes of action induce mutation spectra that are similar. It seems reasonable therefore, to argue that a new mutagen that clusters with mutagens of known mechanism is likely to have a similar mutagenic mechanism. Thus, we have used the clustering of chromium mutations with those causing oxidative damage to suggest that chromium mutagenesis likely involves oxidative-type DNA damage. In the case of exposure to an unknown agent, is it possible to deduce from a mutation spectrum the identity of the primary mutagenic agent? Or, as stated in the introduction, is it possible to identify signatures of carcinogens that allows back-extrapolation? From the analysis presented here, we can certainly conclude that mutation spectrum analysis provides for identification of classes of mutagenic agents. More refinement of these groupings would likely require generation of larger spectra. Present technology of high-throughput sequence analysis would certainly be adequate for this application.

Application of this type of analysis to exposed individuals requires an endogenous mutagenic target for which mutation spectra can be easily determined. The *HPRT* locus affords some of the requirements for this analysis. For example, mutations can be easily selected in peripheral lymphocytes. However, mutations occurring in stem cells can become clonally expanded to dominate the analysis and mutations occurring in later stages can become lost due to turnover of the cell population.[77] The other limitation of this system is in the large size of the target gene; a large group of mutants would have to be screened to allow meaningful analysis. Analysis of genes, such as p53, in the tumor itself cannot provide much information about the carcinogenic exposure because mutations in the tumor are selected and clonally expanded during tumor development. Also, tumor cells appear to exhibit elevated levels of spontaneous mutagenesis. For application in human populations, ideally a mutagenesis target should be small, easily accessible, easily selected, and easily sequenced.

In conclusion, clustering analysis of mutation spectra provides a means of identifying a similar mechanism of action among a broad spectrum of mutagenic agents. In addition, this method allows inferences about the mechanism of action of unknown agents. Refinement of this type of analysis will require generation of larger mutation spectra. Application to studies in human populations is theoretically possible but will require the development of a gene target that is more amenable to analysis than those currently used for such studies.

Acknowledgments

This work was supported by NIH grants R01-ES05400, P42-ES04908, and P30-ES06096.

References

1. Hueper, W.C., Occupational and environmental cancers of the respiratory system, in *Recent Results in Cancer Research*, Springer-Verlag, New York, 1966.
2. Langard, S., One hundred years of chromium and cancer: a review of epidemiological evidence and selected case reports, *Am. J. Ind. Med.*, 17, 189–215, 1990.
3. Smith, A.H. et al., Cancer risks from arsenic in drinking water, *Environ. Health Perspect.*, 97, 259–267, 1992.
4. Abernathy, C.O., Calderone, R.L., and Chappell, W.R., *Arsenic: Exposure and Health Effects*, Chapman & Hall, London, 1997.
5. Ames, B.N., Lee, F.D., and Durston, W.E., An improved bacterial test system for the detection and classification of mutagens and carcinogens, *Proc. Natl. Acad. Sci. USA*, 70, 782–786, 1973.
6. Tennant, R.W., Stratification of rodent carcinogenicity bioassay results to reflect relative human hazard, *Mutat. Res.*, 286, 111–118, 1993.
7. U.S. Department of Health and Human Services, 9th Report on Carcinogens, 2001. http://ehis.niehs.nih.gov/roc/toc9.html.
8. Wang, S.Y., in *Photochemistry and Photobiology of Nucleic Acids*, Academic Press, New York, 1976.
9. Kadlubar, F.F., A transversion mutation hypothesis for chemical carcinogenesis by N2-substitution of guanine in DNA, *Chem. Biol. Interact.*, 31, 255–263, 1980.
10. Lehmann, A.R., Replication of UV-damaged DNA: new insights into links between DNA polymerases, mutagenesis and human disease, *Gene*, 253, 1–12, 2000.

11. Batty, D.P. and Wood, R.D., Damage recognition in nucleotide excision repair of DNA, *Gene*, 241, 193–204, 2000.
12. Featherstone, C. and Jackson, S.P., DNA double-strand break repair, *Curr. Biol.*, 9, R759–R761, 1999.
13. Woodgate, R., A plethora of lesion-replicating DNA polymerases, *Genes Dev.*, 13, 2191–2195, 1999.
14. Albertini, R.J., Nicklas, J.A., and O'Neill, J.P., Somatic cell gene mutations in humans: biomarkers for genotoxicity, *Environ. Health Perspect.*, 101 (Suppl 3), 193–201, 1993.
15. Seidman, M.M. et al., A shuttle vector plasmid for studying carcinogen-induced point mutations in mammalian cells, *Gene*, 38, 233–237, 1985.
16. Harwood, J., Tachibana, A., and Meuth, M., Multiple dispersed spontaneous mutations: a novel pathway of mutation in a malignant human cell line, *Mol. Cell Biol.*, 11, 3163–3170, 1991.
17. Pierce, M.K., Giroux, C.N., and Kunz, B.A., Development of a yeast system to assay mutational specificity, *Mutat. Res.*, 182, 65–74, 1987.
18. Carothers, A.M. et al., DNA base changes and RNA levels in N-acetoxy-2-acetylaminofluorene-induced dihydrofolate reductase mutants of Chinese hamster ovary cells, *J. Mol. Biol.*, 208, 417–428, 1989.
19. Nieboer, E., Rossetto, F.E., and Turnbull, J.D., Molecular biology approaches to biological monitoring of genotoxic substances, *Toxicol. Lett.*, 64–65 Spec No, 25–32, 1992.
20. Seidman, M.M. and Dixon, K., Mutational Analysis of Genotoxic Chemicals, in *Molecular Biology of Toxic Response*, Puga, A. and Wallace, K.B., Eds., Taylor & Francis, Philadelphia, 1998.
21. Cariello, N.F. and Skopek, T.R., *In vivo* mutation at the human HPRT locus, *Trends Genet.*, 9, 322–326, 1993.
22. Munson, P.J. et al., Test of models for the sequence specificity of UV-induced mutations in mammalian cells, *Mutat. Res.*, 179, 103–114, 1987.
23. Brash, D.W. et al., Photoproduct frequency is not the major determinant of UV base substitution hot spots or cold spots in human cells, *Proc. Natl. Acad. Sci. USA*, 84, 3782–3786, 1987.
24. Bredberg, A., Kraemer, K.H., and Seidman, M.M., Restricted ultraviolet mutational spectrum in a shuttle vector propagated in xeroderma pigmentosum cells, *Proc. Natl. Acad. Sci. USA*, 83, 8273–8277, 1986.
25. Seetharam, S. et al., Abnormal ultraviolet mutagenic spectrum in plasmid DNA replicated in cultured fibroblasts from a patient with the skin cancer-prone disease, xeroderma pigmentosum, *J. Clin. Invest.*, 80, 1613–1617, 1987.
26. Adams, W.T. and Skopek, T.R., Statistical test for the comparison of samples from mutational spectra, *J. Mol. Biol.*, 194, 391–396, 1987.
27. Robb, J.A. and Huebner, K., Effect of cell chromosome number on simian virus 40 replication, *Exp. Cell Res.*, 81, 120–126, 1973.
28. Kraemer, K.H. and Seidman, M.M., Use of supF, an *Escherichia coli* tyrosine suppressor tRNA gene, as a mutagenic target in shuttle-vector plasmids, *Mutat. Res.*, 220, 61–72, 1989.
29. Parris, C.N. et al., Proximal and distal effects of sequence context on ultraviolet mutational hotspots in a shuttle vector replicated in xeroderma cells, *J. Mol. Biol.*, 236, 491–502, 1994.

30. McCutchan, J.H. and Pagano, J.S., Enchancement of the infectivity of simian virus 40 deoxyribonucleic acid with diethylaminoethyl-dextran, *J. Natl. Cancer Inst.*, 41, 351–357, 1968.

31. Protic-Sabljic, M. and Kraemer, K.H., One pyrimidine dimer inactivates expression of a transfected gene in xeroderma pigmentosum cells, *Proc. Natl. Acad. Sci. USA*, 82, 6622–6626, 1985.

32. Invitrogen Life Technologies, *Guide to Transfection with Cationic Lipid Reagents*, 2001. http://www.invitrogen.com.

33. Hirt, B., Selective extraction of polyoma DNA from infected mouse cell cultures, *J. Mol. Biol.*, 26, 365–369, 1967.

34. Peden, K.W. et al., Isolation of mutants of an animal virus in bacteria, *Science*, 209, 1392–1396, 1980.

35. Cellux, G. and Govaert, G., A classification EM algorithm for clustering and two stochastic versions, *Comp. Stat. and Data Anal.*, 14, 315–332, 1992.

36. Medvedovic, M. et al., Clustering mutational spectra via classification likelihood and Markov chain Monte Carlo algorithms, *J. Agr., Biol., Environ. Stat.*, 6, 19–37, 2001.

37. Hauser, J. et al., Sequence specificity of point mutations induced during passage of a UV-irradiated shuttle vector plasmid in monkey cells, *Mol. Cell Biol.*, 6, 277–285, 1986.

38. Seidman, M.M. et al., Multiple point mutations in a shuttle vector propagated in human cells: evidence for an error-prone DNA polymerase activity, *Proc. Natl. Acad. Sci. USA*, 84, 4944–4948, 1987.

39. Hauser, J., Levine, A.S., and Dixon, K., Unique pattern of point mutations arising after gene transfer into mammalian cells, *EMBO J.*, 6, 63–67, 1987.

40. Tarone, R.E., Testing for non-randomness of events in sparse data situations, *Ann. Hum. Genet.*, 53 (Pt. 4), 381–387, 1989.

41. Westfall, H.P. and Young, S.S., P-value adjustments for multiple tests in multivariate binomial models, *J. Am. Stat. Assoc.*, 84, 780–786, 1989.

42. Carty, M.P. et al., Analysis of mutations induced by replication of UV-damaged plasmid DNA in HeLa cell extracts, *Environ. Mol. Mutagen.*, 26, 139–146, 1995.

43. Yagi, T. et al., Analysis of point mutations in an ultraviolet-irradiated shuttle vector plasmid propagated in cells from Japanese xeroderma pigmentosum patients in complementation groups A and F, *Cancer Res.*, 51, 3177–3182, 1991.

44. Keyse, S.M., Amaudruz, F., and Tyrrell, R.M., Determination of the spectrum of mutations induced by defined-wavelength solar UVB (313-nm) radiation in mammalian cells by use of a shuttle vector, *Mol. Cell Biol.*, 8, 5425–5431, 1988.

45. Moraes, E.C., Keyse, S.M., and Tyrrell, R.M., Mutagenesis by hydrogen peroxide treatment of mammalian cells: a molecular analysis, *Carcinogenesis*, 11, 283–293, 1990.

46. Moraes, E.C. et al., The spectrum of mutations generated by passage of a hydrogen peroxide damaged shuttle vector plasmid through a mammalian host, *Nucleic Acids Res.*, 17, 8301–8312, 1989.

47. Tsou, T.C., Lin, R.J. and Yang, J.L., Mutational spectrum induced by chromium(III) in shuttle vectors replicated in human cells: relationship to Cr(III)-DNA interactions, *Chem. Res. Toxicol.*, 10, 962–970, 1997.

48. Voitkun, V., Zhitkovich, A., and Costa, M., Cr(III)-mediated crosslinks of glutathione or amino acids to the DNA phosphate backbone are mutagenic in human cells, *Nucleic Acids Res.*, 26, 2024–2030, 1998.

49. Juedes, M.J. and Wogan, G.N., Peroxynitrite-induced mutation spectra of pSP189 following replication in bacteria and in human cells, *Mutat. Res.*, 349, 51–61, 1996.

50. Waters, L.C. et al., Mutations induced by ionizing radiation in a plasmid replicated in human cells, part I: Similar, nonrandom distribution of mutations in unirradiated and X-irradiated DNA, *Radiat. Res.*, 127, 190–201, 1991.

51. Jaberaboansari, A. et al., Mutations induced by ionizing radiation in a plasmid replicated in human cells, part II: Sequence analysis of alpha particle-induced point mutations, *Radiat. Res.*, 127, 202–210, 1991.

52. Szeliga, J. et al., Characterization of DNA adducts formed by anti-benzo[g]chrysene 11,12-dihydrodiol 13,14-epoxide, *Chem. Res. Toxicol.*, 8, 1014–1019, 1995.

53. Szeliga, J. et al., Reaction with DNA and mutagenic specificity of syn-benzo[g]chrysene 11,12-dihydrodiol 13,14-epoxide, *Chem. Res. Toxicol.*, 7, 420–427, 1994.

54. Page, J.E. et al., Mutational spectra for 5,6-dimethylchrysene 1,2-dihydrodiol 3,4-epoxides in the supF gene of pSP189, *Chem. Res. Toxicol.*, 8, 143–147, 1995.

55. Page, J.E. et al., Mutational specificity of the syn 1,2-dihydrodiol 3,4-epoxide of 5-methylchrysene, *Cancer Lett.*, 110, 249–252, 1996.

56. Roilides, E. et al., Mutational specificity of benzo[a]pyrene diolepoxide in monkey cells, *Mutat. Res.*, 198, 199–206, 1988.

57. Endo, H., Schut, H.A., and Snyderwine, E.G., Mutagenic specificity of 2-amino-3-methylimidazo[4,5-f]quinoline and 2-amino-1-methyl-6-phenylimidazo[4,5-b]pyridine in the supF shuttle vector system, *Cancer Res.*, 54, 3745–3751, 1994.

58. Courtemanche, C. and Anderson, A., Shuttle-vector mutagenesis by aflatoxin B1 in human cells: effects of sequence context on the supF mutational spectrum, *Mutat. Res.*, 306, 143–151, 1994.

59. Maccubbin, A.E. et al., Mutations induced in a shuttle vector plasmid exposed to monofunctionally activated mitomycin C, *Environ. Mol. Mutagen.*, 29, 143–151, 1997.

60. Page, J.E. et al., Mutagenic specificities and adduct distributions for 7-bromomethylbenz[a]anthracenes, *Carcinogenesis*, 17, 283–288, 1996.

61. Bigger, C.A. et al., Mutational specificity of the anti 1,2-dihydrodiol 3,4-epoxide of 5-methylchrysene, *Carcinogenesis*, 11, 2263–2265 (1990).

62. Boldt, J. et al., Kinds of mutations found when a shuttle vector containing adducts of 1,6-dinitropyrene replicates in human cells, *Carcinogenesis*, 12, 119–126, 1991.

63. Levy, D.D. et al., Sequence specificity of aflatoxin B1-induced mutations in a plasmid replicated in xeroderma pigmentosum and DNA repair proficient human cells, *Cancer Res.*, 52, 5668–5673, 1992.

64. Mah, M.C. et al., Mutations induced by aminofluorene-DNA adducts during replication in human cells, *Carcinogenesis*, 10, 2321–2328, 1989.

65. Yang, Maher, V.M., and McCormick, J.J., Kinds of mutations formed when a shuttle vector containing adducts of (+/-)-7 beta, 8 alpha-dihydroxy-9 alpha, 10 alpha-epoxy-7,8,9, 10-tetrahydrobenzo[a]pyrene replicates in human cells, *Proc. Natl. Acad. Sci. USA*, 84, 3787–3791, 1987.

66. Yang, J.L., Maher, V.M. and McCormick, J.J., Kinds and spectrum of mutations induced by 1-nitrosopyrene adducts during plasmid replication in human cells, *Mol. Cell Biol.*, 8, 3364–3372, 1988.

67. Bigger, C.A. et al., Mutagenic specificities of four stereoisomeric benzo[c]phenanthrene dihydrodiol epoxides, *Proc. Natl. Acad. Sci. USA*, 89, 368–372, 1992.

68. Bigger, C.A. et al., Mutagenic specificity of a potent carcinogen, benzo[c]phenanthrene (4R,3S)-dihydrodiol (2S,1R)-epoxide, which reacts with adenine and guanine in DNA, *Proc. Natl. Acad. Sci. USA*, 86, 2291–2295, 1989.

69. Sikpi, M.O. et al., Dependence of the mutation spectrum in a shuttle plasmid replicated in human lymphoblasts on dose of gamma radiation, *Int. J. Radiat. Biol.*, 59, 1115–1126, 1991.

70. Liu, S., Medvedovic, M. and Dixon, K., Mutational specificity in a shuttle vector replicating in chromium(VI)-treated mammalian cells, *Environ. Mol. Mutagen.*, 33, 313–319, 1999.

71. Liu, S. and Dixon, K., Induction of mutagenic DNA damage by chromium (VI) and glutathione, *Environ. Mol. Mutagen.*, 28, 71–79, 1996.

72. Yang, J.L. et al., Mutational specificity of chromium(VI) compounds in the hprt locus of Chinese hamster ovary-K1 cells, *Carcinogenesis*, 13, 2053–2057, 1992.

73. Chen, J. and Thilly, W.G., Mutational spectrum of chromium(VI) in human cells, *Mutat. Res.*, 323, 21–27, 1994.

74. Witmer, C. et al., *In vivo* effects of chromium, *Environ. Health Perspect.*, 102 Suppl 3, 169–176, 1994.

75. De Flora, S. et al., Metabolic reduction of chromium, as related to its carcinogenic properties, *Biol. Trace Elem. Res.*, 21, 179–187, 1989.

76. Zhitkovich, A., Voitkun, V., and Costa, M., Formation of the amino acid-DNA complexes by hexavalent and trivalent chromium in vitro: importance of trivalent chromium and the phosphate group, *Biochemistry*, 35, 7275–7282, 1996.

77. Albertini, R.J. et al., *In vivo* somatic mutations in humans: measurement and analysis, *Annu. Rev. Genet.*, 24, 305–326, 1990.

section III

Biomarkers of metal metabolism

chapter fourteen

Learning-induced modulation of neural cell adhesion molecule polysialylation state as a biomarker of prior inorganic lead exposure during the early postnatal period

Keith J. Murphy, Helen C. Gallagher, and Ciaran M. Regan

Contents

Abstract Prospective studies in humans and experimental investigations in animals have correlated elevated perinatal blood lead levels with enduring behavioral and cognitive perturbations. Work in our laboratory has demonstrated prior postnatal lead exposure to result in an enduring impairment of molecular events associated with long-term memory consolidation.[1] In these studies, the influence of prior lead exposure on the transient modulations of hippocampal neural cell adhesion molecule (NCAM) polysialylation state that occur in the 10–12 h posttraining period, a neuroplastic event associated with memory consolidation, were investigated. Direct quantification of polysialylated dentate neurons revealed prior lead exposure to have no effect on their basal number but to significantly delay and blunt the transient increase observed in control animals at the 12 h posttraining time. Thus, lead exposure in the early postnatal period is suggested to result in enduring neuroplastic deficits, and to be most likely associated with the reordering of synaptic connections that occur during memory consolidation. Moreover, it is proposed that human serum expression of NCAM and the associated polysialyltransferase, or its regulator, protein kinase C δ, may serve as a reliable biomarker(s) for the detection of neurobehavioral toxicants.

I. Introduction

Since Pentschew and Garro[2] demonstrated maternal inorganic lead exposure to produce encephalopathy in suckling rodents, there has been continuing concern about the neurodevelopmental sequelae associated with low-level exposure to this toxicant. The majority of prospective studies in humans and experimental investigations in animals have confirmed this concern: perinatal blood lead levels correlate with behavioral and cognitive alterations that endure throughout postnatal development, adolescence and adulthood.[3-9] These findings suggest that perinatal lead burden alters programmed events in the developmental emergence of innate behavior or basic neuroplastic processes necessary for learned responses that guide behavior and provide adaptational potential in the mature individual. Ultimately, such alterations must relate to the synapse that plays a critical role in the consolidation of acquired and learned behaviors.[10,11]

Activity-dependent synapse selection provides a basis to understand neuroplastic events subserving behavioral adaptation in both the developing and adult animal. It is generally accepted that neural activity is translated into enduring synaptic change by a cascade of sequential molecular events involving gene induction, increased protein synthesis and synaptic growth. This process is mediated, in part, by cell recognition systems.[12] For example, the transiently produced hippocampal dentate granule cell synapses are stabilized 6–8 h postlearning and immediately following a period of intense protein synthesis. This stabilization is dependent on the function of neural cell adhesion molecule (NCAM).[13-16] Again, in the 10–12 h posttraining period, synapses are selected to be retained in the memory trace by an NCAM-mediated mechanism. The frequency of dentate neurons with NCAM that are glycosylated with homopolymers of α2,8-linked polysialic acid (PSA) transiently increases in this period.[17,18] In long-term potentiation (LTP), a model of memory-associated neuroplastic change, similar NCAM-mediated events are critical to the induction and stabilization of synaptic modifications underlying this enduring change in neural function.[19-23]

Evidence suggests that exposure to lead alters the outcome of the molecular cascade of events that underlies the enduring synaptic change required for behavioral adaptation. For example, inorganic lead blocks the neuroplastic events necessary for the induction and maintenance of hippocampal LTP following acute exposure *in vitro* and chronic exposure *in vivo*.[24,25] Moreover, low-level exposure to lead induces an abnormal persistence of NCAM polysialylation during postnatal synapse elaboration. This results in a concomitant deficit in dendritic elaboration and spine density in hippocampal CA1 pyramidal cells and dentate granule cells.[26-28] These structural deficits do not arise from lead-induced cell death or altered cell migrations associated with organogenesis.[29,30] However, they seem similar, although less profound, to the dendritic insufficiencies observed in the staggerer (sg/sg) mouse, which are attributed to persistent NCAM polysialylation.[31]

This reduced synaptic complement may result in impaired neuroplastic potential and may account for enduring deficits in avoidance conditioning in adult animals that are no longer being exposed to lead.[32] A recent study, however, has demonstrated compensatory reactive synaptogenesis in cholinergic projections from the septum to the hippocampus following perinatal lead exposure, a pathway known to be involved in acquisition of the passive avoidance paradigm.[33,34] Here, lead-induced loss of septal choline acetyltransferase immunoreactive neurons is enduring, whereas hemicholinium binding, a marker of high affinity choline transport sites, returns to normal after withdrawal from lead exposure. These latter findings imply that learning-associated hippocampal plasticity can be recovered. Postnatal lead exposure might then influence passive avoidance learning and the associated modulations of hippocampal NCAM polysialylation that are necessary for its consolidation. This possibility has been explored in experiments described below.

II. Long-term effects of postnatal inorganic lead exposure on adult neurobehavior

Wistar rat pups were culled to eight at birth and exposed to 400 mg $PbCl_2$/L via their dams' drinking water from postnatal day 1 to 30. This exposure protocol results in pup blood lead levels which increase slowly to 10–15 µg/dl by postnatal day 8, remain constant until postnatal day 16 and subsequently rise to approximately 45 µg/dl at weaning when the pups have direct access to the drinking water.[28,30] At postnatal day 30, the lead was removed from the drinking water and the animals were allowed to reach adulthood (postnatal day 80) at which time blood lead levels ranged from 2–4 µg/dl. This exposure protocol had no effect on body weight gain during the exposure period or thereafter (data not shown). All experimental proce-dures were approved by the Review Committee of the Biomedical Facility of University College, Dublin and were carried out by individuals who held the appropriate licence issued by the Department of Health.

At postnatal day 80, animals were trained in a one-trial, step-through, light-dark passive avoidance paradigm as described previously.[17] A smaller, illuminated compartment was separated from a larger compartment by a shutter with a small entrance. The floor of the training apparatus was a grid of stainless steel bars which could deliver a remotely controlled, scram-bled shock (0.75 mA every 0.5 ms) of 5-s duration when the animal entered the dark chamber. The animals were tested for recall of this inhibitory stimulus at increasing posttraining times by placing them in the light com-partment and noting their latency to enter the dark compartment. A crite-rion period of 600 s was used. Nonparametric statistical comparisons were made using the Mann-Whitney U-test and p values < 0.05 were considered to be significant.

Control animals trained in the passive avoidance paradigm readily acquired and consolidated the task as their mean recall latencies were close to the criterion employed and not significantly different over the 5-day period evaluated (Figure 14.1). When tested within 48 h posttraining, animals exposed postnatally to low levels of lead exhibited recall latencies that were indistinguishable from those of the control group. However, on posttraining day 5 (120 h), significant recall defects were observed in the lead-treated animals. As earlier recall was unaffected, this suggests an enduring impair-ment in long-term memory storage. Moreover, these effects were unrelated to lead-induced alterations in general locomotor activity as no differences were observed in open field measurements for each animal group at all posttraining times.

III. Long-term effects of postnatal inorganic lead exposure on NCAM polysialylation

Transient learning-associated change occurs in neuronal polysialylation in the hippocampal dentate gyrus following training in the passive avoidance paradigm. These changes were investigated in animals postnatally exposed to lead and in control animals. This experiment is designed to determine if lead-induced defects in long-term memory storage involve impaired molec-ular events associated with consolidation.

Following recall at each posttraining time, the animals were sacrificed, the whole brain quickly removed, coated immediately in an optimal cutting temperature compound (Gurr, U.K.), snap-frozen in liquid nitrogen-cooled n-hexane and stored at −80°C until required for further processing. Horizon-tal sections of 12 μm were cut from frozen tissue using a MICROM (Series 500) cryostat. Serial sections were obtained for analysis from a point -5.6 mm from Bregma[35] and thaw-mounted onto 0.1% (w/v) poly-l-lysine coated glass

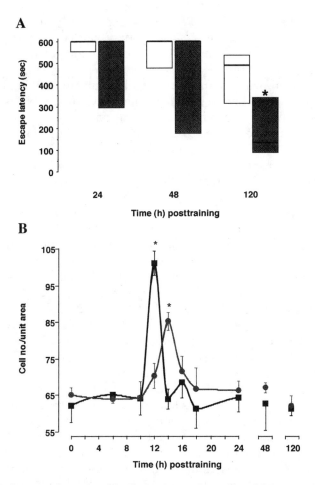

Figure 14.1 Influence of postnatal lead exposure on recall and hippocampal NCAM PSA immunoreactivity following training in the passive avoidance paradigm. In panel A, the data are expressed as the median value (dark horizontal bar) and 25 percent interquartile ranges for control (open boxes) and lead-exposed (shaded boxes) animals at increasing times after training. Each recall latency represents a separate animal group. Values significantly different (p <0.05) from the control are indicated with an asterisk. Panel B shows the influence of postnatal lead exposure on dentate polysialylated neuron frequency at increasing times following passive avoidance training. The data graphed in black and grey represent control and exposed animals, respectively. All values are the mean ± SEM of the number of cells per unit area (3=n=8). Those differing significantly (p < 0.05) from basal expression (0 h) are indicated by an asterisk.

slides. The sections were fixed in 70% (v/v) ethanol for 30 minutes, washed twice for 10 minutes in a washing buffer of 0.1M phosphate buffered 0.9% saline, pH7.4, (PBS) and incubated overnight (20 h) in a humidified chamber at room temperature with antipolysialic acid (PSA)[36] diluted 1:500 in an incubation buffer composed of PBS containing 1% (w/v) bovine serum albumen (Sigma Chemical Co., U.K.) and 1% (v/v) normal goat serum (DAKO, Denmark) in order to eliminate nonspecific staining. The sections were washed again and exposed for 3 h to fluorescein-conjugated goat antimouse IgM (Calbiochem, U.K.) diluted 1:100 with incubation buffer. The sections received a final wash before being mounted in Citifluor® (Agar, UK), a fluorescence-enhancing medium. The staining pattern was observed with a Leitz DM RB fluorescence microscope using an exciting wavelength of 495 nm and an emitting wavelength of 525 nm.

Immunofluorescence was specific as it was eliminated completely by omission of either the primary or secondary antibody; preabsorbing anti-NCAM PSA with colominic acid (1 mg/ml), which contains α2,8 linked homopolymers of sialic acid; or by prior incubation of the sections with 0.3% (v/v) endoneuraminidase-N.[17,18] The total number of PSA-immunoreactive neurons at the dentate granule cell layer and hilar border were counted in seven alternate 12 μm sections commencing -5.6 mm from Bregma, to preclude double counting of the 5–10 μm perikarya. Cell counts were divided by the total area of the granule cell layer and multiplied by the average granule cell layer area which was 0.15 ± 0.01 mm^2 at this level, and the mean $\pm$ SEM calculated. These means were used to establish the overall mean $\pm$ SEM for each animal group. Area measurements were performed using a Quantimet 500 Image Analysis System.

The polysialylated neurons in this region were restricted to the granule cell layer/hilar border and the immunostaining appeared to be located on the cell surface as judged by the silhouette of several appropriately sectioned perikarya (Figure 14.2A). In addition, many of these cells extended immunopositive dendrites through the granule cell layer into the molecular layer. In control animals, the frequency of polysialylated cells was invariant up to and including 10 h posttraining, and was indistinguishable from that observed in the naive animal (64.1 ± 4.5 vs 64.4 ± 3.9, respectively; $p > 0.05$; Figure 14.1B). At 12 h posttraining, the expected transient increase was observed and this returned to the basal level at 14 h posttraining and remained constant for up to 5 days. In contrast, learning associated modulation of NCAM polysialylation was delayed and significantly blunted in lead exposed animals. At 12 h posttraining, the frequency of polysialylated cells in lead-exposed animals was significantly lower than in control animals (Figures 14.1 and 14.2). This appears to be a specific delay in the polysialylation response to learning: the frequency of polysialylated cells did not change between 10 and 12 h posttraining or between 16 and 18 h posttraining in lead exposed animals. Moreover, learning-associated change in frequency of polysialylated cells was observed at 14 h posttraining in lead-exposed animals, but it was lower than in control animals at 12 h posttraining (101 ± 3.3 vs 85.3 ± 2.5, respectively; $p < 0.05$). However, the number of polysialylated cells was not significantly different for lead-exposed and control animals at any other time. This suggests that the effect of lead is specific to learning-associated changes in frequency of polysialylated cells.

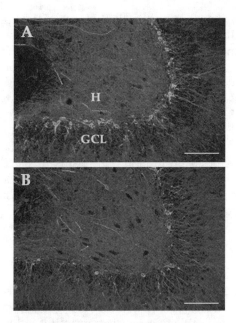

Figure 14.2 Influence of postnatal lead exposure on NCAM PSA immunoreactivity in the rat hippocampal dentate gyrus. Panels A and B represent control and lead-exposed animals, respectively, 12 h following passive avoidance training. The scale bar is 100 μm. GCL: granule cell layer; H: hilus.

IV. Discussion

The neurobehavioral deficits following postnatal lead exposure involve primarily an inability to consolidate memory into long-term storage as described previously.[32] Recall of the aversive stimulus in control animals was robust and avoidance latency did not decline significantly for 5 days. In contrast, recall latency declined significantly by posttraining day 5 in animals postnatally exposed to lead; this effect is specific to the learned response, as no obvious behavioral alterations were observed in open-field studies. Recall was unaffected in lead exposed animals within 48 h posttraining, indicating that the task was acquired successfully but was not processed effectively. This is consistent with impaired modulation of NCAM polysialylation observed in the 10–12 h posttraining period of consolidation. Moreover, the basal frequency of dentate polysialylated neurons was unaffected in the lead exposed groups prior to and following learning-induced modulations of NCAM PSA; this suggests that developmental exposure to inorganic lead results in an enduring inability to fully activate the polysialylation response by mechanisms largely independent of the regulatory polysialyltransferase.

These observations raise a number of issues with respect to the long-term consequences of postnatal lead exposure. It is likely that this study identifies the minimal effects of postnatal exposure to inorganic lead, because the exposure was restricted to the final stages of synapse elaboration. In contrast, perinatal exposure results in enduring cell loss and more severe learning deficits, including impaired acquisition.[32,34] Nevertheless, the exposure protocol was sufficient to induce a significant hippocampal CA1 synaptic deficit in the order of 40%.[27] The same reduction in posttraining polysialylation response was observed in animals with lead exposure; thus, it is tempting to speculate that this reduction reflects enduring deficits in synaptic complement and inability to mount a neuroplastic response of appropriate magnitude during memory consolidation. This is speculative because there are no systematic studies on the long-term consequences of postnatal lead exposure on synaptic complement in all regions of hippocampal formation. However, it is consistent with the fact that synapse formation is required during memory consolidation in the hippocampal dentate gyrus and the CA1 region.[15,16,37] Similarly, impaired synaptic growth has been associated with reduced persistence of LTP in the dentate gyrus of conscious rats exposed developmentally to inorganic lead.[38]

In this study, a temporal delay is observed in the learning-associated polysialylation response that suggests a defect in associated activation mechanisms. Although memory may be committed to the processes of long-term storage it remains malleable to extrinsic transmitter influences up to at least 12 h posttraining. For example, cholinergic antagonism 6–8 h posttraining and D1-mediated dopamine agonism 10–12 h posttraining cause loss of learning-associated NCAM polysialylation activation and amnesia at 24 h posttraining.[39] Lead-induced delay in learning-associated polysialylation may be relevant to the dopamine receptor supersensitivity observed in animals postnatally exposed to lead.[4,40]

These lead-induced learning deficits emerge after a lengthy delay; however; this is not at variance with the period of hippocampal function in memory consolidation. Primate and rodent lesion studies demonstrate that it requires several days for consolidation of associative tasks within the hippocampus.[41,42] Thus, irrespective of the mechanism of action, the minimal long-term consequence of neurodevelopmental lead exposure appears to be an enduring deficit in hippocampal memory consolidation. This view is supported by the inherent ability of the basal forebrain cholinergic system to recover from lead exposure and the belief that this system subserves more general regulatory aspects of cognition rather than specific mnemonic functions.[34,43]

V. Implications of NCAM polysialylation as a biomarker for lead-induced neurobehavioral deficits

Despite the identification of multiple routes through which lead perturbs learning mechanisms in the adult brain, its influence on neural development and structuring events in early life must play a critical role in low-level lead exposure-induced neurobehavioral deficits. In this regard, NCAM PSA may be considered an indicator of neurodevelopmental function as its expression is markedly regulated during development. In the early embryonic period NCAM polypeptides are poorly sialylated, however, in the late embryonic and perinatal period highly polysialylated protein forms predominate and are believed to facilitate cell migration and axon pathfinding and fasiculation in the CNS.[44-46] Thereafter, NCAM PSA is gradually downregulated in all brain areas. This developmental downregulation appears to occur at similar rates in the human and rodent brain although areas in which neurogenesis and synaptic plasticity persist do retain NCAM PSA expression into adulthood.[47-49] Given these central roles in developmental synaptic structuring, perturbation of NCAM polysialylation by neurotoxicants may indeed contribute to the persisting neurobehavioral deficits they are known to induce.

This study indicates that neurobehavioral deficits induced by postnatal lead exposure correlate with perturbations in NCAM PSA function during learning. Although it is unlikely this dysregulation of NCAM polysialylation is the primary causative factor, these observations do provide a rationale for considering NCAM as a sensitive downstream biomarker for toxicant-induced neurobehavioral deficits. An important clinical consideration in the validation of molecular biomarkers is that they be accessible by non-invasive procedures. NCAM PSA fits this criterion, because it can be measured in human serum by immunocapture techniques, and these measurements reflect tissue expression patterns. For example, children have a higher level of serum polysialylated NCAM than adults.[50] An elevated level of serum PSA is also detected in various pathological states such as tumor cells with high metastatic potential.[51,52] In a study of childhood neuroblastoma, serum NCAM PSA levels were directly correlated with the polysialylated NCAM content of tumours and decreased with successful treatment.[50] This suggests that the level of PSA in the serum reflects the rate of release of this molecule from the neural cell surface.

Serum fragments of the NCAM protein may also be detected in humans using standard immunoblotting techniques. Aberrant expression of NCAM serum and/or cerebrospinal fluid fragments has previously been shown to distinguish psychotic patients from healthy controls and, specifically, to discriminate the psychotic disorders schizophrenia and autism, both of which are believed to have a neurodevelopmental origin.[53-56] Hippocampal expression of NCAM PSA is also altered in the human schizophrenic brain[57] and in other disease states including Alzheimer's disease and epilepsy.[58,59] To date there has been no direct attempt to correlate the serum NCAM fragment and brain NCAM PSA expression patterns in human conditions with neurobehavioral manifestations. However, parallel observations that both appear to be perturbed in schizophrenia — a psychotic disorder of neurodevelopmental origin — tentatively suggests that alterations in the pattern of serum NCAM expression may reflect altered functioning and/or turnover of the neuroplastic form of NCAM–NCAM PSA. If this is true then our study, which implicates brain NCAM polysialylation as an indicator of lead exposure, suggests that serum NCAM expression may indeed be a valid biomarker for the detection of developmental toxicant activity.

An alternative strategy for monitoring NCAM polysialylation state is to quantify the polysialyltransferase activity known to directly regulate PSA expression.[60] Previous studies suggest that sialyltransferase activity is both sensitive to neurotoxicants and also reflective of related perturbations in NCAM PSA expression. For example, developmental methylmercury exposure modulates both cerebellar NCAM PSA and associated Golgi sialyltransferase activity.[61] Moreover, in that study, sialyltransferase sensitivity to methylmercury was developmentally regulated, decreasing with age. The same authors also found that in birds perinatally exposed to lead, an age-dependent elevation in NCAM polysialylation and sialyltransferase activity was evident in brain synaptosomes.[62] In these birds, lead exposure during critical periods of development causes neurological deficits that are believed to compromise survival in the wild. We and others obtained similar effects in earlier studies, where the normal developmental regulation of rat brain Golgi sialyltransferase activity was also perturbed by low-level lead exposure.[63,64] Furthermore, this coincided with a delayed conversion of cerebellar sialic acid-rich forms of NCAM to sialic acid-poor forms.[28]

These studies support the view that NCAM polysialylation state, via perturbed sialyltransferase activity, may be a general *in vivo* target of neurotoxicants. This is of particular relevance to the validation of a readily accessible biomarker, as sialyltransferase activity in human serum is believed to reflect, at least in part, the activation state of neural polysialyltransferase enzymes.[65] Furthermore, since the regulation of polysialyltransferase activity is directly regulated by protein kinase C delta (PKCδ),[66,67] monitoring this isozyme is another possible route to the generation of a reliable molecular biomarker. PKC enzymes are known to be exquisitely sensitive to inorganic lead. Just recently secretion of several PKC isoforms, including delta, was shown to occur from human platelets — suggesting that PKCδ may indeed be a sensitive and accessible molecular endpoint.[68–70]

In conclusion, the developmental expression profile of NCAM PSA, together with its known function in memory consolidation and sensitivity to heavy metals, suggests that this neuroplastic substrate may serve as a reliable biomarker for the detection of toxicants that may adversely influence neurobehavior. Further studies directed towards the adaptation of this marker for accessible monitoring of NCAM PSA levels may be required prior to validation. However, analysis of NCAM PSA itself, sialyltransferase activity or PKCδ levels in serum appear to be promising in this regard.

Acknowledgments

CMR wishes to acknowledge the support of University College Dublin and GlaxoSmithKline (Harlow, U.K.) while writing this manuscript during a leave of absence for research.

References

1. Murphy, K.J. and Regan, C.M., Low-level lead exposure in the early postnatal period results in persisting neuroplastic deficits associated with memory consolidation, *J. Neurochem.*, 72, 2099–2104, 1999.
2. Pentschew, A. and Garro, F., Lead encephalo-myelopathy of the suckling rat and its implications for the porphyrinopathic nervous diseases, *Acta Neuropathol.*, 6, 266–278, 1966.
3. Cory-Slechta, D.A., Weiss, B., and Cox, C., Performance and exposure indices of rats exposed to low concentrations of lead, *Toxicol. Appl. Pharmacol.*, 78, 291–299, 1985.
4. Cory-Slechta, D.A., Pokora, M.J., and Widzowski, D.V., Postnatal lead exposure induces supersensitivity to the stimulus properties of a D_2–D_3 agonist, *Brain Res.*, 598, 162–172, 1992.
5. Lilienthal, H. et al., Prenatal and postnatal lead exposure in monkeys: Effects on activity and learning set formation, *Neurobehav. Toxicol. Teratol.*, 8, 265–272, 1986.
6. Bellinger, D. et al., Longitudinal analyses of prenatal lead exposure and early cognitive development, *N. Eng. J. Med.*, 316, 1037–1043, 1987.
7. Needleman, H.L. et al., The long term effects of exposure to low doses of lead in childhood. An 11-year follow up report, *New Eng. J. Med.*, 322, 83–88, 1990.
8. Kuhlmann, A.C., McGlothan, J.L., and Guilarte, T.R., Developmental lead exposure causes spatial learning deficits in adult rats, *Neurosci. Lett.*, 233, 101–104, 1997.
9. Finkelstein, Y., Markowitz, M.E., and Rosen, J.F., Low-level lead-induced neurotoxicity in children, *Brain Res. Rev.*, 27, 168–176, 1998.
10. Greenough, W.T. and Bailey, C.H., The anatomy of a memory: Convergence of results across a diversity of tests, *Trends Neurosci.*, 11, 142–147, 1988.
11. Bailey, C.H. and Kandel, E.R., Structural changes accompanying memory storage, *Annu. Rev. Physiol.*, 55, 397–426, 1993.
12. Bailey, C.H., Bartsch, D., and Kandel, E.R., Toward a molecular definition of long-term memory storage, *Proc. Natl. Acad. Sci. USA*, 93, 13445–13452, 1996.
13. Squire, L.R. and Barondes, S.H., Variable decay of memory and its recovery in cycloheximide-treated mice, *Proc. Natl. Acad. Sci. USA*, 69, 1416–1420, 1972.
14. Doyle, E. et al., Intraventricular infusions of anti-neural cell adhesion molecules in a discrete post-training period impair consolidation of a passive avoidance response in the rat, *J. Neurochem.*, 59, 1570–1573, 1992.
15. O'Malley, A., O'Connell, C., and Regan, C.M., Ultrastructural analysis reveals avoidance conditioning to induce a transient increase in hippocampal dentate spine density in the 6 h post-training period of consolidation, *Neuroscience*, 87, 607–613, 1998.
16. O'Malley, A. et al., Transient spine density increases in the mid-molecular layer of hippocampal dentate gyrus accompany consolidation of a spatial learning task in the rodent, *Neuroscience*, 99, 229–232, 2000.
17. Fox, G.B. et al., Memory consolidation induces a transient and time-dependent increase in the frequency of NCAM polysialylated cells in the adult rat hippocampus, *J. Neurochem.*, 65, 2796–2799, 1995.
18. Murphy, K.J., O'Connell, A.W., and Regan, C.M., Repetitive and transient increases in hippocampal neural cell adhesion molecule polysialylation state following multitrial spatial training, *J. Neurochem.*, 67, 2538–2546, 1996.

19. Geinisman, G., Perforated axospinous synapses with multiple, completely partitioned transmission zones; probable structural intermediates in synaptic plasticity, *Hippocampus*, 3, 417–434, 1993.
20. Lüthi, A. et al., Hippocampal long-term potentiation and neural cell adhesion molecules L1 and NCAM, *Nature*, 372, 777–779, 1994.
21. Rønn, L.C.B. et al., NCAM-antibodies modulate induction of long-term potentiation in rat hippocampal CA1, *Brain Res.*, 677, 145–151, 1995.
22. Becker, C.G. et al., The polysialic acid modification of the neural cell adhesion molecule is involved in spatial learning and hippocampal long-term potentiation, *J. Neurosci. Res.*, 45, 143–152, 1996.
23. Muller, D. et al., PSA-NCAM is required for activity-induced synaptic plasticity, *Neuron*, 17, 413–422, 1996.
24. Altmann, L., Sveinsson, K., and Wiegand, H., Long-term potentiation in rat hippocampal slices is impaired following acute lead perfusion, *Neurosci. Lett.*, 128, 109–112, 1991.
25. Gilbert, M.E., Mack, C.M., and Lasley, S.M., Chronic developmental lead exposure increases the threshold for long-term potentiation in the rat dentate gyrus *in vivo*, *Brain Res.*, 736, 118–124, 1996.
26. Alfano, D.F. and Petit, T.L., Neonatal lead exposure alters the dendritic development of hippocampal dentate granule cells, *Exp. Neurol.*, 75, 275–288, 1982.
27. Kiraly, E. and Jones, D.G., Dendritic spine changes in rat hippocampal pyramidal cells after postnatal lead treatment: A Golgi study, *Exp. Neurol.*, 77, 236–239, 1982.
28. Cookman, G.R., King, W., and Regan, C.M., Chronic low-level lead exposure impairs embryonic to adult conversion of the neural cell adhesion molecule, *J. Neurochem.*, 49, 399–403, 1987.
29. Cookman, G.R. et al., Chronic low-level lead exposure precociously induces rat glial development *in vitro* and *in vivo*, *Neurosci. Lett.*, 86 33–37, 1988.
30. Hasan, F. et al., The effect of low level lead exposure on the postnatal structuring of the rat cerebellum, *Neurotoxicol. Teratol.*, 11, 433–440, 1989.
31. Edelman, G.M. and Chuong, C.M., Embryonic to adult conversion of neural cell adhesion molecules in the normal and staggerer mice, *Proc. Natl. Acad. Sci. USA*, 79, 7036–7040, 1982.
32. Regan, C.M. and Keegan, K., Neuroteratological consequences of chronic low level lead exposure, *Devl. Pharmacol. Therapeut.*, 5, 189–195, 1990.
33. Rashidy-Pour, A., Motaghed-Larijani, Z., and Bures, J., Reversible inactivations of medial septal area impairs consolidation but not retrieval of passive avoidance learning in rats, *Behav. Brain Res.*, 72, 185–188, 1995.
34. Bourjeily, N. and Suszkiw, J.B., Developmental cholinotoxicity of lead: loss of septal cholinergic neurons and long-term changes in cholinergic innervation of the hippocampus in perinatally lead-exposed rats, *Brain Res.*, 771, 319–328, 1997.
35. Paxinos, G. and Watson, C., The rat brain in stereotaxic coordinates, Academic Press, San Diego, 1986.
36. Rougon, G. et al., A monoclonal antibody against meningococcus group B polysaccharides distinguishes embryonic from adult NCAM, *J. Cell Biol.*, 103, 2429–2437, 1986.
37. Moser, M., Trommold, M., and Anderson, P., An increase in dendritic spine density on hippocampal CA1 pyramidal cells following spatial learning in adult rats suggests the formation of new synapses, *Proc. Natl. Acad. Sci. USA*, 91, 12673–12675, 1994.

38. Gilbert, M.E. and Mack, C.M., Chronic lead exposure accelerates decay of long-term potentiation in rat dentate gyrus *in vivo*, *Brain Res.*, 789, 139–149, 1998.
39. Doyle, E. and Regan, C.M., Cholinergic and dopaminergic agents which inhibit a passive avoidance attenuate the paradigm-specific increases in NCAM sialylation state, *J. Neural Transm.*, 92, 33–49, 1993.
40. Cory-Slechta, D.A. and Widzowski, D.V., Low-level lead exposure increases sensitivity to the stimulus properties of dopamine D1 and D2 agonists, *Brain Res.*, 553, 65–74, 1991.
41. Zola-Morgan, S. and Squire, L.R., The primate hippocampal formation. Evidence for a time-limited role in memory storage, *Science*, 250, 288–290, 1990.
42. Kim, J.J. and Fanselow, M.S., Modality-specific retrograde amnesia of fear, *Science*, 256, 675–677, 1992.
43. Everitt, B.J. and Robbins, T.W., Central cholinergic systems and cognition *Annu. Rev. Psychol.*, 48, 649–684, 1997.
44. Hoffman, S. and Edelman, G.M., Kinetics of homophilic binding by embryonic and adult forms of the neural cell adhesion molecule, *Proc. Natl. Acad. Sci. USA*, 89, 5762–5766, 1983.
45. Tang, J., Rutishauser, U., and Landmesser, L., Polysialic acid regulates growth cone behavior during sorting of motor axons in the plexus region, *Neuron*, 8, 405–414, 1994.
46. Wang, C. et al., Functional N-Methyl-D-aspartate receptors on O-2A glial precursor cells: A critical role in regulating polysialic acid-neural cell adhesion molecule expression and migration, *J. Cell Biol.*, 135, 1565–1581, 1996.
47. Miragall, F. et al., Expression of cell adhesion molecules in the olfactory system of the adult mouse: presence of the embryonic form of N-CAM, *Dev. Biol.*, 129, 516–531, 1988.
48. Hu, H. et al., The role of polysialic acid in migration of olfactory bulb interneuron precursors in the subventricular zone, *Neuron*, 16, 735–743, 1996.
49. Ní Dhúill, C.M. et al., Polysialylated neural cell adhesion molecule expression in the dentate gyrus of the human hippocampal formation from infancy to old age, *J. Neurosci. Res.*, 55, 99–106, 1999.
50. Gluer, S. et al., Serum polysialylated neural cell adhesion molecule in childhood neuroblastoma, *Br. J. Cancer*, 78, 106–110, 1998.
51. Jaques, G. et al., Evaluation of serum neural cell adhesion molecule as a new tumor marker in small cell lung cancer, *Cancer*, 72, 418–425, 1993.
52. Kaiser, U. et al., Serum NCAM: a potential new prognostic marker for multiple myeloma, *Blood*, 83, 871–873, 1994.
53. Lyons, F. et al., The expression of an N-CAM serum fragment is positively correlated with severity of negative features in type II schizophrenia, *Biol. Psychiatry*, 23, 769–775, 1988.
54. Plioplys, A.V., Hemmens, S.E., and Regan, C.M., Expression of a neural cell adhesion molecule serum fragment is depressed in autism, *J. Neuropsychiatry Clin. Neurosci.*, 2, 413–417, 1990.
55. van Kammen, D.P. et al., Further studies of elevated cerebrospinal fluid neuronal cell adhesion molecule in schizophrenia, *Biol. Psychiatry*, 43, 680–686, 1988.
56. Vawter, M.P. et al., VASE-containing N-CAM isoforms are increased in the hippocampus in bipolar disorder but not schizophrenia, *Exp. Neurol.*, 154, 1–11, 1998.

57. Barbeau, D. et al., Decreased expression of the embryonic form of the neural cell adhesion molecule in schizophrenic brains, *Proc. Natl. Acad. Sci. USA*, 92, 2785–2789, 1995.
58. Mikkonen, M. et al., Remodeling of neuronal circuitries in human temporal lobe epilepsy: increased expression of highly polysialylated neural cell adhesion molecule in the hippocampus and the entorhinal cortex, *Ann. Neurol.*, 44, 923–934, 1998.
59. Mikkonen, M. et al., Hippocampal plasticity in Alzheimer's disease: changes in highly polysialylated NCAM immunoreactivity in the hippocampal formation, *Eur. J. Neurosci.*, 11, 1754–1764, 1999.
60. McCoy, R.D., Vimr, E.R., and Troy, F.A., CMP-NeuNAc:Poly-α2,8-sialosyl sialyltransferase and the biosynthesis of polysialosyl units in neural cell adhesion molecules, *J. Biol. Chem.*, 260, 12694–12699, 1985.
61. Dey, P.M., Gochfeld, M., and Reuhl, K.R., Developmental methylmercury administration alters cerebellar PSA-NCAM expression and Golgi sialyltransferase activity, *Brain Res.*, 845, 139–151, 1999.
62. Dey, P.M. et al., Developmental lead exposure disturbs expression of synaptic neural cell adhesion molecules in herring gull brains, *Toxicol.*, 146, 137–147, 2000.
63. Breen, K.C. and Regan, C.M., Lead stimulates golgi sialyltransferase at times coincident with the embryonic to adult conversion of the neural cell adhesion molecule, *Toxicol.*, 49, 71–76, 1988.
64. Davey, F.D. and Breen, K.C., Stimulation of sialyltransferase by subchronic low-level lead exposure in the developing nervous system. A potential mechanism of teratogen action, *Toxicol. Appl. Pharmacol.*, 151, 16–21, 1998.
65. Shen, A.L. et al., Alterations in serum sialyltransferase activities in patients with brain tumors, *Surg. Neurol.*, 22, 509–514, 1984.
66. Gallagher, H.C., Odumeru, O.A., and Regan, C.M., Regulation of neural cell adhesion molecule polysialylation state by cell-cell contact and protein kinase C delta, *J. Neurosci. Res.*, 61, 636–645, 2000.
67. Gallagher, H.C. et al., Protein kinase C delta regulates neural cell adhesion molecule polysialylation state in the rat brain, *J. Neurochem.*, 77, 425–434, 2001.
68. Bressler, J. et al., Molecular mechanisms of lead neurotoxicity, *Neurochem. Res.*, 24, 595–600, 1999.
69. Coppi, A.A. et al., The effects of lead on PKC isoforms, *Ann. NY Acad. Sci.*, 919, 304–306, 2000.
70. Hillen, T.J., Aroor, A.R., and Shukla, S.D., Selective secretion of protein kinase C isozymes by thrombin-stimulated human platelets, *Biochem. Biophys. Res. Commun.*, 280, 259–264, 2001.

chapter fifteen

Molecular biomarkers for mercury

Thomas W. Clarkson and Nazzareno Ballatori

Contents ·

Abstract   As our understanding increases of the molecular mechanisms of action of mercury and its compounds, so do possibilities arise for applying molecular markers for exposure, effect, and susceptibility. The underlying chemistry of mercury differs from that of alkylating carcinogens or other toxic organic chemicals that are metabolized to highy reactive intermediates to form stable covalent adducts with macromolecules. Mercury ions bind reversibly to thiol groups present in proteins and smaller molecules. Thus, it is much more difficult to isolate and identify mercury adducts as mercury may dissociate from its original attachment during the isolation procedures. Consequently, most work on molecular biomarkers has focused on induction of changes in gene or protein expression resulting from the biochemical action of mercury.

To date, there has been little need to identify biomarkers of exposure. Methylmercury levels in hair and blood samples have been shown to correlate with levels in the brain, the target organ for methylmercury. Also, inorganic mercury levels in urine appear to correlate with levels in the kidney, the target for inorganic mercury. Nevertheless, there are promising possibilities in this area as we begin to identify specific molecular species of mercury in blood and other body fluids responsible for transport into tissues.

Markers of effect and susceptiblity have so far been directed towards the consequences of mercury-induced reactive oxygen species in the cell, which in turn induce expression of cellular oxidant defence mechanisms. The ability of mercury to induce metallothionein has been known for several decades. More recent work demonstrates that methylmercury and mercury vapor can induce metallothioneins I and II in brain tissue. Unfortunately, metallothionein III, which is uniquely located in neuronal cells, is noninducible. Gamma glutamyl cysteine synthetase (GCS), the rate determining enzyme in glutathione synthesis, is induced in kidney tissues by inorganic mercury and by methylmercury in brain tissue. Both of these inductions of gene expression are the outcome of oxidative stress; thus, they are not specific to the actions of mercury but can be caused by other toxic metals as well as numerous oxidizing organic toxicants such as quinone.

Interesting future possibilities might include those proteins involved in the membrane transport of mercury. For example, the large neutral amino acid transporters (LAT) carry methylmercury as its cysteine complex into the cells including the endothelial cells of the blood–brain barrier. Some of these transporters have recently been cloned, opening up possibilities for their measurement in individual cells. LATs form a super family of glycoprotein associated amino acid transporters, allelic variants of which might play a role in susceptibility of the central nervous system to methylmercury. Methyl and inorganic mercury can be transported out of cells on carriers responsible for glutathione transport. Specifically, the secretion of mercury into bile and its ultimate fecal elimination involves glutathione carriers in the canalicular membrane of liver cells. These proteins have a potential as determinants of susceptibility.

I. Introduction

Strictly speaking, the term biomarkers should include any molecule that is a marker for dose, effect, or susceptibility. However, current usage restricts the term to changes in macromolecules (e.g., adduct formation), the degree of gene expression, or the role of polymorphisms.

The underlying chemistry of mercury differs from that of alkylating carcinogens or other toxic organic chemicals that are metabolized to highly reactive intermediates to form stable covalent adducts with macromolecules. Mercury ions bind reversibly to thiol groups present in proteins and smaller molecules.[1] Thus, it is difficult to isolate and identify mercury adducts as mercury will most likely dissociate from its original attachment during the isolation procedures. Consequently, most work on molecular biomarkers has focused on induction of changes in gene expression or on changes in specific proteins resulting from the biochemical action of mercury.

Table 15.1 The Chemical and Physical Species of Mercury

Inorganic Mercury		
Hg^0	$Hg\text{-}Hg^{++}$	Hg^{++}
Metallic	Mercurous	Mercuric

Organic Mercury	
$C_6H_5\text{-}Hg^+$	$CH_3\text{-}Hg^+$
Phenyl mercury	Methylmercury

The chemical and physical species of mercury are usually divided into two major groupings: inorganic and organic (Table 15.1).

Organic mercury compounds are those in which the mercury atom is covalently linked to at least one carbon atom. Two examples, phenyl and methylmercury, are given in the table. These are monovalent cations that in the body bind primarily to SH ligands on proteins and amino acids. Methylmercury is the generic term that will be used to cover all compounds formed by the methylmercury cation. Methylmercury is the form of mercury to which humans are exposed from consumption of fresh and ocean water fish and other types of seafood.

The inorganic forms of mercury fall into three groups according to their oxidation states. Metallic mercury in the zero or ground oxidation state is the well-known silvery liquid metal often called quicksilver. It releases a monatomic gas usually referred to as mercury vapor with the chemical symbol Hg^0. The inhalation of mercury vapor is a major route of human exposure in the occupational setting and from dental amalgam fillings.

The loss of one electron from the mercury atom results in mercurous compounds formed from mercurous mercury with the chemical symbol Hg_2^{++}. Two atoms of mercury are linked together to form the mercurous ion. The best known compound is mercurous chloride (calomel). This compound had wide medical application up to the middle of the 20th century. Added to teething powders, it was responsible for the childhood disease of Acrodynia, which has almost disappeared.

The loss of two electrons produces the mercuric ion (Hg^{++})that forms the majority of the inorganic compounds of mercury. Mercury in this oxidation state also forms the organic species of mercury discussed above. Compounds of inorganic mercury are still widely used, for example, in mercury batteries. Poisoning either occupational or accidental from inorganic mercury compounds have become rare.

This review will focus on the two principal target organs for the toxic action of mercury and its compounds, namely the kidney and the brain.

II. Kidney

The kidney is the common target for all the inorganic species and for those organic species that are rapidly metabolized to inorganic mercury such as phenyl mercury and the organo-mercurial diuretics.

A. Biomarkers of dose

The kidneys are the main site of accumulation of inorganic mercury in the body. The mechanisms are not completely understood but probably involve the transport of small molecular weight thiol complexes of the inorganic cation.[2] Isotopic studies in humans reveal that the specific activity of mercury in urine is the same as that in kidney tissues and different from that in blood.[3] These data suggest that inorganic mercury in urine comes directly from kidney tissue. If any glomerular filtration takes place, the filtered mercury is first taken up from the tubular fluid into kidney tissues before release into the urine. Thus urinary mercury is an excellent marker for kidney inorganic mercury levels.

Since urine volume undergoes extensive diurnal variation, it is usual to apply the creatinine correction to a measurement of mercury concentrations in urine.[4] The rate of excretion of creatinine in urine is proportional to the lean body mass. In a 70-kg "standard" adult the rate is about 1.6 g/day. Thus, dividing the mercury level in urine (μg/l) by the creatinine concentration (g/l), one obtains an estimate of the urinary excretion rate of mercury (μg Hg/g creatinine) that is independent of urine volume.

Urinary measurement of mercury corrected for creatinine has a long history of application for occupational exposures to mercury vapor and to inorganic aerosols of mercury. They are predictive of effects on the kidney as discussed below. Urinary mercury also correlates with time-weighted exposure to mercury in the breathing zone of the worker.[5] This correlation probably arises because the kidneys are the main depository of inorganic mercury and because kidney levels probably correlate with levels in other tissues after long-term exposures.

B. Biomarkers of effects/susceptibility

A number of markers for kidney damage[6] are now widely used in monitoring workers exposed to inorganic mercury (Table 15.2). Increased excretion of albumin in urine is used as a marker of damage to the filtration mechanism of the glomerular membrane. Damage to the membrane is believed to be mediated by an autoimmue mechanism triggered by inorganic mercury.

The other urinary markers indicate varying degrees of damage to the cells lining the renal tubule, especially the proximal tubules. The appearance

Table 15.2 Molecular Biomarkers of Kidney Function

Target tissue	Marker in urine
Glomerulus	Albumin
Tubular cells	Small proteins: retinol binding
	Cell substrates: porphyrins
	Enzymes: NAG, GGT
	Cell antigens: carbonic anhydrase

Source: Adapted from Roels, H. et al., *Ann. Occup. Hyg.*, 31, 135–145, 1987.

of small molecular weight proteins, such as retinol binding protein, is one example. The failure to reabsorb these proteins completely is a measure of functional damage to the proximal tubular cells. The other urinary markers are measures of damage to the epithelial cells lining the tubules. These markers include enzymes such as NAG (n-acetyl glucosamidase) and GGT (gamma glutamyl transpeptidase). The latter appears to have detected effects at some of the lowest recorded mercury levels.[4] Cellular antigens also appear in urine in elevated amounts. Metabolic changes within kidney cells are indicated by changes in patterns of urinary excretion of porphyrins.[7]

These markers of effects, appearing as they do as molecular species in the urine, are not the type of molecular markers that indicate changes in gene expression. Such markers of effect and susceptibility have so far been directed toward the consequences of mercury-induced reactive oxygen species in the cell, which in turn induce gene expression of the cellular oxidant defense mechanisms.

The ability of mercury to induce metallothionein has been known for several decades.[8] Since this indicator of changes in gene expression is the outcome of oxidative stress, it is not specific to the actions of mercury but can be caused by other toxic metals as well as numerous oxidizing organic toxicants.

Nevertheless, there are promising possibilities in this area as we begin to identify specific molecular species of mercury in blood and other body fluids responsible for transport into tissues. For example, there is evidence that inorganic mercury as the GSH complex enters proximal tubular fluid via filtration of plasma by the glomerulus.[2] Two enzymes located on the surface of proximal tubular cells break down the GSH tripeptide to a dipeptide and then to the three individual amino acids (Figure 15.1). Inorganic mercury attached to cysteine is reabsorbed into the proximal tubular cells on an amino acid carrier. Thus, at least three proteins are involved in the transport of inorganic mercury into kidney cells, such proteins working in a coordinated, directed fashion.

Carrier proteins transporting inorganic mercury from kidney cells to urine are not well characterized. However, studies on other cells indicate that inorganic mercury is transported out of the cell as the glutathione complex on proteins responsible for the membrane transport of glutathione or its conjugates.[9]

In general, mercury appears to enter cells on amino acid or anion carriers and to exit cells on GSH carriers. Thus, the bodily disposition to all tissues is under protein control and, therefore, gene expression. Changes in the expression of these proteins might provide a measure of effects or susceptibility as future studies expand in this direction.

III. Brain

The brain is an important target organ for inhaled mercury vapor and the most important target organ for the short chain alkyl mercurials, especially methylmercury.

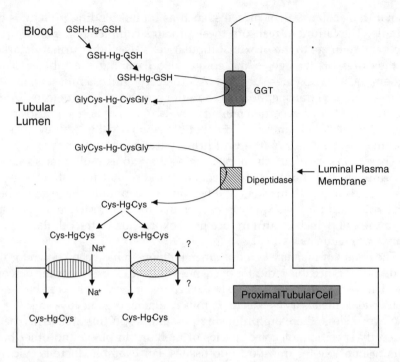

Figure 15.1 A putative mechanism of uptake of inorganic mercury from renal tubular fluid. (Modified from Zalups, R.K., in *Molecular Biology and Toxicology of Metals*, Taylor and Francis, New York, 2000, pp. 234–273.)

A.   Inhaled mercury vapor

1.   Biomarkers of dose

Markers of dose to the brain present a special problem in the case of inhaled mercury vapor. The vapor is a monatomic gas that is highly diffusible and lipid soluble. It can penetrate cell membranes with ease. It diffuses from the alveolar regions in the lung to the blood stream from which it enters all cells and tissues. It passes across the blood–brain barrier (Figure 15.2). Once inside the cell, it is subject to rapid oxidation to mercuric mercury (Hg^{++}) which is believed to be the proximate toxic species.

Ideally, mercury vapor dissolved in the blood stream might be a useful indicator of the amount of vapor entering the brain. However, levels of dissolved vapor in plasma are below the detection limit for current methods of analysis. As the inhaled vapor distributes to tissues by passive diffusion, no protein carriers are involved. Thus the measurement of dose to the brain remains elusive at the present time.

2.   Biomarkers of effect and susceptibility

The mechanisms of action on inhaled mercury vapor on brain tissue is not well understood. The vapor itself is probably nontoxic. Its oxidation product

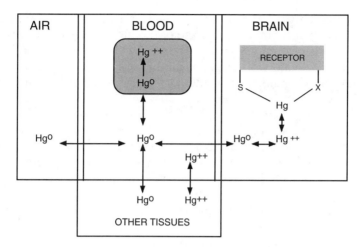

Figure 15.2 A schematic representation of the absorption, distribution, and metabolism of inhaled mercury vapor. For details, see text.

$$\overset{III}{Cat.Fe–OH} + H_2O_2 = \overset{V}{Cat.Fe–OOH} + H_2O$$

$$\overset{V}{Cat.Fe–OOH} + Hg° = \overset{III}{Cat.Fe–OH} + \overset{II}{Hg} + O$$

Figure 15.3 The oxidation of dissolved mercury vapor by the catalase-hydrogen peroxide pathway

(Hg^{++}) is believed to be responsible for the signs and symptoms indicative of disturbances in the central nervous system. Many of these signs and symptoms are psychological in nature, such as changes in mood, excessive shyness, and depression. Inhaled mercury vapor, in general, does not elicit observable morphological effects on brain tissue. Thus, we are not able to identify protein targets.

The only protein known to be involved in the metabolism of inhaled vapor is the enzyme catalase. The oxidation pathway is a two-step reaction first involving the addition of hydrogen peroxide to form catalase complex I. The latter then oxidizes one atom of mercury by extracting two electrons to form mercuric mercury (Figure 15.3).

It is possible that individual differences in the expression of catalase or in the endogenous production of hydrogen peroxide might account for differences in susceptibility to brain damage from inhaled mercury. To date, no studies have been reported on this topic.

B.   *Methylmercury*

1.   *Biomarkers of dose*

Mercury levels in hair and blood samples have served as excellent biomarkers of exposure and have been shown to correlate with levels in the brain, the target organ for methylmercury. This is especially important with regard to the fetal brain as this phase of the life cycle is most sensitive to the toxic effects of methylmercury. Thus, it has been shown that hair and blood levels correlate well between mother and her newly born infant.[10] Most importantly, maternal hair mercury levels parallel those in the new-born infant brain.

Scalp hair has been the sample of choice for the biological monitoring of methylmercury. Not only does it predict brain levels, but depending on the length of the sample, it is able to recapitulate past levels. Hair grows about one centimeter per month. Thus, each centimeter measured from the scalp end will recapitulate past blood and brain levels month by month. It is especially useful for the recapitulation of methylmercury levels during pregnancy. A special technique has been devised[11] that allows measurement of mercury in single strands of hair over intervals corresponding to 3 days of growth. Such detailed recapitulation is desirable, if not essential, if peak mercury levels are to be determined during pregnancy.[12]

2. Biomarkers for effects and susceptibility

A key feature for biomarkers in the brain is that brain neurons generally do not turn over. This raises some unique toxicological aspects, namely that damage can accumulate over long periods of time. This is illustrated by the case of a single exposure to a methylmercury compound where signs of brain damage did not appear until 5 months after exposure.[13] Thereafter, the onset of damage was rapid with the full syndrome of signs and symptoms developing in approximately 2 weeks. Thus, damage must have accumulated in the brain over a 5-month period. The nature of this cumulative damage remains a mystery. There is a need for a molecular marker for early brain effects.

To date, only nonspecific molecular markers have been studied in animals. GSH synthase was affected at subchronic doses of MeHg in mice.[14] Methylmercury can induce metallothioneins in astrocytes.[15] Unfortunately, metallothionein III, which is uniquely found in brain neurons is not inducible.

IV. Future possibilities

Interesting future possibilities might include those proteins involved in the membrane transport of mercury. For example, LATs are believed to carry methylmercury as its cysteine complex into cells, including the endothelial cells of the blood–brain barrier.[16] Some of these transporters have recently been cloned opening up possibilities for their measurement in cells. Allelic variants within the glycoprotein-associated amino acid transporters might play a role in susceptibility of the central nervous system to methylmercury. Methyl and inorganic mercury can be transported out of cells on carriers responsible for glutathione transport. Specifically, the secretion of mercury into bile and its ultimate fecal elimination involves glutathione carriers in the canalicular membrane of liver cells. These proteins have a potential as determinants of susceptibility.

In general, some 9000 genes have nucleotide sequences suggestive of coding for transport proteins.[9] Two thousand of these may be for transport of foreign compounds. The bodily disposition of many other metals is controlled by protein carriers. Genetic errors in one such carrier (ABC7B) are known to cause Wilson's disease, which affects copper disposition in the body.

Besides the transport proteins, enzymes responsible for processing small molecular weight thiols offer potential for assays of susceptibility. As discussed above, two enzymes are involved in the conversion of the glutathione complexes to the transport species, namely, mercury cysteine. Enzymes controlling the plasma levels of cysteine and homocysteine might play a role in susceptibility to brain damage from methylmercury.

The area of molecular markers for effects and susceptibility to mercury and its compounds is still much in its infancy with exciting prospects. Studies conducted mainly in the past decade have indicated the important role of proteins in the transport and metabolism of mercury thiol complexes. From studies in this area, one would expect the new molecular markers to appear.

References

1. Passow, H., Rothstein, A., and Clarkson, T.W., The general pharmacology of the heavy metals, *Pharmacol. Rev.*, 13, 185–224, 1961.
2. Zalups, R.K., Mercury: molecular interactions and mimicry in the kidney, in *Molecular Biology and Toxicology of Metals*, Zalups, R.K. and Koropatnick, J., Eds., Taylor and Francis, New York, 2000, pp. 234–273.
3. Hursh, J.B. et al., Prediction of kidney mercury content by isotope techniques, *Kidney Int.*, 27, 898–907, 1985.
4. Gotelli, A. et al., Early biochemical effects of an organo-mercury fungicide on infants: Dose makes the poison, *Science*, 227, 638–640, 1985.
5. Roels, H. et al., Relationship between the concentrations of mercury in air and in blood or urine in workers exposed to mercury vapour, *Ann. Occup. Hyg.*, 31, 135–145, 1987.
6. Bernard, A. and Lauwerys, R., Epidemiological application of early markers of nephrotoxicity, *Toxicology Letters*, 46, 293–306, 1989.
7. Woods, J.S., Altered porphyrin metabolism as a biomarker of mercury exposure and toxicity, *Can. J. Physiol. Pharmacol.*, 74, 210–215, 1996.
8. Cherian, M.G. and Clarkson, T.W., Biochemical changes in rat kidney on exposure to elemental mercury vapor: Effect of biosynthesis on metallothionein, *Chem. Biol. Interact.*, 12, 109–120, 1976.
9. Ballatori, N., Molecular mechanisms of hepatic metal transport, in *Molecular Biology and Toxicology of Metals*, Zalups, R.K. and Koropatnick, J., Eds., Taylor and Francis, New York, 2000, pp. 336–381.
10. Cernichiari, E. et al., Monitoring methylmercury during pregnancy: Maternal hair predicts fetal brain exposure, *Neurotoxicology*, 16, 705–710, 1995.
11. French, W.R. and Toribara, T.Y., X-ray fluorescence with special application to scanning single hair strands, *Crime Lab. Digest*, 13(1), 4–14, 1986.

12. Cox, C. et al., Dose-response analysis of infants prenatally exposed to methylmercury: An application of a single compartment model to single-strand hair analysis, *Environ. Res.*, 49, 318–332, 1989.
13. Nierenberg, D.W. et al., Delayed cerebellar disease and death after accidental exposure to dimethylmercury, *New Eng. J. Med.*, 338(2), 1672–1675, 1998.
14. Thompson, S.A. et al., Induction of glutamate-cysteine ligase (gamma-glutamylcysteine synthetase) in the brains of adult female mice subchronically exposed to methylmercury, *Toxicol. Lett.*, 110(1–2), 1–9, 1999.
15. Aschner, M., Methylmercury in astrocytes: What possible significance? *Neurotoxicology*, 17(1), 95–106, 1996.
16. Wang, W., Clarkson, T.W., and Ballatori, N., Gamma-glutamyl transpeptidase and L-cysteine regulate methylmercury uptake by HepG2 cells: A human hepatoma cell line, *Toxicol. Appl. Pharmacol.*, 168, 72–78, 2000.

chapter sixteen

Validity of mercury exposure biomarkers

Philippe Grandjean, Poul J. Jørgensen, and Pál Weihe

Contents

Abstract The mercury concentration in scalp hair has been used as a routine biomarker for methylmercury exposure. Although laboratory imprecision of the analysis is low, other factors, such as hair color, may potentially affect the amount of mercury incorporated in the hair. Hair strands from four gray-haired subjects were separated into white hairs and pigmented hair before mercury analysis; pigmented hair contained significantly more mercury than the white hair. To determine the degree of temporal variation of mercury exposure in a population relying on seafood, 15 hair samples with a mercury concentration above 10 µg/g were selected from a cohort study in the Faroe Islands. Most samples showed a coefficient of variation of 10–20% for mercury concentrations in individually analyzed 1.5-cm segments. From this cohort, the full-length hair (mostly 9 cm) obtained at parturition had been analyzed for mercury as a marker of exposure during the pregnancy as a whole. From an additional 683 maternal samples, we then analyzed the proximal 2-cm segment as a biomarker reflecting the exposure during the later periods of the pregnancy. These two hair biomarkers correlated well, and the latter correlated the best with the cord-blood concentration. To determine the diagnostic validity of these exposure biomarkers, their association with neuropsychological test results obtained by the children at age 7 years was calculated. Apart from motor function, which showed a stronger association with the full-length hair mercury concentration, the cord-blood level was found to be the best risk indicator. Thus, while prenatal exposure to methylmercury may be determined from analysis of either maternal hair or cord blood, the latter parameter indicates the amount that has reached the fetal circulation and shows a better association with mercury-related neurobehavioral deficits. The cord-blood mercury level should, therefore, be preferred as exposure biomarker in future studies.

I. Application of exposure biomarkers in the Arctic

Several contaminants biomagnify in marine food chains and result in high concentrations, especially in marine mammals.[1] Exposure to methylmercury and other toxicants is therefore increased in consumers of marine food, especially if the diet includes whale or seal. Thus, increased mercury levels have been documented in Arctic populations[1,2] and in the Faroese who have traditionally eaten pilot whale.[3]

Individual levels of exposure are important in epidemiologic research as well as in health surveillance and medical counseling.[4] However, contaminant intakes may be difficult to estimate from dietary questionnaires because of the substantial variability of the concentrations in food.[1] For this reason, biomarkers of exposure may be important instruments.[4] Formally, a biomarker reflects an event or a sequence of events which occur somewhere in the causal chain between an exposure to a hazardous factor and a related adverse effect.[5] Within the body, an exposure biomarker may be a xenobiotic compound (or a metabolite), an interactive product between the compound (or metabolite)and an endogenous component, or another event related to the exposure.[5] Most commonly, biomarkers of exposures to stable compounds, such as methylmercury, encompass measurements of the elemental concentrations in appropriate samples, such as blood, serum, hair, or urine.

The ideal exposure biomarker should satisfy the following five needs[5]:

1. Sample collection and analysis are simple and reliable.
2. The biomarker is specific for a particular type of exposure with a clearcut relationship to the degree of exposure.
3. The biomarker reflects a subclinical and reversible change only.
4. Relevant intervention or other preventive effort can be considered, if indicated by the biomarker result.
5. Use of the biomarker is regarded as ethically acceptable.

Currently applied exposure biomarkers for mercury particularly need to be evaluated with regard to their validity, i.e., to which degree they reflect risks of adverse health effects.

II. Hair-mercury as exposure biomarker

Methylmercury has a high affinity for sulfur and sulfur-containing amino acids; it is incorporated in hair at relatively high concentrations.[6,7] Sampling of hair is noninvasive and painless; it does not necessitate technical assistance, and storage and transportation also pose advantages over other biological samples.

Human volunteer studies suggest that mercury is incorporated in hair in proportion to the bioavailable concentration in blood.[8] Field studies have confirmed that this parameter is highly feasible and can be applied to assess human exposure levels, e.g., as elegantly demonstrated in the Iraqi poisoning incident.[9] Maternal hair-mercury concentrations during pregnancy have also been utilized as a convenient indicator of fetal exposures from maternal seafood intake.[10,11] Although the mercury concentration may be affected by various factors (see below), the ease of hair sampling and storage has made this parameter highly useful. In addition, the distance of the hair sample from the root will determine the time period reflected by the analysis. This feature will allow assessment of exposures retrospectively, dependent on the length of the hair. For example, in a prospective study in the Seychelles, hair samples were obtained 6 months after parturition, and the hair segments representing the pregnancy period were then analyzed for mercury.[11]

III. Sources of variation

Because of the relatively high mercury concentration in hair, analysis of samples of 5 mg or more can be carried out by atomic absorption methods.[11] For mineralization of the sample, we used microwave digestion in PTFE-lined vessels[12] and mercury analysis by flow-injection cold-vapor atomic absorption spectrometry.[13] The total analytical imprecision was estimated to be 2.3 and 3.2% at mercury concentrations 4.7 and 11.4 µg/g respectively. The accuracy of the mercury determination in human hair was ensured by using the certified reference material CRM 397 (BCR, Brussels, Belgium) as quality control material; the mercury concentration averaged 11.36 µg/g compared to the assigned value of 11.93 ± 0.77 µg/g. Participation in external quality assessment schemes has also rendered satisfactory results. Comparable quality has been obtained using a similar method in another laboratory.[3,11]

In addition to the analytical sources of imprecision, preanalytical factors may also be involved.[6,14] Thus, because hair is exposed to the external environment for several weeks or months before collection of the specimen, contamination from external mercury vapor or hair treatment products may add to the mercury concentration originally incorporated in the hair root. The significance of this factor may be reduced or eliminated by determining the organic mercury component only, but speciation analysis is more cumbersome.

Other sources of variation include hair structure, hair color, weight, and growth rate, which vary considerably between and even within individuals.[6,7] A single strand of scalp hair weighs an average of about 40 µg/cm, but the weight may vary considerably.[15] Thus, a 100-mg hair sample would include an average of about 1,000 2.5-cm long strands, but the number may as low as 500 or as high as 2,000. This factor is of potential importance because the hair mercury concentration is generally expressed in relation to the weight of the sample, not the number of hair roots. Likewise, the hair growth rate is known to vary substantially,[7] and a hair length of 2.5 cm may well correspond to a time period of 10 weeks, possibly as little as 1 month, or as much as 3 months.

The significance of hair color has not been explored in detail and may be difficult to study under properly controlled circumstances. A useful opportunity is to examine gray-haired individuals because they have two different types of hair, i.e., white hair and pigmented hair. All hair roots are exposed to the same environment, internally and externally, and comparison of mercury concentrations would be likely to reveal possible differences in mercury binding.

We obtained hair samples from Faroese whaling foremen who would be expected to have an increased exposure to methylmercury. Four subjects were gray-haired, and the hairs from each subject were then sorted in white hairs and pigmented hairs before analysis. The overall average mercury concentration was 14.2 µg/g for the white hair and 15.3 µg/g for the dark hairs (Table 16.1). One sample clearly exceeded the normal imprecision of the analysis.

While this limited observation is only preliminary, it is in accordance with previous analyses of lead concentrations in hair from gray-haired subjects.[15] Hair color should, therefore, be considered in future studies as a factor that may influence the concentration of mercury incorporated in hair strands. The importance of other factors, such as variations in hair growth rate and hair structure is unknown but may contribute to the variability observed in correlations between blood and hair concentrations (see below).

Table 16.1 Average Mercury
Concentration (µg/g) in White and
Pigmented Hair Strands from Four
Subjects with Gray Hair

Subject	White	Pigmented	Ratio
1	15.1	13.5	1.12
2	27.0	29.6	0.91
3	4.3	4.7	0.91
4	9.1	14.3	0.63

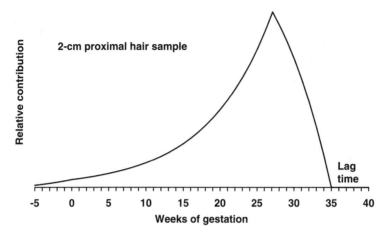

Figure 16.1 Relative contribution of dietary methylmercury intake during pregnancy to the mercury concentration in the proximal 2-cm hair segment obtained at term, assuming a biological half-life of 45 days for methylmercury.

IV. Temporal variation along hair strands

Assuming an average growth rate of about 1 cm per month,[6,7] past methylmercury intake levels can be estimated from mercury concentrations along the hair shaft. The distance from the root indicates the time, but this determination is likely to be less accurate the longer the distance from the root. In addition, a lag time has to be taken into account as the interval between incorporation into the hair root and the appearance in the hair shaft available for sampling. Assuming a biological half-life of 45 days for methylmercury in the body,[16] Figure 16.1 illustrates the time interval represented by a 2-cm hair sample.

When the Faroese birth cohort was established in 1986–1987, maternal hair was cut close to the root at the time of parturition. The hair length varied from about 3 cm to 30 cm. The original analyses involved the proximal 9 cm, when available; most samples were about 9 cm long or more, and only 11% were less than 6 cm (Elsa Cernichiari, pers. comm.). The results previously published,[3] therefore, reflect the average methylmercury exposure during the full duration of the pregnancy.

Because whale meat is the main source of mercury in this population, variations in dietary intake, e.g., related to local availability, could lead to changes in exposure levels. Although a whale kill may result in increased consumption, most whale meat is salted, cured, or frozen for later consumption, whether as a snack or as part of a dinner meal. Further, dietary changes associated with the pregnancy may play a role.

To assess the relative importance of such temporal variability, six maternal hair samples were initially analyzed in 1.1-cm segments of the hair, each segment therefore corresponding to a 4–5 week period. The average coeffi-

Table 16.2 Variation of Mercury Concentrations (μg/g) of 1.5 cm
Segments of 15 Samples of Long Hair Strands with a Known
High Mercury Concentration

Sample No.	No. of Segments	Mean ± s.d.	Coefficient of Variation	Range
1	20	7.29 ± 2.49	34.2	4.1–12.8
2	8	9.22 ± 2.06	22.4	6.4–11.9
3	8	9.37 ± 0.77	8.2	8.4–10.5
4	7	9.60 ± 5.98	62.2	4.2–21.8
5	9	10.37 ± 0.62	6.0	9.3–11.3
6	10	10.50 ± 1.57	14.9	8.5–13.3
7	6	10.51 ± 1.02	9.7	9.2–11.9
8	7	10.60 ± 3.94	37.2	3.3–13.9
9	6	10.87 ± 1.85	17.0	8.5–13.1
10	5	11.38 ± 2.05	18.0	8.5–14.0
11	5	11.75 ± 2.43	20.7	8.5–14.5
12	7	13.31 ± 2.68	20.1	9.1–15.8
13	9	13.59 ± 3.20	23.6	7.5–17.0
14	7	17.38 ± 2.42	13.9	14.5–20.8
15	11	34.02 ± 6.43	18.9	24.0–49.1

cients of variation for 5–18 segments from each woman varied between 8.1% and 23.8%.[3]

In further exploring this variability, additional maternal hair samples were selected for segmental analyses if the mercury concentration of the proximal 9 cm was above 10 μg/g. Given the approximate half-life of 45 days and a hair growth rate of about 1 cm per 30 days, the segment length was chosen to be 1.5 cm, thus corresponding to approximately one half-life. Three of the samples identified were excluded because they contained too few hair strands or did not allow judgment concerning the proximate end of the hair strands. A total of 15 samples were available for segmental analysis.

The results in Table 16.2 show that the variability within three of the 15 hair samples was below 10% and, therefore, not appreciably higher than the analytical imprecision. Most samples had a coefficient of variation between 10 and 20% in accordance with the samples previously analyzed.[3] However, three samples showed a greater variability, and these results for the first six segments are shown in Figure 16.2.

The variations in excess of the analytical imprecision are likely to be due to changes in diet during pregnancy. Records on local whale catches confirm that whale meat was widely available at the time indicated by the hair segments with the highest mercury concentration. Such variations in exposure may be important if the peak concentration determines the risk of toxic effects. However, within this population, exposure levels seem to vary little in most subjects.

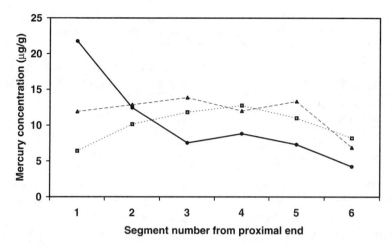

Figure 16.2 Mercury concentration in 1.5-cm hair segments counted from the proximal tip of three hair samples with a large coefficient of variation.

V. Association with other biomarkers

In validating the hair-mercury concentration as an exposure biomarker, its association with other independent biomarkers should be considered. Methylmercury may also be determined in whole blood because most of the methylmercury present in blood is bound to the red blood cells.[6] A blood sample mainly represents third trimester exposure as illustrated in Figure 16.3. Transplacental passage results in fetal blood concentrations that eventually exceed maternal levels.[2,6] The cord blood concentration, therefore, mainly reflects the fetal exposure from maternal diet during the third trimester.

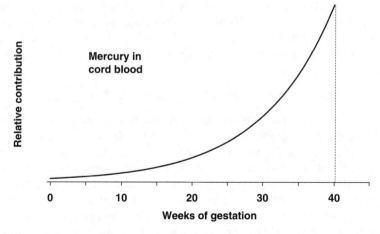

Figure 16.3 Relative contribution of dietary methylmercury intake during pregnancy to the mercury concentration in cord blood (see also Figure 16.1).

Table 16.3 Results of Methylmercury Exposure Biomarker Determinations in 683 Faroese Children with Complete Data on Maternal Hair Concentrations

Mercury Concentration	Geometric Mean	Interquartile Range	Total Range	Correlation[a] with Cord Blood	Correlation[a] with Maternal Hair
Cord blood (µg/l)	22.5	13.0–41.3	0.9–35.1	(1.00)	0.79
Maternal hair (µg/g)	4.26	2.52–7.83	0.3–39.1	0.79	(1.00)
First 2 cm of hair (µg/g)	4.46	2.76–7.76	0.3–40.5	0.84	0.93

[a] After logarithmic transformation of the data.

In the Faroese cohort study, hair for reanalysis was available from 683 mothers of the 1022 cohort members (67%). The proximal 2-cm segment was analyzed as an indication of third trimester exposure.[17] For these cohort members, three biomarkers are now available, i.e., mercury concentrations in cord blood and in 2-cm and full-length maternal hair (Table 16.3).

The two hair mercury measures correlated well. The ranges were similar, and the average ratio between the long-strand and proximal hair concentrations was 1.00 (s.d., 0.35). This finding is remarkable, as the analyses were carried out by two different laboratories with an interval of 10 years. While most paired sample results differed only slightly, 23 samples (3.4%) showed more than a twofold difference. The segmentally analyzed sample with the largest coefficient of variation (Table 16.2) also fell into this category.

Because of possible changes in mercury intake during pregnancy and the estimated half-life of methylmercury of about 45 days,[16] the cord-blood concentration would be expected to correlate better with the mercury concentration recently incorporated in the hair closest to the root. Indeed, the correlation coefficients in Table 16.3 reflect this expectation. The association between the mercury concentrations in cord blood and in the proximal maternal hair is illustrated in Figure 16.4. The average ratio between the hair concentration (in µg/g) and the blood concentration (in µg/ml) was approximately 220 for both hair parameters, but the standard deviation for the proximal hair was smaller (114) than for the long-strand samples (133).

VI. Association with response variables

The exposure biomarkers in current use do not necessarily indicate the amount of the toxic species that has reached the target organ, i.e., the fetal brain. As part of a validation of a biomarker, its usefulness as a predictor of relevant neurobehavioral outcomes must, therefore, be assessed. The Faroese cohort study provides an opportunity to compare the performance of the three exposure biomarkers described above. Although the results discussed above suggest that dietary exposure to methylmercury in most cases did not change substantially during pregnancy, small differences in biomarker uncertainty could affect the predictive validity of the exposure biomarkers.

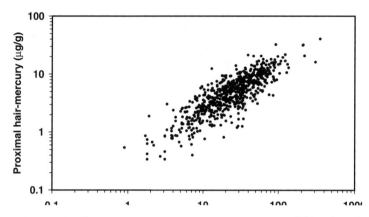

Figure 16.4 Association between mercury concentrations in cord blood and in the proximal 2-cm segment of maternal hair obtained at parturition (r = 0.84).

A total of 917 cohort children were examined by neurobehavioral methods at age 7 years, and mercury-associated deficits were observed in several domains of brain function.[13] A 2-cm maternal hair sample was available for 608 (66%) of these children. The validity of the three biomarkers of prenatal exposure was explored by calculating their association with major outcome variables. Within each domain of brain function, the neuropsychological test outcome that best correlated with the maternal hair mercury concentration[18] was selected. Two of the tests were computer assisted: the Neurobehavioral Evaluation System (NES2) Finger Tapping both hands condition and the Continuous Performance Test reaction time. Also selected were the Boston Naming Test score after cues and the California Verbal Learning Test — Children number correct at short delay. These tests reflect motor speed, attention, language, and verbal memory, respectively. Visuospatial function was not considered, as this parameter appeared to be affected by postnatal exposure levels.[18]

Confounder-adjusted regression coefficients were calculated exactly as before[13,18] within the restricted group of children who had a 2-cm maternal hair result available (Table 16.4). The regression coefficients for cord blood and full-length maternal hair are similar to those previously published for the cohort as a whole.[13,18]

All regression coefficients reflect decreased performance at higher mercury concentrations. As expected from the results for the complete cohort,[13,18] the cord blood was the best predictor of the outcome variables, except for the finger tapping score. The proximal hair concentration showed lower regression coefficients than cord blood in all four instances, but two of the four were higher than those for the long-strand samples.

Table 16.4 Change (Expressed as Percent of SD) in Neuropsychological Test Performance of 608 Faroese Children at Age 7 Years Associated with a Doubling of Three Biomarkers of Prenatal Methylmercury Exposure

Test	Cord Blood	Maternal Hair	First 2 cm of Hair
NES* Finger Tapping			
Both hands	−6.07	−6.94	−5.85
	(0.081)†	(0.057)	(0.120)
NES Continuous Performance Test			
Reaction time	17.70	11.89	13.61
	(<0.001)	(0.007)	(0.003)
Boston Naming Test			
With cues	−9.04	−5.56	−7.19
	(0.006)	(0.110)	(0.045)
California Verbal Learning Test (Children)			
Immediate recall	−7.41	−6.30	−5.44
	(0.037)	(0.089)	(0.153)

* Neurobehavioral Evaluation System.

† Numbers in parentheses, p values.

These results must be interpreted in light of current knowledge concerning vulnerable time periods during fetal development.[7,13,19] Most neuropsychological functions are probably particularly susceptible to neurotoxic exposures toward the end of the gestation period, but it is probable that the development of motor functions may result in vulnerability earlier in gestation. The better correlation of the full-length hair sample with motor speed deficits would be in accordance with this notion. However, overall, the cord-blood concentration seems to be the best risk indicator. Use of the proximal 2-cm segment provides a measure that approaches the predictive value of cord blood, but it correlates less well with motor speed than does the full-length hair concentration.

VII. Validity of mercury exposure biomarkers

Validation of biomarkers must include considerations of preanalytical, analytical, and diagnostic validity.[3,4] The present study has aimed at illuminating some key aspects of these issues.

The preanalytic sources of variability mainly relate to toxicokinetic factors. Thus, the degree to which the biomarker indicates retention in specific body compartments must be considered. While peripheral blood is generally not regarded as a compartment as such, it acts as a transport medium between compartments, and cord blood reflects the availability in the fetal circulation.[11] Thus, toxicokinetic differences must be considered regarding the mercury fractions incorporated into maternal hair roots and the one reaching the fetal circulation. In addition, sample contamination may be an important issue, especially with hair specimens.[14]

The timing of specimen sampling is crucial, especially in studies of prenatal toxicity.[6,11] This issue regards the extent to which the biomarker measurement reflects current or past exposures and accumulated body burdens. In regard to hair samples, it may take 4–6 weeks for metal concentrations to show up in a hair sample cut close to the scalp.[6,11,15] This lag time approaches the half-life of methylmercury in the body. The mercury concentration in the first 2 cm of hair taken at parturition, therefore, mainly reflects exposures in the late second and early third trimester (Figure 16.1).

These considerations are particularly relevant if exposures are not constant. Pregnancy itself may result in changes in diet. In the Faroe Islands, the highest methylmercury concentrations occurs in a single type of food, i.e., pilot whale meat, and some variability would be expected depending on availability and other factors. However, the persistence of methylmercury for several weeks in the body would tend to attenuate the effect of such variations.[7,20] Also, a whale catch in the Faroes provides a supplement to the diet over a substantial time period. Nonetheless, three mothers with a high mercury exposure showed substantial variation in hair-mercury concentrations in clear excess of the analytical imprecision (Table 16.3).

Other factors that may affect hair-mercury concentrations include hair color. The variability seen in the limited study of four gray-haired subjects showed differences that tended to be greater than the analytical imprecision. On the other hand, the differences seemed much lower than the inter-individual variation within the Faroese population, which spanned two orders of magnitude. While such issues may be of minor importance for monitoring studies within a population group, caution must be exerted when comparing data on populations with different hair structure, e.g., Inuit and African populations.

These preanalytical and analytical uncertainties are likely to affect the association between mercury concentrations in cord blood and maternal hair. The results of the present study show less scatter when the proximal segment is used, and this improved correlation is probably due to the proximity of the time periods that the samples reflect. The same conclusion was reached when comparing the variation of the ratio of the two hair concentrations in relation to that of cord blood. Remaining scatter in Figure 16.4 is most likely affected by imprecision of both parameters.

When considering developmental neurotoxicity as the critical issue in risk assessment, assessment of prenatal exposure is essential. Methylmercury is known to pass the placental barrier rather easily, while this is not the case of inorganic mercury. In fish-eating populations, almost all mercury passing into the fetal circulation is thought to be methylmercury. This methylmercury would also be able to penetrate the fetal blood–brain barrier. Accordingly, the cord blood mercury concentration is highly attractive as an exposure biomarker.[13]

When interpreting an exposure biomarker result, the diagnostic validity must be known, i.e., the translation of the biomarker value into the magnitude of possible health risks. However, dose-response relationships are only partially known, and no-effect levels may be controversial.[7] Nonetheless, the present study suggests that cord blood should be the preferred specimen for assessment of individual risk levels in studies of prenatal neurotoxicity risks. For population monitoring purposes, hair has many practical advantages over blood sampling and will improve feasibility in many settings.

VIII. Conclusions

Exposure to methylmercury may be conveniently assessed by analyzing the mercury concentration in scalp hair. Hair-mercury concentrations are affected by several factors, including hair color and variable growth rates, which somewhat limit its usefulness as an indicator of mercury concentrations in the body. However, past exposure levels can be determined by analyzing segments of hair if assuming a constant hair growth rate. Blood concentrations correlate well (r = 0.84) with concentrations in hair close to the root. Prenatal exposure to methylmercury may be determined from analysis of maternal hair or cord blood. The latter parameter shows a better association with mercury-related neurobehavioral deficits, and this exposure biomarker should therefore be preferred in future studies.

Acknowledgments

Supported by grants from the U.S. National Institute of Environmental Health Sciences (ES06112 and ES09797), the European Commission (Environment Research Programme) and the Danish Medical Research Council. The contents of this paper are solely the responsibility of the authors and do not necessarily represent the official views of the NIEHS, NIH or any other funding agency.

References

1. AMAP Assessment Report: Arctic Pollution Issues, Arctic Monitoring and Assessment Programme, Oslo, 1998.
2. Hansen, J.C., Tarp, U., and BohnJ., Prenatal exposure to methylmercury among Greenlandic Polar Inuits, *Arch. Environ. Health*, 45, 355–358, 1990.
3. Grandjean, P. et al., Impact of maternal seafood diet on fetal exposure to mercury, selenium, and lead, *Arch. Environ. Health*, 47, 185–195, 1992.
4. Grandjean, P., Biomarkers in epidemiology, *Clin. Chem.*, 41, 1800–1803, 1995.
5. Grandjean, P. et al., Biomarkers of chemical exposure: state of the art, *Clin. Chem.*, 40, 1360–1362, 1994.
6. Clarkson, T.W. et al., Mercury, in *Biological Monitoring of Toxic Metals*, Clarkson, T.W., et al., Eds., Plenum, New York, 1988, pp. 199–246.
7. National Research Council. *Toxicological Effects of Methylmercury*, National Academy Press, Washington, D.C., 2000.
8. Hislop, J.S. et al., The use of keratinised tissues to monitor the detailed exposure of man to methylmercury from fish, in *Clinical Toxicology and Clinical Chemistry of Metals*, Brown, S.S., Ed., Academic Press, New York, 1983, pp. 145–148.

9. Cox, C. et al., Dose-response analysis of infants prenatally exposed to methylmercury: an application of a single compartment model to single-strand hair analysis, *Environ. Res.*, 49, 318–332, 1989.

10. Kjellström, T., Kennedy, P., and Wallis, S., Physical and mental development of children with prenatal exposure to mercury from fish: Stage 2, Interviews and Psychological Tests at age 6. (Report 3642), National Swedish Environmental Protection Board, Stockholm, 1989.

11. Cernichiari, E. et al., The biological monitoring of mercury in the Seychelles study, *Neurotoxicology*, 16, 613–628, 1995.

12. Pineau, A. et al., Determination of total mercury in human hair samples by cold vapor atomic absorption spectrometry, *J. Anal. Toxicol.*, 14, 235–238, 1990.

13. Grandjean, P. et al., Cognitive deficit in 7-year-old children with prenatal exposure to methylmercury, *Neurotoxicol. Teratol.*, 19, 417–428, 1997.

14. Yamaguchi, S. et al., Factors affecting the amount of mercury in human scalp hair, *Am. J. Publ. Health*, 65, 484–488, 1975.

15. Grandjean, P., Lead content of scalp hair as an indicator of occupational lead exposure, in *Toxicology and Occupational Medicine*, Deichmann, W.M., Ed., Elsevier, Amsterdam, 1979, pp. 311–318.

16. Smith, J.C. and Farris, F.F., Methylmercury pharmacokinetics in man: A reevaluation, *Toxicol. Appl. Pharmacol.*, 137, 245–252, 1996.

17. Steuerwald, U. et al., Maternal seafood diet, methylmercury exposure, and neonatal neurological function, *J. Pediatr.*, 136, 599–605, 2000.

18. Grandjean, P. et al., Methylmercury exposure biomarkers as indicators of neurotoxicity in 7-year-old children, *Am. J. Epidemiol.*, 150, 301–305, 1999.

19. Dobbing, J., Vulnerable periods in developing brain, in *Applied Neurochemistry*, Davison, A.N. and Dobbing, J., Eds., Davis, Philadelphia, 1968, pp. 287–316.

20. Ginsberg, G.L. and Toal, B.F., Development of a single-meal fish consumption advisory for methylmercury, *Risk Anal.*, 20, 41–47, 2000.

chapter seventeen

Urinary arsenic species in relation to drinking water, toenail arsenic concentrations and genetic polymorphisms in GSTM1 in New Hampshire

Margaret R. Karagas, Heather H. Nelson, Karl T. Kelsey, Steven Morris, Joel D. Blum, Tor D. Tosteson, Mark Carey, Xiufen Lu, and X. Chris Le

Contents

Abstract Urinary arsenic is used as a biomarker of occupational or high environmental arsenic exposure and the relative concentrations of urinary metabolites (e.g., inorganic arsenic, monomethylarsonic acid (MMA), and

1-56670-596-7/02/$0.00+$1.50
© 2002 by CRC Press LLC

dimethylarsinic acid (DMA)) are thought to be, at least in part, under genetic control. However, there are limited data on urinary arsenic in populations exposed to environmental levels typical of the U.S. As part of a study of individuals exposed to arsenic through drinking water in New Hampshire, we examined the association between specific urinary fractions of arsenic and drinking water and toenail concentrations and examined whether the distribution of urinary arsenic species differs by *GSTM1* genotype. We recruited 99 individuals between the ages of 25 and 74 years for the study. Water arsenic ranged from undetectable (<0.01 µg/L) to 54.1 µg/L based on high resolution ICP-MS with hydride generation. We tested urine samples using HPLC with hydride generation atomic fluorescence detection and detected arsenic in 77 of the 92 samples (range = less than 1 µg/L to 85.4 µg/L). Only about one fourth of participants had detectable concentrations of inorganic arsenic or MMA, whereas over 80% had detectable DMA. All toenail samples contained at least trace amounts of arsenic using neutron activation analysis (range = 0.01 to 0.53 µg/g). Water and toenail arsenic concentrations correlated with urinary concentrations of MMA and DMA but not inorganic arsenic. Further, water and toenail concentrations were related to the ratio of MMA to DMA but not the ratio of inorganic arsenic to MMA. The percentages or ratios of the urinary fractions did not statistically differ between those with *GSTM1* null or wildtype. Our results have important implications for studies of populations in the U.S. and elsewhere with drinking water containing low but potentially harmful levels of arsenic.

I. Introduction

Urinary concentration of arsenic is frequently used to monitor occupational or high environmental arsenic exposure; however; there are limited data on urinary arsenic in populations exposed to lower environmental levels typical of the U.S. In humans, arsenate (As V) reduces to arsenite (As III) which converts to monomethylarsonic acid (MMA). MMA reduces to the trivalent form and undergoes further methylation to dimethylarsinic acid (DMA). Some organic arsenic compounds, such as arsenobetaine found in seafood, remain unchanged in the body and are excreted in their original state. Arsenosugars, however, are metabolized to DMA and several unidentified arsenic species.[1,2] Within 1–3 days of exposure, 45–85% of ingested arsenic is excreted in urine [3,4] Thus, urine samples can provide an indication of recent arsenic exposure.

The relative concentrations of individual arsenic species may indicate an individual's ability to metabolize arsenic.[5] For this reason, the distribution of arsenic fractions has been explored as a possible marker of susceptibility to arsenic-induced cancers.[6,7] In a study from the northeast of Taiwan (with drinking water concentrations ranging from $\leq$ 50 µg/L to over 300 µg/L of arsenic), the relative concentrations of urinary arsenic were associated with

GSTM1 and *GSTT1* genotypes.[8] Yet, it is unclear whether these effects occur in less exposed populations.

Arsenic accumulates in nail tissue due to its affinity to the sulfhydryls in keratin.[9,10] In prior studies, nail concentrations reflected arsenic exposure via inhalation, drinking water and diet.[4,11] As part of a U.S. study of individuals exposed to drinking water levels of arsenic largely below 50 μg/L, we examined the correlation between specific urinary fractions of arsenic (inorganic, MMA and DMA) and tap water and toenail arsenic levels and examined the association between the urinary fractions and *GSTM1* genotype.

II. Methods

For the current study, we reinterviewed 99 controls who had participated in our nonmelanoma skin cancer case-control study, who reported use of a private well at their current or previous residence and remained a New Hampshire resident.[12] Participants were sent the instructions and materials to collect toenail clippings and a "first morning" urine sample using a protocol similar to the U.S. EPA study.[13] We specifically asked participants to collect and refrigerate the urine sample on the day of the interview. We further requested a three-day diary of seafood and water consumption (i.e., cups per day at home, work, or elsewhere). A study interviewer visited the participant's home to collect the samples and obtain a tap water sample using 125-ml high-density polyethylene bottles (I-CHEM) following a previously described protocol.[14] At the time of the original study interview, we collected a venous blood sample. Within 24 hours, we centrifuged the whole blood at 7000 rpm for 5 minutes, separated the buffy coat into 2 ml aliquots and stored samples at –70°C until shipment to the analysis laboratory.

A.   Laboratory methods

Toenail clipping samples were analyzed for arsenic and other trace elements by Instrumental Neutron Activation Analysis (INAA) at the University of Missouri Research Reactor, using a standard comparison approach as described previously.[15,16] Samples of sufficient weight were split and run in duplicate. If the coefficient of variability (CV, mean divided by the standard deviation) exceeded 15%, the batch was recounted to ensure that the mass and signal were accurately measured and recorded. The detection limit for arsenic measured by INAA is approximately 0.001 μg/g.

Samples of drinking water were analyzed for arsenic concentration using a Finnigan MAT Corp. ELEMENT High Resolution Inductively Coupled Plasma Mass Spectrometer (HR-ICP-MS) equipped with a Precision Glass Blowing Corp. VS-1 Membrane Gas Liquid Separator to allow Hydride Generation (HG) for enhanced sensitivity.[17] All sample preparations and analyses were carried out in a trace-metal clean HEPA-filtered-air environment. Analytical blanks and potential instrumental drifts were carefully monitored, and instrument standardization and reproducibility were performed with Certified Standard Reference Materials. The detection limit using HR-ICP-MS with hydride generation is 0.01 µg/L, two orders of magnitude lower than conventional methods.

Analyses of arsenic speciation in urine samples were carried out by using high-performance liquid chromatography separation with hydride generation atomic fluorescence detection (HPLC-HGAFD).[18] An HPLC system consisted of a Gilson (Middleton, WI) HPLC pump (Model 307), a Rheodyne 6-port sample injector (Model 7725i) with a 20-µL sample loop, and a reversed phase C18 column (ODS-3, 150 x 4.6 mm) with 3-µm particle size packing materials (Phenomenex, Torrance, CA). The column was mounted inside a column heater (Model CH-30, Eppendorf), which was controlled by a temperature controller (Model TC-50, Eppendorf). Column temperature was maintained at 50°C. A mobile phase solution (pH 5.9) contained 5 mM tetrabutylammonium hydroxide, 3 mM malonic acid, and 5% methanol, and its flow rate was 1.2 mL/min. A hydride generation atomic fluorescence detector (HGAFD) (Model Excalibur 10.003, P.S. Analytical, Kent, U.K.) was used for the detection of arsenic. The combination of HPLC and HGAFD has been described previously.[19] A urine sample was filtered through a 0.45 µm membrane and an aliquot (20 µL) was subjected to HPLC-HGAFD analysis. The detection limit was approximately 0.5 µg/L for inorganic arsenite (As III) and MMA and 1.0 µg/L for DMA and inorganic arsenate (As V).

For determining *GSTM1* genotype, we used the methods described by Cheng and colleagues[20] with the CYP1A1 gene as a positive control. Primers hybridizing to the 5′ region of exon 4 and the 3′ region of exon 5 of the *GSTM1* gene were used to amplify a 273-bp fragment in individuals with at least one allele of the *GSTM1* gene. Aliquots of genomic DNA were added to a standard PCR reaction buffer (2.5 mM $MgCl_2$, 200 µM dNTPs, 50 mM KCl, 20 mM Tris HCl, pH 8.6 and 0.1% BSA) that contained 100 ng of each primer (4 total). To this, we added 2.5 units of Taq DNA polymerase and amplified over 40 cycles of PCR using an annealing temperature of 60°C. PCR products were electrophoresed on 2% agarose gels and the products were visualized using ethidium bromide staining to classify individuals as having a *GSTM1* allele or being homozygous null.

B. Statistical analysis

We computed the Pearson correlation between the concentrations of total and specific urinary fractions of arsenic and concentrations of tap water and toenail arsenic. Due to the small number of individuals with either urinary arsenic III or V, we summed the inorganic fractions in the analysis. We examined MMA and DMA along with total organic arsenic (the sum of the two fractions). We summed over all fractions to obtain total urine concentration. Additionally, we evaluated the ratio of the urinary metabolites as an indicator of arsenic metabolism. In these analyses, we excluded the individuals whose urine contained undetectable arsenic.

To assess the association between *GSTM1* genotype and arsenic metabolism, we compared the percent of each urinary arsenic fraction along with inorganic/MMA and MMA/DMA ratios among those with the wild type and null genotypes. Using a regression analysis, we computed an adjusted two-sided p-value for the difference between *GSTM1* and wildtype. In this analysis, we assigned values of half of the detection limit to those with individual fractions in the undetectable range, but still excluded those without any detectable urinary arsenic.

After inspection of scatter plots of the urinary, toenail and water arsenic data, we chose a natural log transformation for each in the analysis. We evaluated the potentially confounding effects of age, gender and seafood intake and found age and seafood intake were significant predictors of urine arsenic concentrations.

III. Results

Of the 99 enlisted participants, 44% were women and 56% were men with an overall mean age of 66.9 years (SD = 10.8). Of these subjects, 96 provided a toenail clipping sample, 92 a urine sample and 96 a water sample. Seventy-one participants had a blood sample genotyped for *GSTM1*.

We found detectable levels of arsenic in all toenails, and all but three water samples. Concentrations in water ranged from undetectable (< 0.01 μg/L) to 54.1 μg/L and in toenails varied from 0.02 μg/g to 0.53 μg/g. Seventy-seven out of 92 subjects (84%) had at least one detectable fraction of urinary arsenic; 26 had detectable inorganic arsenic (26, As III and 3, As V), 22 had MMA, and 75 had DMA. The mean and range values for urinary arsenic are shown in Table 17.1.

As described previously, concentrations of arsenic in water samples correlated with total urinary arsenic.[21] Overall, water arsenic concentrations correlated with MMA and DMA but appeared unrelated to the inorganic fraction. The ratio of MMA to DMA also increased with water arsenic concentrations, unlike the ratio of inorganic arsenic to MMA (Table 17.1). Urinary arsenic was inversely related to age (p = 0.042), and did not differ by gender. Amount of water consumed in the 3 days prior to urine collection did not significantly add to the model. Those who reported seafood intake had significantly higher concentrations of DMA (p = 0.0007). However, adjustment for age and seafood intake had minimal impact on our correlation coefficient estimates (data not shown). Nonetheless, we adjusted correlation estimates for age and seafood intake and separately examined those who did not eat seafood in the 3 days prior to urine collection (Table 17.1).

In an earlier study, we found that toenail concentrations correlated with water concentrations.[22] Thus, as with water, toenail arsenic concentrations correlated with concentrations of MMA and DMA but not inorganic arsenic

Table 17.1 Urinary Fractions of Arsenic and Their Correlation with Tap Water Arsenic in New Hampshire

Urinary Arsenic	N	Mean (μg/L)	Range (μg/L)	All Subjects			Excluding Those Who Ate Seafood		
				N	r*	p-value*	N	r**	p-value**
Inorganic	26	6.94	(0.47, 85.38)	25	−0.01	0.97	13	0.08	0.80
MMA	22	1.50	(0.51, 8.07)	22	0.43	0.055	9	0.77	0.026
DMA	75	7.30	(0.98, 45.50)	74	0.38	0.0011	34	0.57	0.0006
Organic	75	7.74	(0.98, 46.77)	74	0.41	0.0004	34	0.59	0.0003
Total arsenic	77	9.89	(0.98, 85.38)	75	0.40	0.0004	34	0.50	0.0027
Inorganic/MMA* 100	16	102.90	(29.78, 403.38)	16	−0.14	0.64	8	−0.42	0.35
MMA/DMA* 100	22	13.89	(2.08, 34.73)	22	0.26	0.28	9	0.53	0.17

* Adjusted for age and seafood intake.

** Adjusted for age.

Table 17.2 Correlation between Toenail and Urinary Arsenic Concentrations in New Hampshire

Urinary Arsenic	All Subjects			Excluding Those Who Ate Seafood		
	N	r*	p-value*	N	r**	p-value**
Inorganic	26	−0.02	0.91	14	−0.006	0.99
MMA	22	0.75	0.0001	9	0.94	0.0006
DMA	74	0.31	0.0081	35	0.48	0.0037
Organic	74	0.33	0.0042	35	0.50	0.0028
Total arsenic	76	0.29	0.013	36	0.31	0.07
Inorganic/MMA *100	16	−0.36	0.20	8	−0.45	0.32
MMA/DMA*100	22	0.46	0.043	9	0.53	0.18

* Adjusted for age and seafood intake.

** Adjusted for age.

in urine. Likewise, the ratio of MMA to DMA, but not inorganic arsenic to MMA, correlated with toenail arsenic concentrations (Table 17.2).

Overall 47.9% of subjects had the wildtype and 52.1% the null genotype of *GSTM1*. The percentage of inorganic arsenic, MMA and DMA all appeared roughly similar among those with *GSTM1* null and wildtype. There was no observable difference in the ratios of MMA to DMA between the two genotypes (Table 17.3). In the analyses of *GSTM1*, only age was a significant predictor (not seafood); therefore, our models were each age adjusted.

IV. Discussion

In our study, drinking water arsenic concentrations correlated with urinary levels of MMA, DMA and total arsenic. We also found that water levels were related to the ratio of MMA to DMA but not inorganic arsenic to MMA. Results for toenail concentrations of arsenic mirrored those for drinking

Table 17.3 Urinary Concentrations of Arsenic According to *GSTM1* Genotype

Urinary Arsenic	*GSTM1* Null		*GSTM1* Wildtype		
	N	Mean Percent/Ratio	N	Mean Percent/Ratio	p-value*
% Inorganic	37	27.13	34	27.70	0.73
% MMA	37	8.07	34	9.93	0.74
% DMA	37	47.20	34	51.36	0.89
% Organic	37	57.90	34	64.04	0.73
Inorganic/MMA* 100	37	336.18	34	278.88	0.75
MMA/DMA* 100	37	17.10	34	19.34	0.15

* Adjusted for age.

water. We found no statistically significant differences in the fraction of urinary metabolites between those with *GSTM1* null and wild type.

Aside from a limited sample size, a potential limitation of this study is that we collected a first morning void sample, which may not accurately represent exposure over a 24-hour period or the past several days. We based our sampling protocol on results from a study of 96 Utah residents (from 28 families), half of whom had well water arsenic concentrations above 20 µg/L. In this study, "spot" urine samples reflected urine concentrations measured over 24 hours.[13] Further, concentrations varied little over a 5-day period, indicating that arsenic ingestion remained stable. Thus, from a logistical perspective any misclassification from a single urine sample may outweigh the additional time and expense of collecting a 24-hour sample, particularly in a population with relatively constant exposure. However, intra-individual variability in specific urinary metabolites has not been well-characterized, and could have affected our results.

Prior studies have observed a dose-related increase in urinary arsenic with drinking water levels, especially in highly exposed populations.[8,13,23] However, the relationship in less exposed populations is less well studied. Inorganic arsenic and each of the methylated metabolites correlated with water concentrations in data from two Chilean towns combined, one with water concentrations up to 670 µg/L and the other with concentrations of about 15 µg/L (n = 228). But, in the town with lower drinking water levels, correlations were modest for total urinary arsenic ($r = 0.25$; $p < 0.009$), DMA ($r = 0.25$; $p = 0.008$), and MMA ($r = 0.23$; $p = 0.02$) and were poor for inorganic arsenic ($r = 0.11$; $p = 0.27$). Results from our population, with even lower drinking water levels of arsenic, are consistent with the Chilean study; if anything, we found slightly stronger correlations for total urinary arsenic, MMA, and DMA. We likewise found no correlation with urinary inorganic arsenic.

In the previous studies of populations with high drinking water levels of arsenic, the relative contribution of arsenic from food is likely small compared to drinking water. Thus, dietary arsenic probably had only a small or negligible impact on the correlation between urinary arsenic species and drinking water arsenic. However, when the concentrations of arsenic in drinking water are low, as in our study, dietary arsenic ingestion could be a substantial contributor to total arsenic intake. While arsenic levels in drinking water are relatively stable,[21] the amount of arsenic ingested from food varies dramatically with the types of food consumed.[24-27] If a subject had a meal containing relatively high levels of inorganic arsenic during the previous night, the "first void" urine sample collected next morning would contain relatively high levels of residual inorganic arsenic that had not yet been methylated. Therefore, arsenic from food ingestion, which we were unable to measure, could explain the large residual variability in urinary arsenic concentrations and, in particular, the absence of a correlation between urinary inorganic arsenic and drinking water arsenic concentrations.

In addition, temporal fluctuation in drinking water concentrations may contribute to variability when assessing exposure. In a Mexican study, water values varied considerably in a town with high water arsenic (e.g., from 160 µg/L to 590 µg/L (SD = 114 µg/L)) (*28*). Whereas, in a low-exposure town (e.g., mean of 5 µg/L), values appeared more stable (SD = 7 µg/L). Recently, a study conducted in Maryland found significant variation in water samples taken every 2 months for 1 year.[29] Thus, changes in water concentrations may depend on the water supply and its geochemical characteristics. In our New Hampshire study, we retested a subgroup of households' drinking water after 3 to 5 years; levels of arsenic remained consistent (intraclass correlation coefficient = 0.85).[21] Further, in our previous analysis of 217 households from New Hampshire, we found no seasonal pattern in water arsenic levels.[22] In the present study, we only used a single tap water sample from each subject.

In human urine, about 10–30% of total arsenic is in the inorganic form, 10–20% MMA, and 60–70% DMA[5] although wide interperson variability exists. The relative concentrations of arsenic species in urine are suspected of being genetically determined although no genetic trait has been identified. If this is true, the distribution of urinary species of arsenic, in theory, could be a measure of an individual's inherent ability to metabolize arsenic. Indeed, in many prior studies, arsenic concentrations in drinking water had little influence on the percentage of the various fractions. However, we found that drinking water levels were related to the ratio of MMA to DMA. These contrasting results may be due to the lower exposure levels in our population reflected in the large number of participants in whom the levels of MMA were below the detection limit of 0.5 µg/L.

In two small case-control studies of skin lesions (including skin cancers) from an endemic arsenic region in the southwest of Taiwan, the ratio of the urinary arsenic metabolites differed between cancer cases and controls.[6,7] In both studies, cases had a higher percentage of MMA and a higher ratio of MMA to DMA, suggesting impairment of the second methylation among cases. In another study from Taiwan, metabolic forms of urinary arsenic were related to *GSTM1* and *GSTT1* genotypes.[8] Individuals with the null genotype of *GSTM1* had a higher percentage of inorganic arsenic and those with *GSTT1* null had a higher percentage of DMA. Polymorphisms in genes for glutathione-*S*-transferases have been linked to risk of several types of cancers including multiple basal cell skin cancers and bladder malignancies.[30–32] Whether the ratio of the urinary metabolites or polymorphisms in any of the glutathione-*S*-transferase genes affect risk of arsenic-related cancers at low levels of exposure remains to be tested.

In some studies, urinary metabolites of arsenic related to age, gender, and smoking.[5] As mentioned, there is evidence that consumption of some seafoods (containing arsenosugars) results in increases in urinary DMA concentration[1,2,33,34] and, thus, could change the distribution of urinary metabolites. Further, methylation of arsenic involves S-adenosylmethionine.

Diets deficient in methionine could potentially impact methylation, and in turn arsenic metabolism. We did not collect a complete dietary history; however, we did ask about seafood consumption in the 3 days prior to urine collection. In our data, seafood intake and age were positively associated with urinary arsenic concentrations.

Hair and nails have been used forensically to track arsenic poisoning over a period of several months.[9,10] Unlike hair, nails do not appear to be susceptible to external contamination.[22] They are easy to collect and can be analyzed for an array of trace elements (e.g., selenium, mercury, zinc).[35] In the New Hampshire population, we found a significant correlation between drinking water and toenail concentrations among those whose drinking water contained <1μg/L to 180 μg/L of arsenic.[22] We further detected a significant correlation between toenail samples taken 3 to 5 years apart (intraclass correlation coefficient = 0.60).[21] The correlations between total urinary arsenic with tap water and toenail arsenic were all statistically significant.[21] To our knowledge, there are no data from which to compare individuals' urinary species and toenail arsenic concentrations. However, it is not surprising that results based on toenail arsenic, a biomarker of arsenic exposure, parallel those for water arsenic.

In summary, epidemiologic studies have linked environmental arsenic exposure to serious health effects including cancer, cardiovascular disease and diabetes. However, the health consequences of levels found in the U.S., particularly below 50 μg/L remain highly controversial. We conducted our study in a population exposed to arsenic in drinking water at concentrations ranging from <1 μg/L to about 50 μg/L. However, using advanced analytical methods, we were unable to detect inorganic and monomethylated forms of arsenic from first morning void samples of most subjects. Unlike studies in more highly exposed populations, our data suggest that the ratio of MMA to DMA may be exposure (i.e., drinking water) related. Our data have important implications for planning large-scale epidemiologic studies of similar populations in the U.S. and elsewhere.

Acknowledgments

Funding for the study was provided by the National Institutes of Health grants NIEHS ES-07373, NIEHS ES99–001, NCI CA57494, NCI CA50597, NCI CA82354.

References

1. Le, X.C., Cullen, W.R., and Reimer, K.J., Human urinary arsenic excretion after one-time ingestion of seaweed, crab, and shrimp, *Clin. Chem.*, 40, 617, 1994.
2. Ma, M. and Le, X.C., Effect of arsenosugar ingestion on urinary arsenic speciation, *Clin. Chem.*, 44, 539, 1998.

3. Malachowski, M.E., An update on arsenic, *Clin. Lab. Med.*, 10, 459, 1990.
4. Agency for Toxic Substances and Disease Registry, *Toxicological Profile for Arsenic*, U.S. Department of Health and Human Services, Atlanta, 1999.
5. Vahter, M., Genetic polymorphism in the biotransformation of inorganic arsenic and its role intoxicity, *Toxicol. Lett.*, 112–113, 207, 2000.
6. Hsueh, Y.M. et al., Serum beta-carotene level, arsenic methylation capability, and incidence of skin cancer, *Cancer Epidemiol., Biomarkers Prev.*, 6, 589, 1997.
7. Yu, R.C. et al., Arsenic methylation capacity and skin cancer, *Cancer Epidemiol. Biomarkers Prev.*, 9, 1259, 2000.
8. Chiou, H. et al., Arsenic methylation capacity, body retention, and null genotypes of glutathione S-transferase M1 and T1 among current arsenic exposed residents in Taiwan, *Mutat. Res.*, 386, 197, 1997.
9. Henke, G., Nucci, A., and Queiroz, L., Detection of repeated arsenical poisoning by neutron activation analysis of foot nail segments, *Arch. Toxicol.*, 50, 125, 1982.
10. Schroeder, H. and Balassa, J., Abnormal trace metals in man: arsenic, *J. Chron. Dis.*, 19, 85, 1966.
11. Agency for Toxic Substances and Disease Registry, U.S. Department of Health and Human Services, Atlanta, 1993.
12. Karagas, M.R. et al., Skin cancer risk in relation to toenail arsenic concentrations in a US population-based case-control study, *Am. J. Epidemiol.*, 153, 559, 2001.
13. Calderon, R.L. et al., Excretion of arsenic in urine as a function of exposure to arsenic in drinking water, *Environ. Health Perspect.*, 107, 663, 1999.
14. Karagas, M.R. et al., Design of an epidemiologic study of drinking water Arsenic exposure and skin and bladder cancer risk in a US population, *Environ. Health Perspect.*, 106, 1047, 1998.
15. Cheng, T.P. et al., Study of the correlation of trace elements in carpenter's toenails, *J. Radioanal. Nucl. Chem.*, 195, 31, 1995.
16. Nichols, T.A. et al., The study of human nails as an intake monitor for arsenic using neutron activation analysis, *J. Radioanal. Nucl. Chem.*, 236, 51, 1998.
17. Klaue, B. and Blum, J.D., Trace analysis of Arsenic in drinking water by inductively coupled plasma mass spectrometry: High Resolution versus Hydride Generation, *Anal. Chem.*, 71, 1408, 1999.
18. Le, X.C. et al., Speciation of key arsenic metabolic intermediates in human urine, *Anal. Chem.*, 72, 5172, 2000.
19. Le, X.C. and Ma M., Short-column liquid chromatography with hydride generation atomic fluorescence detection for the speciation of arsenic, *Anal. Chem.*, 70, 1926, 1998.
20. Cheng, T.J. et al., Comparison of sister chromatid exchange frequency in perihperal blood lymphocytes in lung cancer cases and controls, *Mutat. Res.*, 348, 75, 1995.
21. Karagas, M.R. et al., Markers of low level arsenic exposure for evaluating human cancer risk in a U.S. population, *Int. J. Occup. Environ. Health*, 14, 171, 2001.
22. Karagas, M.R. et al., Measurement of low levels of arsenic exposure: a comparison of water and toenail concentrations, *Amer. J. Epidemiol.*, 152, 84, 2000.
23. Biggs, M.L. et al., Relationship of urinary arsenic to intake estimates and a biomarker of effect, bladder cell micronuclei, *Mutat. Res.*, 386, 185, 1997.

24. Dabeka, R.W., McKenzie, A.D., and Lacroix, G.M., Dietary intakes of lead, cadmium, arsenic and fluoride by Canadian adults: a 24-hour duplicate diet study, *Food Addit. Contam.*, 4, 89, 1987.

25. Dabeka, R.W. et al., Survey of arsenic in total diet food composites and estimation of the dietary intake of arsenic by Canadian adults and children, *J. AOAC. Int.*, 76, 14, 1993.

26. Schoof, R.A. et al., Dietary arsenic intake in Taiwanese districts with elevated arsenic in drinking water, *Hum. Ecol. Risk Assess.*, 4, 117, 1998.

27. Yost, L.J., Schoof, R.A., and Aucion, R., Intake of inorganic arsenic in the North American diet, *Human Ecol. Risk Assess.*, 4, 137, 1998.

28. Cebrian, M.E. et al., Chronic arsenic poisoning in the north of Mexico, *Human Toxicology*, 2, 121, 1983.

29. Ryan, P.B., Huet, N., and MacIntosh, D.L., Longitudinal investigation of exposure to arsenic, cadmium, and lead in drinking water, *Environ. Health Perspect.*, 108, 731, 2000.

30. Silverman, D.T., Morrison, A.S., and Devesa, S.S., *Cancer Epidemiology and Prevention*, 2nd ed., Schottenfeld, D. and Fraumeni, J.F., Eds., Oxford University Press, New York, 1996, p. 1156.

31. Lear, J.T. et al., Multiple cutaneous basal cell carcinomas: glutathione S-transferase (GSTM1, GSTT1) and cytochrome P450 (CYP2D6, CYP1A1) polymorphisms influence tumour numbers and accrual, *Carcinogenesis*, 17, 1891, 1996.

32. Lear, J. et al., Polymorphism in detoxifying enzymes and susceptibility to skin cancer, *Photochem. Photobiol.* 63, 424, 1996.

33. Le, X.C., Ma, M., and Lai, V., in *Arsenic Exposure and Health Effects*, Abernathy, C.O., Calderon, R.L., and Chappell, W.R., Eds., Elsevier Science, Amsterdam, 1999, p. 69.

34. Buchet, J.P. et al., Assessment of exposure to inorganic arsenic, a human carcinogen, due to the consumption of seafood, *Arch. Toxicol.* 70, 773, 1996.

35. Garland, M. et al., Toenail trace element levels as biomarkers: reproducibility over a 6-year period [published erratum appears in *Cancer Epidemiol Biomarkers Prev.*, 3, 523, 1994], *Cancer Epidemiol. Biomarkers Prev.* 2, 493, 1993.

chapter eighteen

Molecular responses of mammalian cells to nickel and chromate exposure

Max Costa, Konstantin Salnikow, Jessica E. Sutherland, Thomas Kluz, Wu Peng, and Moon-shong Tang

Contents

Abstract Previous studies from our laboratory have developed biomarkers to assess exposure and early toxicological effects to hexavalent chromium (Cr). Hexavalent Cr reacts with the phosphate backbone of DNA forming ternary complexes of protein/Cr/DNA, as well as glutathione and amino acid crosslinks with DNA. Recently, we have utilized the Uvr-ABC system and ligation mediated PCR to study and detect Cr/DNA adducts in exon 5 and 7 of the *p53* gene in human A549 cells. Preliminary results suggest that this technique can detect and map the sites of Cr/DNA adducts at a single nucleotide level in exon 5 and 7 of the *p53* gene. Interestingly, some of these sites appear to be at hotspots for *p53* mutations in human cancer. We are also investigating whether Cr/DNA adducts that were tightly bound to the DNA form these coordinate covalent bonds preferentially at guanines that are neighbored by methylated cytosines since previous studies have suggested that guanines near methylated cytosines are sites for preferential adduct formation.

In addition to these studies, we have been examining the molecular responses of human and rodent cells to nickel exposure. We have found that exposure of human cells to nickel turns on the hypoxia-inducible factor (HIF)-mediated transcriptional pathways which include the glycolytic enzymes, as well as a novel gene we have cloned and named Cap43. An interesting gene in that it is one of the most inducible genes following nickel exposure, Cap43 appears to be induced by nickel through the HIF pathway since it is not induced in HIF-1 deficient cells. We have also found that nickel exposure results in the accumulation of *p53* protein. Previous studies have shown that nickel induces ATF-1-mediated transcription which down-regulates the transcription of thrombospondin 1, a 420,000-dalton antiangiogenic glycoprotein. Thus, exposure of cells to nickel prepares the cell for surviving under low oxygen conditions by inducing the HIF-1 transcription factor, as well as by down-regulating thrombospondin 1 that allows enhanced growth of blood vessels. These events contribute to the ability of nickel to select cells that have *p53* mutations and that are able to grow in low oxygen environments and all these effects probably contribute to nickel's carcinogenic activity.

I. Introduction

It is generally agreed that risk assessment in understanding potential toxicity of chemicals can be greatly improved by the development of biological indicators of exposure, early toxicological effect, and genetic susceptibility. These biological indicators have been called biomarkers. Biological markers are measurements conducted in biological samples that evaluate an exposure or biological effect of that exposure. Biomarkers are generally classified into three groups: biomarkers of exposure, biomarkers of effect, and biomarkers of susceptibility. Typically, biomarkers of exposure include measurements of the actual toxin or its metabolite. An interaction between a xenobiotic or endogenous component can also represent a biomarker of exposure. Frequently, biomarkers of exposure are those that involve DNA or protein adducts. Biomarkers of effect measure the biological response that is mechanistically involved in a pathway leading to injury and disease. Gene mutations and chromosomal rearrangements induced by carcinogen exposure are examples of biomarkers of effect. Most DNA adducts are biomarkers that assess exposure and effect. Biomarkers of susceptibility can be defined as indicators signaling unusually high sensitivity to a toxic exposure. This may include measurements of the activity of enzymes or the ability of the cell to efficiently repair DNA damage.

II. Biomarkers of chromate exposure

Humans are exposed to two major oxidation states of Cr, Cr(III) and Cr(VI).[1] Cr(VI) is the most toxic and carcinogenic form of Cr. The high toxicity of Cr(VI) results from its active accumulation into cells, whereas Cr(III) is much less toxic because it is not able to enter cells well.[2] Inside the cell, all of the Cr(VI) is reduced to Cr(III) and this reduction process and the formation of Cr(III) is responsible for the induction of DNA damage and other cytotoxic effects of Cr(VI).[3] Monitoring of occupational exposure to Cr(VI) compounds is usually based on measurements of Cr in the urine and serum. However, Cr(III) is an element that is found in the diet and background levels in biological fluids can be high masking low levels of Cr(VI) exposure that come from environmental/occupational exposure. Additionally, Cr is readily excreted and redistributed from serum[4] and hence, Cr measurements in human serum and urine reflect relatively recent exposures and are used primarily for heavy industrial exposures.[5] Also, measurement of Cr in erythrocytes is more informative by being indicative of exposure to hexavalent Cr since Cr(III) cannot enter the red blood cell while Cr(VI) readily enters the cells. In fact, Cr-hemoglobin complexes are persistent and therefore a single determination can potentially estimate cumulative Cr(VI) exposure going back in time. Thus, the measurement of Cr in red blood cells is a relatively good indicator of Cr(VI) exposure.[6]

A second indicator of Cr(VI) exposure involves its adduction with DNA. In the cell, the hexavalent form of Cr is reduced to the trivalent form which actually binds to the phosphate backbone of DNA and can cause the formation of ternary complexes involving glutathione, amino acids such as cysteine and histidine, and DNA.[7] Cr(III) can also adduct protein to DNA forming ternary protein-Cr(III)-DNA complexes.[8,9] There has been a substantial amount of work using the DNA-protein crosslinks induced by chromate in white blood cells of humans as a biomarker of chromate exposure and early toxic effects.[10,11] Recently, we have utilized a UvrABC system and ligation-mediated PCR to study and detect Cr-DNA adducts in exons 5 and 7 of the *p53* gene in human A549 cells. Preliminary results suggests that this technique can detect and map the sites of Cr-DNA adducts at a single nucleotide level in exons 5 and 7 of the *p53* gene. Interesting, some of these sites appear to be hotspots for *p53* mutations in human cancer.

We are also investigating whether Cr-DNA adducts that were tightly bound to DNA form these coordinate covalent bonds preferentially on guanines that are neighbored by methylated cytosines since previous studies have suggested that guanines near methylated cytosines are sites for preferential adduct formation.[12] This phenomenon addresses the issue of epigenetic susceptibility with regard to chromate carcinogenesis since the degree of methylation of a promoter region of the gene can vary depending upon the individual and the tissue. Therefore, the sites of Cr adduct formation may also vary depending upon the degree of methylation of the gene's coding region and this may give rise to differences in adduct formation and differences in genetic susceptibility to adduct formation.

The biomarkers available to assess chromate exposure and effect represents useful types of biomarkers. In fact, since a portion of the Cr bound to DNA can be eliminated by exposing the DNA to EDTA,[8] one can address whether the adducts detected were Cr-DNA adducts. There are few adducts that display this unique chemical property of EDTA reversibility, and therefore it is a useful way to identify DNA adducts that contain Cr.

III. Nickel

A second agent that we have interest in developing biomarkers to is nickel (Ni). However, in contrast to Cr, Ni forms weak bonds with DNA and tends to favor protein binding over DNA but, even so, does not form tight bonds with proteins. Thus, it is difficult to develop biomarkers of Ni exposure. One approach to this problem has involved examining the molecular responses of cells to Ni exposure. Are there specific signaling pathways that are turned on by Ni that can be used to indicate Ni exposure. One such pathway that is activated by Ni exposure involves calcium channels and the Cap43 pathway.[13,14] The transcription factors HIF-1α, ATF-1, and p53 are all increased by Ni exposure.[15-18] It is of interest that Ni is a carcinogen and is able to increase the levels of HIF-1α which may be important in allowing cancer cells to survive and proliferate under low oxygen tension. This probably contributes to nickel's carcinogenesis. By examining the cellular pathways that may be uniquely activated by Ni, we hope to identify some that might serve as biomarkers of Ni exposure or effect. However at present, these types of biomarkers are much less promising that those that are being developed for Cr.

References

1. Costa, M., Toxicity and carcinogenicity of Cr(VI) in animal models and humans, *Crit. Rev. Toxicol.*, 27, 431–442, 1997.
2. De Flora, S. and Wetterhahn, K.E., Mechanisms of chromium metabolism and genotoxicity, *Life Chem. Rep.*, 7, 169–244, 1989.
3. Kortenkamp, A., Casadevall, M., and Da Cruz Fresco, P., The reductive conversion of the carcinogen chromium (VI) and its role in the formation of DNA lesions, *Ann. Clin. Lab. Sci.*, 26, 160–175, 1996.
4. Stearns, D.M., Belbruno, J.J., and Wetterhahn, K.E., A prediction of chromium(III) accumulation in humans from chromium dietary supplements, *FASEB J.*, 9, 1650–1657, 1995.
5. Paustenbach, D.J. et al., Urinary chromium as a biological marker of environmental exposure: what are the limitations? *Regul. Toxicol. Pharmacol.*, 26, S23–S34, 1997.
6. Wiegand, H.J., Ottenwalder, H., and Bolt, H.M., Recent advances in biological monitoring of hexavalent chromium compounds, *Sci. Total Environ.*, 71, 309–315, 1988.
7. Zhitkovich, A., Voitkun, V., and Costa, M., Glutathione and free amino acids form stable complexes with DNA following exposure of intact mammalian cells to chromate, *Carcinogenesis*, 16, 907–913, 1995.
8. Zhitkovich, A., Voitkun, V., and Costa, M., Formation of the amino acid-DNA complexes by hexavalent and trivalent chromium *in vitro*: importance of trivalent chromium and the phosphate group, *Biochemistry*, 35, 7275–7282, 1996.

9. Voitkun, V., Zhitkovich, A., and Costa, M., Cr(III)-mediated crosslinks of glutathione or amino acids to the DNA phosphate backbone are mutagenic in human cells, *Nucleic Acids Res.*, 26, 2024–2030, 1998.

10. Costa, M. et al., Monitoring human lymphocytic DNA-protein cross-links as biomarkers of biologically active doses of chromate, *Environ. Health Perspect.*, 104 Suppl 5, 917–919, 1996.

11. Zhitkovich, A. et al., Utilization of DNA-protein cross-links as a biomarker of chromium exposure, *Environ. Health Perspect.*, 106 Suppl 4, 969–974, 1998.

12. Denissenko, M.F. et al., Cytosine methylation determines hot spots of DNA damage in the human P53 gene, *Proc. Natl. Acad. Sci. USA.*, 94, 3893–3898, 1997.

13. Zhou, D., Salnikow, K., and Costa, M., Cap43, a novel gene specifically induced by Ni2+ compounds, *Cancer Res.*, 58, 2182–2189, 1998.

14. Salnikow, K., Kluz, T., and Costa, M., Role of Ca(2+) in the regulation of nickel-inducible Cap43 gene expression, *Toxicol. Appl. Pharmacol.*, 160, 127–132, 1999.

15. Salnikow, K., Wang, S., and Costa, M., Induction of activating transcription factor 1 by nickel and its role as a negative regulator of thrombospondin I gene expression, *Cancer Res.*, 57, 5060–5066, 1997.

16. Salnikow, K. et al., Nickel-induced transformation shifts the balance between HIF-1 and p53 transcription factors, *Carcinogenesis*, 20, 1819–1823, 1999.

17. Salnikow, K. et al., Carcinogenic nickel induces genes involved with hypoxic stress, *Cancer Res.*, 60, 3375–3378, 2000.

18. Salnikow, K. and Costa, M., Molecular mechanisms of nickel carcinogenesis, *J. Environ. Pathol. Toxicol. Oncol.*, 19, 307–318, 2000.

chapter nineteen

Chromium: exposure, toxicity, and biomonitoring approaches

Anatoly Zhitkovich

Contents

Abstract Exposure to Cr(VI) but not other oxidative forms of Cr causes toxic and carcinogenic effects in humans and animals. Cr(VI) is a pro-carcinogen that is reduced intracellularly to form DNA damaging species. Metabolism of Cr(VI) is accomponied by the formation of Cr(V)/Cr(IV) intermediates and finally yields thermodynamically stable Cr(III). Interme-diate Cr species appears to be relatively unreactive although they can cause oxidative DNA damage in the presence of hydrogen peroxide. Cr(III) reacts with DNA phosphates producing mutagenic ternary adducts with amino acids and glutathione. Assessment of human exposure to Cr(VI) has largely been based on measurements of total Cr in urine and plasma. A major concern regarding the use of these analyses for the detection of low exposures to Cr(VI) is a variable and potentially dominant contribution of non-toxic Cr(III). Animal studies with Cr(VI) and inorganic Cr(III) salts have provided evidence that Cr concentrations in erythrocytes could be indicative of Cr(VI) exposures. Measurements of Cr levels in erythrocytes among chrome platers supported the utility of this approach whereas studies on stainless steel welders did not show a clear advantage. Elevated frequencies of chromosomal damage have been detected in several Cr(VI)-exposed populations; however, the results were strongly confounded by smoking and concomitant exposures. Among DNA lesions, the level of DNA-protein crosslinks in lymphocytes was found to be sensitive to low and medium Cr(VI) exposures and these measurements were unaffected by smoking or other common variables. Further developments in methodology are needed to improve Cr-specificity of DNA-protein crosslink assays. Combined use of Cr measurements in biological fluids and DNA-protein crosslink analyses could overcome weaknesses of each individual biomarker and provide a sufficiently sensitive and rapid approach for the initial screening of potentially exposed populations (first-tier biomarkers). Assays for mutagenic Cr(III)-DNA adducts could serve as second-tier biomarkers but the sensitive methodology has yet to be developed.

I. Introduction

Chromium is found in several oxidative and physical forms that differ substantially in their toxicological potency. Cr(0), Cr(III), and Cr(VI) are the major oxidative states of chromium encountered in occupational and environment settings. Cr(0) is present almost exclusively in the metallic form and typically found in alloys with Ni, Fe, and Co. Stainless steel is the largest volume product containing Cr(0). Although Cr(0) is stable to oxidation by atmospheric oxygen under ambient conditions, high-temperature processes such as welding or exposure to corrosive chemicals results in the formation of higher oxidative forms, Cr(III) and Cr(VI). The trivalent state of Cr is stable thermodynamically and Cr(III)-containing compounds typically exhibit slow rates of the ligand exchange. There are hundreds if not thousands of various complexes of Cr(III) currently used in many industrial applications. Many poorly characterized complexes of Cr(III) are present in industrial and residential discharges that sometimes are treated with molasses to reduce Cr(VI) concentrations. The complexity of Cr(III)-ligand interactions is evident even in simple aqueous solutions of Cr(III). Freshly prepared unbuffered solutions of inorganic Cr(III) salts contain hexaaquachromium(III) as the major species (Figure 19.1). Aging of these

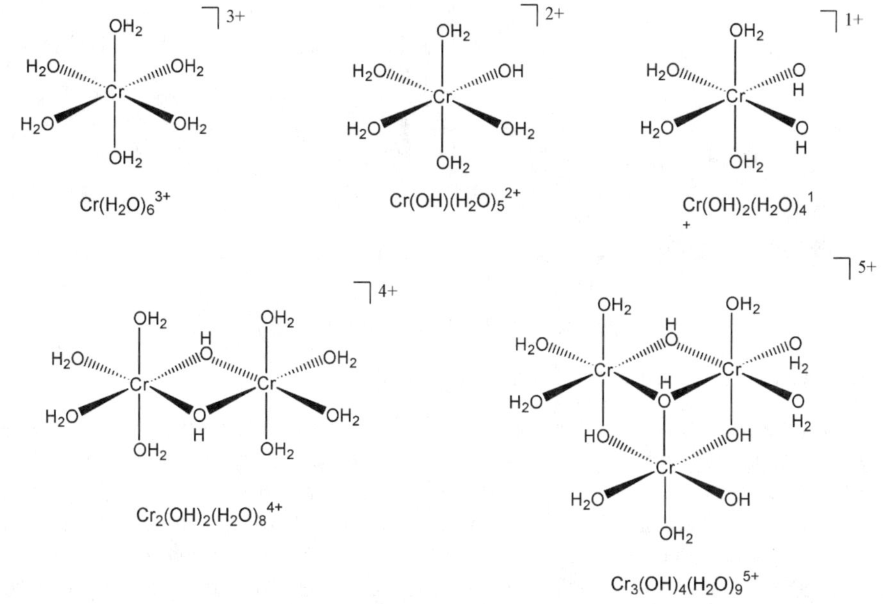

Figure 19.1 Structures of major aqueous complexes of Cr(III).

solutions leads to the conversion of $Cr(H_2O)_6^{3+}$ into $Cr(OH)(H_2O)_5^{2+}$ and $Cr(OH)_2(H_2O)_5^{+}$.[1] The pK of hexaaquachromium(III) is approximately 4. The formation of the Cr(III) hydroxyl species initiates polymerization reactions that yield dimer, trimer, and higher polymeric forms (Figure 19.1). Monomeric and low oligomeric forms of Cr(III) are usually soluble whereas polymeric products form precipitates. The hydrolytic polymerization of Cr(III) is strongly influenced by "aging" time, pH, concentration of Cr(III), and matrix composition of the solution. Complexes of Cr(III) in surface waters are typically represented by a mixture of soluble monomeric and multinuclear hydrolysis products.[2] Waste sites with high Cr(III) concentrations are also expected to contain significant amounts of insoluble hydroxypolymers.

Cr(VI) compounds are most frequently encountered as chromium trioxide (CrO_3) and chromate salts of different solubility. Depending on the pH and metal concentrations of the solution, Cr(VI) can exist as a chromate (CrO_4^{2-}), hydrochromate ($HCrO_4^-$), or dichromate ($Cr_2O_7^{2-}$) ion (Figure 19.2). At neutral pH, chromate is the prevalent form, while lowering pH below 6 leads to the formation of hydrochromate and dichromate. Dissolution of chromium(VI) oxide also produces chromate ion. The solubility of chromate compounds varies dramatically from highly soluble salts with alkali metals to practically insoluble salts with divalent barium and lead. Calcium and magnesium chromates have intermediate solubility. At low pH, Cr(VI) can be reduced by many organic molecules and, therefore, is expected to be a persistent contaminant in aqueous environments with near neutral and alkaline pH.

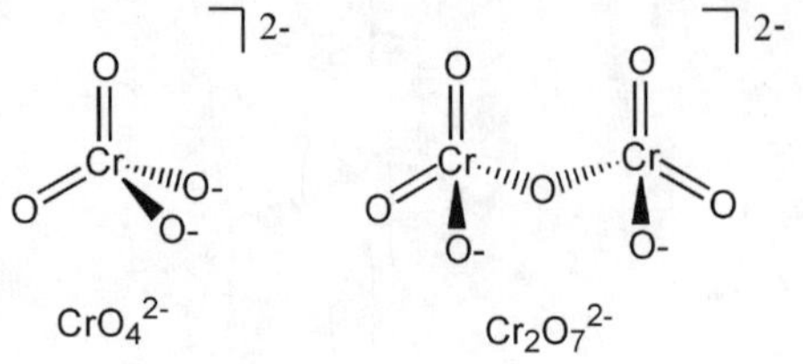

Figure 19.2 Structures of chromate and dichromate ions.

II. Chromium toxicology

A. Oxidative state of Cr and toxicity

Exposure to hexavalent Cr compounds has been consistently found to be associated with an elevated incidence of respiratory cancers and other adverse health effects.[3–5] Squamous cell carcinoma is the most prevalent form of lung cancer among Cr(VI)-exposed workers.[6] The genotoxic potential of Cr(VI) has been confirmed in animal experiments and in several cell-based assays[6a,7] Inhalation of Cr(VI)-containing acid mists in electroplating industry leads to nasal septum ulceration and perforation, as well as impaired lung function such as decreased vital capacity and forced expiratory volume.[8] Other health consequences of exposure to Cr(VI) include pulmonary fibrosis, chronic bronchitis, emphysema, and bronchial asthma.[9] Ingestion of Cr(VI) can result in the irritation of mucous membranes and, in severe cases, intestinal bleeding. High doses of Cr(VI) cause renal tubular necrosis and can be lethal. Animal studies have also detected teratogenic activity of Cr(VI).[9a] Cr(VI) is the second most potent allergen after nickel and chromium contact allergies are frequently found among occupationally exposed workers. The highest incidence of chromium sensitivity is found in chromium plating industry, manufacturing of mineral pigments, shipbuilding, textile industry, and cement-exposed builders.[9]

Epidemiological studies have generally found no association between exposure to Cr(III) compounds and the risk of cancer.[2] The lack of carcinogenic activity of inorganic Cr(III) complexes is further supported by negative results in animal and cellular genotoxicity tests.[7] Trivalent Cr is only weakly allergenic. Metallic Cr is chemically and biologically inert, but exposure to Cr(O)-containing dust may cause irritation in the respiratory tract.

The differential toxicity of Cr(VI) and Cr(III) can be attributed to the dramatically different cellular uptake of these two forms of Cr. Cr(VI) is taken up about 100–1000 times greater than inorganic Cr(III) complexes. As mentioned above, Cr(VI) exists in aqueous solutions as chromate anion that is structurally similar to physiological anions sulfate and phosphate. This structural similarity permits easy entry of chromate through the sulfate channel whereas plasma membranes are essentially impermeable to the trivalent form of Cr. Intracellular Cr(VI) is reduced to Cr(III) that avidly reacts with proteins and DNA producing an assortment of stable coordinate complexes. This uptake-reduction binding process leads to a significant accumulation of Cr inside the cell (up to several hundred-fold) over extracellular concentrations.[8a] Tight binding of Cr(III) by intracellular macromolecules also results in long-term retention of significant amounts of Cr(III). The high stability of Cr(III) complexes with biological ligands is illustrated by the observation that the administration of a strong chelator EDTA did not lead to an increased excretion of Cr from control or highly exposed human subjects.[10,11] Ascorbate and low-molecular weight thiols such as glutathione and cysteine are believed to be the major intracellular reducers of Cr(VI).[12–14]

B. Dietary chromium

The recent popularity of mineral supplements has led to widespread consumption of chromium trispicolinate due to reported beneficial effects of this compound on glucose metabolism and muscle mass.[15] The complexation of Cr(III) with the bidentate ligand 2-pyridinecarboxylate (picolinic acid) is expected to be strong which should ensure a low probability of exchange reactions with the biological ligands (Figure 19.3). Short-term assays generally showed a lack of toxicity of purified chromium trispicolinate.[15] Exposure of cultured cells to unpurified preparations of Cr trispicolinate that likely contained intermediate Cr monopicolinate and Cr bispicolinate complexes led to the induction of chromosomal aberrations.[16] These findings raised the possibility of genetic damage in individuals with the long-term use of Cr trispicolinate,[17] which can potentially occur through the exchange of picolinic ligand by DNA phosphates. At present, there is no evidence to substantiate or exclude this mechanism of delayed toxicity. Mixtures of chromium trispicolinate with ascorbate or hydrogen peroxide have been reported to cause DNA breakage *in vitro*;[18] however, it would be important to address the role of adventitious Fe and Cu in these reactions using Chelex-purified reagents.[30] There are some reports of acute renal toxicity in persons with the recent consumption of the excessively large amounts of chromium picolinate.[19,20] It remains to be established whether these individual cases are directly linked to the consumption of chromium supplements or a coincidence. Complexes of Cr(III) with various organic ligands are also found in many foods and

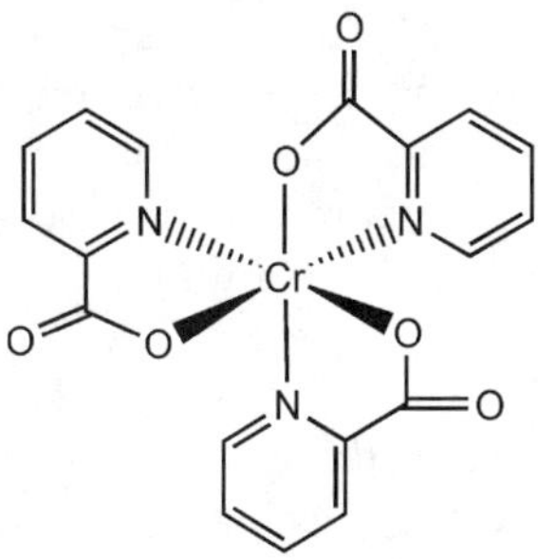

Figure 19.3 Chemical structure of Cr(III) trispicolinate.

beverages.[21] These diet-originated Cr forms do not present any significant health concerns.

C.   *Intracellular metabolism and DNA damage*

Reductive metabolism of Cr(VI) inside the cell generates Cr(III) as the final form. Depending on the nature of the reducing agent and its concentration, this process can generate detectable amounts of unstable Cr(V) and Cr(IV) intermediates, as well as thiyl and carbon-based radicals.[22-25] Reductive reactions with two-electron reducer ascorbate yield Cr(IV) as the major Cr intermediate whereas glutathione-driven metabolism generates significant amounts of Cr(V). Under normal physiological conditions, ascorbate is expected to be the dominant reducing agent for Cr(VI) in many target tissues.[13,14] Generation of Cr(III) inside the cells leads to the rapid formation of stable Cr(III) adducts with proteins and DNA.[26,27]

Ternary DNA complexes formed by Cr(III)-mediated bridging of proteins, glutathione, or amino acids represent the major form of Cr(III)-DNA adducts in Cr(VI)-exposed cells.[17-19] Reduction of Cr(VI) by ascorbate or cysteine also produces Cr(III)-mediated interstrand DNA-DNA crosslinks.[19a,30] Cr-DNA adducts are formed through the attack of DNA phosphates by the freshly formed Cr(III) and Cr(III)-amino acid complexes.[30,31] Coordination to the phosphate group can result in the formation of adducts in either S or R conformation, with only the latter affording additional binding to N^7 of purines (Figure 19.4). Cr-DNA adducts are persistent in treated cells and their repair mechanisms remain unknown.[32] Among oxidative forms of damage, DNA single-strand breaks have frequently been detected following exposure of intact cells to Cr(VI),[33] as well as in some cell-free systems.[34] Ascorbate and glutathione-based *in vitro* reactions containing high concentrations of Cr(VI) also led to the production of abasic sites,[35] but base loss and DNA strand breakage were undetectable when cysteine was used as a reducer.[30] Analysis of DNA from chromate-exposed cells or DNA samples treated *in vitro* under iron-free conditions did not reveal any significant production of 8-hydroxyguanosine and other oxidized DNA bases.[35,36] The formation of DNA strand breaks and abasic sites can be caused by hydroxyl radicals generated through the Fenton-like reactions of Cr(V) and Cr(IV) with hydrogen peroxide.[40] Although synthetic Cr(V) and Cr(IV) complexes can act as direct oxidizing agents, the yield of oxidixed DNA products is quite small.[37-39]

Ternary Cr(III)-DNA adducts formed by crosslinking of glutathione or amino acids have been found to be mutagenic following replication of a shuttle-vector plasmid in human cells.[41] The bulkiness of the crosslinked ligand was found to be associated with a greater mutagenic potency of Cr(III)-DNA complexes. DNA-protein crosslinks have not been tested for their mutagenic activity; however, the size-dependence of mutagenesis by amino acid-Cr(III)-DNA adducts points to a potentially strong mutagenic potential of protein-containing lesions. Shuttle–vector experiments have also

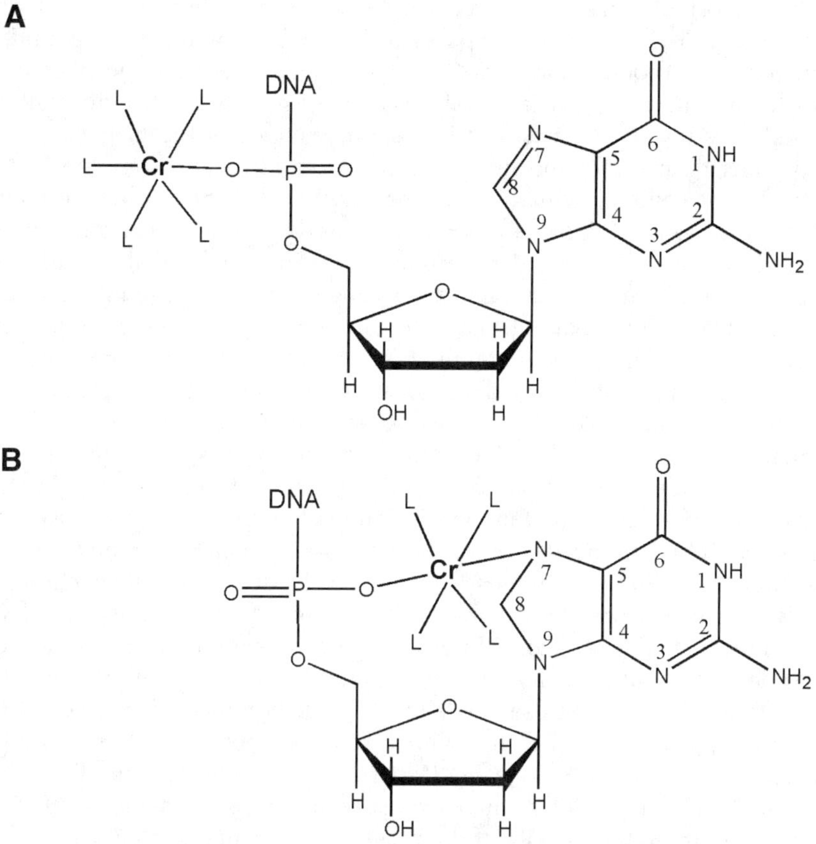

A

B

Figure 19.4 Proposed structures of *S* (A) and *R* (B) stereoisomers of Cr(III)-DNA adducts. Adducts with hydrogen bonding to N⁷-G are also likely.

detected the mutagenic potential of oxidative forms of DNA damage generated in the reaction of Cr(VI) with glutathione in the presence of the trace amounts of iron.[42] The relative importance of Cr(III)-dependent and oxidative pathways in the genotoxicty of Cr(VI) may vary depending on the nature of the dominant reducer in a specific cell type or tissue. Mutational spectra of Cr(VI) and hydrogen peroxide in human lymphoblastoid cells were different,[42a] suggesting that nonradical mechanisms were largely responsible for the induction of mutagenesis by Cr(VI) in intact human cells. The relative roles of oxidative and Cr-dependent mechanisms in the induction of cell death is currently under investigation in various cell models.[43,43a]

III. Human exposure

Workers in several dozen professional groups experience occupational exposure to various forms of chromium. The highest exposure to the toxic Cr(VI) form is usually detected in chrome plating industry, among chromate pro-

duction workers and stainless steel welders. Several million industrial workers worldwide are potentially exposed to Cr and Cr-containing compounds.[44]
Environmental contamination with Cr has also become a significant concern
due to continuous industrial emissions and the presence of heavily contaminated sites in the vicinity of residential areas. One of the major concerns
about environmental exposure to Cr is that children and some other segments of the general population may be particularly sensitive to toxic effects
of Cr. Atmospheric Cr contamination is caused by oil and coal combustion,
steel production, chemical manufacturing, primary metal production,
chrome plating, incinerators, and emissions from cars.[44] Cr-containing discharges from electroplating, leather tanning, cooling towers, and textile
industries lead to water contamination. At present, Cr contamination of
coastal waters has not reached significant levels; however, this may change
because of ongoing ocean dumping of Cr-containing sludge. Disposal of Cr-
containing products and coal ash from electrical utilities and the chemical
industry are the major sources of Cr contamination of soil.[1,44] Approximately
80% of the total Cr released in the environment by the chemical industry are
disposed of by land burial. Cr was found at more than 40% of the toxic sites
designated as Superfund sites by the EPA.[44] Use of inefficient chromate
extraction procedures led to the generation of the large amounts of Cr-
containing slag that was sometimes improperly disposed in landfills located
near densely populated areas. For example, millions of tons of Cr-containing
mine tailings have been used as landfill and in commercial and residential
construction in Hudson County, NJ.[45] Environmental and biomonitoring
studies conducted by New Jersey Department of Environmental Protection
indicated occurrence of human exposure to Cr in residents of Hudson
County living in the vicinity of the most heavily contaminated sites.[46,47]

Cr contamination of soil and water can also result from leaching of this
metal from pressure-treated wood. Treatment of wood with a mixture of
chromate, copper, and arsenate (CCA) is currently the most frequently used
method to preserve wood products from rotting and insect decay. In this
procedure, a solution of Cr(VI), Cu(II) and As(V) is applied under high
pressure, which leads to the binding of these metals up to 5 g/kg.[48] The
treated wood is widely used for construction of decks, playground equipment, utility poles, railway sleepers, marine piles and in many other applications. More than 10 million cubic meters of preservative-treated wood is
produced annually in the U.S. and most of it is CCA-treated wood.[49] It has
been estimated that approximately 52% of Cr in the USA is used for wood
preservation.[50] In the laboratory settings, it was determined that the release
of Cr from CCA-treated wood ranges from 7 to 53% depending on the pH
of the water solution.[49,51] Burning of CCA-treated wood can create highly
concentrated sources of Cr and other toxic metals. At present, there are no
environmental regulations with respect to disposal of CCA-treated wood or
its combustion residues.

Cr has also been found in cosmetics and some household products,
which can lead to allergic reactions. A recent study by Sainio et al.[52] reported

that while the majority of the brands of eye-shadows contained less than 2 ppm of water-soluble Cr, one brand contained more than 300 ppm Cr. High levels of Cr in several brands of detergents and bleaches have been suggested to be responsible for the induction of chromium contact dermatitis among occupationally unexposed individuals.[53]

IV. Biomonitoring approaches

A. Analysis of chromium in biological samples

In order to understand the significance and the limitations of Cr measurements in biological samples, it is important to keep in mind that all measurable Cr is in the trivalent form. Even under conditions of high exposure to Cr(VI), there is a relatively rapid reduction to Cr(III) upon absorption of Cr(VI) into biological fluids or tissues. Measurements of total Cr in urine or plasma are most frequently used to assess occupational exposure to Cr compounds. Ease of the sample collection and the suitability for the direct analysis by graphite furnace atomic absorption spectrometry (GF-AAS) were the major reasons for a broad adoption of Cr measurements in these fluids for biomonitoring purposes. Measurements of Cr in urine are usually adjusted for creatinine concentrations to take into account dilution effects. High levels of Cr in urine and blood were found in chrome platers and stainless steel welders, which are professional groups with a considerable exposure to Cr(VI)-containing fumes.[54-57] One of the important findings in these biomonitoring studies was the detection of significant amounts of absorbed Cr in some occupational groups that had low ambient levels of Cr in the workplace. For example, analysis of Cr in blood and urine samples obtained from various categories of welders found that mild steel welders had significantly elevated Cr levels despite relatively low levels of atmospheric Cr.[54,58]

The toxicological significance of urinary measurements is diminished under conditions of low-level exposure. Urinary and serum measurements can be significantly influenced by Cr intake from dietary sources which is expected to lead to a substantial day-to-day variability. The inability of urinary and serum analyses to differentiate between exposure to toxic Cr(VI) and innocuous organic complexes of Cr(III) makes it difficult to perform risk analysis at environmental exposure levels. The contribution of dietary Cr is not expected to be significant in persons with high industrial exposure, and in these situations serum and urinary Cr measurements can be useful markers of internal dose.

Lewalter et al.[59] proposed to use measurements of Cr content in erythrocytes as a specific index of exposure to the toxic Cr(VI) form. The basis for this suggestion was the well-known ability of Cr(VI) to be readily taken up by cells whereas inorganic Cr(III) complexes have a poor ability to cross plasma membranes. The intracellular reduction of Cr(VI) to Cr(III) in erythrocytes is believed to be followed by a tight binding of Cr(III) to hemoglobin.[60] Since Cr(III)-hemoglobin complexes are unlikely to be excreted and multidentate coordination with protein groups should be stable, the amount of Cr in erythrocytes could potentially be a measure of the cumulative dose of Cr(VI) over the lifespan of red blood cells (120 days). In reality, the half-life of Cr in erythrocytes is expected to be shorter due to exchange reactions with small ligands, which will result in the formation of excretable Cr(III) complexes. Administration of soluble inorganic salts of Cr(III) and Cr(VI) to experimental animals did confirm that erythrocyte measurements of Cr could differentiate between these two oxidative states.[60-62] Cr concentrations in erythrocytes have been found to be highly elevated in chrome platers,[57,63] whereas similar measurements among stainless steel welders showed only moderate increases.[56,64,65] The differences between unexposed controls and chrome platers were greater on the basis of Cr content in erythrocytes than by estimating urinary levels.[57,63] These results provide support for the concept that Cr accumulates in red blood cells and that determinations of the Cr content in these cells can serve as an integrated measure of exposure over an extended period of time. However, studies with stainless welders found a much lower sensitivity of Cr measurements in erythrocytes as compared to urinary determinations.[56,65] Preshift and postshift samples contained similar Cr levels in red blood cells, [56] whereas urinary Cr has been found to be significantly higher in postshift analyses.[56,57] The reduction activity of the blood appeared to be inversely related to Cr accumulation in erythrocytes, and this was caused by a rapid conversion of extracellular Cr(VI) into Cr(III).[66] It is possible that stainless steel welders had a much greater serum reducing activity either due to dietary factors and/or modifying effects of coexposure to other metals.

The toxicological significance of Cr measurements in erythrocytes remains unclear due to the lack of data about the cellular uptake of Cr(III) complexes with lipophilic ligands. Cr(III) complexes with amino acids can cross cell membranes although less readily than the carcinogenic Cr(VI) form.[67] Published animal studies comparing the distribution of Cr between different compartments of the blood typically used commercial chromium(III) chloride that forms $Cr(H_2O)_4Cl_2^+$ ion in a freshly prepared solution. This is not a toxicologically important Cr(III) compound and, therefore, is highly desirable to examine environmentally relevant Cr(III)-aqua ions and dietary Cr(III) complexes with organic ligands. (See introduction (I) and section concerning dietary chromium.) Analysis of the Cr content of biological tissues is not a trivial task even when expensive instrumentation is available because contamination and background problems are frequently encountered. The development of background correction techniques in the early 1980s and strict control of sources of contamination have led to reliable analyses of Cr in biological fluids by the GF-AAS technique. Cr measurements can still be quite variable and profoundly influenced by sample collection practices and laboratory procedures. One of the reasons for a slow adaptation of Cr analyses in erythrocytes was related to the requirement for a high-end GF-AAS instrument utilizing a flow of oxygen to ensure complete mineralization of the organic matrix inside the graphite tubes.

B.   Chromosomal damage

Several investigators have determined the levels of chromosomal damage in occupationally exposed populations in order to assess the biologically active dose of Cr(VI). The frequency of sister chromatid exchanges (SCEs) was unchanged in chrome platers relative to controls,[68] whereas the levels of cytogenetic damage in welders was higher than among unexposed individuals.[65,69,69a] The use of chromosomal damage in the evaluation of biological effects of Cr exposures is complicated by frequent concomitant exposures to other clastogens (i.e., Ni in welders), as well as strong by confounding effects of cigarette smoking. Ni ions have been found to suppress SCE formation by Cr(VI) in cultured human lymphocytes.[70] The frequency of micronuclei in peripheral blood lymphocytes of chrome platers has recently been reported to be significantly elevated and is correlated with internal and external measurements of Cr exposure.[63] The application of the micronuclei assay to individuals with complex exposures will also be confounded by concomitant exposure to other clastogenic chemicals and even by dietary status (i.e., folate intake[70a]).

C.   Measurements of oxidative DNA lesions

Two studies have examined the levels of 8-oxodG in DNA of total white blood cells isolated from chromate production[71] or chromate pigment workers.[72] Although Cr measurements in biological fluids clearly demonstrated a significant internal dose of Cr, the levels of 8-oxodG were unchanged relative to unexposed controls. The negative finding are probably not unexpected considering that 8-oxodG is typically not formed as a result of Cr(VI) exposure. No increase in 8-oxodG levels was found in Cr(VI)-exposed animals[36] or in cell-free systems with the major reducers of Cr(VI), ascorbate or glutathione.[35] *In vitro* reactions utilizing high concentrations of hydrogen peroxide and Cr(VI) caused oxidation of dG but these conditions are unlikely to be encountered under physiological conditions. Elevated levels of single strand breaks in lymphocyte DNA were found among Cr-exposed welders,[69] whereas no increase in DNA breakage was detected in total white blood cells isolated from chromate production workers despite increased Cr concentrations in their blood and urine.[71] Single strand breaks are rapidly repaired, and it is unlikely that they would persist in white blood cells from the time of blood collection at the factory to the final purification step in the laboratory. Measurements of strand breakage in welders were done with the alkaline elution assay in which DNA is typically exposed to alkaline pH over several hours. Prolonged exposure of Cr(III)-adducted DNA to alkali results in strand breakage[30] and, therefore, it seems more likely that an increased rate of elution of DNA from welders resulted from Cr(III)-DNA adduct-dependent cleavage rather than from preexisting breaks. The absence of strand breaks in DNA from white blood cells of chromate workers could have resulted from the use of the alkali unwinding assay that employs only a short exposure to alkali (10–15 min). Additionally, short-lived granulocytes (<24 h) represent the major fraction of white blood cells and this should considerably diminish the sensitivity of measurements in these preparations relative to the analysis of DNA damage in purified lymphocyte populations.

D. DNA-protein crosslinks

DNA-protein crosslinks are known to be produced by exposure to Cr(VI) *in vitro* and *in vivo*.[29,32] Several groups of human subjects who were exposed to Cr(VI) occupationally[57,69,73,74] or lived in Cr-contaminated areas[75] have been examined for the presence of DNA-protein crosslinks in peripheral blood lymphocytes. The levels of DNA-protein crosslinks were found to be elevated in all exposed groups and the dose-response curve showed a good sensitivity at low to medium exposures.[57,69] The saturation of lymphocyte DNA-protein crosslinks was estimated to occur at 7–8 µg/l Cr in erythrocytes.[57] Stainless steel welders from Germany,[69,73] mild steel welders from the U.S.[74] and chrome platers from Bulgaria[57] all exhibited increased levels of DNA-protein crosslinks without confounding effects of smoking. Studies on control and exposed populations have also determined that DNA-protein crosslinks are not affected by body weight, age, race or gender.[57,75,75a] The ability of the DNA-protein crosslink-based biomarker to detect low level human exposures to Cr is probably related to the inefficient repair of these lesions in human lymphocytes.[32,76] Giving that the majority of human lymphocytes have a lifespan of several years,[76a] the inability of these cells to remove DNA-protein crosslinks could lead to the accumulation of these lesions during chronic exposure even to low levels of Cr(VI). The mutagenic potential of DNA-protein crosslinks has never been tested directly but the size-dependence of mutagenesis by other ternary Cr-DNA adducts could be indicative of their high genotoxic activity.[32,41]

A major problem hampering the use of DNA-protein crosslink measurements in biomonitoring is the inability of current methodology to unambiguously establish the chemical nature of the crosslinks. Since the level of DNA-protein crosslinks showed an association with Cr exposure in several studies, it is clear that this biomarker is sensitive to the presence of Cr. The high-temperature version of the K-SDS assay significantly limited other possible crosslinking agents,[32,77] which is illustrated by the absence of an association between smoking and the levels of these adducts. Even inhalation exposure to high levels of formaldehyde has not led to increased DNA-protein crosslinking in lymphocytes of furniture factory workers.[57] A better understanding of the nature of background DNA-protein crosslinks should lead to the development of experimental strategies for the specific determination of Cr-derived lesions. Formation of crosslink-producing malondialdehyde during lipid peroxidation reactions has been suggested to be one of the potential mechanisms for the induction of DNA-protein crosslinks in unexposed individuals.[78] The pattern of crosslinked proteins generated by exposure to malondialdehyde and Cr(VI) is expected to be different, therefore, immunoprecipitation is a promising approach to increase Cr specificity of crosslink determinations.

E. Biomarkers: concluding comments

The major problem encountered in the biomonitoring of exposure to toxic Cr(VI) using measurements of chromium concentrations in biological fluids is the inability to determine whether measurable Cr is derived from the dietary sources or originated from toxic Cr(VI) compounds (Figure 19.5). Since all Cr in the body is trivalent Cr, the assessment of Cr(VI) exposure needs to be based on the mechanistic differences in the interaction of harmless Cr(III) and toxic Cr(VI) forms with cells. Although experiments with animals provided significant support for the idea that Cr concentrations in erythrocytes could preferentially reflect Cr(VI) exposure, uncharged Cr(III) complexes with organic ligands are also likely to enter cells. Measurements of Cr(III)-DNA adducts in lymphocytes or other biological samples should greatly improve risk assessment by providing evidence about the formation of DNA damage which could only result from exposure to toxic Cr(VI). At present, only assays for DNA-protein crosslinks have sufficiently high sensitivity to detect current levels of human exposure to Cr(VI), but the Cr specificity of these determinations still needs to be improved. Inductively coupled plasma mass spectroscopy (ICP-MS) techniques has high analytical sensitivity for the detection of Cr; however, this method suffers from background problems during analysis of biological samples. For example, the current detection limit for ICP-MS is one Cr per 5,000 DNA bases,[79] which means that 10^2 –10^3-fold increase in the sensitivity must be achieved before

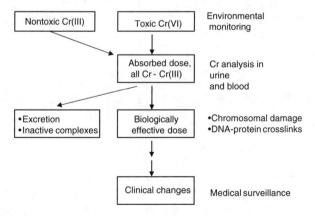

Figure 19.5 A scheme illustrating major approaches for assessment of exposure to chromium compounds.

Table 19.1 Summary of the Major Biomonitoring Approaches

Parameter	Cr in Urine	Cr in RBC*	DPC** Assays
Analytical sensitivity	High	High	High
Sensitivity, samples	High	Medium	High
Chromium specificity	High	High	Medium
Short-term exposure	Yes	No	No
Cumulative exposure	No	Yes	Yes
Dietary Cr	Yes	?	No
Biological damage	No	No	Yes
Sample size	Small	Small	Small
Sample capacity	High	High	High
Personnel training	Moderate	Moderate	Moderate
Equipment	GF-AAS	GF-AAS	Fluorometer
Major limitation	Nontoxic Cr	?Nontoxic Cr	Cr specificity

* Red blood cells.

** DNA–protein crosslinks.

this methodology can be applied to analysis of human samples. One approach for overcoming background problems could be a combination of capillary electrophoresis or HPLC with on-line detection of Cr by ICP-MS.

Since no single biomarker can determine unambiguously exposure to Cr(VI), the most appropriate strategy would be to utilize two biomarkers: one measuring Cr in the biological sample and another quantifying DNA-protein crosslinks. As summarized in Table 19.1, depending on whether chronic or short-term exposure is suspected, Cr determinations in erythrocytes or urine can be used as a biomarker of absorbed dose. Given that the K-SDS crosslink assay[80] and analytical measurements of Cr in blood and urine are relatively inexpensive and have high sample capacity, we envision these two types of analyses could be performed for the initial screening of potentially exposed populations (first tier biomarkers). Additional, more complex Cr-specific analyses (second tier biomarkers), which are being developed for specific Cr(III)-DNA adducts, will be performed only for groups or individuals found positive with the first-tier biomarkers.

Acknowledgments

This work was supported by U.S. Public Health Service Grant ES08786 from the National Institute of Environmental Health Sciences.

References

1. Earley, J.E. and Cannon, R.D., Aqueous chemistry of Cr(III), in *Transition Metal Chemistry*, Carlin, R.L., Ed., Vol. 1, Marcel Dekker, New York, 1965, pp. 33–109.
2. Saleh, F.Y. et al., Ion chromatography-photodiode array UV-visible detection of Cr(III) hydrolytic polymerization products in pure and natural waters, *Anal. Chem.*, 68, 740–745, 1996.

3. IARC Monograph on the Evaluation of Carcinogenic Risk to Humans: Chromium, Nickel and Welding, Vol. 49, IARC, France, Lyon, 1990, p. 677.

4. Langardt, S., One hundred years of chromium and cancer: a review of epidemiological evidence and selected case reports, *Am. J. Ind. Med.*, 17, 189–215, 1990.

5. Sorahan, T. et al., Lung cancer mortality in nickel/chromium platers, 1946–95, *Occup. Environ. Med.*, 55, 236–242, 1998.

6. Ishikawa, Y. et al., Characteristics of chromate workers' cancers, chromium lung deposition and precancerous lesions: An autopsy study, *Br. J. Cancer*, 70, 160–166, 1994.

6a. Biedermann, K.A. and Landolph, J.R., Role of valence state and solubility of chromium compounds on induction of cytotoxicity, mutagenesis and anchorage independence in diploid human fibroblasts, *Cancer Res.*, 50, 7835–7842, 1990.

7. Snow, E., Metal carcinogenesis: mechanistic implications. *Pharmacol. Ther.*, 53, 31–65, 1992.

8. Kuo, H.W., Lai, J.S., and Lin, T.I., Nasal septum lesions and lung function in workers exposed to chromic acid in electroplating factories, *Int. Arch. Occup. Environ. Health*, 70, 272–276, 1997.

8a. Sehlmeyer, U. et al., Accumulation of chromium in Chinese hamster V79-cells and nuclei, *Arch. Toxicol.*, 64, 506–508, 1990.

9. Baruthio, F., Toxic effects of chromium and its compounds, *Biol. Trace Elem. Res.*, 32, 145–153, 1992.

9a. Elbetieha, A. and Al-Hamood, M.H., Long-term exposure of male and female mice to trivalent and hexavalent chromium compounds: effect on fertility, *Toxicology*, 116, 39–47, 1997.

10. Anderson, R.A., Bryden, N.A., and Waters, R., EDTA chelation therapy does not selectively increase chromium losses, *Biol. Trace Elem. Res.*, 70, 265–272, 1999.

11. Kolacinski, Z. et al., Acute potassium dichromate poisoning: A toxicologic case study, *J. Toxicol. Clin. Toxicol.*, 37, 785–791, 1999.

12. Connett, P.H. and Wetterhahn, K.E., Metabolism of the carcinogen chromate by cellular constituents, *Struct. Bonding*, 54, 93–124, 1983.

13. Suzuki, Y. and Fukuda, K., Reduction of hexavalent chromium by ascorbic acid and glutathione with special reference to the rat lung, *Arch. Toxicol.*, 64, 169–176, 1990.

14. Standeven, A.M. and Wetterhahn, K.E., Ascorbate is a principal reductant of chromium(VI) in rat lung ultrafiltrates and cytosols, and mediates chromium-DNA binding *in vitro*, *Carcinogenesis*, 13, 1319–1324, 1992.

15. Preuss, H.G. and Anderson, R.A., Chromium update: Examining recent literature 1997–1998, *Curr. Opin. Clin. Metab. Care*, 1, 509–512, 1998.

16. Stearns, D.M. et al., Chromium picolinate produces chromosome damage in Chinese hamster ovary cells, *FASEB J.*, 9, 1643–1649, 1995.

17. Stearns, D.M., Belbruno, J.J., and Wetterhahn, K.E., A prediction of chromium(III) accumulation in humans from chromium dietary supplements, *FASEB J.*, 9, 1650–1657, 1995.

18. Speetjens, J.K. et al., The nutritional supplement chromium(III) tris(picolinate) cleaves DNA, *Chem. Res. Toxicol.*, 12, 483–487, 1999.

19. Wasser, W.G., Feldman, N.S., and D'Agati, V.D., Chronic renal failure after ingestion of over-the-counter chromium picolinate, *Ann. Intern. Med.*, 126, 410, 1997.

20. Cerulli, R. et al., Chromium picolinate toxicity, *Ann. Pharmacother.*, 32, 428–431, 1998.
21. Garcia, E.M. et al., Chromium levels in portable water, fruit juices and soft drinks: influence on dietary intake, *Sci. Total Environ.*, 241, 143–150, 1999.
22. O'Brien, P., Barret, J., and Swanson, F., Chromium (V) can be generated in the reduction of chromium (VI) by glutathione, *Inorg. Chim. Acta.*, 108, L19-L20, 1985.
23. Goodgame, D.M.L. and Joy, M.A., EPR study of the Cr(V) and radical species produced in the reduction of Cr(VI) by ascorbate, *Inorg. Chim. Acta.*, 135, 115–118, 1987.
24. Stearns, D.M. and Wetterhahn, K.E., Reaction of Cr(VI) with ascorbate produces chromium(V), chromium(IV), and carbon-based radicals, *Chem. Res. Toxicol.*, 7, 219–230, 1994.
25. Shi, X. et al., Chromate-mediated free radical generation from cysteine, penicillamine, hydrogen peroxide, and lipid hydroperoxides, *Biochim. Biophys. Acta.*, 1226, 65–72, 1994.
26. Salnikow, K., Zhitkovich, A., and Costa, M., Analysis of the binding sites of chromium to DNA and protein *in vitro* and in intact cells, *Carcinogenesis*, 13, 2341–2346, 1992.
27. Zhitkovich, A., Voitkun, V., and Costa, M., Glutathione and free amino acids form stable complexes with DNA following exposure of intact mammalian cells to chromate, *Carcinogenesis*, 16, 907–913, 1995.
28. Voitkun, V., Zhitkovich, A., and Costa, M., Complexing of amino acids to DNA by chromate in intact cells, *Environ. Health Perspect.*, 102, Suppl. 3, 251–255, 1994.
29. Costa, M., Analysis of DNA-protein complexes induced by chemical carcinogens, *J. Cell Biochem.*, 92, 127–135, 1990.
30. Zhitkovich, A., Messer, J., and Shrager, S., Reductive metabolism of Cr(VI) by cysteine leads to the formation of binary and ternary Cr-DNA adducts in the absence of oxidative DNA damage, *Chem. Res. Toxicol.*, 13, 2000.
31. Zhitkovich, A., Voitkun, V., and Costa, M., Formation of the amino acid-DNA complexes by hexavalent and trivalent chromium *in vitro*: Importance of trivalent chromium and the phosphate group, *Biochemistry*, 35, 7275–7282, 1996.
32. Zhitkovich, A. et al., Utilization of DNA-protein crosslinks as a biomarker of chromium exposure, *Environ. Health Perspect.*, 106, Suppl. 4, 969–974, 1998.
33. Sugiyama, M., Wang, X.-W., and Costa, M., Comparison of DNA lesions and cytotoxicity induced by calcium chromate in human, mouse, and hamster cell lines, *Cancer Res.*, 46, 4547–4551, 1986.
34. Kortenkamp, A. et al., A role for molecular oxygen in the formation of DNA damage during the reduction of the carcinogen chromium (VI) by glutathione, *Arch. Biochem. Biophys.*, 329, 199–207, 1996.
35. Casavevall, M., da C. Fresco, P., and Kortenkamp, A., Chromium(VI)-mediated DNA damage: oxidative pathways resulting in the formation of DNA breaks and abasic sites, *Chem. Biol. Inter.*, 123, 117–132, 1999.
36. Yuann, J.M. et al., *In vivo* effects of ascorbate and glutathione on the uptake of chromium, formation of chromium(V), chromium-DNA binding and 8-hydroxy-2'-deoxyguanosine in liver and kidney of osteogenic disorder Shionogi rats following treatment with chromium(VI), *Carcinogenesis*, 20, 1267–1275, 1999.

37. Sugden, K.D. and Wetterhahn, K.E., Direct and hydrogen peroxide-induced chromium(V) oxidation of deoxyribose in single-stranded and double-stranded calf thymus DNA, *Chem. Res. Toxicol.*, 10, 1397–1406, 1997.

38. Levina, A. et al., *In vitro* plasmid DNA cleavage by chromium(V) and -(IV) 2-hydroxycarboxylato complexes, *Chem. Res. Toxicol.*, 12, 371–381, 1999.

39. Bose, R.N. et al., Mechanisms of DNA damage by chromium(V) carcinogens, *Nucl. Acids Res.*, 26, 1588–1596, 1998.

40. Shi, X. et al., Reaction of Cr(VI) with ascorbate and hydrogen peroxide generates radicals and causes DNA damage: role of a Cr(IV)-mediated Fenton-like reaction, *Carcinogenesis*, 15, 2475–2478, 1994.

41. Voitkun, V., Zhitkovich, A., and Costa, M., Cr(III)-mediated crosslinks of glutathione or amino acids to the DNA phosphate backbone are mutagenic in human cells, *Nucl. Acids Res.*, 26, 2024–2030, 1998.

42. Lio, S. and Dixon, K., Induction of mutagenic DNA damage by chromium(VI) and glutathione, *Environ. Mol. Mutat.*, 28, 71–79, 1996.

42a. Chen, J. and Thilly, W.G., Mutational spectrum of chromium(VI) in human cells, *Mutat. Res.*, 323, 21–27, 1994.

43. Ye, J. et al., Role of reactive oxygen species and p53 in chromium(VI)-induced apoptosis, *J. Biol. Chem.*, 274, 34974–34980, 1999.

43a. Carlisle, D.L. et al., Apoptosis and p53 induction in human lung fibroblasts exposed to chromium(VI): Effect of ascorbate and tocopherol, *Toxicol. Sciences*, 55, 60–68, 2000.

44. Agency for Toxic Substances and Disease Registry, *Toxicological Profile for Chromium*, U.S. Dept. of Health and Human Services, Washington, D.C., 1993.

45. Burke, T. et al., Chromite ore processing residue in Hudson County, New Jersey, *Environ. Health Perspect.*, 92, 131–138, 1991.

46. Stern, A.H. et al., Residential exposure to chromium waste-urine biological monitoring in conjunction with environmental exposure monitoring, *Environ. Res.*, 58, 147–162, 1992.

47. Stern, A.H. et al., The association of chromium in household dust with urinary chromium in residences adjacent to chromate production waste sites, *Environ. Health Perspect.*, 106, 833–839, 1998.

48. Dawson, B.S.W. et al., Interlaboratory determination of copper, chromium and arsenic in timber treated with wood preservative, *Analyst.*, 116, 339–346, 1991.

49. Warner, J.E. and Solomon, K.R., Acidity as a factor in leaching of copper, chromium and arsenic from CCA-treated dimension lumber, *Environ. Toxicol. Chem.*, 9, 1331–1337, 1990.

50. Barnhard, J., Occurrences, uses, and properties of chromium, *Re.g. Toxicol. Pharm.*, 26, S3-S7, 1997.

51. Aceto, M. and Fidele, A., Rain water effect on the release of arsenic, chromium and copper from treated wood, *Fresenius Environ. Bull.*, 3, 389–394, 1994.

52. Sainio, E.L. et al., Metals and arsenic in eye shadows, *Contact Dermatitis*, 42, 5–10, 2000.

53. Ingber, A., Cammelgaard, B., and David, M., Detergents and bleaches are sources of chromium contact dermatitis in Israel, *Contact Dermatitis*, 38, 101–104, 1998.

54. Bonde, J.P. and Christensen, J.M., Chromium biological samples from low-level exposed stainless steel and mild steel welders, *Arch. Environ. Health*, 46, 225–229, 1991.

55. Kilburn, K.H. et al., Cross-shift and chronic effects of stainless steel welding related to internal dosimetry of chromium and nickel, *Am. J. Ind. Med.*, 17, 607–615, 1990.

56. Stridsklev, I.C. et al., Biologic monitoring of chromium and nickel among stainless steel welders using the manual metal arc method, *Int. Arch. Occup. Environ. Health*, 65, 209–219, 1993.

57. Zhitkovich, A. et al., DNA-protein crosslinks in peripheral lymphocytes of individuals exposed to hexavalent chromium compounds, *Biomarkers*, 1, 86–93, 1996.

58. Edme, J.L. et al., Assessment of biological chromium among steel and mild steel welders in relation to welding process, *Int. Arch. Occup. Environ. Health*, 70, 237–242, 1997.

59. Lewalter, J. et al., Chromium bond detection in isolated erythrocytes: a new principle of biological monitoring of exposure to hexavalent chromium, *Int. Arch. Occup. Environ. Health*, 55, 305–318, 1985.

60. Wiegand, H.J., Ottenwalder, H., and Bolt, H.M., Recent advances in biological monitoring of hexavalent chromium compounds, *Science Total Environ.*, 71, 309–315, 1988.

61. Coogan, T.P. et al., Distribution of chromium within cells of the blood, *Toxicol. Appl. Pharm.*, 108, 157–166, 1991.

62. Merritt, K. and Brown, S.A., Release of hexavalent chromium from corrosion of stainless steel and cobalt-chromium alloys, *J. Biomed. Mat. Res.*, 29, 627–6333, 1995.

63. Vaglenov, A. et al., Genotoxicity and radioresistance in electroplating workers exposed to chromium, *Mutat. Res.*, 446, 23–34, 1999.

64. Angerer, J. et al., Occupational chronic exposure to metals.I. Chromium exposure of stainless steel welders — biological monitoring, *Int. Arch. Occup. Environ. Health*, 59, 503–512, 1987.

65. Jelmert, O., Hansteen, I.-L., and Langard, S., Chromosome damage in lymphocytes of stainless steel welders related to past and current exposure to manual arc welding fumes, *Mutat. Res.*, 320, 223–233, 1994.

66. Miksche, L.W. and Lewalter, J., Health surveillance and biological effect monitoring for chromium-exposed workers, *Regul. Toxicol. Pharmacol.*, 26, S94-S-99, 1997.

67. Kortenkamp, A., Beyersmann, D., and O'Brien, P., Uptake of Cr(III) complexes by erythrocytes, *Toxicol. Environ. Chem.*, 14, 23–32, 1987.

68. Nagaya, T. et al., Sister-chromatid exchanges in lymphocytes from 12 chromium platers: A 5-year follow-up study, *Toxicol. Lett.*, 58, 329–335, 1991.

69. Werfel, U. et al., Elevated DNA single-strand breakage frequencies in lymphocytes of welders exposed to chromium and nickel, *Carcinogenesis*, 19, 413–418, 1998.

69a. Elias, Z. et al., Chromosome aberrations in peripheral blood lymphocytes of welders and characterization of their exposure by biological samples analysis, *J. Occup. Med.*, 31, 477–483, 1989.

70. Katsifis, S.P. et al., Interaction of nickel with mutagens in the induction of sister chromatid exchanges in human lymphocytes, *Mutat. Res.*, 359, 7–15, 1996.

70a. Fenech, M., Aitken, C., and Rinaldi, J., Folate, vitamin B12, homocysteine staus and DNA damage in young Australian adults, *Carcinogenesis*, 19, 1163–1171, 1998.

71. Gao, M. et al., Use of molecular epidemiological techniques in a pilot study on workers exposed to chromium, *Occup. Environ. Med.*, 51, 663–668, 1994.
72. Kim, H., Cho, S.-H., and Chung, M.-H., Exposure to hexavalent chromium does not increase 8-hydroxydeoxyguanosine levels in Korean chromate pigment workers, *Ind. Health*, 37, 335–341, 1999.
73. Popp, W. et al., Investigations of the frequency of DNA strand breakage and cross-linking and of sister chromatid exchanges in the lymphocytes of electric welders exposed to chromium- and nickel-containing fumes, *Int. Arch. Occup. Environ. Health*, 63, 115–120, 1991.
74. Costa, M., Zhitkovich, A., and Toniolo, P., DNA-protein crosslinks in welders: molecular implications, *Cancer Res.*, 53, 460–463, 1993.
75. Taioli, E. et al., Increased DNA-protein crosslinks in lymphocytes of residents living in chromium contaminated areas, *Biol. Trace Element Res.*, 50, 175–180, 1995.
75a. Taioli, E. et al., Normal values of DNA-protein crosslinks in mononuclear cells of a population of healthy controls, *Cancer J.*, 8, 76–78, 1995.
76. Quievryn, G. and Zhitkovich, A., Loss of DNA-protein crosslinks from formaldehyde-exposed cells occurs through spontaneous hydrolysis and an active repair process linked to proteosome function, *Carcinogenesis*, 21, 1573–1580, 2000.
76a. Brazilian, H.E., Schmid, E., and Bauchinger, M., Chromosome aberrations in nuclear power plant workers: the influence of dose accumulation and lymphocyte life-time, *Mutat. Res.*, 306, 197–202, 1994.
77. Costa, M. et al., DNA-protein crosslinks produced by various chemicals in cultured human lymphoma cells, *J. Tox. Env. Health*, 50, 101–116, 1997.
78. Voitkun, V. and Zhitkovich, A., Analysis of DNA-protein crosslinking activity of malondialdehyde *in vitro*, *Mutat. Res.*, 424, 97–106, 1999.
79. Singh, J. et al., Sensitive determination of chromium-DNA adducts by inductively coupled plasma mass spectroscopy with a direct injection high-efficiency nebulizer, *Toxicol. Sciences*, 46, 260–265, 1998.
80. Zhitkovich, A. and Costa, M., A simple, sensitive assay to detect DNA-protein-crosslinks in intact cells and *in vivo*, *Carcinogenesis*, 13, 1485–1489, 1992.

chapter twenty

Using plankton food web variables as indicators for the accumulation of toxic metals in fish

Carol L. Folt, Celia Y. Chen, and Paul C. Pickhardt

Contents

Abstract Recent studies of metal contamination worldwide have shown that even in what appear to be pristine lakes far from industrial centers, certain metals reach high levels in fish. However, there is a great deal of unexplained variation in metal levels in fish from different lakes and aqueous metal concentrations often do not predict metals in fish. The explanation for these observations lies in part in the way the food webs in different lakes influence metal accumulation by fish. Several key characteristics of plankton assemblages are known to be important, yet these variables have rarely been discussed together. The objectives of this paper are to highlight these processes by reviewing: 1) the role played by zooplankton in metal trophic transfer to fish, 2) differences in metal transfer to fish among food webs with different zooplankton assemblages, and 3) the extent to which plankton food web structure is a good predictor of metal burden in fish. We also identify several indicator variables that describe aspects of plankton food webs, which strongly influence metal contamination in fish (e.g., food chain length, horizontal complexity, plankton size-structure, algal productivity). Finally, we discuss the potential for metal contamination in polar lakes in light of these food web variables. Characterized by cold temperatures, high flushing rates, low algal production, and low DOC these lakes are in many ways some of the most pristine in the world. Yet, low productivity, simple trophic structure, long food chains, and large lipid rich zooplankton make fish in many of these lakes strong candidates for accumulating high levels of Hg and contaminants like PCBs. Programs to detect early signs of metal contamination should be designed to include key food web indicator variables in order to better predict rates of metal trophic transfer into fish, mammals and birds.

I. Introduction

Metals enter aquatic food webs in several ways, ranging from atmospheric deposition to surface or groundwater flow of metal enriched water. Once in lakes, metals become incorporated in the food web via direct uptake from the water or the ingestion of metal laden foods.[1] Moreover, metal contamination can result in large changes in species composition, species diversity, biomass, and growth of aquatic organisms.[2,3] The accumulation of metals in fish tissue has been recognized as a human health hazard. Other consumers of fish, such as wildlife, are also at risk from metal contamination in some lakes. Mercury (Hg) is a metal of special concern because elevated levels of Hg have been found in fish from a large number of lakes in remote areas, including high latitude lakes.[4] These lakes are not associated with local sources of contamination but receive Hg and other contaminants via long-range atmospheric transport from anthropogenic and natural sources.[5,6]

Complex biological processes in lakes make it impossible to simply predict the concentration of metals in fish from the concentration of metals in water. One of the primary reasons for this poor correlation is that diet appears to be responsible for $\geq 50\%$ of the burden of many metals in fish.[7-9]Therefore, it is critical to investigate the metal uptake and transfer properties of the organisms comprising fish diets. Metals are taken up directly from the water and the food via bioaccumulation and some metals also biomagnify. Biomagnification describes a systematic increase in the concentration of contaminants in tissue as the chemical moves up a food web (i.e., from algae to zooplankton to fish). Metals such as lead (Pb) and cadmium (Cd) bioaccumulate, but may not biomagnify. In contrast, contaminants such as

Hg, DDT, and PCBs biomagnify in lake food webs, which can result in high concentrations of these contaminants in fish from lakes with low contaminant levels in the water. To follow the movement of such chemicals requires an appreciation of the entire food web.

Past studies have shown that there is a great deal of unexplained variation in metal levels in fish of the same species found in different lakes, even if background metal levels are similar.[10,11] The explanation for the variation among lakes lies in part in the way that food webs in different lakes influence metal accumulation by fish. The most well-studied food web effects include age and trophic position of the fish. Both variables are positively correlated with the burdens of biomagnifying contaminants (e.g., Hg, PCBs) in fish.[12,13] For metals that biomagnify, predatory fish have higher body burdens of Hg than planktivorous fish of equal body size. Older and larger fish also tend to have higher burdens. Geographic differences in metal burdens arise if fish occupy different feeding niches or feed on prey at different levels on the food chain in different lakes. These processes have been reviewed numerous times in the literature so are not covered again here.[12,14,17]

However, other aspects of the food web influence metal movement through aquatic food webs, and these processes have been less well studied. Principally, research has been lacking to link metals in the water to metals in the lower levels of the food web. This need was highlighted in the Mercury Study Report to Congress where it states, "Ongoing efforts to understand mercury bioaccumulation in aquatic systems continue to be focused in trophic levels three and four. Additional emphasis should be placed on research at the lower trophic levels. In particular, there is a need to understand the determinants of mercury accumulation in phytoplankton and zooplankton and how rapid changes in plankton biomass impact these values." [18] This knowledge is needed to evaluate the trophic pathways of different metals. The necessity for such research is widely recognized[19,20] because these pathways ultimately determine the bioavailability of these chemicals to humans or to other predators that consume fish.

There are other compelling reasons to focus on the processes regulating metal in zooplankton. For example, research suggests that the efficiencies with which planktonic food webs transfer metals from algae to fish relate to properties of specific zooplankton taxa (e.g., individual metal burdens, body size, growth rates, and susceptibility to predators). Zooplankton are also especially sensitive to environmental stress,[19,21] so distinct assemblages typify certain contaminants. Therefore, categorizing and elucidating unique abilities of plankton in different food webs to transfer metals to fish will increase our ability to predict metal contamination in fish across different types of lakes.

The aims of this paper are to discuss current understanding for the role of plankton in the bioaccumulation and biomagnification of toxic metals to fish in lakes. Metals with different potentials for bioaccumulation and biomagnification will be contrasted, and the importance of different zooplankton types and assemblages will be explained. Our study focuses on nonessential potentially toxic metals (e.g., Hg, MeHg, As, Cd, Pb) in freshwater systems. Another important feature of this paper, is that we examine influences of the plankton food web on metal transfer rates that occur independently from differences in overall metal inputs across systems. So, for example, this study offers explanations for why two lakes with similar metal inputs but different plankton food webs can produce fish with different metal burdens. Under natural conditions, food web structure and metal inputs influence fish burdens. Processes responsible for the transport and chemical transformation of metals in lakes are not covered here.

We also identify several indicator variables for the accumulation of metals in fish. These variables characterize features of the food web (e.g., food chain length, horizontal complexity, zooplankton mean body size, and algal productivity) that are likely to result in either an increase or decrease in the trophic transfer of metals through the plankton to fish. The paper is divided into sections wherein we discuss the role played by zooplankton in the trophic transfer of metal into fish, the differences in metal transfer rates to fish among food webs with different zooplankton communities, and the extent to which zooplankton food web structure is a good predictor of metal burden in fish. Finally, as part of the proceedings for the International Conference on Arctic Development, Pollution, and Biomarkers of Human Health, we apply these ideas about food web structure to examining metal contamination in arctic and subarctic lakes.

II. The role of plankton in metal trophic transfer to fish

A. Trophic transfer through the food web

To understand metal movement through food webs (i.e., trophic transfer) to fish requires beginning with the phytoplankton. Metals enter the food web by passively adsorbing to algal cells and other negatively charged particles.[22] Algae play a vital role in trophic transfer because they concentrate metals from the water by 10–1000X.[23,26] These concentrations are determined by algal cell size, density,[27] cell wall properties,[27,29] and physiological condition.[30]

The next step is to follow metals as they move from phytoplankton to zooplankton. Metal burdens for particular zooplankton species derive from the net balance between their metal uptake via food and water, and elimination. Metals such as Hg, MeHg, As, Cd, and Zn can enter algal cytoplasm. This is extremely important because there is a strong positive (and predictive) linear relationship between assimilation efficiency (AE) of zooplankton for particular metals and the percentage of each of these metals found in algal cytoplasm.[31] Once ingested, metals accumulate in zooplankton tissues at the site of uptake (gill for water and intestine for food).[32,33] Metals that adhere to the surface of ingested particles can be desorbed in the zooplankton gut leading to uptake and in some cases, toxicity.[22,34]

The last step in trophic transfer is to follow the metal from zooplankton to fish via dietary uptake. Four critical processes determine the bioaccumulation of metals by fish from water and food (i.e., weight-specific ingestion, AE, elimination, and growth.[30] Large differences in fish metal burdens can arise if fish consume prey items that vary with respect to any of these vital processes (i.e., AE, ingestion rates). For instance, metals associated with the chitinous exoskeleton of some species are poorly assimilated by fish relative to those associated with polar elements in other species.[35] The extent to which any species of zooplankton is an important determinant of fish burden is decided by its own metal uptake kinetics and how frequently it is included in fish diets. Therefore, we expect that fish will accumulate metals at different rates from different species of zooplankton.

B.   Taxa-specific information

Currently, little is known about metal burdens and transfer potential of individual zooplankton taxa. In most studies, metal burdens of zooplankton are determined by pooling taxonomically complex samples. Investigations of particular taxa have been limited to a few species. However, some relevant published information exists. For instance, larger zooplankton species usually have greater total burdens of metals.[36] This is especially important because larger zooplankton are also generally the preferred species in fish diets. The large cladoceran, *Daphnia*, has been shown to have greater mass-specific burdens than other zooplankton taxa (including predatory copepods) for at least some metals (e.g., As, MeHg, Cd; Folt unpublished data).[9,37,38] This result suggests that these large cladocerans are likely to be stronger conduits of metal to fish than many smaller species (copepods, rotifers, ciliates).

Fish also consume numerous large macroinvertebrate predators that have been more thoroughly examined as conduits of metal and other contaminants to fish.[12] Their presence in a lake indicates a high potential for fish dietary metal uptake because these invertebrates feed higher on the food web than many other zooplankton, accumulate large metal or contaminant burdens, and are prime fish food. The midge larvae, *Chaoborus*, has been even suggested as a biomonitoring tool because its uptake of Cd is proportional to the concentration of the free metal ion in the water after controlling for free H^+.[39,40] Nevertheless, even within the large macroinvertebrate taxa, metals burdens vary, making generalizations across these taxa problematic.

III. The role of the food web structure in metal transfer to fish

Building upon the premise that individual zooplankton differ in the extent to which they serve as conduits of metal to fish, it becomes obvious that food webs comprising different zooplankton taxa will differ significantly in the way they transfer metal through the food web. In this section, we cover four key aspects of the food web in a lake that can be used as potential indicators for the extent to which the food web is likely to efficiently transfer metals to fish (all things other than food web structure being equal). These four aspects are food chain length, horizontal complexity, zooplankton size-structure, and algal productivity.

A. Food chain length and horizontal complexity

Figure 20.1[41] depicts an aquatic community comprising the full range of zooplankton, macroinvertebrates and fish. Trophic levels of each group are shown; small-bodied species include many herbivores though omnivorous and predaceous rotifers and crustaceans are also present. The length of the food chain is simply the number of trophic levels present. Many studies pool zooplankton in the food web within trophic levels two and three, putting *Mysis* and *Chaoborus* into a single trophic category.[42-44] This has led to the idea that aquatic webs are simple four-level chains.[45] Horizontal complexity is the number of connections between species within a trophic

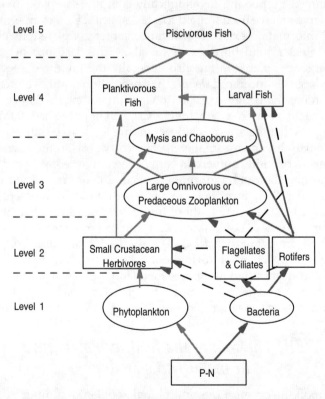

Figure 20.1 An aquatic community comprising the full range of zooplankton, macroinvertebrates and fish.[41] Dashed and solid lines depict likely feeding links; dashed lines delineate links that increase as the composition of the food web shifts toward smaller-bodied zooplanton.

level. Food webs with more horizontal complexity have more links among zooplankton species.

It is well known that an increase in food chain length (the number of trophic levels from algae to fish) results in an increase in MeHg and PCBs in fish via biomagnification.[12,14,46,47] Biomagnification of Zn and Cd has also been measured in specific cases[11,30] but is less well understood for Hg. In contrast, metals, like As and Pb, biodiminish in food webs, so that an increase in trophic levels actually results in a reduction in the transfer of these metals to fish (Table 20.1).[48] *Mysis* and *Chaoborus* (Figure 20.1) add trophic complexity by increasing food chain length from algae to fish. Their presence in lakes has also been found to correlate strongly with the burden of various biomagnifying contaminants in fish.[12]

Recent work has also shown that increasing the number of feeding connections within zooplankton trophic levels results in a reduction in movement of metals to fish.[10,11] This is a new and complementary paradigm for food chain effects. Horizontal complexity probably reduces metal movement into fish by changing food web structure, from large zooplankton with higher mass-specific metal burdens that fish prefer to smaller less preferred species that have lower burdens. Hence, to predict the way metal will transfer to fish in particular lakes requires knowledge of chain length and horizontal complexity. Further, to understand horizontal complexity requires a deeper examination of the zooplankton species composition that characterizes food webs that are more or less horizontally complex.

B. Zooplankton size-structure

To fully characterize aquatic food webs is difficult and time consuming. However, these webs are often distinguished by the mean body length of the zooplankton assemblage, which is a variable that can be more easily measured.[49] For the purpose of predicting the way food webs in different lakes are likely to transfer metal to fish, this may also be a suitable indicator variable. Food webs characterized by small mean size tend to consist of plankton taxa (e.g., rotifers, small cladocerans, copepods) that have lower mass-specific body burdens of contaminants like Hg than larger-bodied taxa, e.g., large cladocerans.[8,31-33] Moreover, small-bodied species tend to be much less favored foods for fish.[50-52] In fact, as the abundance of zooplankton eating fish increases, the dominance of large cladocerans declines, and the relative abundance of smaller species increases.[53-55] Finally, there is evidence that horizontally complex communities are frequently small bodied.[10,21,56] Hence, we suggest that metal transfer to fish should be reduced in lakes characterized by small-bodied plankton food webs that are horizontally complex.

Small-bodied food webs have also been associated with a number of factors in addition to zooplanktivory including high temperatures,[57,58] contaminant stress,[57,59] metal stress,[2,3] changes in the N:P stoichiometry of the algae[60] and extreme nutrient enrichment.[45] Therefore, estimating the mean

Table 20.1   Variables That Have Been Positively (+) or Negatively (−) Correlated with Metal (Mercury, Arsenic or Lead) in Water and Fish in Prior Studies

	Mercury		Arsenic		Lead	
	Fish	Water	Fish	Water	Fish	Water
Water Chemistry						
ANC	−[a,b]					
pH	−[d,e,f]	−[j]			−[i]	
DOC	+[k]			+[h]	+[a]	+[h]
Cl⁻					+[a]	
Total N				+[h]		+[h]
Total P					+[a]	
N:P	+[g]					+[h]
Physical Variables						
Depth			+[a]			
Watershed, Lake area	+[a,b,c]					
Temperature		−[h]			+[g]	
Lake latitude					−[a,g]	
Lake Trophic Status						
Chlorophyll *a*					+[a,g]	
Secchi depth						−[h]
Land Use Variables						
Agricultural				+[h]		
Road density			+[a]		+[a]	+[h]
Food Web Variables						
Maximum chain length	+[a,g]		−[a]			
Linkages	−[a]		−[a]			

[a] Data from Reference 10.   [g] Data from Reference 60.

[b] Data from Reference 94.   [h] Data from Reference 11.

[c] Data from Reference 95.   [i] Data from Reference 99.

[d] Data from Reference 96.   [j] Data from Reference 9.

[e] Data from Reference 97.   [k] Data from Reference 100.

[f] Data from Reference 98.

body size of the plankton usually provides useful information about other vital processes that affect lake water quality. Other variables that are also associated with large-bodied vs. small-bodied zooplankton food webs are summarized in Table 20.2.

Unfortunately, direct tests of the effect of zooplankton size-structure on metal accumulation in fish are not presently available. Still, size-structure has been shown to influence trophic transfer of other elements (e.g., C, N, and P). For example, Stockner and Porter[61] demonstrated that in horizontally complex food webs dominated by small-bodied zooplankton (hereafter, termed small, more complex webs or SMC webs), the transfer of carbon to fish was much less than in less horizontally complex webs dominated by large zooplankton (large, less complex webs or LLC webs). Piscivorous salmon were also significantly smaller in SMC lakes than in LLC lakes, indicating that less carbon was being moved up the food chain in SMC lakes. Lower feeding efficiency

Table 20.2   Summary of Key Descriptors of Conditions Characterizing Food Webs That Consist of Small-Bodied Zooplankton and That Are More Complex (more Linkages) Versus Food Webs That Consist of Large-Bodied Zooplankton and That Are Less Complex (Fewer Linkages)*

Descriptors	Small-Bodied/More Complex Food Webs (SMC)	Large-Bodied/Less Complex Food Webs (LLC)
Environmental conditions	shallower, warmer[a]	deeper, cooler
	more eutrophic[a,b]	less eutrophic
	lower N:P ratio[b]	higher N:P ratio
	lower DO in bottom water[a,b]	greater DO in bottom water
Biological associations	smaller algal cells[c,d,e]	larger algal cells
	lower water clarity[d]	greater water clarity
	lower relative biomass bacteria:phytoplankton[f]	higher relative biomass bacteria:phytoplankton
	higher biomass zooplanktivorous fish[a,b]	lower biomass zooplanktivorous fish
Zooplankton characteristics[g]	rotifers (more), protozoans (more), small cladocerans — *Bosmina* sp., small herbivorous calanoid copepods, smaller predatory cyclopoid copepods smaller average body size[h] more linkages among zooplankton[i]	rotifers (fewer), protozoans (fewer), large cladocerans — *Daphnia* sp., large omnivorous calanoid copepods, larger predatory cyclopoid copepods larger average body size fewer linkages among zooplankton
Macroinvertebrates	some *Chaoborus* littoral, benthic invertebrates	other *Chaoborus* *Mysis relicta, Leptodora*

* See text for further details.
[a] Environmental data from Reference 21.
[b] Nutrient data from Reference 60.
[c] Data from Reference 53.
[d] Data from Reference 54.
[e] Data from Reference 101.

[f] Data from Reference 69.
[g] Zooplankton characteristic data from Reference 21, 49, 53, 54, 56, 101 through 104.
[h] Zooplankton body size data from Reference 54, 101, 102.
Linkage data from Reference 10, 21.

of fish on the smaller species and losses of carbon at each step up the trophic pathway were proposed to explain the pattern. Similar relationships for metals such as MeHg appear plausible and should be tested.

C.   Algal productivity

Another important process that can alter the way different food webs transfer metals is the algal productivity. Growing evidence suggests that increasing nutrients and algal productivity may reduce the transfer of metals that biomagnify from algae to fish. If true, this relationship may explain in part why certain contaminants are found at harmful levels in fish from more pristine systems.[47,62]

Algal productivity could affect metal trophic transfer in lake in several ways. First, nutrient enrichment or eutrophication could alter web structure (e.g., chain length, dominance of particular taxa) by favoring particular algal species.[45,63] Chain length does not appear to increase with enrichment except in a few ultraoligotrophic lakes.[45] However, taxonomic trends suggest that large zooplankton (cladocerans) are less dominant in more eutrophic lakes. Their decline may be due to increased planktivory[64-66] or to an increase in the dominance of inedible blue-green algae.[67] Nevertheless, these shifts in zooplankton size-structure or taxonomic composition under high algal productivity could result in lower metals in fish for reasons discussed earlier. Second, low productivity lakes tend to be dominated by algal cells with high lipid levels. They are also found at lower abundances and have lower mass-specific growth rates. These features all tend to concentrate metals (on a mass-specific basis) in algal cells at low productivity, which in turn increases trophic transfer to their consumers.[68-70] Third, increasing algal productivity can result in metal growth dilution (equal to less metal per cell in rapidly growing populations) and increased sedimentation rates.[68,71-73] This evidence also supports the likelihood of reduced metal trophic transfer to fish in more eutrophic lakes. Finally, is important to note that while many lines of evidence suggest that increasing nutrients and algal productivity may result in lower rates of metal transfer, clear unambiguous evidence is still lacking.

IV. Factors predicting metal transfer to fish

Although numerous studies have investigated burdens of particular metals in fish from various locations, there are few that have combined these measures with estimates of metals in the rest of the food web. Research combining food web variables with other factors that influence metal levels (e.g., limnological conditions, land use patterns) is also uncommon. As a result, there is not a compelling consensus on the relative importance of all the various factors that drive metal burdens in fish across lakes. Toward this goal, we constructed a table comparing key variables that have been correlated with concentrations of certain metals (Hg, As and Pb) in water and fish (Table 20.1). Based on this table, we can compare several metals with respect to the prime factors that are likely to increase or decrease aqueous metal levels and trophic transfer rates. Moreover, we can determine for each metal, whether the factors that predict high levels in water also predict high levels in fish tissue. For instance, Hg appears to be found at higher aqueous levels in lakes with low pH and low temperature. However, high DOC, long food chain length and lower horizontal complexity (low linkage) are the variables that predict high Hg levels in fish tissue. Mercury is clearly a metal for which understanding the food web is critical.[10,11,60] Aqueous levels of Arsenic and Pb are found in high nutrient and high productivity lakes. Like Hg, As and Pb in fish tissue appear to be correlated with high DOC. Arsenic levels in fish tissue are also lower in lakes with longer food chains and high linkage, indicating the influence of food web variables on trophic movements for this metal also. Unlike arsenic, Pb in fish tissue does not appear to be correlated with food web structure but rather with chemical characteristics associated with productivity (DOC, chloride ion, total P, and Chl *a*). Taken together, these studies suggest that Hg concentrations in water and fish are higher from cold, low productivity lakes whereas As and Pb are higher in water and fish from high productivity lakes. Similar comparisons for additional metals would be especially useful for making broad inferences and designing sampling programs.

V. Application to evaluating contamination in polar food webs

In this final section, we consider in brief various food web variables as indicators of the potential for metal contamination in arctic lakes. This is offered as a means to generate hypotheses and to contribute to sampling programs in the region, rather than as a comprehensive review or evaluation of the current status of metals in arctic freshwaters. Although most arctic and subarctic lakes are far from industrial centers, there is mounting evidence that anthropogenically derived contaminant concentrations are elevated above background levels throughout the region. Moreover, arctic biomagnification factors are an order of magnitude per trophic level, which is higher than many more southern lakes.[20,74]

Processes including long-range transport, regional temperatures, and cold condensation appear to concentrate contaminants in the polar regions. Cold condensation (when contaminants in warmer locations are volatized and then condense in cooler environments) results in an accumulation of contaminants at high latitude driven by regional temperature differentials.[75,76] Cold arctic temperatures also result in greater cryoconcentration, which collects contaminants in the water during the long periods of ice cover.[20,75,77]

However, polar lakes have been reported to act as conduits of contaminants rather than sinks.[77] Although snowmelt brings in large pulses of both nutrients[20,78] and contaminants, much of the load is exported downstream in these characteristic flushing events.[79] Arctic lakes also have lower inputs of organic matter and DOC (Table 20.3) and tend to have low algal productivity. Low DOC is usually associated with lower levels of metals such as Hg and Pb,[80] which may explain in part why contaminants in northern lake sediments tend to decrease with increasing latitude.[81] Using the classification systems of Duff et al.[82] and Pienitz et al.,[83] arctic lakes can categorized as tundra lakes, forest/tundra lakes, and forest lakes (Table 20.3). Based upon DOC levels alone, the tundra lakes could be predicted to have lower contaminant levels than forest or forest/tundra lakes. They would also flush more thoroughly each year.

Once contaminants such as Hg and PCBs have been deposited in the polar regions, bioaccumulation and biomagnification appear likely to be

Table 20.3   Ranges and Average Values for Basic Limnological Parameters
Measured in Three Vegetation Categories[82]

	Vegetation of Lake Watershed		
Limnological Parameter	Tundra	Forest/Tundra	Forest
Depth	Very shallow[a]	Shallow[a]	Deeper[a]
Temperature	Coldest[d]	Cold[b,c]	4°C warmer[b,c]
Secchi depth (m)	4.1–7[a,b,c]	≈1.7[b]	≈4.4[b]
Chl. *a* (µg liter^{-1})	≈2[d]	≈4[b,c]	≈2.5[b]
DO (mg · liter^{-1})	Saturated[b,d]	11.1–14.6[b,c]	11.1–14.6[b,c]
DOC (mg · liter^{-1})	7.9–9.4[b]	≈15.9[b]	15.9+[b,c]
Total N:Total P	33:1–42:1[b]	33:1–42:1[b]	60:1[b]
pH	7.85[b]	7.61[b]	8.25[b]
Dissolved ion concentration	Lowest[a,d]	Low[b,c]	Moderate[b,c]

[a] Data from Russian lakes.[82]   [c] Averages/ranges from 24 Canadian
[b] Averages/ranges from 59 Canadian lakes.[105]   lakes.[83]
                                                  [d] Data from Alaskan lakes.[106]

enhanced by some conditions in arctic lakes and features of polar food webs. First, polar lakes are generally cold (Table 20.3). Fish grow more slowly at lower temperatures, but slower growth rates result in greater bioaccumulation of contaminants such as PCBs and Hg.[20,70] Second, zooplankton and fish in colder climates have higher lipid contents. High lipid in zooplankton increases the bioaccumulation and trophic transfer of lipophilic contaminants like Hg from plankton to fish[84-86] and lipid levels in fish are positively associated with levels of persistent contaminants in their tissues.[87] Third, these lakes also tend to have fairly low algal productivity, which can increase the mass-specific burden of metals and other contaminants in the phytoplankton and zooplankton. Fourth, macrozooplankton communities in arctic lakes are considered depauperate compared to temperate lakes.[88,89] However, the low diversity of crustacean zooplankton tends to be associated with lower horizontal food web complexity,[90] which is predicted to increase the trophic transfer of Hg.[10] Fifth, recent evidence on the length of the pelagic food web above phytoplankton revealed by zooplankton stable nitrogen enrichment suggests that the vertical food web structure in many northern lakes is longer than previously hypothesized.[91] The range in mean α15N from primary producers to top predators can be as high as 13 ppt although most food webs average 10 ppt.[17,92] Taken together, these findings underscore the great potential for vertical movement of metals and other contaminants up the pelagic food web in arctic lakes.

In summary, it appears that global climate conditions transport and accumulate anthropogenically-produced contaminants such as Hg, DDT and PCBs in polar regions. Characterized by cold temperatures, high flushing rates, relatively low algal production, DOC and low species diversity, these lakes are in many ways some of the most pristine in the world. Yet, arctic lakes that have relatively low productivity, simple trophic structure, long food chains, and large lipid rich zooplankton are strong candidates for rapid transfer of certain contaminants to fish. These food web processes appear likely to explain why fish in arctic lakes have been found to accumulate significant burdens of metal and other contaminants.[13,17,33,74,86,93] Inclusion of key food web indicator variables (chain length, horizontal complexity, zooplankton size structure, and productivity) in programs to detect early signs of contamination may better predict rates of metal trophic transfer into fish, mammals, and birds.

Acknowledgments

Our research is supported by NIEHS Superfund ES07373. We also thank the organizers of the International Conference on Arctic Development, Pollution and Biomarkers of Human Health (April 2000) wherein this manuscript was presented by C.L.F.

References

1. Wiener, J.G. and Spry, D.J., in *Environmental Contaminants in Wildlife: Interpreting Tissue Concentrations*, Beyer, W.N., Heinz, G.H., and Redmond-Norwood, A.W., Eds., CRC Press, Boca Raton, 1996, pp. 297–339.
2. Yan, N.D. and Strus, R., Crustacean zooplankton communities of acidic, metal-contaminated lakes near Sudbury, Ontario, *Can. J. Fish. Aquat Sci.*, 37, 2282–2293, 1980.
3. Yan, N.D. et al., Recovery of crustacean zooplankton communities from acid and metal contamination: comparing manipulated and reference lakes, *Can. J. Fish. Aquat. Sci.*, 53, 1301–1327, 1996.
4. Meili, M., The coupling of mercury and organic matter in the biogeochemical cycle towards a mechanistic model for the boreal forest zone, *Wat. Air Soil Pollut.*, 56, 333–347, 1991.
5. Allen-Gil, S.M. et al., Organochlorine pesticides and polychlorinated biphenyls (PCBs) in sediments and biota from four U.S. Arctic lakes, *Arch. Environ. Contam. Toxicol.*, 33, 378–387, 1997.
6. Rognerud, S. et al., Concentrations of trace elements in recent and preindustrial sediments from Norwegian and Russian Arctic lakes, *Can. J. Fish. Aquat. Sci.*, 55, 1512–1523, 1998.
7. Rodgers, D.W., in *Mercury Pollution: Integration and Synthesis*, Watras, C.J. and Huckabee, J.W., Eds., Lewis Publishers, Boca Raton, 1994, pp. 427–439.
8. Kraal, M.H. et al., Uptake and tissue distribution of dietary and aqueous cadmium by carp *(Cyprinus-carpio)*, *Ecotoxicol. Environ. Saf.*, 31, 179–183, 1995.
9. Watras, C.J. et al., Bioaccumulation of mercury in pelagic freshwater food webs, *Sci. Total Environ.*, 219, 183–208, 1998.
10. Stemberger, R.S. and Chen, C.Y., Fish tissue metals and zooplankton assemblages of northeastern U.S. lakes, *Can. J. Fish. Aquat. Sci.*, 55, 339–352, 1998.
11. Chen, C.Y. et al., Accumulation of heavy metals in food web components across a gradient of lakes, *Limnol. Ocean.*, 45, 1525–1536, 2000.
12. Cabana, G. et al., Pelagic food-chain structure in Ontario lakes — a determinant of mercury levels in lake trout *(Salvelinus-namaycush)*, *Can. J. Fish. Aquat. Sci.*, 51, 381–389, 1994.

13. VanderZanden, M.J. and Rasmussen, J.B., A trophic position model of pelagic food webs: impact on contaminant bioaccumulation in lake trout, *Ecolog. Monog.*, 66, 451–477, 1996.

14. Rasmussen, J.B. et al., Food-chain structure in Ontario lakes determines PCB levels in lake trout *(Salvelinus-namaycush)* and other pelagic fish, *Can. J. Fish. Aquat. Sci.*, 47, 2030–2038, 1990.

15. Wiener, J.G. et al., Factors influencing mercury concentrations in walleyes in northern Wisconsin lakes, *Trans. Am. Fish. Soc.*, 119, 862–870, 1990.

16. Jackson, L.J. and Schindler, D.E., Field estimates of net trophic transfer of PCBs from prey fishes to Lake Michigan salmonids, *Environ. Sci. Technol.*, 30, 1861–1865, 1996.

17. Kidd, K.A. et al., Effects of trophic position and lipid on organochlorine concentrations in fishes from subarctic lakes in Yukon Territory, *Can. J. Fish. Aquat. Sci.*, 55, 869–881, 1998.

18. USEPA, "Mercury study report to Congress," Vol. 1, U.S. Environmental Protection Agency, Washington, D.C., 1997.

19. Schindler, D.W., Detecting ecosystem responses to anthropogenic stress, *Can. J. Fish. Aquat. Sci.*, 44, 6–25, 1987.

20. Schindler, D.W. et al., The effects of ecosystem characteristics on contaminant distribution in northern fresh-water lakes, *Sci. Total Environ.*, 161, 1–17, 1995.

21. Stemberger, R.S. and Lazorchak, J.M., Zooplankton assemblage responses to disturbance gradients, *Can. J. Fish. Aquat. Sci.*, 51, 2435–2447, 1994.

22. Taylor, G., Baird, D.J. and Soares, A., Surface binding of contaminants by algae: consequences for lethal toxicity and feeding to *Daphnia magna* Straus, *Environ. Toxicol. Chem.*, 17, 412–419, 1998.

23. Lunde, G., The synthesis of fat and water soluble arseno organic compounds in marine and limnetic algae, *Acta. Chemica Scandinavica*, 27, 1586–1594, 1973.

24. Maeda, S. et al., Bioaccumulation of arsenic by freshwater algae *(Nostoc* sp.*)* and the application to the removal of inorganic arsenic from an aqueous phase, *Appl. Organometallic Chem.*, 1, 363–370, 1987.

25. Maeda, S., in *Arsenic in the Environment, Part I: Cycling and Characterization*, Nriagu, J.O., Ed., John Wiley & Sons, Inc., New York, 1994, pp. 155–187.

26. Connell, D.B. and Sanders, J.G., Variation in cadmium uptake by estuarine phytoplankton and transfer to the copepod *Eurytemora affinis, Marine Biol.*, 133, 259–265, 1999.

27. Wang, X.L. et al., Modelling the bioconcentration of hydrophobic organic chemicals in aquatic organisms, *Chemosphere*, 32, 1783–1793, 1996.

28. Timmermans, K.R. et al., Trace-metals in a littoral foodweb — concentrations in organisms, sediment and water, *Sci. Total Environ.*, 87–88, 477–494, 1989.

29. Wang, W.X. and Fisher, N.S., Excretion of trace elements by marine copepods and their bioavailability to diatoms, *J. Marine Res.*, 56, 713–729, 1998.

30. Reinfelder, J.R. et al., Trace element trophic transfer in aquatic organisms: a critique of the kinetic model approach, *Sci. Total Environ.*, 219, 117–135, 1998.

31. Reinfelder, J.R. and Fisher, N.S., The assimilation of elements ingested by marine copepods, *Science*, 251, 794–796, 1991.

32. Playle, R.C., Dixon, D.G., and Burnison, K., Copper and cadmium-binding to fish gills — estimates of metal gill stability-constants and modeling of metal accumulation, *Can. J. Fish. Aquat. Sci.*, 50, 2678–2687, 1993.

33. Kock, G., Hofer, R. and Wograth, S., Accumulation of trace metals (Cd, Pb, Cu, Zn) in Arctic char *(Salvelinus alpinus)* from oligotrophic alpine lakes: relation to alkalinity, *Can. J. Fish. Aquat. Sci.*, 52, 2367–2376, 1995.

34. Sick, L.V. and Baptist, G.J., Cadmium incorporation by the marine copepod *Pseudodiaptomus-coronatus, Limnol. Ocean.*, 24, 453–462, 1979.

35. Reinfelder, J.R. and Fisher, N.S., Retention of elements absorbed by juvenile fish *(Menidia-menidia, Menidia-beryllina)* from zooplankton prey, *Limnol. Ocean.*, 39, 1783–1789, 1994.

36. Visman, V., Mierle, G. and McQueen, D.J., Uptake of aqueous methylmercury by larval *Chaoborus-americanus, Wat. Air Soil Pollut.*, 80, 1007–1010, 1995.

37. Tsalkitzis, E., Methylmercury in golden shiners *(Notemigonus crysoleucas)* and zooplankton from Mouse Lake (Ontario), M.S. thesis, York University, 1995.

38. Back, R.C. and Watras, C.J., Mercury in zooplankton of northern Wisconsin lakes — taxonomic and site-specific trends, *Wat. Air Soil Pollut.*, 80, 931–938, 1995.

39. Hare, L. and Tessier, A., Predicting animal cadmium concentrations in lakes, *Nature*, 380, 430–432, 1996.

40. Croteau, M.N., Hare, L., and Tessier, A., Refining and testing a trace metal biomonitor *(Chaoborus)* in highly acidic lakes, *Environ. Sci. Technol.*, 32, 1348–1353, 1998.

41. Porter, K.G., in *Food Webs: Integration of Patterns and Dynamics*, Polis, G.A. and Winemiller, K.O., Eds., Chapman & Hall, New York, 1996, pp. 51–59.

42. Hairston, N.G., Cause-effect relationships in energy-flow, trophic structure, and interspecific interactions, *Am. Naturalist*, 142, 379–411, 1993.

43. Jackson, L.J. and Carpenter, S.R., PCB concentrations of Lake-Michigan invertebrates — reconstruction based on PCB concentrations of alewives *(Alosa-pseudoharengus)* and their bioenergetics, *J. Great Lakes Res.*, 21, 112–120, 1995.

44. Paterson, M.J., Rudd, J.W.M., and St. Louis, V., Increases in total and methylmercury in zooplankton following flooding of a peatland reservoir, *Environ. Sci. Tech.* 32, 3868–3874, 1998.

45. Persson, L. et al., in *Food Webs: Integration of Patterns and Dynamics* Polis, G.A. and Winemiller, K.O., Eds., Chapman & Hall, New York, 1996, pp. 396–434.

46. Bloom, N.S., On the chemical form of mercury in edible fish and marine invertebrate tissue, *Can. J. Fish. Aquat. Sci.*, 49, 1010–1017, 1992.

47. Lawson, N.M. and Mason, R.P., Accumulation of mercury in estuarine food chains, *Biogeochemistry*, 40, 235–247, 1998.

48. Chen, C.Y. and Folt, C.L., Bioaccumulation and diminution of arsenic and lead in a freshwater food web, *Environ. Sci. Tech.*, 34, 3878–3884, 2000.

49. Cottingham, K.L., Nutrients and zooplankton as multiple stressors of phytoplankton communities: evidence from size-structure, *Limnol. Oceanogr.*, 44(3), 810–827, 1998.

50. Hrbacek, J. et al., Demonstration of the effect of the fish stock on the species composition of zooplankton and the intensity of metabolism of the whole plankton association, *Proc. Intl. Assoc. Theoret. Appl. Limnol.*, 14, 192–195, 1961.

51. Brooks, J.L. and Dodson, S.I., Predation, body size and composition of plankton: the effect of a marine planktivore on lake plankton illustrates theory of size, competition, and predation, *Science*, 150, 28–35, 1965.

52. Vanni, M.J., Layne, C.D., and Arnott, S.E., "Top-down" trophic interactions in lakes: effects of fish on nutrient dynamics, *Ecology*, 78, 1–20, 1997.

53. Mazumder, A., Patterns of algal biomass in dominant odd-link vs. even-link lake ecosystems, *Ecology*, 75, 1141–1149, 1994.
54. Ramcharan, C.W., France, R.L., and McQueen, D.J., Multiple effects of planktivorous fish on algae through a pelagic trophic cascade, *Can. J. Fish. Aquat. Sci.*, 53, 2819–2828, 1996.
55. Schindler, D.E. et al., Influence of food web structure on carbon exchange between lakes and the atmosphere, *Science*, 277, 248–251, 1997.
56. PerezFuentetaja, A., McQueen, D.J., and Demers, E., Stability of oligotrophic and eutrophic planktonic communities after disturbance by fish, *Oikos*, 75, 98–110, 1996.
57. Moore, M. and Folt, C., Zooplankton body size and community structure — effects of thermal and toxicant stress, *Trends Ecol. Evol.*, 8, 178–183, 1993.
58. Moore, M.V., Folt, C.L. and Stemberger, R.S., Consequences of elevated temperatures for zooplankton assemblages in temperate lakes, *Archiv. Hydrobiol.*, 135, 289–319, 1996.
59. Havens, K.E., Yan, N.D., and Keller, W., Lake acidification — effects on crustacean zooplankton populations, *Environ. Sci. Technol.*, 27, 1621–1624, 1993.
60. Stemberger, R.S. and Miller, E.K., A zooplankton-N: P-ratio indicator for lakes, *Environ. Monitor. Assess.*, 51, 29–51, 1998.
61. Stockner, J.G. and Porter, K.G., in *Complex Interactions in Lake Communities*, Carpenter, S.R., Ed., Springer-Verlag, New York, 1988, pp. 69–84.
62. Mason, R.P., Reinfelder, J.R., and Morel, F.M.M., Uptake, toxicity, and trophic transfer of mercury in a coastal diatom, *Environ. Sci. Tech.*, 30, 1835–1845, 1996.
63. Vanni, M.J., in *Food Webs: Integration of Patterns and Dynamics* Polis, G.A. and Winemiller, K.O., Eds., Chapman & Hall, New York, 1996, pp. 81–95.
64. McCauley, E. and Murdoch, W.W., Cyclic and stable-populations — plankton as paradigm, *Am. Naturalist*, 129, 97–121, 1987.
65. Persson, L. et al., Shifts in fish communities along the productivity gradient of temperate lakes — patterns and the importance of size-structured interactions, *J. Fish Biol.*, 38, 281–293, 1991.
66. Persson, L. et al., Trophic interactions in temperate lake ecosystems — a test of food-chain theory, *Am. Naturalist*, 140, 59–84, 1992.
67. MacKay, N.A. and Elser, J.J., Factors potentially preventing trophic cascades: food quality, invertebrate predation, and their interaction, *Limnol. Ocean.*, 43, 339–347, 1998.
68. Currie, R.S. et al., Influence of nutrient additions on cadmium bioaccumulation by aquatic invertebrates in littoral enclosures, *Environ. Toxicol Chem.*, 17, 2435–2443, 1998.
69. Wallberg, P., Bergqvist, P.A., and Andersson, A., Potential importance of protozoan grazing on the accumulation of polychlorinated biphenyls (PCBs) in the pelagic food web, *Hydrobiologia*, 357, 53–62, 1997.
70. Taylor, W.D. et al., Organochlorine concentrations in the plankton of lakes in southern Ontario and their relationship to plankton biomass, *Can. J. Fish. Aquat. Sci.*, 48, 1960–1966, 1991.
71. Andersson, A. et al., Effect of nutrient enrichment on the distribution and sedimentation of polychlorinated biphenyls (PCBs) in seawater, *Hydrobiologia*, 377, 45–56, 1998.
72. Millard, E.S. et al., Effect of primary productivity and vertical mixing on PCB dynamics in planktonic model-ecosystems, *Environ. Toxicol. Chem.*, 12, 931–946, 1993.

73. Baines, S.B. and Pace, M.L., Relationships between suspended particulate matter and sinking flux along a trophic gradient and implications for the fate of planktonic primary production, *Can. J. Fish. Aquat. Sci.*, 51, 25–36, 1994.
74. AllenGil, S.M. et al., Heavy metal accumulation in sediment and freshwater fish in U.S. Arctic lakes, *Environ. Toxicol. Chem.*, 16, 733–741, 1997.
75. Mackay, D. and Wania, F., Transport of contaminants to the Arctic — partitioning, processes and models, *Sci. Total Environ.*, 161, 25–38, 1995.
76. Blais, J.M. et al., Accumulation of persistent organochlorine compounds in mountains of western Canada, *Nature*, 395, 585–588, 1998.
77. Freitas, H. et al., Contaminant fate in high Arctic lakes: Development and application of a mass balance model, *Sci. Total Environ.*, 201, 171–187, 1997.
78. Hobbie, J.E., Corliss, T.L., and Peterson, B.J., Seasonal patterns of bacterial abundance in an Arctic lake, *Arctic Alpine Res.*, 15, 253–259, 1983.
79. Rouse, W.R. et al., Effects of climate change on the freshwaters of arctic and subarctic North America, *Hydrological Processes*, 11, 873–902, 1997.
80. Lucotte, M. et al., Anthropogenic mercury enrichment in remote lakes of northern Quebec (Canada), *Wat. Air Soil Pollut.*, 80, 467–476, 1995.
81. Muir, D.C.G. et al., Spatial trends and historical deposition of polychlorinated biphenyls in Canadian midlatitude and Arctic lake sediments, *Environ. Sci. Technol.*, 30, 3609–3617, 1996.
82. Duff, K.E. et al., Limnological characteristics of lakes located across arctic treeline in northern Russia, *Hydrobiologia*, 391, 205–222, 1999.
83. Pienitz, R., Smol, J.P., and Lean, D.R.S., Physical and chemical limnology of 24 lakes located between Yellowknife and Contwoyto Lake, Northwest Territories (Canada), *Can. J. Fish. Aquat. Sci.*, 54, 347–358, 1997.
84. Sargent, J.R. and Falkpetersen, S., The lipid biochemistry of calanoid copepods, *Hydrobiologia*, 167, 101–114, 1988.
85. Farkas, T., Adaptation of fatty-acid compositions to temperature — study on planktonic crustaceans, *Comp. Biochem. Physiol. B-Biochem. Molec. Biol.*, 64, 71–76, 1979.
86. Campbell, L.M. et al., Organochlorine transfer in the food web of subalpine Bow Lake, Banff National Park, *Can. J. Fish. Aquat. Sci.*, 57, 1258–1269, 2000.
87. Brown, M.P. et al., Polychlorinated-biphenyls in the Hudson River, *Environ. Sci. Technol.*, 19, 656–667, 1985.
88. Kling, G.W., Fry, B., and Obrien, W.J., Stable isotopes and planktonic trophic structure in Arctic lakes, *Ecology*, 73, 561–566, 1992.
89. Gu, B.H., Schell, D.M., and Alexander, V., Stable carbon and nitrogen isotopic analysis of the plankton food-web in a sub-Arctic lake, *Can. J. Fish. Aquat. Sci.*, 51, 1338–1344, 1994.
90. Sprules, W.G. and Bowerman, J.E., Omnivory and food-chain length in zooplankton food webs, *Ecology*, 69, 418–426, 1988.
91. France, R. et al., Vertical foodweb structure of freshwater zooplankton assemblages estimated by stable nitrogen isotopes, *Res. Popul. Ecol.*, 38, 283–287, 1996.
92. Kidd, K.A. et al., Correlation between stable nitrogen isotope ratios and concentrations of organochlorines in biota from a fresh-water food-web, *Sci. Total Environ.*, 161, 381–390, 1995.
93. Kidd, K.A. et al., Bioaccumulation of organochlorines through a remote freshwater food web in the Canadian Arctic, *Environ. Pollut.*, 102, 91–103, 1998.
94. Hakanson, L., Nilsson, A., and Andersson, T., Mercury in fish in Swedish lakes, *Environ. Pollut.*, 49, 145–162, 1988.

95. Suns, K. and Hitchin, G., Interrelationships between mercury levels in yearling yellow perch, fish condition and water quality, *Wat. Air Soil Pollut.*, 650, 255–265, 1990.

96. Lathrop, R.C., Rasmussen, P.W. and Knauer, D.R., Mercury concentrations in walleyes from Wisconsin (U.S.) lakes, *Wat. Air Soil Pollut.*, 56, 295–307, 1991.

97. Hudson, R.J.M., Gherini, S.A., and Watras, C.J., in *Mercury Pollution Integration and Synthesis*, Watras, C.J. and Huckabee, J.W., Eds., Lewis Publishers, Boca Raton, 1994, pp. 473–523.

98. Simonin, H.A. et al., in *Mercury Pollution Integration and Synthesis*, Watras, C.J. and Huckabee, J.W., Eds., Lewis Publishers, Boca Raton, 1994, pp. 457–469.

99. Heit, M. et al., Trace-element concentrations in fish from 3 Adirondack lakes with different pH values, *Wat. Air Soil Pollut.*, 44, 9–30, 1989.

100. Driscoll, C.T. et al., The role of dissolved organic-carbon in the chemistry and bioavailability of mercury in remote Adirondack lakes, *Wat. Air Soil Pollut.*, 80, 499–508, 1995.

101. Cottingham, K.L. et al., Response of phytoplankton and bacteria to nutrients and zooplankton: a mesocosm experiment, *J. Plankton Res.*, 19, 995–1010, 1997.

102. Pace, M.L. and Vaque, D., The importance of Daphnia in determining mortality-rates of protozoans and rotifers in lakes, *Limnol. Ocean.*, 39, 985–996, 1994.

103. Wickham, S.A. and Gilbert, J.J., The comparative importance of competition and predation by Daphnia on ciliated protists, *Arch. Hydrobiol.*, 126, 289–313, 1993.

104. Gasol, J.M. and Vaque, D., Lack of coupling between heterotrophic nanoflagellates and bacteria — a general phenomenon across aquatic systems, *Limnol. Ocean.*, 38, 657–665, 1993.

105. Pienitz, R., Smol, J.P. and Lean, D.R.S., Physical and chemical limnology of 59 lakes located between the southern Yukon and the Tuktoyaktuk Peninsula, Northwest Territories (Canada), *Can. J. Fish. Aquat. Sci.*, 54, 330–346, 1997.

106. Miller, M.C. et al., Primary production and its control in Toolik Lake, Alaska, *Arch. Hydrobiol., Supp.*, 74, 97–131, 1986.

section IV

Organ and systems biomarkers

chapter twenty-one

Validation of exposure and risk biomarkers: aflatoxin as a case study

John D. Groopman, Peta E. Jackson, Paul Turner,
Christopher P. Wild, and Thomas W. Kensler

Contents

Abstract   The use of biomarkers in molecular epidemiology studies for identifying stages in the progression of development of the health effects of environmental agents has the potential for providing important information for critical regulatory, clinical, and public health problems. Investigations of aflatoxins probably represents one of the most extensive data sets in the field and this work may serve as a template for future studies of other environmental agents. The aflatoxins are naturally occurring mycotoxins found on foods such as corn, peanuts, various other nuts, and cottonseed and they have been demonstrated to be carcinogenic in many experimental models. As a result of nearly 30 years of study, experimental data and epidemiological studies in human populations, aflatoxin B_1 was classified as carcinogenic to humans by the International Agency for Research on Cancer (IARC). The long-term goal of the research described herein is the application of bio-markers to the development of preventative interventions for use in human populations at high-risk for cancer. Several of the aflatoxin specific bio-markers have been validated in epidemiological studies and are now being used as intermediate biomarkers in prevention studies. The development of these aflatoxin biomarkers has been based upon the knowledge of the bio-chemistry and toxicology of aflatoxins gleaned from experimental and human studies. These biomarkers have subsequently been utilized in experimental models to provide data on the modulation of these markers under different situations of disease risk. This systematic approach provides encouragement for preventive interventions and should serve as a template for the development, validation and application of other chemical-specific biomarkers to cancer or other chronic diseases.

I. Introduction

The use of biomarkers for identifying stages in the progression of development of the health effects of environmental agents has the potential for providing important information for critical regulatory, clinical, and public health problems.[1,2] Since the development of a paradigm for molecular bio-markers by a committee of the National Research Council over a decade ago, some progress has been made in applying such chemical biomarkers to specific environmental situations that may be hazardous to humans, as exemplified by the study of aflatoxins. The major goals of environmental chemical-specific biomarker research are to develop and validate biomarkers that reflect specific exposures and predict disease risk in individuals. Presumably, after an environmental exposure, each person has a unique response to dose and time to disease onset. These responses will be affected by intrinsic (genetic) and by extrinsic (such as dietary) modifiers. It is assumed that biomarkers that reflect the mechanism of action of an environmental chemical will be strong predictors of an individual's risk of disease. It is also expected that these biomarkers can more clearly classify the status of exposure of individuals, local communities, and larger populations. These studies should also help to elucidate the molecular processes of chemically induced human disease and underlying susceptibility factors. A conceptual model for this work is shown in Figure 21.1.

The molecular epidemiology investigations of aflatoxins probably represents one of the most extensive data sets in the field and this work may serve as a template for future studies of other environmental agents. The aflatoxins are naturally occurring mycotoxins found on foods such as corn, peanuts, various other nuts and cottonseed. They have been demonstrated to be carcinogenic in many animal species including rodents, nonhuman primates and fish. They were also initially suspected to contribute to human hepato-cellular carcinoma (HCC).[3] As a result of nearly 30 years of study, experi-

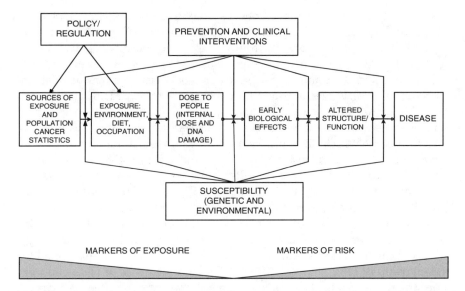

Figure 21.1 Conceptual basis for the development of biomarkers for use in molecular epidemiology and applications in preventive interventions.

mental data and epidemiological studies in human populations, aflatoxin B_1 (AFB$_1$) was classified as carcinogenic to humans by the IARC in 1993.[4]

II. Molecular epidemiological studies of aflatoxin and human liver cancer

HCC is one of the most common cancers worldwide, and there is a striking geographic variation in incidence. For example, in the People's Republic of China, HCC accounts for over 300,000 deaths annually, the third leading cause of cancer mortality.[5] During the 1960s and 1970s, several epidemiological studies were conducted in Asia and Africa that showed there was an association between high aflatoxin exposure, estimated by sampling foodstuffs or by dietary questionnaires, and increased incidence of HCC.[6] While these early studies could not account for additional factors such as hepatitis B virus (HBV) or hepatitis C virus (HCV) infection, this information provided a strong motivation to further investigate the circumstantial relationship between aflatoxin ingestion and liver cancer incidence.

In general, the most rigorous test of an association between an agent and disease outcome is found in prospective epidemiological studies, in which healthy people are recruited, questionnaires and biological samples taken, and the cohort followed until significant numbers of cases are obtained. A nested study within the cohort can then be designed to match cases and controls. Since the controls were recruited at the same time and with the same health status as the cases, they are better matched than in traditional case-control studies.

There have been two major cohort studies to address the relationship of aflatoxin exposure to HCC incidence reported to date. The first comprises over 18,000 people in Shanghai from whom urine and blood samples were collected.[7,8] After a 7-year follow-up, cases and controls were age and residence matched to examine the association between markers of aflatoxin exposure and HBV infection and the development of HCC. The data revealed a highly significant increase in the relative risk (RR = 3.4) for those liver cancer cases where urinary aflatoxin biomarkers were detected. The relative risk for people who tested positive for HBsAg was 7.3, but individuals with urinary aflatoxins and positive HBsAg status had a relative risk for developing HCC of about 59. These results strongly supported a causal relationship between two major HCC risk factors, HBV and AFB_1 exposure. Finally, when individual aflatoxin metabolites were stratified for liver cancer outcome, the presence of the aflatoxin-nucleic acid adduct (AFB-N^7-gua) in urine always resulted in a two- to threefold elevation in risk of developing HCC.

Subsequent cohort studies carried out in Taiwan have also examined the relationship of HBV status, AFB_1 exposure and incidence of HCC and have confirmed the results of the Shanghai investigation.[9,10] A nested case-control study from a cohort of over 15,000 people in Taiwan found that in HBV-infected males there was an adjusted odds ratio (OR) of 2.8 for detectable compared with nondetectable aflatoxin–albumin adducts and 5.5 for high compared with low levels of aflatoxin metabolites in urine.[9] A second cohort study in Taiwan observed a dose-response relationship between urinary AFM_1 levels and HCC in chronic HBV carriers.[10] Similar to the Shanghai data, the HCC risk associated with AFB_1 exposure was more striking among the HBV carriers with detectable AFB–N^7-gua in urine.

The use of biomarkers in cohort studies has clearly shown the chemical-viral interaction in the induction of HCC. However, aflatoxin exposure in the absence of chronic hepatitis B infection is also etiologically associated with liver cancer. These findings provide the compelling basis to increase efforts in HBV immunization programs and in the development of concerted programs to lower dietary aflatoxin exposure as means of lowering human cancer risk.

III. Aflatoxin exposure and HBV infection in children

The use of biomarkers in cohort studies has clearly shown the chemical-viral interaction in the induction of HCC. However, aflatoxin exposure in the absence of chronic hepatitis B infection is also etiologically associated with liver cancer. These findings provide the compelling basis to increase efforts in HBV immunization programs and in the development of concerted programs to lower dietary aflatoxin exposure as means of lowering human cancer risk. These epidemiological studies have also compelled the case to understand if a mechanistic interaction between HBV infection and enhanced aflatoxin metabolism. Early studies in adults infected with HBV and exposed to aflatoxin showed no interaction between these two risk factors and the levels of aflatoxin biomarkers[11,12]; however, a different picture has emerged in children.

Aflatoxin–albumin adduct levels were measured in serum samples obtained from a group of Gambian children. Aflatoxin–albumin adduct was found in nearly all serum samples collected during a survey performed at the end of the dry season and levels of adduct were generally high (up to 720 pg aflatoxin–lysine equivalent/mg albumin). Higher levels of aflatoxin–albumin adduct were detected in Wollof children than in children of other ethnic groups and marked variation in mean adduct levels between villages was observed. Aflatoxin–albumin adduct levels were higher in children who were HBsAg positive than in controls. Much lower levels of aflatoxin–albumin adduct were detected in repeat samples obtained at the end of the rainy season. There was poor correlation between dry and rainy season levels of adduct in individual children. Thus, these researchers showed that Gambian children are exposed to high levels of aflatoxin and HBV infection appeared to enhance albumin adduct formation.[13]

There have been continuing studies demonstrating that the strongest interaction for HBV infection and aflatoxin exposure biomarkers occurs in children. In a recent cross-sectional analysis of unvaccinated (against HBV) children in rural Gambia, sera from 444 children aged 3–4 yrs, selected to be representative of their communities, were analyzed for aflatoxin–albumin adducts and markers of HBV infection. There was a large interindividual variation in adduct levels between children (2.2 to 459 pg AF–lysine eq./mg albumin). Adduct level was strongly correlated with season, with an approximately twofold higher mean level in the dry season than the wet. Geometric mean adduct levels in uninfected children, chronic carriers and acutely infected children were 31.6 (n = 404), 44.9 (n = 34), and 96.9 (n = 6) pg/mg, respectively. The relationship of adduct level to ethnicity, month of sampling and HBV status was examined in a multiple regression model. Month of obtaining the blood sample (p = 0.0001) and HBV status (p = 0.0023) each made a highly significant contribution to the model; the high aflatoxin–albumin levels were particularly associated with acute infection. The effect of seasonality on adducts was also observed in a previous study of 347 Gambian adults; however, no correlation between adduct level and HBV status was observed in that population.[14] This difference between children and adults may reflect a more severe effect of HBV infection, particularly acute infection, in childhood on hepatic AF metabolism. Thus, childhood may be a particularly critical period to which intervention strategies should be addressed.[15]

IV. Aflatoxin exposure and mutations in the p53 tumor suppressor gene

In addition to the evidence from epidemiological studies and the use of biomarkers of biologically effective dose, further support for the involvement of aflatoxin in HCC incidence in certain parts of the world has come from investigations of mutations in the *p53* tumor suppressor gene. The *p53* gene is found mutated in a majority of human cancers and there is a large variation in number and type of mutations between cancers of different tissues.[16,17] Such diversity lends itself to the analysis of the mutational spectrum within the gene with a view to determining information about the etiology of the tumor and potential risk assessment.[18,19] One of the most striking examples of a molecular fingerprint in the *p53* gene is a characteristic G→T transversion at the third base of codon 249 observed in liver cancer patients from regions of high aflatoxin exposure.

Initial reports of a specific mutation in the *p53* gene of HCCs in populations exposed to high levels of aflatoxin came from two independent studies in southern Africa and Qidong, China, in which a G→T transversion in codon 249 of the *p53* was observed in approximately 50% of HCCs.[20,21] In contrast, no codon 249 mutations were detected in areas with low aflatoxin exposure such as Japan, Europe, and the U.S.[22]

Since these early findings, numerous investigations to determine the frequency of *p53* mutations, in particular the prevalence of G→T transitions at codon 249, in populations with high and low levels of aflatoxin exposure have been conducted. Li and coworkers[23] analyzed samples from two areas of China and found that 9/20 (45%) HCCs from Qidong, an area of high AFB$_1$ exposure, had a *p53* mutation and all of these were codon 249 mutations. In contrast, in HCCs from Shanghai, an area of intermediate aflatoxin exposure, 3/18 (17%) cases had a *p53* mutation and only 1/3 (33%) was located in codon 249. All but one of the *p53* mutations determined in this study were found to be positive for HBV infection. A recent study of the HCC cases from the same regions found 50%) and 50%) samples from Qidong and Shanghai, respectively, to have a *p53* mutation. Of these, 100% (7/7) and 60% (3/5) were codon 249 transversions, respectively.[24] Senegal is another country which has an extremely high incidence of liver cancer and in which there is a high exposure to dietary aflatoxin. A study conducted on 15 HCC samples from this country detected 10 (67%) cases with a mutation at codon 249 of the *p53* gene, the highest frequency reported to date.[25]

Aguilar et al.[26] investigated the mutagenesis of the *p53* gene in nontumorigenic liver DNA from HCC cases in the United States, Thailand and Qidong, areas of negligible, low and high aflatoxin exposure. They found that the frequency of codon 249 mutations paralleled the level of aflatoxin exposure. Similar observations were made in a comparison of HCCs from Qidong and Beijing, an area of low aflatoxin exposure, in which a codon 249 G→T transversions were seen in 52% (13/25) and 0% (0/9) of HCCs respectively.[27] However, the total frequency of mutations and loss of heterozygosity in the *p53* gene in the two regions (both of which have a high incidence of HBV) were similar indicating that alterations in *p53* independent of the 249 mutation may play an important role in HBV-associated HCC.

The implication from these studies in populations of varying exposure to aflatoxin that the G→T mutation at the third base of codon 249 is aflatoxin specific has been supported by studies in bacteria which have shown that aflatoxin exposure causes almost exclusively G→T transversions.[28] It has also been shown that the aflatoxin-epoxide can bind to codon 249 of *p53* in a plasmid *in vitro*, providing further indirect evidence for a putative role of aflatoxin exposure in *p53* mutagenesis.[29] A recent study has mapped AFB_1 adduct formation to codon 249.[30] However, this was not the major site of adduction by AFB_1 and adducts were detected in several other codons in exons 7 and 8. It was also observed that the rate of adduct removal from codon 249 was relatively high suggesting that additional mechanisms are involved in mutagenesis at this site.

In vitro studies to measure the frequency of *p53* mutations in HepG2 human hepatocarcinoma cells incubated with AFB_1 in the presence of rat liver microsomes found that G→T transversions at the third base of codon 249 were preferentially induced.[31,32] G→T and C→A transversions were also observed in adjacent codons in cells exposed to AFB_1, although at lower frequencies. These findings also suggest that there is some additional selection of cells with a G→T transversion at the third base of codon 249, perhaps due to an altered function of the mutant 249 serine p53 protein.

While the evidence for a role of aflatoxin in the incidence of codon 249 mutations in *p53* from the above studies is persuasive, the level of exposure to aflatoxin has been classified by factors such as geographic residence rather than measurement of aflatoxin exposure at the individual level. The limitations of this approach have been reviewed by Lasky and Magder.[33] Only a few studies have been conducted to date which measure aflatoxin adduct levels and *p53* mutational status in HCC cases. The largest and most recent study to assess the relationship between the presence of AFB_1 adducts, HBV status and *p53* mutations was reported by Lunn and coworkers[34] who investigated 105 HCC cases in Taiwan. Mutations in the *p53* gene were detected in 28% (29/105) of cases by single strand conformation polymorphism and DNA sequencing revealed that 12 of these were specific codon 249 mutations. Mutant p53 protein was detected in 35% (37/105) of HCC cases by immunohistochemisty. AFB_1–DNA adducts were detected in 66% (69/105) of tumor tissues and were associated with *p53* DNA (OR = 2.9 p = 0.082) and protein mutations (OR = 2.9, p = 0.054) with only borderline significance. There was no statistically significant association between AFB_1–DNA adducts and codon 249 mutations. All of the codon 249 mutations (n = 12) occurred in HBsAg-seropositive carriers, resulting in an OR of 10.0 (P < 0.05), suggesting that HBV may be involved in the selection of these mutations.

While the detection of specific *p53* mutations in liver tumors has provided insight into the etiology of certain liver cancers, the application of these specific mutations to the early detection of cancer offers great promise for prevention.[35] A recent report by Kirk et al.[36] reported for the first time the detection of codon 249 *p53* mutations in the plasma of liver tumor patients from the Gambia; however, the mutational status of the tumors were not known. These authors also reported a small number of cirrhosis patients having this mutation and given the strong relation between cirrhosis and future development of HCC, the possibility of this mutation being an early detection marker needs to be explored.

In a recent paper by Jackson et al.[37] Short Oligonucleotide Mass Analysis (SOMA) was compared with DNA sequencing in 25 HCC samples for specific *p53* mutations. Mutations were detected in ten samples by SOMA in agreement with DNA sequencing. Analysis of another 20 plasma and tumor pairs showed 11 tumors containing the specific mutation and this change was detected in six of the paired plasma samples. Four of the plasma samples had detectable levels of the mutation; however, the tumors were negative suggesting possible multiple independent HCCs. Ten plasma samples from healthy individuals were all negative. This molecular diagnostic technique has implications for prevention trials and the early diagnosis of HCC.

In summary, studies of the prevalence of codon 249 mutations in HCC cases from patients in areas of high or low exposure to aflatoxin suggest that a G→T transition at the third base is associated with aflatoxin exposure and *in vitro* data would seem to support this hypothesis. A majority of codon 249 mutations are found in patients with an HBV infection implicating an association. However, in comparisons of codon 249 mutations in regions of high HBV infection but with varying levels of AFB_1 exposure, the mutation only occurs in areas of high AFB_1 exposure. HBV evidently plays an important role in mutagenesis, perhaps by causing preferential selection of cells harboring the mutation. The use of the codon 249 mutation as a marker of exposure to aflatoxin must be done with caution until evidence has been obtained from studies measuring AFB_1 adducts and mutations in the same individual.

V. Biomarkers and liver cancer prevention

Several approaches can be considered for the prevention of liver cancer. A first approach is vaccination against HBV. Unfortunately, many people living in high-risk areas for liver cancer acquire the HBV infection before age three. Thus, an immunization program for total population protection would have to occur over several generations, provided that mutant strains of HBV do not arise, thereby eliminating the utility of current vaccines. Despite these problems, vaccination programs for HBV have been implemented in many areas of Africa and Asia. A second approach for cancer prevention would be the elimination of aflatoxin exposures. Primary prevention of aflatoxin exposures could be accomplished through large expenditures of resources for proper crop storage and handling; however, this approach is not economically feasible in many areas of the world. Secondary prevention measures using chemopreventive agents, which block the activation and enhance the detoxification of AFB_1 are being investigated in high-risk populations.

Cancer prevention trials that use biomarkers as intermediate endpoints provide the ability to assess the efficacy of promising chemopreventive agents in an efficient manner by reducing sample size requirements as well as the time required to conduct the studies compared to trials that have cancer incidence or mortality as endpoints.[38] These intermediate markers are particularly valuable when investigating chemopreventive agents such as oltipraz that may have an effect at early, preclinical stages of carcinogenesis. The key issue in trials that use biomarkers as the outcome of interest is to have a marker directly associated with the evolution or development of neoplasia. These biomarkers are typically organ site-specific genomic, proliferation, and differentiation markers that reflect different intermediary stages of the neoplastic process. As an adjunct to these process-dependent markers it may also be possible to devise markers specific to interventions in selected groups at high risk for carcinogen exposure. These agent-dependent approaches would be based upon knowledge of the etiologic agent(s) in the study population.

It has been shown that several agents can provide some level of protection against aflatoxin-induced liver cancer in experimental systems (Kensler et al.[39]). One of the most promising of these agents is oltipraz (4-methyl-5-(2-pyrazinyl)-1,two-dimensionalithiole-3-thione), which has been shown to inhibit AFB_1 hepatocarcinogenesis in rats when administered 1 week prior to and throughout carcinogen exposure.[40] This and subsequent studies have also shown that hepatic AFB–DNA adducts, urinary AFB–N^7-gua, and serum albumin adducts are all reduced in animals given oltipraz during aflatoxin administration indicating the utility of these biomarkers in intervention studies.[41]

A double-blind Phase IIa clinical chemoprevention trial with oltipraz was conducted in Qidong, China, in 1995.[42] Healthy individuals were randomized into groups receiving 125 mg of oltipraz daily, 500 mg of oltipraz weekly, or placebo. Blood and urine specimens were collected biweekly over the 8-week intervention period and an 8-week follow-up period. Levels of aflatoxin-albumin adducts in serum and AFM1 and AFB-NAC excreted in the urine were examined as primary biomarker endpoints in the study. There were no consistent changes observed in levels of aflatoxin-albumin adducts in the placebo arm or the arm receiving 125 mg of oltipraz daily. However, there was a significant decline in aflatoxin-albumin levels beginning 1 month into intervention with 500 mg of oltipraz weekly which continued for 1 month after treatment was stopped.[43]

Urinary levels of AFM_1, the primary oxidative metabolite of AFB_1, were found to be reduced by 51% in individuals receiving weekly doses of 500 mg oltipraz as compared to the placebo controls.[42] No significant differences were observed in the levels of AFM_1 in the arm receiving 125 mg of oltipraz daily. In contrast, levels of AFB–NAC, a detoxification product of AFB_1, were increased 2.6-fold in the 125 mg group, but were unchanged in the 500 mg group. An increase in AFB–NAC in the 125 mg group confirms the ability of oltipraz to induce phase II enzymes to increase aflatoxin conjugation. The lack of an effect of 500 mg of oltipraz on AFB–NAC probably reflects masking due to diminished substrate formation through the inhibition of cytochrome p450 activities seen in this group (measured as a reduction in AFM1 levels). Overall, these results highlight the use of biomarkers in chemoprevention studies to determine the efficacy of such agents.

VI. Summary

The long-term goal of the research described herein is the application of biomarkers to the development of preventative interventions for use in human populations at high-risk for cancer. Several of the aflatoxin specific biomarkers have been validated in epidemiological studies and are now being used as intermediate biomarkers in prevention studies. The development of these aflatoxin biomarkers has been based upon the knowledge of the biochemistry and toxicology of aflatoxins gleaned from experimental and human studies. These biomarkers have subsequently been utilized in experimental models to provide data on the modulation of these markers under different situations of disease risk. This systematic approach provides encouragement for preventive interventions and should serve as a template for the development, validation, and application of other chemical-specific biomarkers to cancer or other chronic diseases.

Acknowledgments

We acknowledge the significant contributions of many of our colleagues to the development of aflatoxin biomarkers and our experimental and clinical studies on chemoprevention of aflatoxin hepatocarcinogensis. Financial support for this work has been provided by grants R01 CA39416, P01 ES06052, NIEHS Center P30 ES03819 and contract N01-CN-25437

References

1. Anonymous, Biological markers in environmental health research, *Environ. Health Perspect.*, 74, 3–9, 1987.
2. Wogan, G.N., Molecular epidemiology in cancer risk assessment and prevention: Recent progress and avenues for future research, *Environ. Health Perspect.*, 98, 167–178, 1992.
3. Busby, W.F.J. and Wogan, G.N., Aflatxoins, in *Chemical Carcinogens*, 2nd ed., Searle, C.B., Ed., American Chemical Society, Washington, D.C., 1984, pp. 945-1136.
4. IARC, Aflatoxins, *Monogr. Eval. Carcinog. Risks. Hum.*, 56, 1993, pp. 245-395.
5. National Cancer Office of the Ministry of Public Health, P.R.C., *Studies on Mortality Rates of Cancer in China*, People's Publishing House, Beijing, 1980.
6. Groopman, J.D., Scholl, P., and Wang, J.S., Epidemiology of human aflatoxin exposures and their relationship to liver cancer, *Prog. Clin. Biol. Res.*, 395, 211–222, 1996.
7. Ross, R.K. et al., Urinary aflatoxin biomarkers and risk of hepatocellular carcinoma, *Lancet*, 339(8799), 943–946, 1992.
8. Qian, G.S. et al., A follow-up study of urinary markers of aflatoxin exposure and liver cancer risk in Shanghai, People's Republic of China, *Cancer Epidemiol. Biomarkers. Prev.*, 3(1), 3–10, 1994.
9. Wang, L.Y. et al., Aflatoxin exposure and risk of hepatocellular carcinoma in Taiwan, *Intl. J. Cancer*, 67(5), 620–625, 1996.

10. Yu, M.W. et al., Effect of aflatoxin metabolism and DNA adduct formation on hepatocellular carcinoma among chronic hepatitis B carriers in Taiwan, *J. Hepatol.*, 27(2), 320–330, 1997.

11. Groopman, J.D. et al., Molecular dosimetry of aflatoxin-N⁷-guanine in human urine obtained in the Gambia, West Africa, *Cancer Epidemiol. Biomarkers, Prev.*, 1, 221–228, 1992.

12. Wild, C.P., et al., Correlation of dietary intake of aflatoxins with the level of albumin bound aflatoxin in peripheral blood in the Gambia, West Africa, *Cancer Epidemiol. Biomarkers Prev.*, 1, 229–234, 1992.

13. Allen, S.J. et al., Aflatoxin exposure, malaria and hepatitis B infection in rural Gambian children, *Trans. R. Soc. Trop. Med. Hyg.*, 86, 426–430, 1992.

14. Wild, C.P. et al., Environmental and genetic determinants of aflatoxin-albumin adducts in the Gambia, *Int. J. Cancer*, 86: 1–7, 2000.

15. Turner, P.C. et al., Hepatitis B infection and aflatoxin biomarker levels in Gambian children, *Trop. Med. Int. Health*, 5(12), 837–841, Dec. 2000.

16. Hollstein, M. et al., p53 mutations in human cancers, *Science*, 253(5015), 49–53, 1991.

17. Greenblatt, M.S. et al., Mutations in the p53 tumor suppressor gene: Clues to cancer etiology and molecular pathogenesis, *Cancer Res.*, 54(18), 4855–4878, 1994.

18. Vogelstein, B. and Kinzler, K.W., Carcinogens leave fingerprints, *Nature*, 355(6357), 209–210, 1992.

19. Harris, C.C., p53 tumor suppressor gene: at the crossroads of molecular carcinogenesis, molecular epidemiology, and cancer risk assessment, *Environ. Health Perspect.*, 104 Suppl 3, 435–439, 1996.

20. Hsu, I.C. et al., Mutational hotspot in the p53 gene in human hepatocellular carcinomas, *Nature*, 350(6317), 427–428, 1991.

21. Bressac, B. et al., Selective G to T mutations of p53 gene in hepatocellular carcinoma from southern Africa, *Nature*, 350(6317), 429–431, 1991).

22. Challen, C. et al., Analysis of the p53 tumor-suppressor gene in hepatocellular carcinomas from Britain, *Hepatology*, 16(6),1362–1366, 1992.

23. Li, D. et al., Aberrations of p53 gene in human hepatocellular carcinoma from China, *Carcinogenesis*, 14(2), 169–173, 1993.

24. Rashid, A. et al., Genetic alterations in hepatocellular carcinomas: Association between loss of chromosome 4q and p53 gene mutations, *Brit. J. Cancer*, 80, 59–66, 1999.

25. Coursaget, P. et al., High prevalence of mutations at codon 249 of the p53 gene in hepatocellular carcinomas from Senegal, *Brit. J. Cancer*, 67(6), 1395–1397, 1993.

26. Aguilar, F. et al., Geographic variation of p53 mutational profile in nonmalignant human liver, *Science*, 264(5163), 1317–1319, 1994.

27. Fujimoto, Y. et al., Alterations of tumor suppressor genes and allelic losses in human hepatocellular carcinomas in China, *Cancer Res.*, 54(1), 281–285, 1994.

28. Foster, P.L., Eisenstadt, E., and Miller, J.H., Base substitution mutations induced by metabolically activated aflatoxin B1, *Proc. Nat. Acad. Sci. USA*, 80(9), 2645–2648, 1983.

29. Puisieux, A. et al., Selective targeting of p53 gene mutational hotspots in human cancers by etiologically defined carcinogens, *Cancer Res.*, 51(22), 6185–6189, 1991.

30. Denissenko, M.F. et al., The p53 codon 249 mutational hotspot in hepatocellular carcinoma is not related to selective formation or persistence of aflatoxin B1 adducts, *Oncogene, 17(23),* 3007–3014, 1998.

31. Aguilar, F. Hussain, S.P., and Cerutti, P., Aflatoxin B1 induces the transversion of G→T in codon 249 of the p53 tumor suppressor gene in human hepatocytes, *Proc. Nat. Acad. Sci. USA,* 90(18), 8586–8590, 1993.

32. Cerutti, P. et al., Mutagenesis of the H-ras protooncogene and the p53 tumor suppressor gene, *Cancer Res.,* 54(7 Suppl.), 1934s–1938s, 1994.

33. Lasky, T. and Magder, L., Hepatocellular carcinoma p53 G > T transversions at codon 249: the fingerprint of aflatoxin exposure?, *Environ. Health Perspect.,* 105(4), 392–397, 1997.

34. Lunn, R.M. et al., p53 mutations, chronic hepatitis B virus infection, and aflatoxin exposure in hepatocellular carcinoma in Taiwan, *Cancer Res.,* 57(16), 3471–3477, 1997.

35. Sidransky, D. and Hollstein, M., Clinical implications of the *p53* gene, *Annu. Rev. Med.,* 47, 285–301, 1996.

36. Kirk, G.D. et al., Ser-249 *p53* mutations in plasma DNA of patients with hepatocellular carcinoma from the Gambia, *J. Natl. Cancer Inst.,* 92, 148–153, 2000.

37. Jackson, P.E. et al., Specific *p53* mutations detected in plasma and tumors of hepatocellular carcinoma patients by electrospray ionization mass spectrometry, *Cancer Res.,* 61, 33–35, 2001.

38. Kensler, T.W., Groopman, J.D., and Wogan, G.N., Use of carcinogen-DNA and carcinogen-protein adduct biomarkers for cohort selection and as modifiable end points in chemoprevention trials, *Principles of Chemoprevention,* 139, 237–248, 1996.

39. Kensler, T.W., Groopman, J.D., and Roebuck, B.D., Use of aflatoxin adducts as intermediate endpoints to assess the efficacy of chemopreventive interventions in animals and man, *Mutat. Res.,* 402, 165–172, 1998.

40. Roebuck, B.D. et al., Protection against aflatoxin B1-induced hepatocarcinogenesis in F344 rats by 5-(2-pyrazinyl)-4-methyl-1,two-dimensionalithiole-3-thione (oltipraz): predictive role for short-term molecular dosimetry, *Cancer Res.,* 51(20), 5501–5506, 1991.

41. Egner, P.A. et al., Levels of aflatoxin-albumin biomarkers in rat plasma are modulated by both long-term and transient interventions with oltipraz, *Carcinogenesis,* 16(8), 1769–1773, 1995.

42. Jacobson, L.P. et al., Oltipraz chemoprevention trial in Qidong, People's Republic of China: study design and clinical outcomes, *Cancer Epidemiol. Biomarkers Prev.,* 6(4), 257–265, 1997.

43. Kensler, T.W. et al., Oltipraz chemoprevention trial in Qidong, People's Republic of China: Modulation of serum aflatoxin albumin adduct biomarkers, *Cancer Epidemiol. Biomarkers, Prev.,* 7(2), 127–134, 1998.

chapter twenty-two

Biomarkers of early effect in the study of cancer risk

Laura Gunn, Luoping Zhang, Matthew S. Forrest,
Nina T. Holland, and Martyn T. Smith

Contents

Abstract   Cancer is an abnormal genetic phenomenon, involving multiple steps of somatic mutation. Genetic damage can occur at the level of the gene (e.g., point mutations and deletions) or the chromosome (e.g., aneuploidy, translocations). During the last two decades, a wide spectrum of biomarkers of genetic damage has been developed to detect early mutational and chromosomal effects of carcinogenic exposure in humans. Historically, biomarkers have tended to measure mutations in surrogate genes or use cytogenetics to assess overall changes in chromosome structure and number. These biomarkers have been shown to be associated with a wide range of carcinogenic exposures, but they are not truly biomarkers of early effect as they are not on the causal pathway of disease. Identification of early causal genetic events in cancer has led to the recent development of novel biomarkers of early effect in high-risk populations. These novel biomarkers measure changes frequently observed among cancer patients, including point mutations in genes such as *p53* and *RAS*, altered gene methylation, and aneuploidy (chromosome loss or gain), including monosomy 7 and trisomy 8, and specific chromosome rearrangements such as translocations. Future technologies will measure >50,000 endpoints from a drop of blood using proteomics and array technologies and identify all genetic polymorphisms related to susceptibility using high-throughput genomics. Application of these biomarkers to study individuals who may be at risk, but who do not yet have cancer, will result in improved early detection, as well as a better understanding of the risk factors for cancer itself.

I. Introduction

Carcinogenesis is a complex, multistage process, which involves the accumulation of a variety of mutations within a particular cell and its progeny.[1] Although carcinogenesis depends on a number of different factors including environmental exposure, diet, genetics, and target tissue, certain general characteristics of cancers are known. The identification of the role of particular genes in cancer has opened a new avenue of research over the past two decades. Oncogenes and tumor suppressor genes have taken center stage with their respective roles in cancer. Alterations in these genes ranging from small insertions, deletions, point mutations, and aberrant methylation, to gross chromosomal aberrations, like translocations, and gene amplification, either enhance or inactivate the normal function of the gene and lead to abnormal proliferation, lack of cell cycle control, genomic instability, and eventually cancer. Mutations in these genes provide telltale signs of genetic changes or damage and possible cancer risk, often long before the onset of cancer. Particular genes, chromosomal regions or entire chromosomes are vulnerable to mutation at variable points in carcinogenesis.[1] This suggests that certain mutations play a specific role in the ability of a cell to survive and continue to the next step of this multistep process, as well as potentially determining what the next mutation will be. These mutations, particularly early events, may provide markers, which are indicative of genetic damage and potential cancer risk.

During the last two decades, a wide spectrum of biomarkers of genetic damage has been developed to detect mutational and chromosomal effects of carcinogenic exposure in humans.[2] Historically, biomarkers have tended to measure mutations in surrogate genes, including hypoxanthine phosphoribosyltransferase and glycophorin A[3] or use cytogenetics to assess overall changes in chromosome structure and number.[4-6] These biomarkers have been shown to be associated with a wide range of carcinogenic exposures, but they are not true biomarkers of early effect as they are not on the causal pathway of disease. In addition, early biomarkers were not specific enough to identify small or gene-specific alterations, but rather indicated generalized chromosome damage or instability.

As our understanding of the multistage process of carcinogenesis has improved over the years, biomarkers have evolved into more sensitive, specific, predictive markers of cancer risk. Current biomarkers measure changes frequently observed among cancer patients, including point mutations in key genes such as *p53* and *RAS,* aberrant gene methylation, altered gene expression, aneuploidy (chromosome loss or gain), and specific chromosome rearrangements such as translocations.

The evolution of biomarker research has included the identification of causal genetic events as well as the means to study these events. A key contributing factor to the evolution of cancer biomarkers has been the parallel evolution of biotechnology. Technological advances, such as PCR and fluorescent microscopy, have revolutionized the field of cancer research by providing new levels of sensitivity and specificity. Technology will continue to fuel the development of genetic biomarkers as well as enabling the elucidation and analysis of complex phenotypes. These phenotypic markers will soon be analyzed by extremely high-throughput methods, requiring minimal amounts of biological material. This paper will review recent progress in the development of biomarkers of early effect for carcinogenesis and illustrate their use in our own work. Finally, we will describe recent technological advances that are likely to be responsible for the biomarkers of the future.

II. Molecular cytogenetics

A. Fluorescence in situ hybridization

Molecular cytogenetics employing fluorescence *in situ* hybridization (FISH) has been around for more than a decade and its application in biomarker research continues to grow. Recent technological developments, most significantly rolling circle amplification and automated spot counting, are likely to dramatically increase the utility of FISH in the next few years. FISH has several advantages over conventional cytogenetics, perhaps the most important of which is its rapid ability to detect specific chromosome changes on the

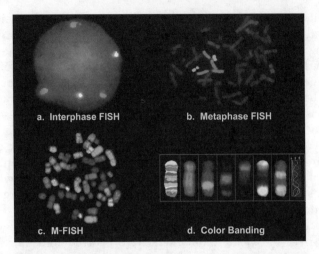

Figure 22.1 (See color insert following page 392.) Different applications of FISH in human cells: (a) two color interphase FISH shows aneuploidy of chromosome 7 in a human lymphocyte with 3 spots of chromosome 7 (red) and 2 copies of chromosome 8 (green); (b) three color painting of metaphase FISH with chromosomes 1, 16, and 19; (c) M-FISH (multicolor FISH); (d) color banding of chromosome 5. (Images (a) and (b) were obtained in the laboratory of the authors. Images (c) and (d) are from MetaSystems, Inc., Belmont, MA. With permission.)

causal pathway to cancer. Interphase FISH, in particular, offers several advantages over classical cytogenetics.[7] First, interphase FISH allows analysis of nondividing cells (Figure 22.1a). Second, a much larger number of cells, at least 1000 or more, may be analyzed. Third, the detection of aneuploidy is facilitated by simply counting the number of labeled regions representing a particular chromosome of interest within the isolated interphase nucleus. By contrast, metaphase FISH can readily detect structural rearrangements in addition to aneuploidy (Figure 22.1b). Furthermore, because metaphase FISH, like classical cytogenetics, analyzes dividing cells, the results from these two methods may be directly compared. A number of studies have determined that FISH is more sensitive and convenient than classical cytogenetics.[8-10] Therefore, FISH appears to be the more suitable method for large-scale population biomonitoring. However, one of the drawbacks of FISH is its high cost due to a restrictive patent and monopoly of probe distribution.

B.   Applications of FISH

One example of a specialized FISH assay primarily employed in radiation research is that developed by Tucker and coworkers.[11,12] This assay uses single-color FISH by painting the chromosome pairs 1, 2, and 4 (or 3, 5 and 6) the same color, which allows for the detection of numerical and structural chromosome aberrations among these painted chromosomes and structural rearrangements between these and other untargeted chromosomes. This assay has been applied *in vitro* and *in vivo* in animal and human studies.[11,13,14]

Since radiation is thought to cause equal levels of damage across all chromosomes,[15] and chromosomes 1 through 6 (the largest chromosomes) make up 40% of the genome,[16] it is hypothesized that measurement of damage in these large chromosomes can be extrapolated to the whole genome.[11] This may not be true for chemical exposures as certain chemicals may have selective or preferential effects on certain chromosomes.[17] For example, we showed that epoxide metabolites of 1,3-butadiene had more effect on certain chromosomes than others.[18] Indeed, the hypothesis of equal levels of damage across the genome may not hold true even for low doses of radiation, as inversion of chromosome 10 has been shown to be highly sensitive to low intensity radiation exposure.[19] Interestingly inv(10) rearranges the *RET* gene and is associated with thyroid cancer, potentially caused by linear energy transfer (LET) radiation.

Our laboratory is currently employing FISH to examine the cytogenetic changes in human blood cells caused by exposure to the established leukemogen, benzene. Our plan is to examine all 24 chromosomes and in particular to examine for chromosome changes associated with the development of leukemia. This study is being performed along with Drs. Rothman and Hayes of the National Cancer Institute (NCI), and Drs. Li and Yin at the Chinese Academy of Preventive Medicine in Beijing. We have already applied various FISH techniques in this collaborative study of 43 Chinese workers highly exposed to benzene (median exposure level = 31 ppm, 8 hr TWA) and 44 frequency-matched controls. To date, five chromosomes (1, 5, 7, 8, and 21) have been examined by metaphase FISH in these highly exposed Chinese workers and their matched controls. Frequencies of monosomy 5, 7, and 8, but not 1 or 21, increased with elevated exposure levels, whereas a significant trend was observed for trisomy of all five chromosomes.[17,20] The most striking dose-dependent increases were found in monosomy 7 and trisomy 7, 8 and 21. The most common structural changes detected among chromosomes 1, 5, 7, 8, and 21 were t(8;21), t(8;?) (translocation between chromosome 8 and another unidentified chromosome), breakage of chromosome 8, and deletions of the long (q) arms of chromosomes 5 and 7. A significant trend was observed for all these changes.[17,20] The loss and long arm deletion of chromosomes 5 and 7, two of the most common cytogenetic changes in therapy and chemical-related leukemia, were significantly increased in benzene-exposed workers over controls.[17]

C. Recent developments in FISH technology

Recently, multicolor FISH (mFISH) and color banding have been developed. The mFISH method involves painting each of the 24 different chromosomes a different color using four or five fluorophores with combined binary ratio labeling, which allows the entire karyotype to be screened for chromosome aberrations (Figure 22.1c).[21] Since the human eye cannot effectively distinguish the 24 colors, this method requires the use of an automated imaging system. In color banding, which is based on traditional banding techniques, each chromosome is labeled by subregional DNA probes in different colors, resulting in a unique chromosome bar code (Figure 22.1d).[22] This method allows the rapid identification of chromosomes and chromosome rearrangements. These techniques are at present relatively new and have not been employed as widely or extensively as FISH, but their potential is high.

Of great promise in its application to FISH is rolling circle amplification (RCA), which is a molecular methodology that allows visualization by FISH of small changes previously undetectable by microscopy. It relies upon the isothermal linear amplification of a single stranded circle of DNA. A variation of RCA uses a probe with two 3′ ends, a target specific sequence (40–50 nt) attached by a poly-T linker and $(CH_2)_{18}$ spacer to a primer (24–28 nt), which will initiate an RCA.[23] When the probe is added to cells fixed onto microscope slides, the target specific sequence anchors the probe to a region of DNA and stops it from being removed during washing. Thus, when RCA occurs the product is attached to the genomic DNA; when decorator probes and biotin labeled dUTPs are added, the linear repeat can be collapsed by avidin to give a localized area of fluorescence.[23] This technique means that much smaller fragments of DNA than would be apparent by traditional FISH can be visualized, including small deletions, duplications, point mutations, and even single nucleotide polymorphisms.

III. PCR-based methods for detecting somatic mutations and aberrant gene methylation

Chromosome translocations and other structural rearrangements produce novel fusion genes or products that can be detected at the DNA or RNA level by the polymerase chain reaction (PCR) or reverse-transcriptase PCR (RT-PCR) as well as by FISH. PCR holds a number of advantages over FISH, including: the ability to detect rare events (1 copy/10^{6-7} cells vs. $1/10^{3-4}$ cells by FISH) and the ability to study large numbers of people easily and at low cost. These potent advantages are accompanied, however, by two disadvantages. First, the high sensitivity of PCR makes it prone to false-positive results caused by sample contamination. However, contamination artifacts can be overcome with extremely rigorous lab procedures[24] as well the use of dUTP and uracil glycosylase in PCR reactions to prevent carryover contamination. Second, until recently, quantitation was difficult, especially for RT-PCR. Quantitation has also become feasible through recent advances in exonuclease-dependent real-time PCR. This quantitative PCR assay, now generally called real-time PCR, allows for the absolute number of novel sequences to be quantified in a cell population. Real-time PCR is more sensitive than conventional PCR, where a sensitivity of 1 in 10^7 can be reached if a stochastic multitube approach is taken.[25,26] This technology has, therefore, paved the way for a new generation of biomarkers to be developed. Though no methods yet exist which employ PCR to measure rare aneuploidies or genome-wide structural damage, real-time and conventional PCR techniques which measure specific chromosome rearrangements, such as translocations, inversions, and the methylation status of genes, have become available.

A. Conventional PCR detection of chromosome rearrangements

RT-PCR and PCR have previously been used to detect a number of translocations including t(14;18), t(8;21), t(9;22), and t(4;11). Using these techniques, t(9;22) and t(14;18) have been detected in unexposed individuals of different ages and in smokers.[27-29] Both translocations were found to increase with age and the t(14;18) translocation was increased in cigarette smokers.[30] Studies from our laboratory showing detectable t(8;21) by RT-PCR in an otherwise healthy benzene exposed worker[20] clearly demonstrate the potential of RT-PCR for monitoring specific aberrations in populations exposed to suspected or established leukemogens. Because many of these translocations have multiple breakpoints or translocation partners, multiplex assays have also been developed to detect multiple or unknown rearrangements. Despite recent improvements in sensitivity and applicability, conventional PCR methods remain semiquantitative. However, with the recent advent of real-time PCR, quantitation is no longer an obstacle. Now that quantitation problems can be overcome, a new avenue of biological monitoring for early detection of cancer has been opened. PCR-based procedures therefore hold further promise for detecting specific chromosome aberrations, especially when used in combination with FISH.

B. Quantitative real-time PCR

Real-time PCR is comparable to conventional PCR in that it uses sense and antisense primers to amplify a targeted sequence of DNA. However, real-time PCR techniques such as TaqMan, molecular beacons, and Scorpion probes, rely upon fluorescent signals to quantitate the rate of amplification as the reaction progresses. TaqMan employs an additional, nonextendable oligonucleotide probe, which is positioned between the two primers during the annealing phase of amplification (Figure 22.2).[31] The oligonucleotide probe is labeled with a fluorescent reporter dye, such as FAM (6-carboxy-fluorescein), at the 5′ end and a quencher fluorescent dye, such as TAMRA (6-carboxy-tetramethyl-rhodamine), at the 3′ end. When the probe is intact, fluorescence resonance energy transfer (FRET) to TAMRA quenches the FAM emission. During the extension phase of amplification, the *Taq* polymerase extends the primer to the region of the probe, at which point the 5′ exonuclease property of *Taq* cleaves the reporter dye from the probe. This results in an increase in fluorescent signal that is proportional to the amount of amplification product. The increase in reporter molecules is measured in real time by the ABI Prism 5700 or 7700 Sequence Detection Systems (PE Applied Biosystems). After each cycle, a fluorescent signal is measured resulting in an amplification plot, in which the point at which the fluorescence crosses a defined threshold, Ct, that is proportional to the starting copy number. Cts

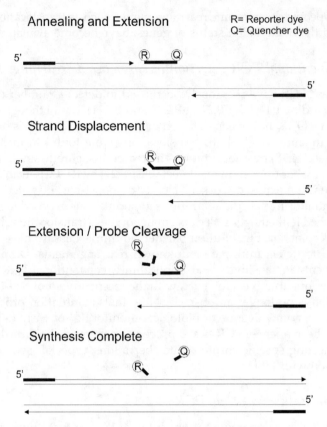

Figure 22.2 Diagram of TaqMan technology in quantitative PCR. (Adapted from PCR Applications, 1999.[31])

of positive control samples are used to generate a standard curve. From this standard curve, it is possible to calculate copy number of unknown samples. Methods for the quantitative detection of translocations using the above TaqMan technology have recently been reported. For example, methods for the analysis of t(14;18), t(8;21), t(9;22) and other translocations have been presented or published.[25,32–35]

We have further developed and refined these real-time RT-PCR methods so that they permit the quantification of t(8;21), t(15;17), inv(16), t(11q23), t(14;18), t(12;21), and t(1;19) fusion gene expression at low levels (1 transcript per 100,000 cells). This has enabled us to examine the levels of these translocations in adult and cord blood of control individuals, in workers exposed to benzene and in patients given the flavonoid quercetin therapeutically. We have found that a significant number of control individuals express measurable levels of the specific translocations and that mitogen-stimulated blood culture enhances the expression of several translocations. Data from benzene-exposed workers and quercetin treated patients are currently under analysis, but *in vitro* studies in cord blood have demonstrated the ability of the benzene metabolite hydroquinone to produce t(14;18), providing a mechanistic basis for the production of lymphoma by benzene.[36]

C.   Another potential application of real-time PCR: measurement of aberrant gene methylation

In addition to the different types of genetic damage involved in carcinogenesis, epigenetic mechanisms, such as DNA methylation, have gained attention as potential key players in certain cancer types. Aberrant methylation, which may be induced by environmental exposures, may result in altered carcinogen metabolism, cell cycle regulation, and DNA repair. For example, in leukemia and lymphoma, translocations cause the formation of novel fusion genes that produce excessive growth,[37-38] and other genes undergo transcriptional silencing by methylation, which causes aberrant cell cycle control.[39] Aberrant methylation and transcriptional silencing appears to be an early event in both solid tumors, including lung,[40] colon,[41] hepatocellular,[42] and bladder,[43] as well as hematologic malignancies.[39] A number of different methods have been developed to detect aberrant methylation of genes, including the use of methylation sensitive restriction enzymes, bisulfite sequencing, and methylation-specific PCR.

Perhaps one of the most interesting targets of aberrant methylation is the tumor suppressor gene p16^{INK4a}, which is a key component in the G1/S cell cycle checkpoint and has been shown to be involved in colon cancer, leukemia, and lung cancer. Recently, Lo et al. have developed a real-time methylation-specific PCR protocol[44] and applied it to bone marrow samples of patients with multiple myeloma as well as cell lines with known methylation status. The authors demonstrated that the real-time method had high concordance with the conventional method, however, with the added sensitivity and specificity of the real-time technology. In addition, the authors correlated methylation status with p16 mRNA expression and observed transcription was inversely correlated with methylation status. As with other real-time methods, this application shows great potential for future studies involving methylation of key genes in carcinogenesis as well as other biological processes.

D.   Multiplex real-time PCR with molecular beacons

Yet another novel application of real-time PCR technology is the multiplex amplification of DNA by Vet et al.[45] Using molecular beacons rather than TaqMan probes, the authors successfully performed multiplex PCR reactions which were monitored in real-time. Although similar to TaqMan probes, molecular beacons do not require the 5′ exonuclease properties of Taq polymerase. Instead, when unbound, the fluorescent moiety is kept in physical proximity to the quencher in a loop conformation. When the probe anneals to a target sequence, the loop is linearized and the reporter and quencher are separated in space, and the reporter emits fluorescence. In a multiplex reaction, different colored fluorescent moieties are used for each amplicon. The potential applications of this assay are promising. Because research potential is often limited by the amount of material available for analysis, the possibility of real-time multiplex PCR provides the opportunity to examine multiple points of interest simultaneously with the same amount of material previously required for one point.

E.   The use of Scorpion™ primers

An alternative to molecular beacons and TaqMan are Scorpion primers.[46] While TaqMan and molecular beacons rely upon a bimolecular approach, with separate probes and primers, Scorpion primers use a unimolecular method with an integral tail that is used to probe an extension product of the primer. Scorpion primers consist of a normal PCR primer linked to a nonamplifiable monomer, such as hexethylene glycol, which prevents copying of elements further in the 5′ direction, a nonfluorogenic (dark) quencher of the fluorophore, a probe element flanked by self-complementary stems (hairpin), and a fluorophore. In its unextended form, the Scorpion is nonfluorescent (the fluorophore is quenched). During PCR, the primer element of the Scorpion is extended at its 3′ end and the Scorpion becomes a full PCR product. The recognition sequence of the Scorpion then hybridizes to its complementary target sequence within the same strand of the PCR product and fluorescence is emitted. A recent paper suggests that this means Scorpions perform better than TaqMan or molecular beacons, especially under fast cycling conditions.[47]

F.   PCR-based detection of point mutations

The use of PCR technology has vastly improved detection and identification of mutations in cancers. Increased sensitivity and reproducibility has provided the possibility of utilizing these mutation assays as biomarkers of early effect, detection of minimal residual disease, or precursors to relapse. Because of the low frequency of many of these mutations in the normal population, the normal background levels and variability have not yet been established. Recently, a number of assays have been developed which improve sensitivity orders of magnitude over previously used methods. Many of these assays employ methods to selectively amplify the relative number of mutants in a large pool of wildtype in order to increase the sensitivity of detecting rare mutant alleles, a method referred to as genotypic selection.

One recently published assay used genotype selection for the detection of mutations in the *H-RAS* gene. By combining two previously published methods,[48,49] the Mut-Ex + ACB-PCR technique[50] is one of the most sensitive genotypic selection methods. This assay begins with the denaturation of a heterogeneous sample of mutant and wildtype double stranded DNA. When reannealing, mutant DNA forms heteroduplex DNA with normal strands, while normal DNA strands form homoduplexes. Mut S, a thermostable protein, is added. This binds to the mispaired sequence of the heteroduplex, which protects the short sequence of mutant DNA from digestion from 3'–5' exonuclease activity of T7 DNA polymerase, whereas the wildtype DNA is digested. This Mut-Ex step results in a 1000-fold enrichment of mutant alleles relative to wildtype. To further increase sensitivity, the next step utilizes an additional selection technique, ACB-PCR: (allele-specific competitive blocker PCR). This genotypic selection method is based on preferential amplification by allele specific primers. The first primer has more mismatches to wildtype than mutant, resulting in preferential amplification of mutant DNA. The second primer is a blocker primer which preferentially anneals to the wildtype sequence, but is modified with a 3' dideoxyguanosine residue, which prevents extension. ACB-PCR therefore results in preferential amplification of mutant DNA with a sensitivity of as few as ten mutant alleles detected in the presence of 10^8 copies of the wildtype allele.

As one of the most sensitive methods available for mutation detection, the MutEX+ ACB-PCR technique has many potential applications. This method is based on increasing the ratio of mutant DNA relative to wildtype and is therefore a sensitive method for the detection of rare mutations. However, this method is not appropriate for unknown mutations, as the sequence of the mutated region is necessary for the design of ACB-PCR primers.

Genotypic selection methods have also been applied to *p53* mutation detection. Sites which are commonly mutated in the *p53* gene, referred to as mutational hotspots, have been targeted as potential biomarkers of early effect. Assays utilizing allele specific PCR have been designed to detect and preferentially amplify mutations in these hotspots. These assays either used alone or in combination with single strand conformational polymorphism and sequencing result in a considerable improvement in sensitivity over conventional methods of mutation detection.[51]

IV. PCR-independent molecular technologies

Traditional mutation detection and genotyping, be it RFLP, sequencing (normal, mini-, or pyro-), SSCP, etc., have relied upon one technology, PCR. However there are two new methodologies which break free from the restrictions of PCR: rolling circle amplification and the Invader® assay.

A. Rolling circle amplification

Rolling circle amplification has a variety of potential uses, all of which rely upon the isothermal linear amplification of a single stranded circle of DNA. Padlock probes use rolling circle amplification to identify small sequences, polymorphisms, or mutations. A linear probe is designed to have target-complementary segments of 15–20 nucleotides at both ends, separated by a linker of approximately 50 nucleotides. The linker can include sequence elements that allow for amplification or identification of individual probes. Upon hybridization to a target DNA or RNA sequence, the two ends of the probe become juxtaposed and can be joined by a DNA ligase, ensuring a similar level of specificity to that of PCR. Circularized probes can then be detected in a number of ways. For example, the primer can be designed to hybridize to the padlock with which primer extension creates a linear tandem repeat of the circle that can be probed with labeled decorator oligos. Hyper-branched RCA occurs if a second primer is added which copies the RCA product, resulting in faster replication than normal RCA. Digestion with a restriction enzyme, known to cut the padlock only once, allows the resolution of the product on an agarose gel.[52] It is also possible to use PCR to amplify a padlock probe if the primers used amplify across the ligation point.

This method can also be used for *in situ* allelic discrimination. For allelic discrimination, there are two probes which only have 20 nucleotide target specific sequences which differ at the 3' nucleotide. A single anchor probe is designed to hybridize adjacent to the rolling circle probes. When ligase is added, the anchor probe will ligate to the allele discriminating probe only if the 3' nucleotide matches the target sequence. Amplification can then proceed as described above. Genotype can then be determined by using different labels for each decorator probe.[23]

B. The Invader assay

The Invader assay (Third Wave Technologies, Inc., Madison, WI) is a plat-form-independent homogenous single-tube assay that does not involve PCR, restriction digests, or gel electrophoresis. Invader assays rely upon the creation of a unique substrate for Cleavase®, a structure-specific 5' nuclease, by the annealing of a probe and an upstream oligonucleotide to a target sequence. The probe contains two regions, an analyte-specific region that forms a duplex with the target and a noncomplementary 5' arm region, which serves as a reporter molecule precursor. Cleavage of the probe only occurs when the probe and upstream oligonucleotide overlap[53,54]; therefore, two target DNAs differing by a single nucleotide may be dis-criminated. The cleaved arm of the probe then forms cleavage structures with FRET (fluorescence resonance energy transfer) cassettes, labelled with reporter and quencher dyes. Cleavage results in the release of the reporter dye generating an increase in fluorescent signal. Therefore, Invader can be used as an alternative to real-time PCR as well as for allelic discrimination and mutation detection.

V. The future application of arrays and nanobarcodes

A wide variety of cDNA microarrays are now available for studying gene expression. To date, these have found only limited application in biomarker research. This is probably due to at least two reasons. The first is expense. Academic researchers have been hindered in the application of this technology because of its high cost. However, competition, consortia, and new technolo-gies (such as Operon's 70mer oligonucleotides) are driving prices downward.

The second is the need for large amounts of high-quality RNA. cDNA microarrays rely on the RT-PCR reaction, which we have recently shown is relatively inefficient using commercial enzymes (Curry J., McHale C., and Smith M.T., manuscript in preparation). Thus, large quantities of RNA are required as starting material, somewhere in the order of 50 micrograms. Again new advances are driving this amount down, but a second problem is that most epidemiologists do not collect biological material in such a fashion that RNA is preserved. RNA is inherently a highly unstable molecule and thus its isolation and storage are challenging. We are currently working on protocols for the field collection of RNA from samples obtained in biomarker studies.

Investigators also realize that gene expression is not necessarily phenotype. Thus, protein arrays and proteomics have become of particular interest. Protein arrays hold great potential in biomarker research as they may identify a pattern of protein expression associated with a particular exposure or the early onset of disease. There are also large banks of biological material available for analysis by such arrays.

Possibly of even greater potential is the future application of nanotechnology, including Nanobarcodes™. A company called Surromed (California) is attempting to measure over 50,000 endpoints in a single drop of blood. In order to do this, it is developing several new technologies including Surroscan to replace flow cytometry and Nanobarcodes labelled with antibodies so that thousands of proteins can be recognized in a drop of plasma. Nanobarcodes are freestanding, cylindrically shaped metal nanoparticles whose composition can be varied along length, such that more than seven metals can be used to produce a nanoscale supermarket-like bar code. Their identification is based on differential reflectivity using microscopy and they can be functionalized with proteins, oligonucleotides, and organics. They can also be made magnetic. This allows for a wide variety of applications including multiplexed analysis of gene expression and multiplexed immunoassays with 10–20,000 flavors. Molecular epidemiologists would benefit by recognizing this coming technology and begin storing small amounts of cryopreserved whole blood as well as the usual serum, buffy coat, etc., that are conventionally stored.

VI. Conclusion

This is an exciting time in biomarker research. Technologies are emerging which will make a significant impact on the field. The potential for improved early detection of cancer is apparent. In addition to enhanced sensitivity and specificity, many new technologies emphasize functional aspects of mutation which require RNA, intact cryopreserved cells, and undamaged proteins. It is therefore important to recognize that biological samples will have to be processed in a more sophisticated manner to ensure their preservation and utility in future biomarker research. Further, it is clear that a significant effort is necessary to lower the cost of these new technologies if they are to be applied in developing countries where they are most needed.

Acknowledgments

Our work on biomarkers is supported by the National Institute of Environmental Health Sciences through grants P42ES04705, P30ES01896 and RO1ES06721 and by the National Foundation for Cancer Research. LG is supported by training grant T32ES07075.

References

1. Vogelstein, B. and Kinzler, K.W., The multistep nature of cancer. *Trends Genet.*, 9, 138–141, 1993.
2. Toniolo, P. et al., Eds., *Application of Biomarkers in Cancer Epidemiology*, International Agency for Research on Cancer, Lyon, France, 1997.
3. Albertini, R.J. and Hayes, R.B., Somatic cell mutations in cancer epidemiology, *IARC Sci. Publ.*, 6, 159–184, 1997.
4. Hagmar, L. et al., Cancer risk in humans predicted by increased levels of chromosomal aberrations in lymphocytes: Nordic Study Group on the Health Risk of Chromosome Damage, *Cancer Res.*, 54, 2919–2922, 1994.
5. Hagmar, L. et al., Chromosomal aberrations in lymphocytes predict human cancer: A report from the European Study Group on cytogenetic biomarkers and health (ESCH), *Cancer Res.*, 58, 4117–4121, 1998.
6. Sorsa, M., Wilbourn, J., and Vainio, H., Human cytogenetic damage as a predictor of cancer risk, *IARC Sci. Publ.*, 18, 543–554, 1992.
7. Eastmond, D.A., Schuler, M., and Rupa, D.S., Advantages and limitations of using fluorescence *in situ* hybridization for the detection of aneuploidy in interphase human cells, *Mutat. Res.*, 348, 153–162, 1995.
8. Kadam, P. et al., Combination of classical and interphase cytogenetics to investigate the biology of myeloid disorders: Detection of masked monosomy 7 in AML, *Leuk. Res.*, 17, 365–374, 1993.
9. Kibbelaar, R.E. et al., Detection of monosomy 7 and trisomy 8 in myeloid neoplasia: A comparison of banding and fluorescence *in situ* hybridization, *Blood*, 82, 904–913, 1993.
10. Poddighe, P.J. et al., Interphase cytogenetics of hematological cancer: Comparison of classical karyotyping and *in situ* hybridization using a panel of eleven chromosome specific DNA probes, *Cancer Res.*, 51, 1959–1967, 1991.
11. Matsumoto, K. et al., Persistence of radiation-induced translocations in human peripheral blood determined by chromosome painting, *Radiat. Res.*, 149, 602–613, 1998.
12. Tucker, J.D. and Senft, J.R., Analysis of naturally occurring and radiation-induced breakpoint locations in human chromosomes 1, 2 and 4, *Radiat. Res.*, 140, 31–36, 1994.
13. Tucker, J.D. et al., Persistence of radiation-induced translocations in rat peripheral blood determined by chromosome painting, *Environ. Mol. Mutagen.*, 30, 264–272, 1997.
14. Tucker, J.D. et al., Biological dosimetry of radiation workers at the Sellafield nuclear facility, *Radiat. Res.*, 148, 216–226, 1997.
15. Sachs, R.K., Chen, A.M., and Brenner, DJ., Review: Proximity effects in the production of chromosome aberrations by ionizing radiation, *Intl. J. Radiat. Biol.*, 71, 1–19, 1997.

16. Morton, N.E., Parameters of the human genome, *Proc. Natl. Acad. Sci. USA*, 88, 7474–7476, 1991.

17. Zhang, L. et al., Increased aneusomy and long arm deletion of chromosomes 5 and 7 in the lymphocytes of Chinese workers exposed to benzene, *Carcinogenesis*, 19, 1955–1961, 1998.

18. Xi, L. et al., Induction of chromosome-specific aneuploidy and micronuclei in human lymphocytes by metabolites of 1,3-butadiene, *Carcinogenesis*, 18, 1687–1693, 1997.

19. Scarpato, R. et al., FISH analysis of translocations in lymphocytes of children exposed to the Chernobyl fallout: Preferential involvement of chromosome 10, *Cytogenet. Cell Genet.*, 79, 153–156, 1997.

20. Smith, M.T. et al., Increased translocations and aneusomy in chromosomes 8 and 21 among workers exposed to benzene, *Cancer Res.*, 58, 2176–2181, 1998.

21. Schröck, E. et al., Multicolor spectral karyotyping of human chromosomes, *Science*, 273, 494–497, 1996.

22. Müller, S. et al., Toward a multicolor chromosome bar code for the entire human karyotype by fluorescence *in situ* hybridization, *Hum. Genet.*, 100, 271–278, 1997.

23. Zhong, X.B. et al., Visualization of oligonucleotide probes and point mutations in interphase nuclei and DNA fibers using rolling circle DNA amplification, *Proc. Natl. Acad. Sci. USA*, 98, 3940–3945, 2001.

24. Kwok, S. and Higuchi, R., Avoiding false positives with PCR published erratum appears in Nature 1989 Jun 8;339(6224):490], *Nature*, 339, 237–238, 1989.

25. Luthra, R. et al., Novel 5′ exonuclease-based real-time PCR assay for the detection of t(14;18)(q32;q21) in patients with follicular lymphoma, *Am. J. Path.*, 153, 63–68, 1998.

26. Doelken, L., Schueler, F., and Doelken, G., Quantitative detection of t(14;18)-positive cells by real-time quantitative PCR using fluorogenic probes, *Biotechniques*, 25, 1058–1064, 1998.

27. Liu, Y. et al., BCL2 translocation frequency rises with age in humans, *Proc. Natl. Acad. Sci. USA*, 91, 8910–8914, 1994.

28. Biernaux, C. et al., Detection of major bcr-abl gene expression at a very low level in blood cells of some healthy individuals, *Blood*, 86, 3118–3122, 1995.

29. Fuscoe, J.C. et al., Quantification of t(14;18) in the lymphocytes of healthy adult humans as a possible biomarker for environmental exposures to carcinogens, *Carcinogenesis*, 17, 1013–1020, 1996.

30. Bell, D.A., Liu, Y., and Cortopassi, G.A., Occurrence of bcl-2 oncogene translocation with increased frequency in the peripheral blood of heavy smokers, *J. Natl. Cancer Inst.*, 87, 223–224, 1995.

31. Innis, M., Gelfand, D.H., and Sninsky, J.J., Eds. PCR applications: protocols for functional genomics, Academic Press, San Diego, CA, 1999.

32. Bories, D. et al., Real-time quantitative RT-PCR monitoring of chronic myelogenous leukemia, *Blood*, 92, 73A, 1998.

33. Krauter, J. et al., AML1/MTG8 real time RT-PCR for the detection of minimal residual disease in patients with t(8;21) positive AML, *Blood*, 92, 74A-75A, 1998.

34. Preudhomme, C. et al., Detection of BCR-ABL transcripts in chronic myeloid leukemia (CML) using a novel "real time" quantitative RT-PCR assay: A report on 15 patients, *Blood*, 92, 93A, 1998.

35. Saal, R.J. et al., Quantitation of AML-ETO transcripts using real time PCR on the ABI7700 sequence detection system, *Blood*, 92, 74A, 1998.
36. Turakulov, R.I. et al., Use of real-time PCR to show that the benzene metabolite, hydroquinone, induces the BCL-2 translocation(14;18) found in non-Hodgkin's lymphoma, AACR Annual Meeting, *Proceedings of the AACR*, 42, 283(abstr. 1524), 2001.
37. Ong, S.T. and Beau, M.M.L., Chromosomal abnormalities and molecular genetics of non-Hodgkin's lymphoma, *Semin. Oncol.*, 25, 447–460, 1998.
38. Rowley, J.D., The critical role of chromosome translocations in human leukemias, *Annu. Rev. Genet.*, 32, 495–519, 1998.
39. Issa, J.P., Baylin, S.B., and Herman, J.G., DNA methylation changes in hematologic malignancies: Biologic and clinical implications, *Leukemia*, 11, S7–S11, 1997.
40. Belinsky, S.A. et al., Aberrant methylation of p16INK4a is an early event in lung cancer and a potential biomarker for early diagnosis, *Pro. Natl. Acad. Sci. USA*, 95, 11891–11896, 1998.
41. Hsieh, C.-J. et al., Hypermethylation of the p16INK4a promoter in colectomy specimens of patients with long-standing and extensive ulcerative colitis, *Cancer Res.*, 58, 3942–3945, 1998.
42. Kanai, Y. et al., Aberrant DNA methylation on chromosome 16 is an early event in hepatocarcinogenesis, *Japan. J. Cancer Res.*, 87, 1210–1217, 1996.
43. Jones, P.A. et al., DNA methylation in bladder cancer, *Europ. Urol.*, 33, 7–8, 1998.
44. Lo, Y.M.D. et al., Quantitative analysis of aberrant p16 methylation using real-time quantitative methylation-specific polymerase chain reaction, *Cancer Res.*, 59, 3899–3903, 1999.
45. Vet, J.A.M. et al., Multiplex detection of four pathogenic retroviruses using molecular beacons, *Proc. Natl. Acad. Sci. USA*, 96, 6394–6399, 1999.
46. Whitcombe, D. et al., Detection of PCR products using self-probing amplicons and fluorescence, *Nat. Biotechnol.*, 17, 804–807, 1999.
47. Thelwell, N. et al., Mode of action and application of Scorpion primers to mutation detection, *Nucleic Acids Res.*, 28, 3752–3761, 2000.
48. Parsons, B.L. and Heflich, R.H., Evaluation of MutS as a tool for direct measurement of point mutations in genomic DNA, *Mutat. Res.*, 374, 277–285, 1997.
49. Parsons, B.L. and Heflich, R.H., Detection of a mouse H-ras codon 61 mutation using a modified allele-specific competitive blocker PCR genotypic selection method, *Mutagenesis*, 13, 581–588, 1998.
50. Parsons, B.L. and Heflich, R.H., Detection of basepair substitution mutation at a frequency of 1×10^{-7} by combining two genotypic selection methods, MutEx enrichment and allele-specific competitive blocker PCR, *Environ. Mol. Mutagen.*, 32, 200–211, 1998.
51. Wada, H. et al., Clonal expansion of p53 mutant cells in leukemia progression *in vitro*, *Leukemia* 8, 53–59, 1994.
52. Lizardi, P.M. et al., Mutation detection and single-molecule counting using isothermal rolling-circle amplification, *Nat. Genet.*, 19, 225–232, 1998.
53. Lyamichev, V. et al., Comparison of the 5′ nuclease activities of taq DNA polymerase and its isolated nuclease domain, *Proc. Natl. Acad. Sci. USA*, 96, 6143–6148, 1999.
54. Kaiser, M.W. et al., A comparison of eubacterial and archaeal structure-specific 5′-exonucleases, *J. Biol. Chem.*, 274, 21387–21394, 1999.

chapter twenty-three

Proteomics in toxicology: an experience with gentamicin

Sandy Kennedy, M.D. Kelly, C. Moyses, R. Amess, and R. Parekh

Contents

Abstract The potential value of the use of proteomic technology in the discovery of biomarkers will be described. The example of an experiment to demonstrate this utility using the antibiotic, gentamicin, in rats is discussed. In the discovery of biomarkers in this way, experimental rigor, and the comparison with conventional methods, e.g., histopathology, is paramount. The sensitivity of the proteomic approach is shown by the identification of potential biomarkers of gentamicin toxicity in the serum at lower doses than any changes were seen by conventional means.

I. Introduction

Proteomics is a novel technology that has the potential to rapidly change our approach to safety assessment. This technology can potentially identify all or nearly all of the proteins (the proteome) of any biological system or sample. An individual protein or a group of proteins can be associated with a disease state or toxic response and this information can be used to develop biomarkers for a disease or toxic response. Proteomic analysis involves the systematic separation, identification, and characterization of all of the proteins, or a subset of the proteins, in a biological sample. This analysis is done using high-throughput, automated techniques involving digital imaging of proteins fractionated on a two-dimensional gel, robotic excision of proteins from the gel, and identification of the proteins by mass spectrometry.[1,2]

Examples of experimental designs using proteomics are as follows: Proteins are identified and compared in samples from animals treated or not treated with a test material; proteins are identified whose expression changes after exposure to the test material which may be involved in a toxic response to the test material in the target tissue. A similar experiment could be carried out with normal or diseased tissue, identifying proteins that are involved in etiology or progression of a particular disease.

Proteomics is viewed as a sister technology to genomics and there is a strong synergy between the two approaches. However, in some cases gene expression is abnormal in a diseased tissue, but comparable change is not observed at the protein level, and protein expression in the diseased tissue correlates poorly with gene expression. This is especially true when a xenobiotic chemical interferes with protein synthesis and/or posttranslational modification of a protein. Therefore, the information that can be derived about a pathological process from gene expression is less complete than the information that can be derived from proteomics. Proteomics has the potential to reveal the full phenotypic picture of secreted proteins.

Proteomic analysis of body fluids can be of particular value in identifying potential noninvasive biomarkers.[3,4] This capability is enhanced by the ability to remove high abundance proteins from the gel such as albumin, IgG, haptoglobulin, and transferrin. When an immunoaffinity-based enrichment technique is used to remove these proteins, hundreds of proteins can be identified that would not have been detected due to strong signals from the high abundance proteins.

When using any novel technology, scientific rigor is paramount. It is crucial to have a sound scientific protocol based on good toxicological principles for generating samples.[5] There is also a need for a comprehensive and careful analysis of the results. These principles will be demonstrated in the experiments described below, in which the proteome of the rat was studied in animals dosed with the antibiotic gentamicin.[3] The proteomics analysis was compared to more conventional techniques including clinical chemistry, haematology, and histopathology, and the proteomics approach compared favorably. The study was designed to answer the following questions:

1. Can a dose response be identified in the proteome?
2. Is proteomic analysis more sensitive than classical endpoints?
3. Is the drug-induced expression of up-regulated or down-regulated proteins reversed after a drug-free withdrawal phase?
4. Can protein markers of toxicity be identified?
5. Can additional mechanistic information be acquired?

The following protocol was designed to answer some of these questions. The protocol also took into account available information concerning renal toxicity of gentamicin.

Table 23.1 **Seven Days of Treatment with Gentamicin Sulphate Followed by 14-day recovery period**

Route of administration:	intravenous
Dose levels:	0, 0.1, 1.0, 10.0, 40.0, 60.0 mg/kg/day
Group size:	10 male rats per treated group
	20 male rats in control group
Blood and urine samples:	2, 3, and 8 days
Blood Parameters:	BUN and creatinine
Urine parameters:	NAG, GGT, volume, and specific gravity

II. Kidney histopathology

Urine biochemistry was analyzed in animals dosed with 60 mg/kg/day. The results indicated an increased level of n-acetyl glucosaminidase and gamma glutamyl transferase at days 2, 3, and 8, which is consistent with previously reported renal toxicity of gentamicin. Urine biochemistry was normal in animals dosed at 40mg/kg/day or lower, and serum biochemistry was normal at all doses and time points. Mild histological lesions were observed in the proximal tubule epithelial cells of the kidney in animals dosed at 60 mg/kg/day for 7 days. These lesions were characterized by a loss of the brush border indicative of an early necrotic change. In animals dosed at 40 and 60 mg/kg/day followed by a 14-day drug-free recovery phase, the proximal tubule epithelium began to regenerate. However, the lesion was relatively minor, involving a small portion of the kidney cortex.

Four hundred and twenty samples were taken from kidney cortex, urine, and serum for proteomic analysis. Analysis of kidney cortex samples produced 114 digitised images from which 95,524 data points were resolved. This large amount of data can be distilled into a smaller amount of meaningful information using good scientific practice and a pragmatic approach. For example, appropriate dataset comparisons should be made (such as high dose versus control) and a tiered approach can be used. In this study, 11 dataset comparisons were made resulting in 50 annotations or protein identifications. In the kidney cortex samples, treatment-related changes in expression were observed involving proteins that participate in many cellular functions (i.e., amino acid metabolism, the urea cycle, oxidative stress, mitochondrial and structural proteins). In the urine, structural proteins were identified such as cytokeratins, which could be nonspecific indicators of renal damage.

One protein was consistently overexpressed in serum samples from rats dosed at 40 mg/kg/day for 7 days. This protein is involved in activating the alternate pathway of complement and it also binds to human renal proximal tubule epithelial cells. This protein has potential as a noninvasive serum marker for gentamicin toxicity. Routine clinical pathology showed increases in this protein in the serum as early as day three of treatment. However, the level of expression returned to normal by the end of the treatment withdrawal phase.

This example demonstrates that proteomics is more sensitive than some conventional toxicological assays, detecting altered protein expression at lower doses and earlier time points than the other methods. This study also identified potential protein biomarkers of renal toxicity in kidney tissue, urine, and serum. Thus it is evident that proteomic analysis could facilitate identification of biomarkers of toxicity, drug efficacy, or exposure to a xenobiotic compound.[6] Biomarkers can be analyzed using a noninvasive methodology, in which biological samples are collected that are representative of the affected tissue, fluid, or organ. Immunoassays can also be developed to screen for a biomarker(s).

In the near future, proteomics could provide us with libraries of biomarkers and immunoassay screens for monitoring exposure to xenobiotics. This kind of research will inevitably identify new proteins and protein-protein interactions. This new information may lead to an increasingly complex picture of biological systems. Appreciation of this complexity should not deter us from moving forward but should rapidly increase our understanding in many scientific fields including toxicology.[7]

References

1. Celis, J.E. and Gromov, P., Two-dimensional protein electrophoresis: Can it be perfected? *Curr. Opin. Biotechnol.*, 10, 16–21, 1999.
2. Moller, A. et al., Two-dimensional gel electrophoresis: A powerful method to elucidate cellular responses to toxic compounds, *Toxicology*, 160, 129–38, 2001.
3. Kennedy, S., Proteomic profiling from human samples: The body fluid alternative, *Toxicol. Lett.*, 120, 379–384, 2001.
4. Page, M.J. et al., Proteomic definition of normal human luminal and myoepithelial breast cells purified from reduction mammoplasties, *Proc. Natl. Acad. Sci. USA*, 96, 12589–12594, 1999.
5. Pennie, W.D. et al., The principles and practice of toxicogenomics: Applications and opportunities, *Toxicol. Sci.*, 54, 277–283, 2000.
6. Holt, G. and Sistare, F., Proteomics and safety biomarker discovery, FDA docket 00N-0930, www.fda.gov/ohms/dockets/ac/cder00.htm#Pharmaceutical Science, 2000.
7. Moyses, C., Pharmacogenetics, genomics, proteomics: The new frontiers in drug development, *Intl. J. Pharm. Med.*, 13, 197–202, 1999.

chapter twenty-four

Xenopus laevis genomic biomarkers for environmental toxicology studies

Perry J. Blackshear

Contents

Abstract   *Xenopus laevis* has long been an important model system for experimental embryology and developmental biology, but it has not been used extensively for genetic studies because the *Xenopus* genome is pseudotetraploid. In the past several years, large-scale projects to sequence *Xenopus laevis* expressed sequence tags (ESTs) have been undertaken and the amount of DNA sequence data for *Xenopus* has increased rapidly. Recent studies indicate that these ESTs can be heterogeneous and this could potentially cause problems in large-scale genomic analyses of *Xenopus* using cDNA or oligonucleotide microarrays. One recent study showed that approximately 87% of ESTs for commonly expressed mRNAs have duplicate alleles. No inbred strains are available, which increases the probability of high frequencies of less dramatic polymorphisms. Nonetheless, microarrays can be developed to hybridize selectively to one or the other of two duplicate alleles, or conversely, to hybridize to both allelic variants with equal avidity. *Silurana tropicalis* is closely related to *Xenopus* and it has been recognized for its reported diploidy, small size, rapid growth to sexual maturity, and large clutch size. It is possible that *Xenopus* microarrays can be used to analyze gene expression in *Silurana*. Another area for potential future development is a *Xenopus* Virtual Genome Project, which would use cluster analysis to assemble contigs covering a large fraction of expressed *Xenopus* alleles whether or not the genes are duplicated. These future molecular studies on *Xenopus* will extend the extensive existing body of data on morphological endpoints and perturbations of *Xenopus* development.

I. Introduction

Xenopus laevis, the South African clawed frog, has been recognized for decades as one of the premier model organisms for studying early developmental biology. It has been extremely useful, not only to study normal developmental processes, but also to study abnormal development in embryos exposed to environmental agents, surgical manipulation, exogenous gene expression, antisense inhibitors of endogenous gene expression and other perturbations. *Danio rerio* is a species that offers some of the advantages of *Xenopus*, but the wealth of information on normal and perturbed *Xenopus* development makes it one of the most useful experimental systems for developmental biology.

Xenopus has been used extensively to study how external conditions and agents influence developmental processes. Many studies characterize abnormal development at the level of gross morphology, and developmental abnormalities in *Xenopus* have been codified to some extent in the Frog Embryo Teratogenesis Assay–*Xenopus* (FETAX) system. However, it is clear that developmental morphological abnormalities must be preceded and accompanied by major changes in gene expression in the whole embryo, in specific tissues and cell types or in specific developmental systems. Current efforts to sequence *Xenopus* expressed sequence tags (ESTs) have provided enough *Xenopus* DNA sequence data for construction and evaluation of DNA microarrays, Thus, it should soon be possible to examine expression changes in *Xenopus* genes and correlate gene expression patterns with developmental perturbations in the embryo.

This chapter covers three areas related to *Xenopus* genomic biomarkers: a brief review of the current status of the *Xenopus* EST sequencing efforts; a discussion of the advantages and disadvantages of *Xenopus* for this type of analysis, particularly in relation to the pseudotetraploid genome of *Xenopus*; and a discussion of recent comparisons between ESTs from *Xenopus laevis* and *Silurana tropicalis* (also known as *Xenopus tropicalis*; for convenience, in this chapter, this species will be referred to as *Silurana*). For interested readers, a comprehensive guide to FETAX was published by the American Society for Testing and Materials, and a companion atlas is also available.[1] A review of current research using FETAX was presented at a recent conference: A summary of the conference is available on the Internet at http://iccvam. niehs.nih.gov/methods/fetax.htm.

II. Xenopus EST sequencing projects

In 1999, the National Institutes of Health (NIH) named *Xenopus* as one of five nonmammalian models of human development and disease. In March 2000, an NIH-sponsored workshop was held on the topic of "Identifying the genetic and genomic needs for *Xenopus* research." The summary of this workshop, available at www.nih.gov/science/models/xenopus/reports/xenopus_report.pdf, describes in detail the advantages (as well as some of the disadvantages) of using *Xenopus* as a model system in developmental biology. The highest priority recommendation of this workshop was to begin large-scale EST sequencing projects on a variety of *Xenopus* embryonic stages and adult tissues. Library construction for this project was supported by the National Cancer Institute, the National Institute of Child Health and Development, the National Institute of Environmental Health Sciences, others in this country, the Welcome Trust in the U.K. and the RIKEN Institute in Japan. Most of the sequencing for this NIH-supported effort was performed at the Consortium sequencing laboratory at the Washington University of St. Louis (see http://genome.wustl.edu/est/ xenopus_esthmpg.html). However, other large sequencing efforts were carried out by the NIH Intramural Sequencing Center at the National Human Genome Research Institute. At the time of this writing (August 22, 2001), this and other *Xenopus* EST sequencing efforts had generated approximately 101,762 EST sequences that had been deposited in GenBank. This represents an enormous explosion in the amount of publicly available *Xenopus* sequence data compared to several years ago. By comparison, there were approximately 4016 non-EST *Xenopus laevis* mRNAs in GenBank on August 22, 2001; there were also a total of 7747 *Silurana tropicalis* sequences (including ESTs and mRNAs) and a total of 121 other *Xenopus* sequences (e.g., *borealis*).

The libraries used in this effort are quite varied with respect to cloning vector, mRNA preparation, developmental stage and tissue, etc.; rate of success in generating useful sequence data from the libraries has also varied significantly. In addition, many libraries were normalized in an attempt to remove particularly abundant mRNAs while others were not. Still others were selectively subtracted for previously cloned sequences in an attempt to enrich for poorly represented genomic regions. Nonetheless, the aggregate collection of sequences is highly redundant for highly enriched genes. This is not entirely a drawback since such redundancy can provide assurance that sequences are correct, provide sequence information over the complete length of long transcripts, provide information on allelic variants, provide additional information on nonallelic variants such as polymorphisms among strains, etc. Nonetheless, in an attempt to quantify this redundancy, we subjected all of the *Xenopus* sequence data in GenBank on August 23, 2000 to a fragment assembly analysis using SeqMerge (Genetics Computer Group,

Madison, WI), as described previously.[2] This program will merge fragments of closely related sequences into contigs. For *Xenopus* mRNA sequences found in the GenBank nr database, there were 2237 entries, and there were 31939 entries in the *Xenopus* GenBank EST database. After vector, poly A and poly T sequences were masked, these 34,176 individual sequence entries could be merged into 17,148 contigs using SeqMerge. Thus, on this date, approximately 17,148/34,176*100% = 50% of the total sequences in GenBank probably represented individual alleles, or pairs of alleles, since the allelic variants in general were similar enough to be merged with the primary allele by the program. This was also true of more minor non-allelic polymorphic variants and sequencing errors, which were in general combined into a single contig per locus. Of the 17,148 contigs generated in this way, 10,828 were represented by a single sequence in GenBank, whereas the remaining 6,320 represented more than one sequence, ranging as high as 30 individual entries in two cases. Thus, we estimated that the number of sequenced alleles (i.e., primary alleles plus allelic variants) present in GenBank was approximately 50% of the total number of sequences.

Estimates of how many *Xenopus* ESTs should be sequenced are complicated by previous widely disparate estimates of the size of the *Xenopus* genome. One calculation is based on the predicted size of the pseudotetraploid *Xenopus* genome, which is estimated to include between 20,000 and 46,000 expressed genes[2] considering each allele and allelic variant pair as a single expressed gene. This would mean that at least 80,000 ESTs would need to be sequenced to approach near saturation of expressed genes assuming that the correct number of genes is 40,000. However, our preliminary analysis suggested that at least 87% of expressed *Xenopus* genes have two expressed alleles, which for the purpose of this discussion are called an allele and a corresponding allelic variant. If this percentage were applied to the total projected size of 40,000 expressed genes, then an estimated 80,000 + 0.87 * 80,000 = 149,600 ESTs would need to be sequenced before we could expect to represent a copy of each allele and its variant in the database. Given the other aspects of EST redundancy mentioned above, as well as the scarcity or other underrepresentation of some expressed genes in *Xenopus* libraries, it might be reasonable to continue sequencing EST libraries until approximately 200,000 EST sequences have been deposited in GenBank. At the current rate of EST sequencing, this goal should be achieved within one year or so. The question of whether a genomic sequencing effort for *Xenopus* should be undertaken to fill in the gaps that will remain after this large scale EST sequencing effort is complete is beyond the scope of this review. However, the answer to this question will probably depend on whether the scientific community comes to the consensus that *Silurana*, the diploid relative of *Xenopus*, is an even more useful experimental animal than *Xenopus* itself.

The current status of *Xenopus laevis* EST redundancy was also evaluated as follows. The 101,859 ESTs contributed by the Washington University Sequencing group were clustered on approximately September 15, 2001, using the program estcluster (Mike Dante, pers. comm.; see http://www.nematode.net/Programs/ for this and the other icatools programs used in this analysis). This program does not produce contigs or consensus sequences and can use each EST more than once as it attempts to cluster them. On that date, there were only 4944 clusters with a single EST sequence; there were 26,128 with two ESTs; 15,277 with 3–10 ESTs; 3864 with 11–100 ESTs; 236 with 101–1000 ESTs; and, somewhat remarkably, eight clusters with more than 1000 ESTs each. The latter eight clusters included 1013, 1044, 1172, 1223, 1227, 1262, 1333, and 1362 ESTs, respectively. These presumably represent common mRNAs since many of the libraries used in the EST sequencing project were not normalized. It is unclear at this time how this clustering technique approaches the phenomenon of allelic variants, that is, if they are generally clustered together as in the SeqMerge analysis or if they form individual clusters.

The relative paucity of singletons (only 4944 out of 101,859 ESTs; ≈5%) suggests that the yield of novel sequences through continued sequencing is likely to be quite low; as noted above, it is not clear whether allelic variants are treated as different sequences in this analysis. When the probability that the next sequence would be novel was estimated using the program estfreq, the remarkable prediction was made that the probability that the next EST sequenced would be novel was zero. In addition, it was predicted that the expected number of novel sequences that would be obtained by an EST sequencing program of similar size was also zero.

III. Allelic variants in Xenopus and their relevance to microarray analyses

Despite the utility of *Xenopus* as an experimental animal in studies of vertebrate development, including toxicology studies and the widespread use of its oocytes as cellular expression factories for mRNAs and proteins, the genetics of *Xenopus* have made it unpalatable as a useful test species. However, the advent of large-scale EST sequencing projects like the ones outlined above produced large amounts of expressed sequence information, which makes it possible to carry out simultaneous analyses of many thousands of transcripts during normal *Xenopus* development and during development perturbed by environmental toxins. However, the problem of allelic variants, which we found to be present in 87% of 100 expressed genes[2] in addition to the presence of more limited polymorphisms in individual animals, make certain types of microarray platforms more useful than others for specific types of analyses. In addition, as *Silurana* becomes more widely used as an experimental animal, it would seem to be advantageous to make microarrays that could be useful in analyzing expression data from either species. The following two sections of this chapter provide examples of some of the difficulties involved in developing transcript data for these analysis platforms.

As a test sequence, we used the cDNA sequence encoding the highly expressed maternal *Xenopus* protein XC3H-4 (GenBank accession number AF061983). This protein is the fourth member of a small family of tandem

CCCH zinc finger proteins expressed in *Xenopus*[3]; the first three members of this family, XC3H-1,-2, and -3, have known orthologs in mammals, whereas the XC3H-4 is only known to occur in frogs and fish[3-5] as of this writing. The best understood member of this family, tristetraprolin (TTP), is involved in the physiological regulation of the stability of mRNAs encoding the important mammalian cytokines tumor necrosis factor alpha and granulocyte-macrophage colony stimulating factor;[6,7] it is possible that XC3H-4 is involved in mRNA stability and turnover during early, maternally controlled development in *Xenopus*.

To evaluate the status of ESTs from this gene in GenBank, the nonhuman nonmouse EST databases were searched on August 5, 2001 with the parent sequence AF061983 with the additional limitation *Xenopus laevis*. All ESTs that matched this query sequence as the top hit were evaluated, and *Silurana* sequences that were mistakenly included were removed. This yielded 44 ESTs; one of these was removed from further analysis because it appeared to contain a large deletion that was not shared by either the query sequence or the other ESTs. Approximately half of these EST sequences aligned with the 5' end of the cDNA and half with the 3'-end.

Figure 24.1 shows the alignment of 19 ESTs with the query sequence over the region representing bp 41–141 of AF061983. The query sequence is the bottom sequence in the alignment. This alignment covers a segment of the query sequence near the amino terminus of the protein coding region. Several things are apparent from an inspection of this alignment. First, the sequences are highly related, but they fall into at least three major categories. For example, the top seven sequences all have the sequence TGG at positions 9–11 in the alignment, whereas the 13 lower sequences (which include the search sequence AF061983) have the sequence CAA at those positions. Similarly, the top seven ESTs have TG at positions 16 and 17, whereas the bottom 13 have CA at that site. This is what would be expected if there were two allelic variants at this site or two closely related alleles. However, this pattern is not observed throughout the region of the alignment. At position 186, the lower 13 sequences suddenly diverge into two distinct groups: the bottom 4 sequences have T at this position, whereas the middle 8 sequences have G. The bottom 4 sequences are not identical throughout: at position 88 the middle 2 sequences have a T whereas the top and bottom sequences have a C at that position. Other single nucleotide differences can tentatively be ascribed to either sequencing errors or true individual animal or library polymorphisms. However, it is clear from even this simple analysis that the sequences do not correspond to two different allelic variants of a single locus but instead correspond to at least three distinct alleles.

The three major groups described above were each reduced to a consensus sequence starting with the initiator ATG, shown in Figure 24.2A; the translations of these three consensus sequences are shown in Figure 24.2B. The bottom sequence in the DNA alignment, labeled be506162, represents the group of ESTs that correspond to the search sequence, AF061983. It is clear that the three consensus DNA consensus sequences represent individ-

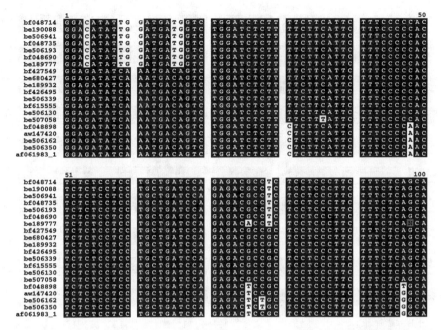

Figure 24.1 Alignment of 19 *Xenopus* ESTs with the 5'-end of AF061983, bp 41–141. Shown is an alignment using Pileup from GCG of the *Xenopus* cDNA clone AF061983 with the same 100 bp from 19 *Xenopus* ESTs that were highly related. The accession numbers of the ESTs are to the left of the sequences. Base differences are in white; "N" shaded gray means the sequence of the EST was ambiguous at that point. See the text for further details.

ual alleles, and encode distinct proteins; nonetheless, the amino termini of their encoded proteins shown in Figure 24.2B are similar, with no gaps, and all three contain the four CCCH zinc fingers that are the hallmark of this protein in *Xenopus*.[3] Nonetheless, it possible to identify regions in which three consensus sequences are identical (Figure 24.2A); these regions of identity could be used to generate allele-independent oligonucleotide probes for microarray experiments.

Fifteen ESTs were also aligned with a more 3' region of the query sequence AF061983. This alignment revealed more diversity, particularly in the 3'-untranslated region, suggesting that there are more independent alleles than the three major alleles identified by alignment of sequences near the 5'-end of the coding region. Figure 24.3 shows an alignment of these EST sequences in the 838–938 bp region of AF061983. There are at least four independent alleles in this group of aligned sequences, and numerous individual variants are detected within each group of alleles. The four consensus sequences, and their translated regions, are shown in Figures 24.4A and 24.4B, respectively. The most likely interpretation of these data is four independent alleles are in this collection of ESTs, possibly representing two different genes with two allelic variants for each gene; however, the single

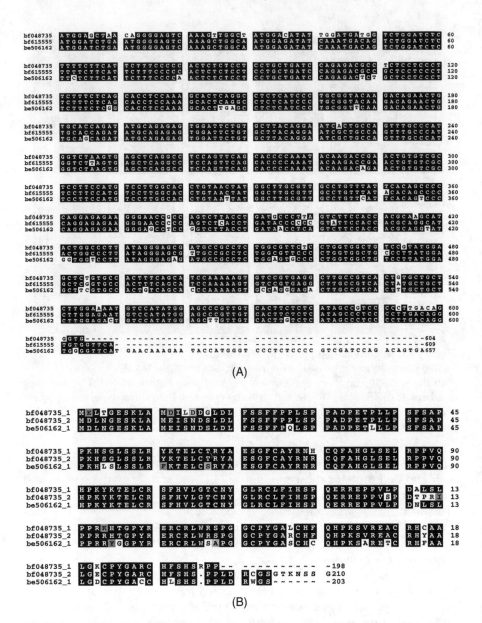

Figure 24.2 Alignment of three consensus EST sequences from Figure 24.1. In A., representative ESTs from the three major groups of ESTs, shown in Figure 24.1, were aligned with ClustalW. The accession numbers of the ESTs are on the left. All three ESTs were aligned beginning at the initiator ATG. Base differences are indicated by lack of shading at a position. In B., the translations of the same three ESTs are shown. Gray shading indicates a conservative amino acid substitution at that position; white shading indicates a nonconservative substitution. Loss of a triplet codon is indicated by a dot.

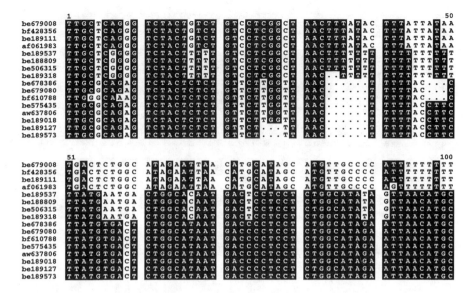

Figure 24.3 Alignment of 15 *Xenopus* ESTs with a more 3′ region of AF061983, bp 838–938. All details of this alignment are as in the legend to Figure 24.1.

base differences may be sequencing errors or individual polymorphisms. As shown in Figures 24.3 and 24.4, it would be difficult to develop oligonucleotides of significant length that would hybridize efficiently to all four alleles in this part of the transcript; however, it should be relatively easy to develop allele-specific oligonucleotides based on these alignments.

This example underscores some of the potential difficulties in developing transcript arrays for *Xenopus*. It is clear that a single full-length cDNA on a chip representing this locus or pair of loci would hybridize to its cognate cDNA with highest affinity, and hybridize to the other three allelic variants with varying affinity, making interpretation of this kind of data almost impossible. However, using 5′ probes of relatively short length, such as single or multiple synthetic oligonucleotides of 25–50 bp, it should be possible to develop filter or chip probes that would hybridize with equal efficiency to each of the four transcripts. It should also be possible to develop oligonucleotides that hybridize to the more diverse regions of the 3′-end of the transcript that identify specific transcripts with reasonable reliability.

Could the purported diploid species *Silurana tropicalis* simplify this analysis by alleviating the problems introduced by the *Xenopus* allelic variants? We investigated this question by searching the database of *Silurana* sequence for ESTs corresponding to the 3′-end of AF061983. An alignment was made of seven ESTs and the *Xenopus* search sequence starting at position 666 of AF061983 (Figure 24.5). As predicted from the previous data, the ESTs were similar in the initial protein coding region of the transcript; however, the similarity weakens in the 3′-noncoding region (position 201 and higher of

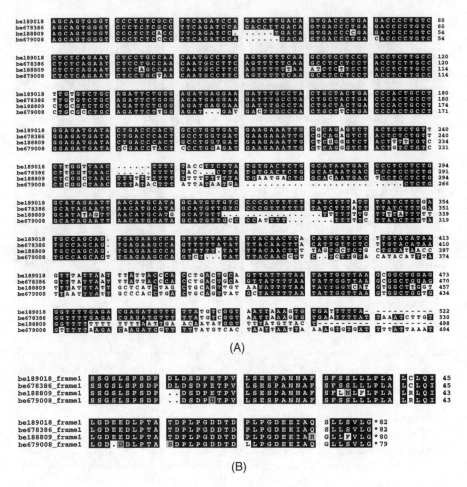

Figure 24.4 Alignment of four consensus EST sequences from Figure 24.3. In A., representative ESTs from the four major groups of ESTs, shown in Figure 24.3, were aligned with Pileup. The accession numbers of the ESTs are on the left. Base differences are indicated by lack of shading at a position. In B., the translations of the same four ESTs are shown. Gray shading indicates a conservative amino acid substitution at that position; white shading indicates a nonconservative substitution. Loss of a triplet codon is indicated by a dot; the positions of stop codons are indicated by an asterisk.

sequence AF061983; Figure 24.5). This alignment shows that there are two major alleles of this locus in *Silurana*, supporting the theory discussed above that this diversity represents two individual gene products rather than two variant alleles of the same gene. Close to the extreme 3'-end of the sequences, there is more heterogeneity in the *Silurana* sequences. For example, the EST bi347377 is identical to three other ESTs (bg514806, bg514705, bg514690) until position 182 of AF061983, when it suddenly diverges from the other three

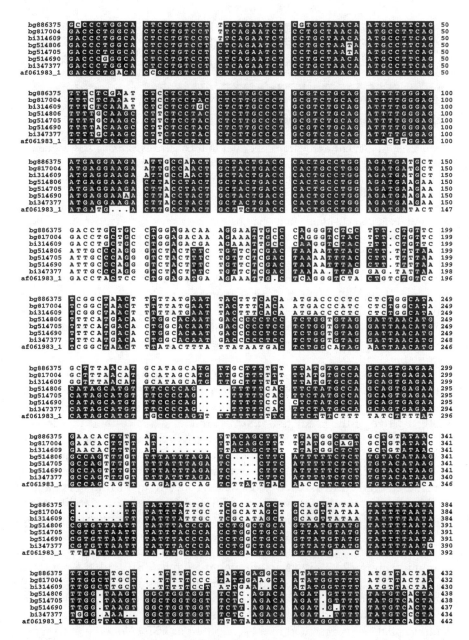

Figure 24.5 Alignment of seven *Silurana* ESTs with a 3' portion of the *Xenopus* cDNA AF061983, starting at bp 666. Shown is an alignment using Pileup from GCG of the *Xenopus* cDNA clone AF061983 with the same region of seven *Silurana* ESTs that were highly related. The accession numbers of the ESTs are to the left of the sequences. Base differences are in white; "N" shaded gray means the sequence of the EST was ambiguous at that point. See the text for further details.

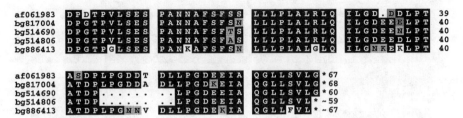

Figure 24.6 Alignment of the translations of four consensus *Silurana* EST sequences from Figure 24.5. Representative ESTs from the four major groups of ESTs shown in Figure 24.5 were translated and then aligned with Pileup. The accession numbers of the ESTs are on the left; the translation of the *Xenopus* cDNA clone af061983 is at the top of the alignment as indicated. Gray shading indicates a conservative amino acid substitution at that position; white shading indicates a nonconservative substitution. Loss of a triplet codon is indicated by a dot; the positions of stop codons are indicated by an asterisk.

sequences for 10–15 bp, then regains the consensus. Similar differences are observed close to the 3′end of other members of this subgroup. Similarly, there are a number of differences between the EST bg514690 and four other closely related ESTs; can these all be attributed to sequencing errors or polymorphisms, or does this represent evidence for more allelic diversity in *Silurana*?

These ESTs were reduced to four consensus sequences and translated into predicted amino acid sequences (Figure 24.6). However, the protein sequences did not help clarify the significance of the sequence variation in the ESTs. Two of the predicted *Silurana* protein sequences and the *Xenopus* protein terminate in a glycine, but the other two *Silurana* sequences do not end in glycine. Nonetheless, the two sequences with a similar terminus are not otherwise closely related, as is obvious from Figure 24.6. Therefore, it is considered premature to reach a conclusion concerning the number of alleles of this protein represented as *Silurana* ESTs; the sequence variability of *Silurana* ESTs may pose a significant problem in performing expression analyses in this species.

There is one other point that should be made concerning *Silurana* genomics. Figure 24.7 shows an alignment *Silurana* ESTs with a 5′-region of the *Xenopus* sequence af061983 (shown as gi 1_20116). Base identities with the *Xenopus* sequence are shown as dots, and differences at an individual position are indicated. This alignment also suggests that more than one *Silurana* allele is represented in this collection of ESTs. Nonetheless, there is a high level of overall similarity between the *Silurana* sequences and the *Xenopus* sequence, which would permit hybridization by a single probe if a longer EST is used in a *Xenopus* microarray. The affinity for the cognate *Xenopus* cDNA might be higher, but the level of similarity between the two species suggests that array patterns could be used in cross-species hybridization experiments.

```
1_20116     10   tggatctgaatggggagtcaaagctggcaatggagatatcaaatgacagtctggatctct 69
13483542    25   ...............a...c.....atg........g........g..c.......... 84
13483556    48                                     ........g..c.......... 69
14186846    38                                              .......... 47

1_20116     70   tctcttcattctttccccaactctctcctcctgctgatccagagactccgctcctccctt 129
14187961     5                   ...........t........t.tt-.............. 43
13483542    85   ....c..t..t.......c...........t.............t.............. 144
13483556    70   ....c..t..t.......c...........t.............t.............. 129
14186846    48   ....c..t..........c...........t.............t.............. 107
14186507    39          ..........c...........t.............t......... 90

1_20116    130   ctttctcggcacctccaaagcacttgagcctctcatccctgcggtacaagacagaactgt 189
14187961    44   .......a.................c...............t..c.............. 103
13483542   145   .......a.................c...............t..c.............. 204
13483556   130   .......a.................c...............t..c.............. 189
14186846   108   .......a.................................t..c.............. 167
14186507    91   .......a.................................. 130

1_20116    190   gcagcagatatgcaga-gagtggattctgtgcttacaggaatcgctgccagtttgcccat 248
14187961   104   ...c...........-.........a....c........................... 162
13483542   205   ...c...........-................c......................... 263
13483556   190   ...c...........-................c......................... 248
14186846   168   ...c..........ttt.t.............c......................... 227
14186507   162      ....-....c..c...c..c.....................cc 208

1_20116    249   ggtctaagtgagctcaggcctccagttcagcaccccaaatacaagacagaactgtgtcgc 308
14187961   163   ........................................t................. 222
13483542   264   ........................................t................. 323
13483556   249   ........................................t................. 308
14186846   228   ........................................t................. 287
14186507   209   cc..c...........c...ct...............t....... 251

1_20116    309   tccttccatgtccttggcacctgtaactatggcttgcgttgcctgttcattcacagtccc 368
14187961   223   ........c..g.....t...........a.......tt.............c..a 282
13483542   324   ........c..g.....t...........a.......tt.............c..a 383
13483556   309   ........c..g.....t...........a.......tt.............c..a 368
14186846   288   ........c........t...........a.......tt.............c..a 347

1_20116    369   caggagagaagagagcctccggtcttacctgataacctcagtcttccaccacgcaggtat 428
14187961   283   ..........g...t....a....cc......gc..ctc.......a......a.... 342
13483542   384   ..........g...t....a....cc......gc..ctc.......a......a.... 443
13483556   369   ..........g...t....a....cc......gc..ctc.......a......a.... 428
14186846   348   ..........g.........a....cc......gc..ctg........a.......... 407

1_20116    429   ggtggtccttatagggagagatgccgcctctggagtgccctggtggctgtccttatgga 488
14187961   343   .c...c..........c.g.........c..t.t..................c.. 402
13483542   444   .c...c..........c.g.........c..t.t.................. 503
13483556   429   .c...c..........c.g.........c..t.t.................. 488
14186846   408   .c...c..c.........ca...........c.ct.a.................... 467

1_20116    489   gctcggtgccactttcagcacccaaaaagtgccagggagacttgccgtcactttgctgct 548
14187961   403   ...................agtc.....gt.................. 462
13483542   504   ...................agtc.....gt.................. 563
13483556   489   ...................agtc.....gt.................. 548
14186846   468   ...................t......g..tt...a...g.................. 527
15040654   606   ...................agtc.....gt................. 558
```

Figure 24.7 Alignment of *Silurana* ESTs with a 5′ region of AF061983. An alignment of a 5′ region of the *Xenopus* cDNA AF061983 with available *Silurana* ESTs is shown; the *Xenopus* sequence is at the top, and the numbers to the left and right of each alignment are the bp numbers in each sequence. Dots indicate base identity with the top *Xenopus* sequence; other base differences are indicated. The GI numbers of each EST are indicated to the left of each alignment.

IV. Implications for Xenopus genomics

Because *Xenopus laevis* is an important model system for developmental biology and aquatic toxicology, it has been suggested that a full-scale genomic sequencing effort should be undertaken, i.e., a *Xenopus* Genome Project. As indicated above, this would be a complex undertaking for a variety of reasons, including the variability among individual frogs of this noninbred strain, the possibility of allelic variants for most expressed genes, etc. *Silurana* has been suggested as a potential substitute species since it appears to be a closely related diploid; other experimental factors, such as egg clutch size, time to sexual maturity, etc., make it more useful than *Xenopus* for some types of experiments.[8] Still, there is variability among *Silurana* transcripts as well, that perhaps can be accounted for by individual polymorphisms but may also indicate nondiploidy. It seems likely that a full-scale sequencing effort will be accomplished someday, for one or both organisms, but it also seems likely that this will be only after a large number of other diploid organisms, pathogenic microorganisms, and others have been sequenced to completion.

In the meantime, *Xenopus* ESTs are being sequenced at a high rate and improved tools for clustering sequences have led us to propose the Virtual *Xenopus* Genome Project. The goal of this project will be to condense the raw EST sequence traces into individual contigs that represent full-length transcripts for *Xenopus* expressed genes. Because of the likelihood that a high proportion of *Xenopus* genes will be represented by at least two alleles, it will be necessary to improve the resolution of currently available clustering tools so that individual alleles can be distinguished; these alleles can also be aligned with each other so that common and distinct regions are identified for use in microarray and other types of transcript analyses. A method should also be developed to identify and curate single nucleotide and other polymorphisms in the *Xenopus* EST sequences. The data from the EST collections should also be available for comparison and clustering with archival genomic and cDNA sequence data in GenBank for which individual sequence traces are not available. Such a data repository should be kept current by adding and updating contigs as new ESTs and other *Xenopus* sequences are added to GenBank. Because of the expected frequency of allelic variants and polymorphism, a significant amount of manual curation will be required to keep such a database current and to resolve database management issues such as establishing rules for the labeling of allelic variants, isotypes, and polymorphisms and developing a systematic nomenclature. However, we believe that such an undertaking will maximize the utility of the large *Xenopus* EST database, and the result might be a catalogue of all expressed *Xenopus* alleles and their common variants.

Acknowledgments

I am grateful to Mike Dante for the cluster analysis and many helpful discussions.

References

1. Bantle, J.A., Finch, R.A., and Burton, D.T., FETAX interlaboratory validation study: Phase III–Part 1 testing, *J. Appl. Toxicol.*, 16, 517–528, 1996.
2. Blackshear, P.J. et al., The NIEHS *Xenopus* Maternal EST Project: Interim analysis of the first 13,879 ESTs from unfertilized eggs, *Gene*, 267, 71–87, 2001.
3. De, J. et al., Identification of four CCCH zinc finger proteins in *Xenopus*, including a novel vertebrate protein with four zinc fingers and severely restricted expression, *Gene*, 228, 133–145, 1999.
4. Stevens, C.J. et al., Blastomeres and cells with mesendodermal fates of carp embryos express cth1, a member of the TIS11 family of primary response genes, *Int. J. Dev. Biol.*, 42, 181–188, 1998.
5. te Kronnie, G. et al., Zebrafish CTH1, a C3H zinc finger protein, is expressed in ovarian oocytes and embryos, *Dev. Genes Evol.*, 209, 443–446, 1999.
6. Carballo, E., Lai, W.S., and Blackshear, P.J., Feedback inhibition of macrophage tumor necrosis factor α production by tristetraprolin, *Science*, 281, 1001–1005, 1998.
7. Carballo, E., Lai, W.S., and Blackshear, P.J., Evidence that tristetraprolin (TTP) is a physiological regulator of granulocyte-macrophage colony-stimulating factor (GM-CSF) mRNA deadenylation and stability, *Blood*, 95, 1891–1899, 2000.
8. Amaya, E., Offiedl, M.F., and Grainger, R.M., Frog genetics: *Xenopus tropicalis* jumps into the future, *Trends Genet.*, 14, 253–255, 1998.

chapter twenty-five

Molecular biomarkers targeting signal transduction systems: the AH receptor pathway

Timothy R. Fennell, Frederick J. Miller, and William F. Greenlee

Contents

Abstract Biomarkers provide tools for assessing exposure to environmental toxicants and for elucidating the intermediate steps between exposure and outcome. Thus, the development of biomarkers is an important component of an integrated approach to informed human health risk assessments for xenobiotics. This chapter discusses some of the basic elements and concepts that drive this process using the Ah receptor signal transduction pathway as a paradigm.

I. Introduction

The measurement of biomarkers in people and in laboratory animals has much to offer in improving our understanding of the adverse health effects and safety of chemicals. Much of the philosophy and terminology employed in the area of biomarkers has been extensively described by the Committee on Biological Markers of the National Research Council. Two reports of the use of biological markers in pulmonary toxicology and reproductive toxicology were published in the late 1980s.[1,2] These initial reports were followed by reports on biological markers in the immune system and in urinogenital system in response to chemical exposure.[3,4] Biological markers (i.e., biomarkers) are changes measured in a biological system that can be related to exposure to, or effects of, a foreign chemical or toxic substance.

The validation of biomarkers prior to their use in molecular epidemiology studies has been a subject of concern.[5,6] In a decision model for biomarkers of exposure, Stevens et al.,[7] identified a number of critical considerations. The specificity and mechanism of the biomarker response should be known. A dose response relationship must be established if the biomarker is to have any application to the assessment of potential human risk from exposure to a given compound. The temporal response kinetics should be known, and the response kinetics should be relatable to chemical kinetics or pharmacodynamics. There should be appropriate analytical methodology, and the marker should be stable between sampling and analysis. Field validation of the method for humans is requisite for acceptance of the biomarker as a useful research or risk assessment tool.

A.   Biomarker categories

Biomarkers can be classified into a number of categories according to the type of information obtained.

Biomarkers of exposure indicate whether exposure to an agent has taken place, and they include measurement of specific metabolites and/or adducts formed by reaction of the compound or its metabolites with macromolecules, such as DNA, RNA, and protein. Measures of a biologically effective dose include DNA adducts, and as a surrogate indicator of the amount of the reactive chemical or metabolite circulating in the blood, hemoglobin adducts and albumin adducts. Biomarkers of exposure can also aid in performing interspecies extrapolations when data are available from experimental animals and humans.[8] In addition, in some circumstances, biomarkers can provide insights into the potential shape of the dose response curve below levels where tumors have been observed.[9,10] One of the challenges in developing a biomarker of exposure is demonstrating that the marker is indeed specific and does not reflect a nonspecific response that could arise from a number of different compounds.

Markers of effect provide an indication of early events in the etiology of disease or in toxicologic or carcinogenic processes. Markers of effect for

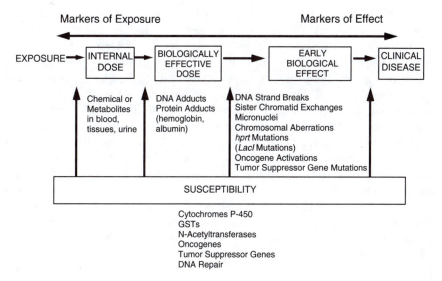

Figure 25.1 Biological markers of exposure, effect, and susceptibility that are important in chemically induced carcinogenesis. (From Fennell, T.R., *CIITAct.*, 16, 11, 1996.
With permission.

carcinogens include endpoints that can be measured in the tissue of concern
with regard to particular stages of carcinogenesis. In most cases, measurements are made in surrogate cells such as blood cells, which can be readily
obtained from human subjects. The markers may include sister chromatid
exchanges, micronuclei, chromosomal aberrations, and mutations at several
genes such as *hprt* and glycophorin A. A confounding factor in measuring
these markers in people is that they are not chemical specific and may be
elevated by lifestyle factors. More specific measures of effect that may be
critically linked to the carcinogenic process include the measurement of
mutations in oncogenes and tumor supressor genes. Mutations in the *ras*
oncogene and the *p53* tumor suppressor gene are frequently detected in
human tumors, and with polymerase chain reaction for sensitive detection
may provide early markers of events in human carcinogenesis.[11]

Markers of susceptibility can be used to identify specific individuals at
greater risk than the general population due to a genetic or other predisposition to the effects of exposure to a particular compound. These can involve
enzymes involved in the metabolism of carcinogens, activation and detoxi-
cation, DNA repair enzymes, oncogenes, and tumor supressor genes. Markers can be measured at various stages in the progression from exposure to
the end effect, and represent a continuum of the events involved (Figure
25.1). The biomarkers presented in Figure 25.1 are particularly relevant to
chemical carcinogenesis.[12,13] However, biomarkers are of interest in the
molecular epidemiology of other diseases.[6]

There are many issues in toxicology that can be addressed with biomarkers. Among these are the following: the comparison of routes of expo-

sure (e.g., dermal vs. inhalation exposure to chemicals); the sources of exposure (e.g., lifestyle vs. workplace); endogenous generation of chemicals vs.
exogenous exposure; comparisons between species; interindividual differences in sensitivity and susceptibility; and gaining insights on the mode of
action of a given chemical. Making effective and efficient use of biomarkers
to address the above issues requires that we understand the relationship
among exposure, effects, and a biomarker, the specificity and sensitivity of
a biomarker, and the role of timing between exposure and measurement on
the level of a biomarker. For humans, exposure to mixtures and the potential
for exposure by multiple routes and sources adds a significant degree of
complexity to the development and interpretation of biomarkers. In subsequent section of this chapter, we examine the potential dual role that the Ah
Receptor (*AHR*) can play in establishing the mode of action of dioxin, an
ubiquitous compound arising as a contaminant of a variety of industrial
processes and for which considerable controversy exists concerning potential
effects in humans.

II. The AHR *signal transduction pathway —* *mode of action insights*

A. *Background*

The AHR is a member of the bHLH/PAS subfamily of transcription factors
that have been linked to development, oxygen homeostasis, and circadian
rhythm.[14,15] AHR is the only member of this family demonstrated to require
ligand-dependent activation for heterodimerization and transcriptional activity. No endogenous ligand for *AHR* has been identified; however, a large group
of environmental aromatic hydrocarbons and dietary indolecarbinols (e.g.,
ICZ) have been shown to bind and activate the *AHR*.[16,17] TCDD, an unwanted
environmental contaminant and one of the most potent AhR ligands, has been
utilized as a molecular probe to study *AHR*-dependent signal transduction.
Adverse outcomes observed in animals treated with TCDD include a wasting
syndrome, tumor promotion in rodent liver and skin, immune dysfunction,
and reproductive and developmental anomalies.[18–20]

B. AHR-*dependent gene transcription*

In the presence of a ligand, AHR partners with ARNT to bind to dioxin-
responsive enhancer (DRE) motifs (5′-GCGTG-3′), transcriptionally regulating a diverse set of gene encoding proteins involved in the metabolism of
xenobiotics, fatty acids, and steroid hormones, and a second set of AhR-
dependent genes postulated to play a role in growth regulation.[15,19] Two
target genes, CYP1A1 and CYP1B1, have been demonstrated to metabolize
the mitogen 17β-estradiol (E_2)[21,22] to form 2-hydroxy- and 4-hydroxy-E_2,
respectively. To date, none of the identified AhR-regulated genes has been

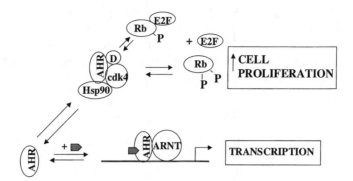

Figure 25.2 Proposed model for AHR ligand-dependent control of cell proliferation and gene transcription. In the absence of ligand, AhR is part of an active cdk4 complex (not all of the proteins comprising this complex are shown in the model) facilitating the phosphorylation of Rb that in turn supports cell proliferation.[32] In the presence of ligand (gray symbol), AHR partners with ARNT to enhance gene transcription.

conclusively linked to decreased proliferation or enhanced differentiation for a particular cell and/or tissue type.

C. Interaction of ligand-activated AHR and oxygen signaling pathways

The AHR signal transduction pathway mediating ligand-dependent gene expression recently has been linked to the pathway regulating transcriptional responses to hypoxia (Figure 25.2). ARNT, the AHR heterodimerization partner, has been identified as Hif-1β, the heterodimerization partner for hypoxia-inducible factor 1α (HIF-1α).[23] Like ARNT (Hif-1β) and AHR, HIF-1α contains contiguous bHLH and PAS domains. HIF-1α/ARNT heterodimers bind to asymmetric E-box core motifs (5′-TACGTG-3′ or 5′-GACTGT) found within hypoxia response elements (HREs) controlling the hypoxia-induced transcription of erythropoietin (EPO), vascular endothelial growth factor (VEGF), and a number of glycolytic enzymes.[24]

D. AHR regulation of cell proliferation

Several studies support the involvement of AHR in the proliferative processes associated with development. AHR mRNA and protein are detected in most tissues at gestation days 10–16, and expression is associated with phases of rapid proliferation and differentiation in these tissues.[25] AHR is expressed in preimplantation embryos, and when cultured in the presence of AHR antisense oligonucleotides, a decrease in embryo cell proliferation and in blastocyte formation are observed.[26] AHR-deficient cell lines demonstrate a reduced proliferation rate.[27] Ligand-dependent activation of AHR has been linked to decreased proliferation in mammary, uterine, and liver tumor cell lines, enhanced terminal differentiation in skin cells,[28-30] and the palatal epithelium.[31] Taken together, these observations suggest that in the absence of exogenous ligand, AHR contributes to proliferation in developing tissues and possibly tumors. A potential model for the proposed dual actions of the AHR is shown in Figure 25.2.

III. Value of biomarkers

The use of biological markers can accomplish a number of objectives in improving epidemiological studies. Measurement of markers of exposure can integrate exposure from all routes, can measure the dose delivered to the critical molecular target, and can measure the total change from different chemicals eliciting the same effect. Markers of effect can help us understand critical biological events involved in the generation of toxic effects. Markers of susceptibility can elucidate sources of individual variability. Epidemiological studies of the case-control type are beginning to provide some insight into markers of susceptibility and cancer. The expectation from the use of biomarkers in industrial health would be to provide better methods of exposure assessment and to identify early effects prior to the development of cancer or other adverse health effects in exposed people, resulting in disease prevention. The broad use of biomarkers in epidemiological studies and industrial medicine requires the development of cost effective analytical methods.

Further potential value of biomarkers in understanding human health effects from exposure to chemicals can be derived from their use in the risk assessment process. The quantitative estimation of the risk associated with exposure to chemicals requires exposure characterization, hazard characterization, and dose-response assessment. Biomarkers can provide a means to reduce or to make explicit some of the uncertainties present in the risk assessment process. Biomarkers can be used to provide a dose metric as an alternative to external exposure concentration for comparison with effect. The dose level at critical targets is controlled by a number of processes, including absorption, metabolism, excretion, and DNA repair. These processes can be saturable, resulting in nonlinear behavior at low-exposure levels. Relating effect to critical dose measures should improve high- to low-dose extrapolation as well as interspecies extrapolation. The measurement of biomarkers in exposed rodents can provide an understanding of events in the development of adverse effects and an improved context for the interpretation of measurements made in people. Biomarkers of exposure can be used to improve the interspecies scaling of dose. There is a considerable parallel between the measurements made using biomarkers in understanding the early events linking exposure with adverse health effects and in developing an understanding of the mode of action of toxicants. As illustrated in Figure 25.3, biomarkers for dioxin exposure and effects can be integrated into the biomarkers continuum and can be used in the risk assessment process, as has been conducted by the Environmental Protection Agency in the draft risk assessment for dioxin (http://www.epa.gov/ncea/

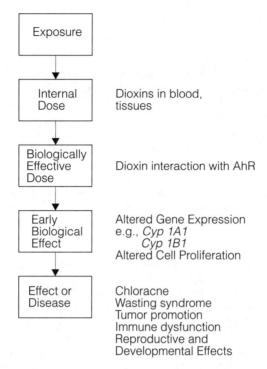

Figure 25.3 Examples of biomarkers for dioxin and the AhR signal transduction pathway.

pdfs/dioxin/dioxreass.htm). Many controversial issues remain in the risk assessment of dioxin, including whether the effects generated in rodents are generated in exposed people, the doses that may cause effects in people, and whether the early effects are linked to the generation of adverse outcomes. Biomarkers can provide an understanding of mode of action as part of the risk assessment process, and we have illustrated this potential by examining the dual role the AhR signal transduction pathway could be used to establish the mode of action of dioxin, an ubiquitous environmental contaminant.

References

1. National Research Council, *Biologic Markers in Pulmonary Toxicology,* National Academy of Sciences, Washington D.C., 1989.
2. National Research Council, *Biologic Markers in Reproductive Toxicology,* National Academy of Sciences, Washington D.C., 1989.
3. National Research Council, *Biologic Markers in Urinary Toxicology,* National Academy of Sciences, Washington D.C., 1995.
4. National Research Council, *Biologic Markers in Immunotoxicology,* National Academy of Sciences, Washington D.C., 1992.
5. Schulte, P.A. and Talaska, G., Validity criteria for the use of biological markers of exposure to chemical agents in environmental epidemiology, *Toxicology,* 101, 73–88, 1995.

Biomarkers of environmentally associated disease

6. Schulte, P.A. and Perera, F.A., *Molecular Epidemiology*, Academic Press, San Diego, 1993.
7. Stevens, D.K. et al., Decision model for biomarkers of exposure, *Regul. Toxicol. Pharmacol.*, 14, 286–296, 1991.
8. Hatch, G.E. et al., Ozone dose and effect in humans and rats: A comparison using oxygen-18 labeling and bronchoalveolar lavage, *Am. J. Respir. Crit. Care Med.*, 150, 676–683, 1994.
9. Poirier, M.C. and Beland, F.A., DNA adduct measurements and tumor incidence during chronic carcinogen exposure in animal models: Implications for DNA adduct-based human cancer risk assessment, *Chem. Res. Toxicol.*, 5, 749–755, 1992.
10. Choy, W.N., A review of the dose-response induction of DNA adducts by aflatoxin B1 and its implications to quantitative cancer-risk assessment, *Mutat. Res.*, 296, 181–198, 1993.
11. Perera, F.P. and Weinstein, I.B., Molecular epidemiology: recent advances and future directions, *Carcinogenesis*, 21, 517–524, 2000.
12. Perera, F.P. and Weinstein, I.B., Molecular epidemiology and carcinogen-DNA adduct detection: New approaches to studies of human cancer causation, *J. Chronic. Dis.*, 35, 581–600, 1982.
13. Perera, F.P. and Whyatt, R.M., Biomarkers and molecular epidemiology in mutation/cancer research. *Mutat. Res.*, 313, 117–129, 1994.
14. Crews, S.T., Control of cell lineage-specific development and transcription by bHLH-PAS proteins. *Genes Dev.*, 12, 607–620, 1998.
15. Schmidt, J.V. and Bradfield, C.A., Ah receptor signaling pathways, *Ann. Rev. Cell Dev. Biol.*, 12, 55–89, 1996.
16. Bjeldanes, L.F. et al., Aromatic hydrocarbon responsiveness-receptor agonists generated from indole-3-carbinol *in vitro* and *in vivo*: Comparisons with 2,3,7,8-tetrachlorodibenzo-p-dioxin, *Proc. Natl. Acad. Sci. USA*, 88, 9543–9547, 1991.
17. Kafafi, S.A. et al., Affinities for the aryl hydrocarbon receptor, potencies as aryl hydrocarbon hydroxylase inducers and relative toxicities of polychlorinated biphenyls: A congener specific approach, *Carcinogenesis*, 14, 2063–2071, 1993.
18. Vanden Heuvel, J.P. and Lucier, G., Environmental toxicology of polychlorinated dibenzo-p-dioxins and polychlorinated dibenzofurans, *Environ. Health Perspect.*, 100, 189–200, 1993.
19. Greenlee, W.F., Sutter, T.R., and Marcus, C., Molecular basis of dioxin actions on rodent and human target tissues, *Prog. Clin. Biol. Res.*, 387, 47–57, 1994.
20. Safe, S.H., Modulation of gene expression and endocrine response pathways by 2,3,7,8-tetrachlorodibenzo-p-dioxin and related compounds, *Pharmacol. Ther.*, 67, 247–281, 1995.
21. Hayes, C.L. et al., 17 Beta-estradiol hydroxylation catalyzed by human cytochrome P450 1B1, *Proc. Natl. Acad. Sci. SA*, 93, 9776–9781, 1996.
22. Spink, D.C. et al., The effects of 2,3,7,8-tetrachlorodibenzo-p-dioxin on estrogen metabolism in MCF-7 breast cancer cells: Evidence for induction of a novel 17 beta-estradiol 4-hydroxylase, *J. Steroid Biochem. Mol. Biol.*, 51, 251–258, 1994.
23. Wang, G.L. et al., Hypoxia-inducible factor 1 is a basic-helix-loop-helix-PAS heterodimer regulated by cellular O2 tension, *Proc. Natl. Acad. Sci. USA*, 92, 5510–5514, 1995.

24. Semenza, G.L., Transcriptional regulation by hypoxia-inducible factor 1: Molecular mechanisms of oxygen homeostasis, *Trends Cardiovasc. Med.*, 6, 151–157, 1996.
25. Abbott, B.D., Birnbaum, L.S., and Perdew, G.H., Developmental expression of two members of a new class of transcription factors: I. Expression of aryl hydrocarbon receptor in the C57BL/6N mouse embryo, *Develop. Dyn.*, 204, 133–143, 1995.
26. Peters, J.M. and Wiley, L.M., Evidence that murine preimplantation embryos express aryl hydrocarbon receptor, *Toxicol. Appl. Pharmacol.*, 134, 214–221, 1995.
27. Ma, Q. and Whitlock, Jr., J.P., The aromatic hydrocarbon receptor modulates the Hepa 1c1c7 cell cycle and differentiated state independently of dioxin, *Mol. Cell Biol.*, 16, 2144–2150, 1996.
28. Hushka, D.R. and Greenlee, W.F., 2,3,7,8-Tetrachlorodibenzo-p-dioxin inhibits DNA synthesis in rat primary hepatocytes, *Mutat. Res.*, 333, 89–99, 1995.
29. Safe, S. et al., 2,3,7,8-Tetrachlorodibenzo-p-dioxin (TCDD) and related compounds as antioestrogens: characterization and mechanism of action, *Pharmacol. Toxicol.*, 69, 400–409, 1991.
30. Osborne, R. and Greenlee, W.F., 2,3,7,8-Tetrachlorodibenzo-p-dioxin (TCDD) enhances terminal differentiation of cultured human epidermal cells, *Toxicol. Appl. Pharmacol.*, 77, 434–443, 1985.
31. Abbott, B.D. and Birnbaum, L.S., TCDD alters medial epithelial cell differentiation during palatogenesis, *Toxicol. Appl. Pharmacol.*, 99, 276–286, 1989.
32. Greenlee, W.F., Hushka, L.J., and Hushka, D.R., Molecular basis of dioxin actions: evidence supporting chemoprotection, *Toxicol. Pathol.*, 29, 6–7, 2001.

chapter twenty-six

Measuring oxidative damage caused by environmental stressors: potential for reversal and abatement

Robert A. Floyd

Contents

1-56670-596-7/02/$0.00+$1.50
© 2002 by CRC Press LLC

Abstract   Many environmental stressors, when they interact with biological systems, enhance production of reactive oxygen species (ROS). ROS may help mediate the successful response to environmental stressors and conversely, they may be involved in deleterious events in many instances. In this chapter, we offer the novel concept that the deleterious effects of many damaging consequences caused by some environmental stressors can be abated and/or reversed by agents that quell oxidative stress. Specific agents expected to be able to carry out this useful purpose include certain dietary antioxidants, plant constituents, and other synthetic agents such as nitrones. The background information reinforcing the validity of this concept are presented with specific data illustrating its high probability of the success.

I. Introduction

Reactive oxygen species (ROS) are implicated at various stages at which environmental stressors influence biological systems. This chapter proposes that some environmental insults may be abated and/or reversed by agents that quell oxidative stress. This is a relatively novel concept and is presented schematically in Figure 26.1. The plausibility of this concept rests on two observations. First, many environmental insults act in part by enhancing ROS, oxidative stress, oxidative damage and their biological consequences. Second, several agents are known which inhibit the deleterious effects of oxidative stress. Agents that may prevent oxidative damage include some dietary antioxidants, plant constituents, and nitrones.

Suppressing biological oxidative stress is much more complex than it appears. This observation was made during studies on nitrones, some of which are free radical trapping agents and may act as antioxidants in biological systems. Our extensive study of the action of nitrones in neurodegenerative diseases yielded surprising perspectives showing that their biological action is much more complex than previously thought and that their action in signal transduction is much more important than their direct free radical scavenging activities.[1] It is important to note that inhibitors of oxidative stress may exert their effect at various stages of disease. Therefore, the model and stage of development at which such agents are used should be carefully considered.

There are many observations in the literature showing that antioxidants have anticarcinogenic potential. At first glance, it would appear that these observations prove the concept advanced above. However, it is not known at what stage of disease these agents act, and it is not known which, if any, of the cancers being treated, were caused by an environmental stressor.

This chapter is not a comprehensive review but simply presents basic information to support these notions. In addition, the limitations of the major ideas are outlined in reference to some environmental stressors and the biological damage they cause. Sensitive methods will be needed to assess oxidative stress and oxidative damage caused by environmental stressors and to test these ideas; methods will also be needed to monitor signal

Reactive Oxygen Species Contribution to the Action of Environmental Stressors

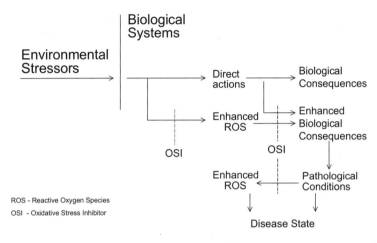

Figure 26.1 Scheme showing that some environmental stressors act by enhancing production of reactive oxygen species (ROS) and enhancing biological damage. Pathological conditions may develop, including cancer. In theory, oxidative stress inhibitors (OSIs) should mitigate this process and reduce production of ROS.

transduction, gene induction, and pathological change. Numerous observations indicate that environmental insults can cause oxidative damage and that their effects are mediated by oxidative events. Table 26.1 lists chemical and environmental agents that stimulate oxidative effects in biological systems. The effects of many cytokines and growth factors are mediated by ROS intermediates or produce ROS intermediates as byproducts of their action. In addition, H_2O_2 can trigger signal transduction processes. Therefore, ROS cause oxidative damage and can interfere with normal signal transduction

Many environmental insults trigger signal transduction processes or set off signaling events. The complexity of these interactions are shown diagrammatically in Figure 26.2. For many environmental stressors, oxidative stress enhances other types of cellular damage and exacerbates the biological consequences of the exposure (Figures 26.1 and 26.2). Pathological change can result from such exposure, including sustained oxidative stress. It, therefore, becomes important to quantify the amount of oxidative damage caused by an environmental stressor.

There have been recent advances in understanding of the etiology of diseases in which oxidative stress plays a significant role and in developing treatment for these diseases. This progress shows that novel therapeutic approaches are possible.[1] These results and several epidemiological studies suggest that it may be possible to either reverse or abate damage caused by environmental insults using agents that suppress oxidative damage. For example, some dietary components, including antioxidants, other plant con-

Table 26.1 Some Representative Stressors Where Consequences and Oxidative Stress Are Clearly Linked

Environmental Stressor	Consequences/Mechanism	References
Kainic Acid, Domoic Acid	Epilepsy/Excitotoxity, ROS, NO	22, 24
Malonate, 3-nitropropionate	Huntington's Disease/mitochondria inhibitor, ROS	25, 27
MPTP, Mn, Organochlorides	Parkinson's Disease/mitochondria inhibitors, ROS	33–37
Acrylonitrile	Actrocytomas/glutathione conjugation, ROS	38, 39
Trinitrotoulene	Cataracts/ROS production	41
Ultraviolet Radiation	Photocarcinogenesis/ROS, DNA lesions	42
Arsenite	Cancer/ROS, ?	52
Trimethyltin	Hearing Loss/ROS, ?	53
Carbon Monoxide/Noise	Hearing Loss/ROS, ?	54,55

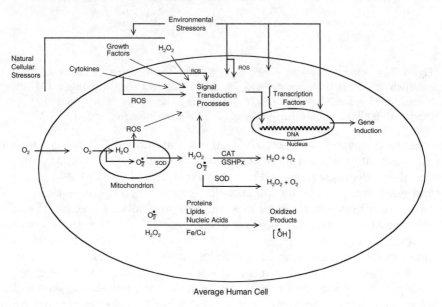

Average Human Cell

Figure 26.2 The complexity of environmental stressors acting on biological systems. Some environmental stressors act directly on DNA and other stressors induce signal transduction processes either by increasing ROS production or by enhancing the action of growth factors and cytokines. Endogenous ROS production from mitochondria is expected to contribute to the action of environmental stressors. Oxidative damage to proteins, lipids and nucleic acids is also considered important in some environmental stressor action.

stituents, and agents such as nitrones may be quite effective in abating or reversing the damage caused by some environmental stressors. Many of these ideas have not yet been evaluated thoroughly.

Basic information on oxidative stress, oxidative damage, signal transduction, and novel therapeutic approaches will be presented in this chapter.

These ideas will be used to synthesize concepts that may be useful in thinking about reversal and/or abatement of environmental stress. Additionally, recent data may help explain why the consequences of exposure to specific environmental stressors vary between individuals and populations. For instance, the individual-to-individual variation in response to exposure to an environmental stressor may be due in part to the different exogenous antioxidant defenses for different individuals. Clearly, the best approach to minimize an environmental insult is to remove the environmental stressor completely. This should be the first approach before other approaches are considered. Nevertheless, in some cases this is not possible and abatement and reversal may be useful mechanisms to minimize the deleterious consequences of exposure to an environmental stressor.

II. Oxidative stress and oxidative damage

In general terms, all aerobic organisms experience oxidative stress simply because ROS are continuously produced as a byproduct of normal oxygen metabolism (Figure 26.2). The amount of ROS produced is equal to about 2% of the oxygen consumed. A human cell, on average, consumes about 10 million O_2 molecules per second and produces about 0.2 million ROS molecules per second.* Thus, continuous oxidative stress is exerted upon aerobic cells at all times. As much as 90% of the total ROS is produced as superoxide in the mitochondria at Complexes I and III of the electron transport chain. The superoxide is normally dismutated to H_2O_2 by manganese superoxide dismutase (MnSOD) in the mitochondrial matrix. H_2O_2 readily penetrates membranes and diffuses into the cytosol. H_2O_2 is reduced to water by catalase or glutathione peroxidase; alternatively, if it is not reduced, H_2O_2 mediates oxidative damage and/or triggers signal transduction (Figure 26.2).

Oxidative processes damage biomolecules at a low level at all times in aerobic cells as a result of normal oxidative metabolism. Oxidative damage is often not detrimental because the damaged molecules are either repaired or removed. Oxidative damage affects protein, lipid, or nucleic acid (Figure 26.2). Most oxidative damage is caused by H_2O_2 or other peroxides acting via a transition metal (Fe or Cu) catalyst ligated on a reactive surface or at a specific site on the biomolecule.[3] Oxidative damage to proteins has been studied extensively and reviews are available.[3-5] It is likely that most of the oxidative damage is targeted to proteins.[3] Oxidized proteins are degraded by specific proteases in the proteosomal system.[6] Oxidized proteins are degraded and formed at nearly equal rates because the total amount of oxidized protein in a cell remains relatively constant in the absence of an

---

* The average oxygen consumption for a human cell was calculated from oxygen utilization values as follows: 0.05 L oxygen per 10 s at rest; and 1 L per 10 s after completing a sprint.[2] It is assumed that oxygen consumption is slightly higher during "normal" activity than at rest. The number of cells per 70 kg human was assumed to be 6×10^{13} and the size of each hypothetical average cuboidal cell has 10 μ unit dimension with a density of 1. See Tabatabaie and Floyd.[3]

abnormally large oxidative insult. However, the oxidized protein content in cells increases with increasing age of an organism.[5,7] It is not known if oxidized proteins contribute to any specific pathological state that might develop due to a high level of oxidative stress.[8]

Lipids are also damaged under conditions of increased oxidative stress. Many studies demonstrate an increase in malondialdehyde, a lipid oxidation product produced when oxidative stress increases. The definitive nature of the malondialdehyde products are not yet known, but it is clear that lipid oxidation occurs under many conditions of increased oxidative stress. Recent rigorous characterization of lipid oxidation products shows that 4-hydroxy-2-nonenal (HNE) is a lipid oxidation breakdown product, which has potent biological action. HNE mediates glutathione consumption and inactivates glutathione peroxidase in cultured cells. HNE reacts with lysine residues on proteins[9] and HNE-protein adducts are present in normal human kidney, in human kidney cancer,[10] and in rat kidney cells following iron overload.[11] Humphries and Szweda[12] demonstrated that HNE selectively inactivates α-ketoglutarate dehydrogenase (α-KGDH). This enzyme is critical as a rate-limiting complex in mitochondrial energy metabolism.[12] When isolated mitochondria are exposed to low levels of HNE, they rapidly lose NADH-linked respiratory capacity.[13]

DNA and RNA suffer continuous oxidative damage in aerobic cells. Many types of oxidative lesions have been identified, and repair systems have evolved to remove oxidative lesions in nuclear and mitochondrial DNA. One of the most widely studied oxidative DNA lesions is 8-hydroxy-2'-deoxyguanosine (8-OHdG). This lesion represents at least 5% of all oxidative lesions formed in DNA. It has been studied most extensively using a highly sensitive HPLC-electrochemical method for its detection and quantification.[14] Refinements in methods of DNA extraction[15] have led to measurements of lower 8-OHdG levels, which may have been elevated in early studies by Fe ligation during DNA isolation. The 8-OHdG content varies from one lesion in one million to one lesion in ten million guanine bases, depending on the tissue, age of the animal, and oxidative stress state of the system. Higher levels of 8-OHdG are detected in cancer tissues and under conditions that have an oxidative stress component.[16,17] 8-OHdG is mutagenic[18] and there are at least three enzyme systems that either remove it from DNA or prevent it from being incorporated into DNA. Many studies have investigated the role of 8-OHdG in mutagenesis and carcinogenesis. High levels of 8-OHdG are detected in brain, especially in older individuals,[19] and 8-OHdG may have effects in post-mitotic cells. It is possible that 8-OHdG may play a role in the development of conditions or pathological states in the brain, but this possibility is speculative at present.

III. Environmental stressors

Table 26.1 presents a small list of approximately 100 pathological conditions closely linked to exposure to environmental stressors that cause oxidative damage. Some of these conditions are discussed below.

A. Kainic acid-induced epilepsy

Epilepsy syndromes can be induced rapidly in rats by a small dose of kainic acid (KA), an environmental contaminant closely related to the chemical domoic acid present in some shellfish. In animals injected with 1–2 mg of KA, there is drastic remodeling of the brain involving the death of neurons in the hippocampus and changes in gene expression 1 year after injection.[20] KA interferes with glutamate uptake so neuronal death is due largely to excitotoxicty, which involves ROS and nitric oxide.[21] Superoxide intermediates are likely to be involved in KA-mediated neuronal death because scavengers protect cerebellar neurons from KA.[22] *In vivo* studies also demonstrate that phenidone protects against KA-mediated neurotoxicity, suggesting that ROS play an important role.[23] As discussed below, α-phenyl-tert-butyl nitrone (PBN) also protects from KA-induced seizures and death[24]; in contrast to phenidone, PBN provides protection against KA when it is administered after KA.

B. Mitochondrial toxins and Huntington's Disease

Huntington's Disease (HD) is a neurologic disorder that is strongly associated with exposure to agents that cause oxidative stress.[25] Animal models of HD can be produced by administration of mitochondrial toxins 3-nitropropionic acid[26] or malonate.[27] 4-hydroxybenzoic acid was used as a trap for ROS to demonstrate that mice treated with 3-nitropropionic acid have a higher level of ROS than untreated animals.[26]

C. Environmental agents and Parkinson's Disease

The cause of Parkinson's Disease (PD) is thought to have a strong environmental component. The strength of this argument has increased significantly in the last few years. Langston and colleagues[28] discovered that a synthetic chemical (1-methyl-4-phenyl-1,2,3,6-tetrahydropyridine, or MPTP), formed as a byproduct in elicit drug manufacture, was the cause of PD-like tremors in young individuals in a south San Francisco neighborhood in the early 1980s. These observations showed for the first time that a chemical could induce a PD-like syndrome. Subsequent studies proved that MPTP causes symptoms resembling PD in several animal models. Genetic linkage studies also support the idea that an environmental agent(s) can cause PD; studies of monozygotic and dizygotic twin pairs showed unequivocally that PD with onset after 50 years of age does not have a significant hereditary component.[29] In contrast, PD with onset before age 50 is likely to have a strong hereditary component.[29] Several environmental chemicals have been implicated in PD. These include heavy metals (especially manganese, a known neurotoxin), and organochloride pesticides. Consumption of rural well water is also correlated with PD and this result supports the idea that PD has an environmental component.[28]

The etiology of PD may also involve ROS. For example, it is possible that H_2O_2 is produced during dopamine metabolism by monoamine oxidase, which would put dopaminergic neurons under increased oxidative stress.[30,31] Brain mitochondrial inhibitors have also been implicated in PD. These include manganese, which inhibits mitochondrial aconitase,[32] salsolinol (1-methyl-6–7-dihydroxy-1,2,3,4-tetrahydroisoquinoline) which is toxic to dopaminergic SH-5Y5Y cells and impairs energy production,[33] and MPP+ (1-methyl-4-phenylpyridinuim, the active MPTP metabolite) which inhibits complex I of mitochondria and increases ROS production.[34] MPP+ is selectively transported into dopaminergic neurons.[35]

Studies with MPP+ support and illustrate the main concepts of this chapter. Chun et al.[36] demonstrated that dopaminergic cell death induced by MPP+ was mediated by ROS (specifically H_2O_2). They also showed that MPP+-mediated apoptosis was inhibited by the antioxidant trolox as well as by a caspase inhibitor. Interestingly, they also showed in the same system that H_2O_2 caused the same effects as MPP+.[36]

Some studies also suggest that organochloride pesticides may cause PD. Heptachlor increases dopamine transport into dopamine neurons but inhibits its sequestration into dopamine containing vesicles.[37,38] The mechanism of this effect involves heptachlor-mediated change in expression of the genes for two dopamine transport proteins.[37]

D. Brain tumors and environmental agents

In animal models, astrocytomas (glial cell tumors) are induced by chronic treatment with acrylonitrile (ACN) by a nongenotoxic mechanism, and there is strong evidence that ROS and oxidative damage are involved.[39,40] ACN was chronically administered to rats and oxidative damage was measured in the brain (the target tissue) and the liver (a nontarget tissue). Oxidative damage increased in a dose-dependent manner in the brain but not in the liver.[39] Levels of 8-OHdG, malondialdehyde, and trapped hydroxyl free radicals increased in the brain but not in the liver. Catalase activity, which is low in the brain, decreased with increasing dose of ACN in brain but not in the liver; in contrast, small changes in superoxide dismutase (SOD) and glutathione peroxidase were noted.[39] Cultured rat astrocytes but not hepatocytes had increased oxidative damage after exposure to ACN.[40] These experiments support the notion that oxidative damage plays a significant role in astrocytoma initiation and development.

E. Environmental agents and cataracts

Cataract prevalence increases with increasing age and with increased exposure to sunlight. However, there is strong evidence that environmental agents especially those that increase ROS, such as UV radiation or 2,4,6-trinitrotoluene (TNT), enhance cataract formation.[41] A mechanistic study by Kumagai et al.[41] showed that cataract formation increases in workers exposed to TNT. They showed that TNT is reduced by a lens protein called ζ-crystallin which leads to a large increase in superoxide when reduced TNT is re-oxidized. This lens protein has a low level of antioxidants and is sensitive to ROS-induced damage.

F. UV carcinogenesis

The role played by free radicals in carcinogenesis has been debated for many years. There is little doubt that carcinogenesis involves free radical events and changing these events can effect the rate if not the final outcome of carcinogenesis. However, it is not clear how or at what stage antioxidants or other agents that affect free radical processes interfere with carcinogenesis. This discussion will focus on the early phases of carcinogenesis. Skin tumors are a good example because they are largely due to UV radiation and the ROS-mediated processes caused by UV.[42] UV-mediated tumor formation in rodents was studied by Black and colleagues in the 1970s under the basic premise that ROS are important mechanistically.[43-46] They determined that a cholesterol epoxide formed by ROS intermediates was important in the carcinogenic process.[43] They showed that UV-mediated tumor formation decreased significantly in animals fed a diet containing antioxidants (i.e., 1.2% ascorbate, 0.5% butylated hydroxytoluene (BHT), 0.2% DL-α-toco-pherol acetate, and 0.1% reduced glutathione).[44-46] In contrast, a higher level of unsaturated fat in the diet led to a higher rate of tumors.[46] These studies suggested that BHT was the most effective anticarcinogenesis antioxidant in the diet.[46]

Slaga and colleagues also studied skin carcinogenesis in rodent models and show that ROS play a role in this process and antioxidants are protective.[47-49] Their studies show that benzyl peroxide[48,49] or lauroyl peroxide[49] are effective skin tumor promoters but not initiators when applied directly to the skin. In addition, BHT was an effective anticarcinogenic agent against skin tumors initiated with dimethylbenzanthracene (DMBA).[47] They show that BHT prevents the formation of DNA–DMBA adducts,[47] which presumably is the primary early neoplastic event. It is likely that BHT inhibits DMBA-induced skin carcinogenesis because it is a competitive inhibitor of P450 metabolism, which reduces the rate of DMBA activation and reduces the number of DNA–DMBA adducts; thus, BHT may not inhibit skin carcinogenesis by scavenging free radicals per se. These results demonstrate that caution is needed in interpreting the results obtained with antioxidants in biological systems.

The anticarcinogenic activities of BHT and butylated hydroxyanisole (BHA), a related antioxidant, have been demonstrated in various animal models.[50,51] The results obtained support the notion that their anticarcinogenic activity is not due to their free radical scavenging activity per se. BHT was effective in preventing hepatomas induced by N-2-fluorenylacetamide and N-hydroxy-N-fluorenylacetamide but was not effective in preventing esophageal tumors induced by diethylnitrosoamine.[50] In the same study, diphenyl-p-phenylenediamine, a well-known antioxidant, was ineffective in all tumor models studied.[50] BHA was effective in preventing methyla-zoxymethanol (MAM)-induced colon tumors in mice.[51] Its anticarcinogenic activity in this model was shown not to be due to its antioxidant activity, but to its inhibition of the enzyme NAD-dependent alcohol dehydrogenase, the enzyme that metabolizes MAM.[51]

G. Arsenic carcinogenesis

Arsenic is present in various forms in the environment. It is closely linked to several human diseases including blackfoot, diabetes, hypertension and cancers of the skin, lung, bladder and liver. Much research has shown that metabolism of arsenic increases the level of ROS.[52] Wang et al.[52] demonstrated that arsenite enhances production of ROS and causes ROS-mediated apoptosis in Chinese hamster ovary cells and that antioxidants could prevent the ROS-mediated effects. N-acetyl-cysteine was quite effective, whereas Trolox and Tempo (2,2,6,6-tetramethyl-1-piperidinyloxy) were less effective. Tempo is a stable nitroxide compound.

H. Hearing loss: trimethyltin/carbon monoxide

The mechanisms of hearing loss remain poorly understood and are difficult to study because the cochlea is so small. Nevertheless, there is good evidence that ROS are involved in hearing loss especially when it is caused by exposure to trimethyltin (TMT) or carbon monoxide (CO). TMT damages neural tissue in the cochlea via an ROS-mediated process.[53] Administration of superoxide dismutase (SOD) ligated to polyethylene glycol protected guinea pigs from TMT-mediated hearing loss.[53] Exposure to CO or to CO and noise causes hearing loss.[54,55] Fechter and colleagues used allopurinol and PBN to suppress ROS-mediated processes. Allopurinol blocks xanthine oxidase-mediated superoxide production. Both agents were effective in preventing CO-mediated hearing loss; however, PBN was more effective.[54] When noise is combined with CO, hearing loss is more severe.[55] In a study using CO and noise together, PBN was effective in preventing hearing loss. In addition, PBN provided partial protection against hearing loss caused by noise.[55]

IV. Antioxidant action

A. Ascorbate and α-tocopherol

Most of our understanding of the action of antioxidants is based on historical discoveries in nutrition research on the action of vitamins E and C — essential because they suppress oxidative stress in biological systems. The molecular action of these compounds in biological systems can be explained by the chemistry of free radicals, which has been recently reviewed.[56] Figure 26.3 is a summary of key ideas in this area. One important concept is that free radical reactions can be propagated from one chemical entity to another through as many as 20 sequential reactions; in addition, oxygen changes the inherent reactivity and damage potential of free radicals in biological systems. Antioxidants such as α-tocopherol are important because they can terminate these reactions. The ability of α-tocopherol to act as a free radical scavenger has been investigated in the bilayer phase in liposomes dispersed in aqueous systems. These studies show that the radical scavenging action of α-tocopherol in the membrane phase is coupled to the radical scavenging action of ascorbate in the aqueous phase; hence, these compounds are likely to work in concert in dispersed liposomal aqueous systems and in biological systems (Figure 26.3).

The action of α-tocopherol and ascorbate in quelling lipid peroxidation in biological membranes is complex and is not a simple one-to-one scavenging of free radicals in a mass action one phase model. For example, the physiological concentrations of ascorbate (<1 mM) and α -tocopherol (<100 μM) are low, which suggests that the mechanism of action of these antioxidants is complex and not based on scavenging. Therefore, it should not be surprising that other agents that quell oxidative damage also act in a complex manner. This is also true for nitrone antioxidants which quell oxidative damage processes in several pathological conditions where ROS events are implicated.

B.   Nitrones as suppressors of biological oxidative damage

The chemical structures of nitrones are shown in Figure 26.4. Nitrones can be used to spin-trap free radicals in chemical, biochemical and biological systems[57-59] and they stabilize highly reactive free radicals so that they can be characterized (Figure 26.4). Because the spin of the original radical is preserved during its reaction with a nitrone, they were termed spin-traps.[60] Nitrones began to be used in analytical chemistry for this purpose in 1969,[61] and they began to be used in biochemical and biological systems in the 1970s and 1980s. It was not until 1985, when Novelli discovered that PBN was protective in a rat model of traumatic shock[62] that the pharmacological potential of these compounds were even considered. PBN has neuroprotective activity during stroke[59,63] and PBN and other nitrones have pharmacological potency against a wide range of age-related diseases and pathological conditions.

The potent pharmacological action of nitrones is certainly not a result of their direct mass action free radical trapping properties. The primary rationale for this statement is that in situations where nitrones act potently, their target tissue concentration can be as low as 100 nM; in some cases, nitrones make cells resistant to oxidative insult for an extended time, long after the nitrone concentration is undetectable. These ideas are discussed further.[64-67] Nitrones are active in tissues when their concentration is too low for effective trapping of radicals; thus, the neuroprotective activity of the nitrones may be to suppress oxidative damage in the brain that is due to aging or rapid injury. It is possible that nitrones mediate this effect by

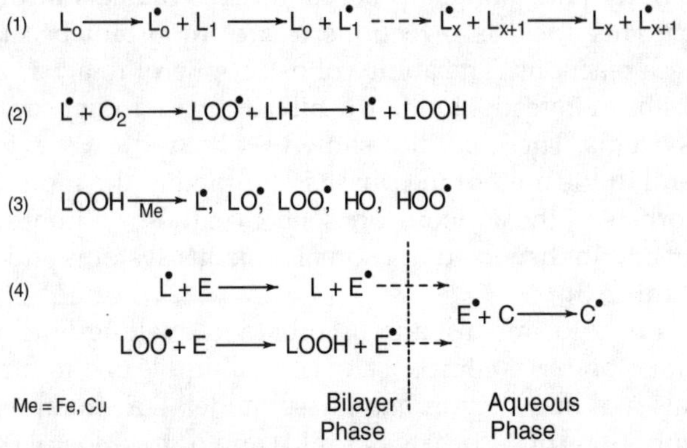

(1) $\quad L_0 \longrightarrow \overset{\bullet}{L}_0 + L_1 \longrightarrow L_0 + \overset{\bullet}{L}_1 \dashrightarrow \overset{\bullet}{L}_x + L_{x+1} \longrightarrow L_x + \overset{\bullet}{L}_{x+1}$

(2) $\quad \overset{\bullet}{L} + O_2 \longrightarrow LOO^{\bullet} + LH \longrightarrow \overset{\bullet}{L} + LOOH$

(3) $\quad LOOH \xrightarrow[Me]{} \overset{\bullet}{L}, LO^{\bullet}, LOO^{\bullet}, HO^{\bullet}, HOO^{\bullet}$

(4)
$$\overset{\bullet}{L} + E \longrightarrow L + E^{\bullet} \dashrightarrow$$
$$LOO^{\bullet} + E \longrightarrow LOOH + E^{\bullet} \dashrightarrow$$

$$E^{\bullet} + C \longrightarrow C^{\bullet}$$

Me = Fe, Cu   Bilayer   Aqueous
　　　　　　　　　Phase   Phase

Figure 26.3 This figure shows the chemistry of free radical reactions, the importance of oxygen in free radical reactions as in biological membranes, and the antioxidant action of α-tocopherol in biological membranes and the interaction of vitamin E (E) with vitamin C (C) in the aqueous phase. Many biological molecules can become free radicals simply by losing an electron or by gaining an electron in their outer molecular orbital. As noted in equation 1, free radicals react readily with many other biological molecules because the reaction is an electron exchange with low activation energy. Lipids (L) in biological membranes are susceptible to free radical reactions involving loss of a hydrogen atom from the methylene carbon of a carbon–carbon double bond. Free radical reactions often propagate for several reaction cycles in membranes. Oxygen is more soluble in the lipid phase than in the aqueous phase, and will rapidly add to lipid free radicals (equation 2) to form an alkoxy free radical LOO•; this molecule can abstract hydrogen from a nearby lipid to form a lipid hydroperoxide (LOOH) and another lipid radical, which can propagate other free radical reactions. This illustrates the importance of oxygen in free radical reactions in a membrane. The lipid hydroperoxide is reactive (equation 3) in the presence of catalytic amounts of a metal (Me) such as Fe or Cu. Oxygen supports a cascade with more free radicals leading to auto-oxidation. α-tocopherol (E) resides in the membrane where it terminates chain reactions involving lipid radicals or lipid alkoxy free radicals (equation 4) by forming α-tocopherol free radical (E•). The radical quenching portion of α-tocopherol resides at the bi-layer interface where, E• can be reduced to α-tocopherol by ascorbate (C), forming an ascorbate free radical in the aqueous phase. The ascorbate free radical can be reduced back to ascorbate enzymatically using NADH as reductant. Additionally, the lipid hydroperoxide (LOOH) can be catalytically decomposed by a selenium containing glutathione peroxidase system with NADPH as reductant. Thus, α-tocopherol and ascorbate work in tandem to control lipid peroxidation and glutathione peroxidase suppresses oxidative damage by decomposing LOOH.

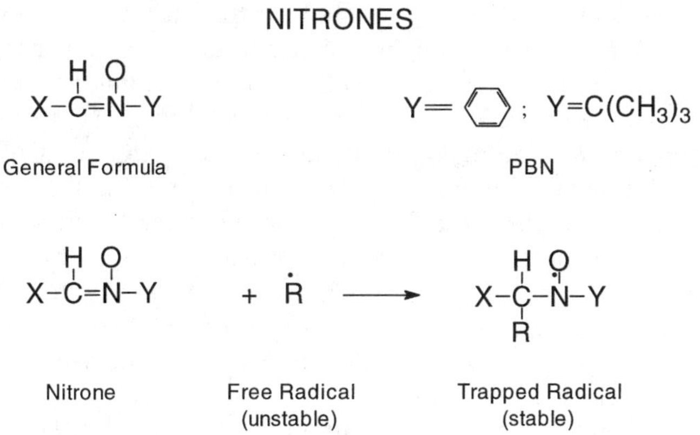

NITRONES

General Formula

PBN

Nitrone   **Free Radical**
(unstable)   **Trapped Radical**
(stable)

Figure 26.4 The general formulas for nitrones, PBN (α-phenyl-*tert*-butylnitrone), and a general reaction where an unstable free radical is trapped by a nitrone to yield a stable free radical (nitroxide).

preventing the upregulation of genes that produce toxic products.[64–66] Nitrones may interfere with the signal transduction cascade which is activated in brain injury. Studies using cultured brain cells[67,68] and in the brain of experimental animals[24] provide data that supports these conclusions.

C. Other antioxidants

As stated above, the mechanism of the action of ascorbate, α–tocopherol, and nitrones, agents that suppress oxidative damage in biological systems, is complex. Simple mass action free radical trapping in a one-phase system does not explain the action of any of these compounds, and therefore, it is unreasonable to assume that this concept would strictly apply to other potential antioxidants. On the other hand, it is reasonable to expect that agents that suppress oxidative damage in biological systems may limit or abate the oxidative stress associated with environmental stressors. More research is needed in order to understand the mechanism(s) and potential of these and other antioxidants and to examine the validity of the concepts and approaches discussed above.

V. ROS-mediated signaling

Recent studies of the mechanisms of signal transduction have established several facts about this process. Signaling processes are activated as a response to a change in the cellular environment. The mechanism of signaling often involves MAP-kinase mediated protein phosphorylation cascades, which lead to *de novo* protein synthesis or other adaptive responses. ROS (H_2O_2) and ROS-mediated events play an important role in some signaling processes.[69–71] Some stressors induce mitochondrial production of H_2O_2, which may act as a second messenger.

In 1991, Baeuerle and colleagues showed that ROS was involved in NF-κB mediated signal transduction, in which cytokine stressors or H_2O_2 upregulate the HIV-1 gene in human T-cells.[72] Other studies show that H_2O_2 activates signal transduction in many cell types[68,73–77] and that H_2O_2 may act as a messenger in epidermal growth factor (EGF) signaling.[78,79] Bae and colleagues inhibited EGF signaling in A431 human epidermal carcinoma cells by electroporating catalase into the cells. They showed that the EGF receptor kinase activity was necessary to induce production of H_2O_2, but that the 126 amino acid carboxyl terminal tail of EGF receptor was not necessary. Thus, H_2O_2 signaling may involve inactivation of a protein tyrosine phosphatase.[78] Other studies strongly implicate H_2O_2-mediated phosphatase inactivation in signaling in primary astrocytes.[67,68]

These observations highlight an important concept in H_2O_2-mediated signaling which has implications in the larger context of redox active signaling. Figure 26.5 presents a scheme illustrating the central importance of phosphatase inactivation in signaling. Several previous studies provide clear-cut biochemical mechanisms for this process. Denu and Tanner[80] rigorously demonstrated that the active sites of some phosphatases have an acid cysteine sulfhydryl residue which is rapidly converted to a sulfenic acid residue (-SOH) by H_2O_2. The sulfenic acid residue is then converted back, at a slower rate, to the sulfhydryl by glutathione. Thus, H_2O_2 rapidly inacti-

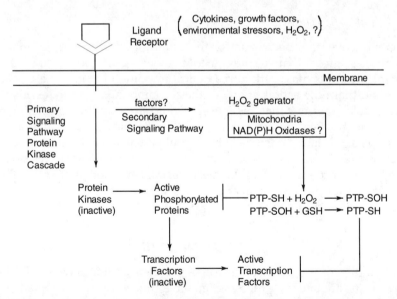

Figure 26.5 Inactivation and reactivation of phosphatases by H_2O_2 and glutathione (GSH), respectively, and the importance of this process in regulating stressor-mediated signal transduction. PTP is phosphotyrosine phosphatase. H_2O_2 may be generated by mitochondria or a plasma membrane oxidase. This scheme was adapted from Figure 26.2.[69]

vates the phosphatase allowing its target protein to remain phosphorylated and stimulating the kinase cascade. Glutathione eventually reactivates the phosphatase activity, tyrosine phosphate residues are removed from the activated kinases and the phosphorylation cascade ceases (Figure 26.5).

VI. Quantitation of oxidative stress and oxidative damage

If the effects of environmental stressors are mediated by ROS, then it should be possible to quantify parameters that are indicators of this process. Table 26.2 presents a list of such parameters, and a brief discussion of them is presented below.

If environmental stressors increase oxidative stress, then it is likely that signal transduction components involving ROS will be activated and pro-inflammatory cytokines and chemokines will increase. Signal transduction pathways involving NF-κB involves ROS; AP-1 signaling is not linked to ROS-mediated processes. The activation state of the MAP-kinase p38 has been useful in our laboratory.[67,68]

It is possible, although difficult in most cases to measure the level of ROS intermediates. In the case of H_2O_2 efflux from cells or mitochondria, a horseradish peroxidase catalyzed dichlordihydroflorescein diacetate oxidation method is useful and accurate.[68,80] For tissue values of hydroxyl free radical flux, salicylate trapping[81] has been used by many laboratories.

Increasingly, it is becoming clear that mitochondria play a central role in producing cellular ROS, and are important in apoptosis. For example, cytochrome c release from mitochondria is used as a diagnostic marker of apoptosis. As mitochondrial function decreases due to increased oxidative stress, proteins are oxidized. Oxidative damage to mitochondrial cis-aconitase, adenine nucleotide translocase and α-ketoglutarate dehydrogenase can be quantified as a measure of this process. DNA damage in mitochondria can be measured, including point mutations, deletions and 8-OHdG.

Table 26.2   Quantifiable Parameters Useful to Assess Environmental Stressor Effect on Oxidative Stress and Oxidative Damage

- Activation State of Signal Transduction Components
- Level of Pro-inflammatory Cytokines and Chemokines
- Level of ROS Intermediates
- Level of Mitochondrial-mediated ROS Production
- Level of Oxidized Mitochondrial Protein Components
- Amount of Lipid Oxidation Products
- Amount of Oxidatively Damaged Proteins
- Amount of Nucleic Acids Oxidation Products
- Content of Antioxidants and Activity of Antioxidant Enzymes

Oxidative damage is also assessed by measuring oxidized protein. In this assay, oxidation of amino acid residues to carbonyls is quantified using dinitrophenylhydrazine (DNP). DNP-modified proteins can be separated by column chromatography and identified on a Western blot or by optical quantification of the chromophore.

Oxidized DNA is an important parameter even though oxidative lesions are subject to DNA repair. In addition, oxidative lesions can be introduced into DNA during isolation. The oxidative lesion 8-OHdG is widely used to assess oxidative DNA damage using the HPLC-electrochemical method. Care must be taken to avoid exposure to Fe during DNA isolation, by using NaI for example.

Malondialdehyde has been used to measure lipid oxidation and F2-isoprostanes have recently been measured as a specific peroxidation product of arachidonate. The level of 4-hydroxynonenal (HNE) has also been a useful parameter, although it should be considered that. HNE binds to protein.

Overall, the methods currently available for assessing oxidative stress and oxidative damage to cells and tissue are problematic; this should be kept in mind when these methods are used. These methods require careful evaluation and standardization and it is best to compare the results obtained using several different methods to assess oxidative damage.

VII. Conclusions

This chapter poses the following question: if an environmental stressor(s) acts through an oxidative mechanism to damage a biological system, will agents that minimize oxidative processes abate or reverse the damage? In some cases, circumstantial evidence suggests a positive answer, but the question requires more rigorous study. Antioxidative mechanisms may be the only practical means to minimize the damage caused by some environmental stressors. Oxidative stress causes different amounts of damage to individuals who have different antioxidant capacity; interindividual variation in the effects of environmental stressors may be partially explained by this fact. More study in this area will help explain the action of certain environmental stressors and help evaluate the validity of the questions posed in this chapter.

Acknowledgments

Research discussed here was funded by NIH grants NS 35747 and CA 082506.

References

1. Floyd, R.A., Antioxidants, oxidative stress, and degenerative neurological disorders, *Proc. Soc. Exp. Biol. Med.*, 222, 236–245, 1999.

2. Lehninger, A., *Principles of Biochemistry,* Worth Publishers, New York, 1982, pp. 719.
3. Tabatabaie, T. and Floyd, R.A., Protein damage and oxidative stress, in *Cellular Aging and Cell Death*, Holbrook, N.J., Martin, G.M., and Lockshin, R.A., Eds., John Wiley & Sons, New York, 1996, pp 35–49.
4. Stadtman, E.R. and Oliver, C.N., Metal-catalyzed oxidation of proteins, *J. Biol. Chem.*, 266 (4), 2005–2008, 1991.
5. Stadtman, E.R., Metal ion catalyzed oxidation of proteins: Biochemical mechanism and biological consequences, *Free Radic. Biol. Med.*, 9, 315–325, 1990.
6. Stadtman, E.R., Oxidation of proteins by mixed-function oxidation systems: implication in protein turnover, aging and neutrophil function, *Trends Biochem. Sci.*, 2, 11–12, 1986.
7. Stadtman, E.R., Protein oxidation and aging, *Science*, 257, 1220–1224, 1992.
8. Grune, T., Reinheckel, T., and Davies, K., Degradation of oxidized proteins in mammalian cells, *FASEB J.*, 11, 526–534, 1997.
9. Berlett, B.S. and Stadtman, E.R., Protein oxidation in aging disease, and oxidative stress, *J. Biol. Chem.*, 272, 20313–20316, 1997.
10. Oberley, T.D., Toyokuni, S., and Szweda, L.I., Localization of hydroxynonenal protein adducts in normal human kidney and selected human kidney cancers, *Free Radic. Biol. Med.*, 27, 695–703, 1999.
11. Zainal, T.A. et al., Localization of 4-hydroxy-2-nonenal-modified proteins in kidney following iron overload, *Free Radic. Biol. Med.*, 26, 1181–1193, 1999.
12. Humphries, K. and Szweda, L.I., Selective inactivation of α-ketoglutarate dehydrogenase and pyruvate dehydrogenase: Reaction of lipoic acid with 4-hydroxy-2-nonenal, *Biochem.*, 37, 15835–15841, 1998.
13. Cooney, G.J., Taegtmeyer, H., and Newsholme, E.A., Tricarboxylic acid cycle flux and enzyme activities in the isolated working rat heart, *Biochem.*, 200, 701–703, 1981.
14. Floyd, R.A. et al., Hydroxyl free radical adduct of deoxyguanosine: Sensitive detection and mechanisms of formation, *Free Radic. Res. Commun.*, 1, 163–172, 1986.
15. Nakae, D. et al., Improved genomic/nuclear DNA extraction for 8-hydroxydeoxyguanosine analysis of small amounts of rat liver tissue, *Cancer Lett.*, 97, 233–239, 1995.
16. Floyd, R.A., The role of 8-hydroxyguanine in carcinogenesis, *Carcinogenesis*, 11, 1447–1450, 1990.
17. Denda, A. et al., Prevention by acetylsalicylic acid of liver cirrhosis and carcinogenesis as well as generations of 8-hydroxydeoxyguanosine and thiobarbituric acid-reactive substances caused by a choline-deficient, L-amino acid-defined diet in rats, *Carcinogenesis*, 15, 1279–1283, 1994.
18. Cheng, C.K. et al., 8-Hydroxyguanine, an abundant form of oxidative DNA damage, causes G T and A C substitutions, *J. Biol. Chem.*, 267, 166–172, 1992.
19. Nakae, D. et al., Age and organ dependent spontaneous generation of nuclear 8-hydroxydeoxyguanosine in male Fischer 344 rats, *Lab. Invest.*, 80, 249–261, 2000.
20. Bing, G. et al., A single dose of kainic acid elevates the levels of enkephalins and activator protein-1 transcription factors in the hippocampus for up to 1 year, *Proc. Natl. Acad. Sci. USA*, 94, 9422–9427, 1997.
21. Schulz, J.B. et al., Involvement of free radicals in excitotoxicity *in vivo*, *J. Neurochem.*, 64, 2239–2247, 1995.

22. Dykens, J.A., Stern, A., and Trenkner, E., Mechanism of kainate toxicity to cerebellar neurons *in vitro* is analogous to reperfusion tissue injury, *J. Neurochem.*, 49, 1222–1228, 1987.

23. Kim, H.C. et al., Phenidone prevents kainate-induced neurotoxicity *via* antioxidant mechanisms, *Brain Research*, 874, 15–23, 2000.

24. Floyd, R.A., Hensley, K., and Bing, G., Evidence for enhanced neuro-inflammatory processes in neurodegenerative diseases and the action of nitrones as potential therapeutics, *J. Neural. Trans.*, 60, 337–364, 2000.

25. Browne, S.E., Ferrante, R.J., and Beal, M.F., Oxidative stress in Huntington's disease, *Brain Pathol.*, 9, 147–163, 1999.

26. Bogdanov, M.B. et al., Increased vulnerability to 3-nitropropionic acid in an animal model of Huntington's disease, *J. Neurochem.*, 71, 2642–2644, 1998.

27. Beal, M.F. et al., Age-dependent striatal excitotoxic lesions produced by the endogenous mitochondrial inhibitor malonate, *J. Neurochem.*, 61, 1147–1150, 1993.

28. Langston, J.W. et al., Chronic Parkinsonism in humans is due to a product of meperdine analogue synthesis, *Science*, 219, 279–280, 1983.

29. Tanner, C.M. et al., Parkinson disease in twins an etiologic study, *JAMA*, 281, 341–346, 1999.

30. Zeevalk, G.D., Bernard, L.P., and Nicklas, W.J., Role of oxidative stress and the glutathione system in loss of dopamine neurons due to impairment of energy metabolism, *J. Neurochem.*, 70, 1421–1430, 1998.

31. Han, J. et al., Inhibitors of mitochondrial respiration, iron (II), and hydroxyl radical evoke release and extracellular hydrolysis of glutathione in rat striatum and substantia nigra: potential implications to Parkinson's disease, *J. Neurochem.*, 73, 1683–1695, 1999.

32. Zheng, W., Ren, S., and Graziano, J.H., Manganese inhibits mitochrondial aconitase: a mechanism of manganese neurotoxicity, *Brain Res.*, 799, 334–342, 1998.

33. Storch, A. et al., 1-Methyl-6,7-dihydroxy- 1,2,3,4-tetrahydroisoquinoline (salsolinol) is toxic to dopaminergic neuroblastoma SH-SY5Y cells *via* impairment of cellular energy metabolism, *Brain Res.*, 855, 67–75, 2000.

34. Adams, Jr., J.D., Klaidman, L.K., and Leung, A.C., MPP+ and MDPD+ induced oxygen radical formation with mitochondrial enzymes, *Free Radic. Biol. Med.*, 15, 181–186, 1993.

35. Shen, R.S., et al., Serotonergic conversion of MPTP and dopaminergic accumulation of MPP+, *FEBS Lett.*, 189, 225–230, 1985.

36. Chun, H.S. et al., Dopaminergic cell death induced by MPP+, oxidant and specific neurotoxicants shares the common molecular mechanism, *J. Neurochem.*, 76, 1010–1021, 2001.

37. Kirby, M.L., Barlow, R.L., and Bloomquist, J.R., Neurotoxicity of the organochlorine insecticide heptachlor to murine striatal dopaminergic pathways, *Toxicological Sci.*, 61, 100–106, 2001.

38. Miller, G.W. et al., Heptachlor alters expression and function of dopamine transporters, *Neurotoxicology*, 4, 631–637, 1999.

39. Jiang, J., Xu, Y., and Klaunig, J.E., Induction of oxidative stress in rat brain by acrylonitrile (ACN), *Toxicological Sci.*, 46, 333–341, 1999.

40. Kamendulis, L.M. et al., Induction of oxidative stress and oxidative damage in rat glial-cells by acrylonitrile, *Carcinogenesis*, 20, 1555–1560, 1999.

41. Kumagai, Y. et al., ζ-Crystallin catalyzes the reductive activation of 2,4,6-trinitrotoluene to generate reactive oxygen species: a proposed mechanism for the induction of cataracts, *FEBS Lett.*, 478, 295–298, 2000.

42. Scharffetter-Kochanek, K. et al., UV-Induced reactive oxygen species in photocarcinogenesis and photoaging, *Biol. Chem.*, 378, 1247–1257, 1997.
43. Chan, J.T. and Black, H.S., Skin carcinogenesis: Cholesterol-5 α, 6 α-epoxide hydrase activity in mouse skin irradiated with ultraviolet light, *Science*, 186, 1216–1217, 1974.
44. Black, H.S., Effects of dietary antioxidants on actinic tumor induction, *Res. Commun. Chem. Pathol. Pharmacol.*, 7, 783–786, 1974.
45. Black, H.S. and Chan, J.T., Suppression of ultraviolet light induced tumor formation by dietary antioxidants, *J. Invest. Dermatol.*, 65, 412–414, 1975.
46. Black, H.S. et al., Relation of antioxidants and level of dietary lipid to epidermal lipid peroxidation and ultraviolet carcinogenesis, *Cancer Res.*, 45, 6254–6259, 1985.
47. Slaga, T.J. and Bracken, W.M., The effects of antioxidants on skin tumor initiation and aryl hydrocarbon hydroxylase, *Cancer Res.*, 37, 1631–1635, 1977.
48. Slaga, T.J. et al., Skin tumor-promoting activity of benzoyl peroxide, a widely used free radical-generating compound, *Science*, 213, 1023–1025, 1981.
49. Klein-Szanto, A.J.P. and Slaga, T.J., Effects of peroxides on rodent skin: Epidermal hyperplasia and tumor promotion, *J. Invest. Dermatol.*, 79, 30–34, 1982.
50. Ulland, B.M., Antioxidants and carcinogenesis: Butylated hydroxytoluene, but not diphenyl-*p*-phenylenediamine, inhibits cancer induction by N-2-fluorenyl acetamide and by N-hydroxy-N-2-fluorenylacetamide in rats, *Food Cosmetol. Toxicol.*, 11, 199–207, 1973.
51. Wattenberg, L.W. and Sparnins, V.L., Inhibitory effects of butylated hydroxyanisole on methylazoxymethanol acetate-induced neoplasia of the large intestine and on nicotinamide adenine dinucleotide-dependent alcohol dehydrogenase activity in mice, *J. Natl. Cancer Inst.*, 63, 219, 1979.
52. Wang, T.S. et al., Arsenite induces apoptosis in Chinese hamster ovary cells by generation of reactive oxygen species, *J. Cell. Physiol.*, 169, 256–268, 1996.
53. Clerici, W.J., Effects of superoxide dismutase and U74389G on acute trimethyltin-induced cochlear dysfunction, *Toxicol. Appl. Pharmacol.*, 136, 236–242, 1996.
54. Fechter, L.D., Liu, Y., and Pearce, T.A., Cochlear protection from carbon monoxide exposure by free radical blockers in the guinea pig, *Toxicol. Appl. Pharmacol.*, 142, 47–55, 1997.
55. Rao, D. and Fechter, L.D., Protective effects of phenyl-N-tert-butylnitrone on the potentiation of noice-induced hearing loss by carbon monoxide, *Toxicol. Appl. Pharmacol.*, 167, 125–131, 2000.
56. Floyd, R.A., Basic free radical biochemistry, in *Free Radicals in Aging*, Yu, B.P., Ed., CRC Press, Boca Raton, FL, 1993, pp. 39-55.
57. Floyd, R.A. and Soong, L.M., Spin trapping in biological systems: Oxidation of the spin trap 5,5-dimethyl-1-pyrroline-1-oxide by a hydroperoxide-hematin system, *Biochem. Biophys. Res. Commun.*, 74, 79–84, 1977.
58. Oliver, C.N. et al., Oxidative damage to brain proteins, loss of glutamine synthetase activity, and production of free radicals during ischemia/reperfusion-induced injury to gerbil brain, *Proc. Natl. Acad. Sci. USA*, 87, 5144–5147, 1990.
59. Floyd, R.A., Role of oxygen free radicals in carcinogenesis and brain ischemia, *FASEB J.*, 4, 2587–2597, 1990.
60. Janzen, E.G., Spin trapping, *Acc. Chem. Res.*, 4, 31–40, 1971.

61. Janzen, E.G. and Blackburn, B.J., Detection and identification of short-lived free radicals by electron spin resonance trapping techniques (spin trapping): Photolysis of organolead, tin, and -mercury compounds, *J. Am. Chem. Soc.*, 91, 4481–4490, 1969.

62. Novelli, G.P. et al., Phenyl-T-butyl-nitrone is active against traumatic shock in rats, *Free Radic. Res. Commun.*, 1, 321–327, 1985.

63. Carney, J.M. and Floyd, R.A., Phenyl Butyl Nitrone Compositions and Methods for Treatment of Oxidative Tissue Damage, U.S. Patent 5,025,032, 1991.

64. Floyd, R.A., Protective action of nitrone-based free radical traps against oxidative damage of the central nervous system, in *Advances in Pharmacology*, Sies, H., Ed., Academic Press, San Diego, 1996, pp. 361–378.

65. Carney, J.M. and Floyd, R.A., Protection against oxidative damage to CNS by α-phenyl-*cert*-butyl nitrone (PBN) and other spin trapping agents: a novel series of nonlipid free radical scavengers, *J. Mol. Neurosci.*, 3, 47–57, 1991.

66. Floyd, R.A. and Hensley, K., Nitrone inhibition of age-associated oxidative damage, in *Reactive Oxygen Species from Radiation to Molecular Biology*, Chiueh, C.C., Ed., New York Academy of Sciences, New York, 2000, pp. 222–237.

67. Robinson, K.A. et al., Basal protein phosphorylation is decreased and phosphatase activity increased by an antioxidant and a free radical trap in primary rat glia, *Arch. Biochem. Biophys.*, 365, 211–215, 1999.

68. Robinson, K.A. et al., Redox-sensitive protein phosphatase activity regulates the phosphorylation state of p38 protein kinase in primary astrocyte culture, *J. Neurosci. Res.*, 55, 724–732, 1999.

69. Hensley, K. et al., Reactive oxygen species, cell signaling, and cell injury, *Free Radic. Biol. Med.*, 28, 1456–1462, 2000.

70. Gabbita, S.P. et al., Redox regulatory mechanisms of cellular signal transduction, *Arch. Biochem. Biophys.*, 376, 1–13, 2000.

71. Janssen-Heininger, Y.J.W., Poynter, M.E., and Baeuerle, P.A., Recent advances towards understanding redox mechanisms in the activation of nuclear factor κB, *Free Radic. Biol. Med.*, 28, 1317–1327, 2000.

72. Schreck, R., Rieber, P., and Baeuerle, P.A., Reactive oxygen intermediates as apparently widely used messengers in the activation of the NF-kB transcription factor and HIV-1, *EMBO J.*, 10, 2247–2258, 1991.

73. Meyer, M., Schreck, R., and Baeuerle, P.A., H_2O_2 and antioxidants have opposite effects on activation of NF-kB and AP-1 in intact cells: AP-1 as secondary antioxidant-responsive factor, *EMBO J.*, 12, 2005–2015, 1993.

74. Sundaresan, M. et al., Requirement for generation of H_2O_2 for platelet-derived growth factor signal transduction, *Science*, 270, 296–299, 1995.

75. Guyton, K.Z. et al., Activation of mitogen-activated protein kinase by H_2O_2, *J. Biol. Chem.*, 271, 4138–4142, 1996.

76. Rao, G.N., Hydrogen peroxide induces complex formation of SHC-Grb2-SOS with receptor tyrosine kinase and activates Ras and extracellular signal-regulated protein kinases group of mitogen-activated protein kinases, *Oncogene*, 13, 713–719, 1996.

77. Goldkorn, T. et al., H_2O_2 acts on cellular membranes to generate ceramide signaling and initiate apoptosis in tracheobronchial epithelial cells, *J. Cell. Sci.*, 111, 3209–3220, 1998.

78. Bae, Y.S. et al., Epidermal growth factor (EGF)-induced generation of hydrogen peroxide, *J. Biol. Chem.*, 272, 217–221, 1997.

79. Goldkorn, T. et al., EGF-Receptor phosphorylation and signaling are targeted by H_2O_2 redox stress, *Am. J. Respir. Cell Mol. Biol.*, 19, 786–798, 1998.
80. Denu, J.M. and Tanner, K.G., Specific and reversible inactivation of protein tyrosine phosphatases by hydrogen peroxide: evidence for a sulfenic acid intermediate and implications for redox regulation, *Biochem.*, 37, 5633–5642, 1998.
81. Floyd, R.A. et al., Use of salicylate with high pressure liquid chromatography and electrochemical detection (LCED) as a sensitive measure of hydroyxl free radicals in adriamycin treated rats, *J. Free Radic. Biol. Med.*, 2, 13–18, 1986.

chapter twenty-seven

Biomarkers of male reproductive health

Anne L. Golden

Contents

Abstract Concerns that environmental exposures have been detrimental to male sexual development and fertility have been heightened by reports of declining sperm counts over the past 50 years. Marked geographic variation has been found in semen quality and in the incidence of testicular cancer and certain urogenital defects. Debate continues over the existence, magnitude and significance of these trends, and how best to evaluate the hypothesis that *in utero* and childhood exposures to estrogenic chemicals may be to blame.

   Epidemiologic methods for assessing the impact of hazardous exposures on male reproductive health have been developed mainly in the area of occupational medicine, and this paper will review the tools that are currently available. These include questionnaires to determine reproductive history and sexual function; reproductive hormone profiles; and semen analyses such as sperm concentration, motility, and morphology. Recently developed biomarkers that show significant promise from the fields of clinical reproductive medicine and reproductive toxicology will be discussed as possible additions to epidemiologic studies, including assays of sperm function and genetic integrity, and biomarkers of DNA damage. For population-based studies involving occupational groups or communities with environmental exposures, issues related to the cost, validity, precision, and utility of these methods must be carefully considered.

I. Introduction

In 1977, a striking relationship between duration of occupational exposure to the nematocide 1,2-dibromo-3-chloropropane (DPCP) and diminished fertility was documented among men working in a California pesticide factory.[1] This study by Whorton and colleagues provided unprecedented evidence of the potential for chemically-induced injury to male reproductive health.[1,2] More recently, threats to male fertility have received increased attention following several reports of a decline in sperm counts over the past 50 years in some[3-10] but not all[11-13] populations, and evidence of marked geographic variation in semen quality.[10,14] The incidence of testicular cancer has progressively increased in many countries over the last century and other disorders of the male reproductive tract such as hypospadias and cryptorchidism may have increased in some populations.[15-19] Hypotheses have been put forward that environmental chemical exposures, especially *in utero* and childhood exposures to estrogenic compounds, may be contributing to these observed changes in male reproductive health and fertility.[20,21] Consequently, in recent years an increased emphasis has been placed on determining the frequency and origins of reproductive dysfunction in females and males.[22-26] The significance of possible trends in indicators of reproductive impairment, and the influence of occupational and environmental hazards, are active areas of international research.[27-32]

Reproductive capacity in the male is analogous to fecundability in the female, i.e., the physiologic capacity of an individual to produce a pregnancy, whether or not that capacity has been fulfilled. Because of the multifactorial etiologies of infertility and of adverse pregnancy outcomes in humans, determining the proportion of cases that are attributable to impaired male reproductive capacity is a considerable challenge.[33] Recent estimates suggest that a male factor is present in at least 50% of infertile couples and that 30% may be caused by a pure male factor. Paradoxically, advances in the availability and success of assisted reproductive technologies, such as intracytoplasmic sperm injection (ICSI) for treating male factor infertility, may reduce the chance that a couple will receive a complete infertility work-up to determine the underlying cause(s).[34,35]

Table 27.1 Chemical and Physical Toxicants with Established or Suspected Human Reproductive Effects

Occupation/Environment	
• Anesthetic gases	• Lead[a]
• Benzene	• Manganese
• Cadmium[a]	• Methylmercury[a]
• Carbon disulfide	• Napthyl methylcarmamate (Carbaryl)
• Chlordecone (Kepone)[a]	• Organic solvents
• Cytotoxic drugs[a]	• Polybrominated biphenyls (PBBs)
• DDT	• Polychlorinated biphenyls (PCBs)[a]
• Dibromochloropropane (DBCP)[a]	• Polycyclic aromatic hydrocarbons (PAHs)
• Dioxins	• Vinyl chloride
• Ethylene dibromide	
• Ethylene oxide	**Lifestyle**
• Ethylene glycol ethers[a]	• Alcohol[a]
• Fluoride	• Illicit drugs (cocaine, heroin, marijuana)
• Heat[a]	• Tobacco smoke[a]
• Ionizing radiation[a]	

[a] Established effects.

Many chemical and physical agents found in the workplace and environment are suspected reproductive toxicants, as summarized in Table 27.1.[36,37] However, only four exposures (ionizing radiation, lead, the pesticide DBCP, and the disinfectant and fungicide ethylene oxide) are regulated in the United States by the Occupational Safety and Health Administration (OSHA) in part due to their reproductive effects. The National Institute for Occupational Safety and Health (NIOSH) has designated the reproductive hazards of occupational exposures as a priority area in need of further study under its National Occupational Research Agenda (NORA).[38] But exposure to environmental hazards is not limited to the workplace; other potential sources of exposure include food, air, water, soil, hobbies, and lifestyle choices (Table 27.1). Individuals may have multiple exposures and, in many cases, to chronic low doses of a variety of agents. The reproductive health implications of chronic exposures to reproductive toxicants are not well documented and the mechanisms of toxicity are either poorly understood or unknown.

Researchers and clinicians interested in male reproductive health and fertility are utilizing increasingly sophisticated methodologies from the fields of toxicology, reproductive medicine, environmental and occupational medicine, and epidemiology.[39–41] This chapter aims to review the methods that have contributed to our current understanding of the impact of hazardous substances on male reproductive capacity. The second objective is to describe recently developed biomarkers that may become feasible for use in future studies of environmental and occupational toxicants.

II. Experimental models of male reproductive health

Experimental animal models using the test protocols specified by regulatory agencies play a central role in hazard identification, risk assessment, and regulation of exposures.[42,43] However, animal models can also provide valuable information for reproductive risk assessment on many other fronts. Animal studies can be designed to confirm reproductive toxicity when initial observations in exposed humans or wildlife are suggestive of an adverse effect.[44] Furthermore, such observations can be extended across a wide range of exposures in animals, using any route of exposure and any specified dose versus time scenario.[45,46] For example, when human exposure is likely to be acute or intermittent, animal models are ideal for defining critical exposure windows based on developmental stage or to reveal the pathogenesis of an effect at various times after exposure. This is particularly important with respect to male reproductive effects since alterations in semen quality or fertility may not become evident until some time after the exposure, particularly if the target is an early stage of spermatogenesis.

A rodent model is most commonly used for the study of reproductive and developmental toxicity.[42,43,47] In order to use toxicology data derived from animal studies in human risk assessment, it is critical to identify and understand species-specific differences in physiology and metabolism that may affect the response to the toxicant in question. It is also important to recognize that the genetic homogeneity of rodents, while advantageous in its lack of potential confounding factors, makes it difficult to discern susceptible subpopulations unless different strains are studied. Nevertheless, rodent studies provide valuable information about hazard identification, dose response, and critical thresholds for fertility, and are often helpful for developing paradigms for human studies. Rodent models have, for example, been used to determine the relationship between sperm parameters and fertility.[47,48]

Determining that a substance is toxic to the male reproductive system is only the first step. The next step is to examine its mechanisms of toxicity. Mechanistic information allows for predictions about the potential toxicity of individual compounds or complex mixtures in humans; for better understanding of the windows of vulnerability in the development of the male reproductive system; and for development of possible preventive or curative measures.

Acute short-term exposure models combined with serial exposure models give a complete picture of the range of effects.[47] Exposing animals over a long period of time allows for the detection of transgenerational effects from chemicals, such as male-mediated developmental effects in offspring. If developmental effects appear, researchers can go back and administer a dose during that critical period of development to refine knowledge about how such problems occur. Early developmental endpoints measurable at birth in animal research include litter gender ratio, anogenital distance, testis position, genital malformations, vaginal opening, preputial separation, and serum hormone levels. Acute, short-term exposures, on the other hand, can be useful for identifying critical windows of exposure. Acute exposures followed over time can help identify the pathogenesis of a lesion, isolate the cell type that is susceptible to damage (germ cells, spermatocytes, or spermatid), and determine genetic effects, including the repair capability of affected genes. Serial sacrifice studies are best used for identifying the earliest detectable pathological changes in target organs, cells, or processes. Multi-generational studies, in particular continuous breeding studies, yield the most thorough assessment of the many complex processes that result in reproductive and developmental toxicity.[47,49]

III. Epidemiologic methods of assessing male reproductive health

Epidemiologic investigation of male reproductive health requires the use of several complementary methods ranging from questionnaires that elicit information on reproductive history (e.g., pubertal development, sexual functioning, paternity), as well as urogenital disorders, to sophisticated tests of semen quality and neuroendocrine function. A full battery of evaluations is most realistic when individuals are motivated by infertility or by another clinical reason to seek diagnosis and treatment. For population-based studies involving occupational groups or communities with environmental exposures, issues related to the cost, validity, precision, and acceptability of the available methods are more critical.[22-24]

Table 27.2 provides a list of measures that are most commonly used to assess male reproductive capacity and the reproductive functions that they assess. The choice of appropriate methodologies to employ is predicated by the investigators' understanding of several factors: the nature of the exposed population; the source levels and known routes of exposure; the organ systems in which a toxicant exerts its actions; the hypothesized mechanisms of a toxicant's actions; and the tests available to assess the effects of toxicants in the relevant organ systems. As shown in Table 27.2, the principal targets for male reproductive toxicants are the neuroendocrine system, the testes, the accessory sex glands, and sexual function.[50,51]

There are several characteristics of the male reproductive system that make it simpler to evaluate than the female system. Male reproductive function is not dependent on a cycle, male germ cells can be obtained by the millions, and the male gonads are more accessible for examination to diagnose or rule out abnormalities.[52] In humans, the total duration of spermatogenesis, the process that results in the formation of spermatozoa from stem cells, is approximately 74 days. Therefore, is it feasible to conduct a prospective study with the expectation that recent exposures can be related to current measures of sperm quantity, quality, fertilizing capacity, and germ cell mutations.[53] On the other hand, studying chronic or historical exposures may be more problematic. With the exception of live births, men may be less likely than women to accurately recall prior experiences such as waiting time to pregnancy, adverse pregnancy outcomes, or disorders in their children.[54,55] These are important considerations when the only source of information on these outcomes is self-report, although retrospective data can be improved through confirmation by medical records, by vital records, or by the female partner.

Table 27.2   Assessment of Male Reproductive Capacity

Method of Assessment	Endocrine System	Testes	Post-Testicular Events[a]	Sexual Function
Neuroendocrine hormone levels[b]	✓			
Sperm density (concentration)		✓		
Sperm morphology and morphometry		✓		
Sperm motility (% motile and velocity)		?	✓	
Sperm viability (vital stain and HOS[c])			✓	
Semen volume			✓	
Semen pH			✓	
Marker chemicals from accessory glands			✓	
Sperm function assays[d]		✓	✓	
Sperm chromosome analyses[e]		✓		
Nocturnal penile measurements				✓
Personal reproductive history[f]	✓	✓		✓

[a] Includes production of seminal plasma components and capacitation of sperm in the epididymis, vas deferens, and accessory sex glands.

[b] Includes testosterone (total and bioavailable), sex hormone-binding globulin (SHBG), follicle stimulating hormone (FSH), luteinizing hormone (LH), prolactin, and inhibin-B.

[c] HOS = Hyperosmotic swelling.

[d] Includes acrosome reaction, hemizona assay (HZA) of sperm binding, and sperm penetration assays (SPA).

[e] Includes sperm chromatin stability assay (SCSA), Comet assay, and assessment of chromosomal aneuploidy and nuclear microdeletions.

[f] Includes pubertal development, paternity (time to pregnancy and outcomes), sexual function (erection, ejaculation, orgasm, and libido).

Sources: Schrader, S.M. and Kesner, J.S., in *Occupational and Environmental Reproductive Hazards,* Williams and Wilkins, Baltimore, 1993; Schrader, S.M., in *Handbook of Human Toxicology,* CRC Press, Boca Raton, FL, 1997.

Research studies of male reproductive health should determine each participant's history of urogenital disorders including infections, injuries, or surgeries. When feasible, physical examinations should be done by a trained clinician, to assess overall male habitus and maturation. Routine biochemistry and blood counts should be performed to rule out medical conditions associated with fertility problems, e.g., abnormal renal or liver function.

Although any of the commonly used epidemiological study designs can be used to study male reproductive health (e.g., case-control studies of occupational risk factors for congenital defects or other adverse pregnancy outcomes, and retrospective cohort studies for cancers of the genitourinary tract), the most promising current and future initiatives utilize prospective cohort studies that are designed to test specific exposure-outcome hypotheses.[53] Currently, several ongoing studies of male reproductive health are addressing relationships between hazardous exposures and measures of

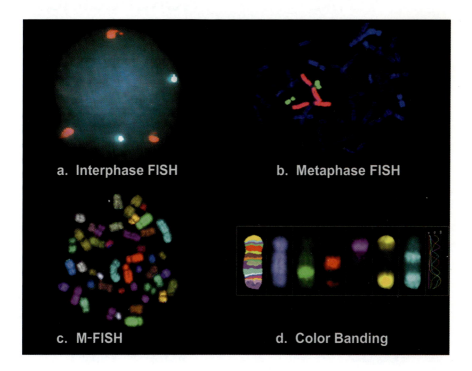

Figure 22.1 Different applications of FISH in human cells: (a) two color interphase FISH shows aneuploidy of chromosome 7 in a human lymphocyte with 3 spots of chromosome 7 (red) and 2 copies of chromosome 8 (green); (b) three color painting of metaphase FISH with chromosomes 1, 16, and 19; (c) M-FISH (multicolor FISH); (d) color banding of chromosome 5. (Images (a) and (b) were obtained in the laboratory of the authors. Images (c) and (d) are from Meta Systems, Inc., Belmont, MA. With permission.)

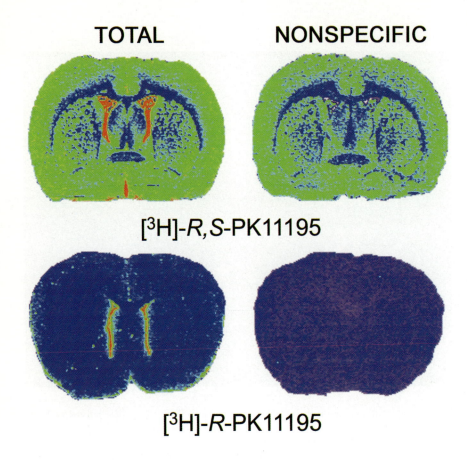

TOTAL　　　　　　**NONSPECIFIC**

[³H]-*R,S*-PK11195

[³H]-*R*-PK11195

Figure 28.1 Comparison of total and nonspecific binding for the racemic (*R,S*) and pharmacologically active (*R*) enantiomer of [³H]-PK11195. Total and nonspecific binding were performed in adjacent coronal rat brain slices at the level of the corpus striatum. Red represents high levels of binding; yellow-green, intermediate levels; and blue, low levels. The pharmacologically active (*R*) enantiomer of [³H]-PK11195 produces a dramatic decrease in nonspecific binding levels.

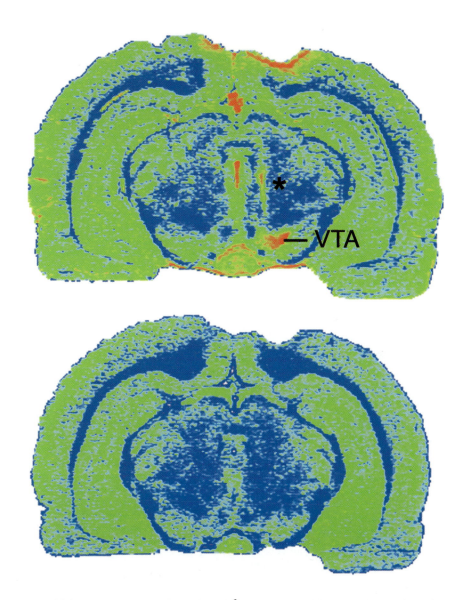

Figure 28.2 Total and nonspecific binding of [³h]-*(R,S)*-PK11195 to coronal rat brain sections at the level of the ventral tegmental area (VTA). In this study, adult rats were stereotactically injected with the dopaminergic neurotoxicant 6-hydrodopamine into the VTA. The dark band to the left of the star (*) represents increased binding of [³H]-*(R, S)*-PK11195 to peripheral benzodiazepine receptors associated with the physical trauma of inserting a cannula guide for stereotactic injection. The dark area pointed by the line represents increased levels of [³H]-*(R, S)*-PK11195 binding resulting from the 6-hydroxydopamine-induced degeneration of dopaminergic neurons at the site of injection. Red represents high levels of binding; yellow-green, intermediate levels; and blue, low levels.

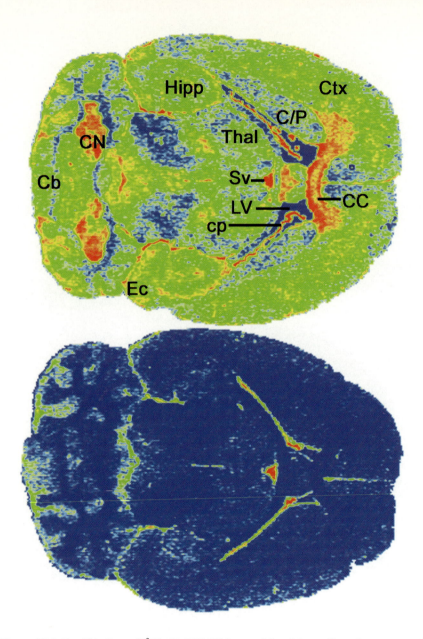

Figure 28.3 Total binding of [³H]-(*R*)-PK11195 to peripheral benzodiazepine receptors in horizontal rat brain sections of cuprizone treated (top) animals (0.2% in diet for 4 weeks) and controls (bottom). Note the high levels of binding in the cuprizone-treated brain in the corpus collosum (CC) and deep cerebellar nuclei (CN). Increased binding was also present in other brain structures such as the hippocampus (Hipp), cerebral cortex (ctx), caudate/putamen (C/P), entorhinal cortex (Ec), and thalamus (Thal). A high level of [³H]-(*R*)- PK11195 to peripheral benzodiazepine receptors is normally found in the choroid plexus (cp) and ventricles (3v, third ventricle). The cerebellum (Cb) is noted as an anatomical landmark. Red represents high levels of binding; yellow-green, intermediate levels; and blue, low levels.

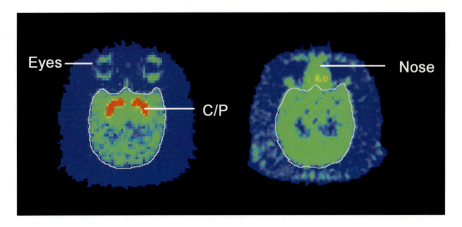

Figure 28.4 PET studies of $[^{11}C]$-(R)-PK11195 to peripheral benzodiazepine receptors (right) and $[^{11}C]$-WIN 35,428 to dopamine transporters (left) in a normal nonhuman primate (baboon) brain (transaxial view). Note the highly concentrated levels of dopamine transporters in the caudate/putamen (C/P). A much more homogeneous distribution of peripheral benzodiazepine receptors in the brain neuropil is observed in the same animal. The brain is delineated by the white boundaries. Red represents high levels of binding; yellow-green, intermediate levels; and blue, low levels.

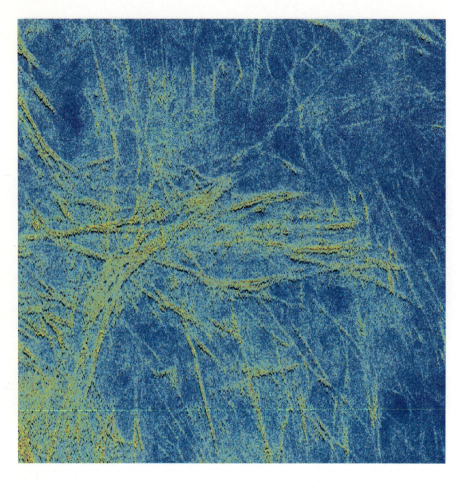

Figure 37.4 A video-rate image of the elastin fiber structures in the dermal layer (Zeiss Fluar 40 oil).

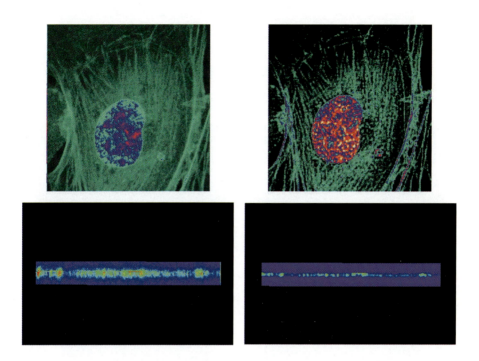

Figure 37.5 Two-photon fluorescence (left) and blind deconvoluted (right) images of a bovine pulmonary endothelial cell (green F-actin, blue nucleus). Top: lateral view; bottom: axial section. (Zeiss Fluar 100 oil).

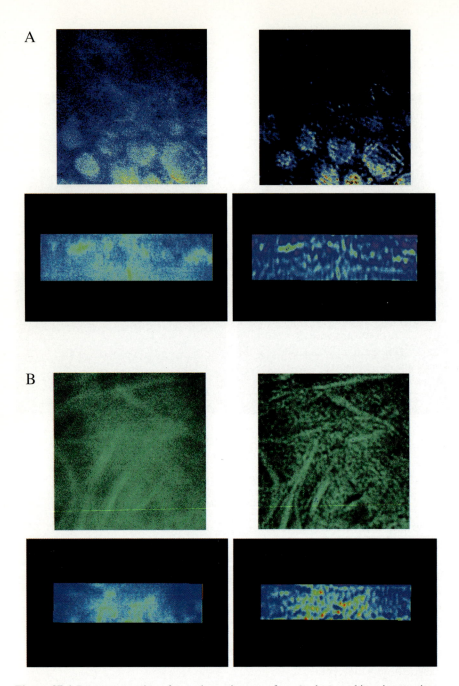

Figure 37.6 Image restoration of two-photon images of *ex vivo* human skin using maximum likelihood approach: (A) auto-fluorescence (left) and blind deconvoluted (right) images of human basal layer. Top: lateral view; bottom, axial section; (B) auto-fluorescence (left) and blind deconvoluted (right) images of human basal layer. Top: lateral view; bottom: axial section. (Zeiss Fluar 40 oil).

semen quality, fertility, and pregnancy outcomes. These investigations have been designed to take into account potential confounders such as geographic variability; factors related to the female partner including abnormal menstrual cycle, previous treatment for infertility and parity; contraception status; and abstinence interval.[30,56,57]

A.   Reproductive history from questionnaires and vital records

The most common method of assessing fertility in epidemiologic studies involves interviewing the man and his current partner about their individual reproductive histories, including prior marriages and sexual partners. These questionnaires elicit information about all previous reproductive events, specifically the number of pregnancies, waiting time to pregnancy, interpregnancy intervals, and pregnancy outcomes for each partner. Clearly, bias can occur if the analysis of these data fails to take into account potential confounding due to time period, age, race, marital status, parity, frequency of intercourse, sterilization, and contraceptive use.[50]

The most common applications of this type of data involve computing indirect epidemiological measures of reduced fertility or increased incidence of adverse events in comparison to a standard population or across a range of exposure levels. These measures include the standardized fertility ratio (SFR) which compares the observed number of live births to the expected number of live births based on person-years of observation.

More recently, the average time to pregnancy or the risk of failure to conceive within one year has been used to study environmental exposures or other risk factors in males and females.[58-62] In the strictest sense, using delayed time to pregnancy captures the probability of nonconception. In truth, subclinical embryonic losses will contribute misclassified delays in time to pregnancy.

Women respondents have remarkable long-term recall of the time to conception of recognized (clinical) pregnancies. A recent study found that retrospective data from a questionnaire conducted after a median duration of 14 years were almost identical to data obtained at the time of pregnancy.[54] Male respondents are able to give values for time to pregnancy, but data collected from men tend to be less complete and may be less reliable.[55,56,63] Among the limitations of this method are that people who have never achieved a pregnancy are excluded from the analysis, as are all unplanned pregnancies.[61] Further, errors in recalling time to pregnancy, although likely to be nondifferential across exposure groups, will lead to underestimation of associations. Potential confounding by each of the factors listed above must also be taken into consideration.[58,64]

Indirect measures of fertility continue to be widely used as functional parameters that are relatively easy to obtain from vital records or by surveys in large populations. Despite their limitations, they have been shown to correlate with biological markers of reproductive capacity.[29,65] They can be used to monitor male-mediated and female-mediated effects as long as potentially confounding characteristics of both partners can be measured and controlled.[63] Data from population-based surveys and from existing medical or vital records can also be used to explore differences and temporal or geographic variation in rates of clinically recognized spontaneous abortion, sentinel phenotypes including congenital defects or cancers in offspring, and various other reproductive outcomes (e.g., offspring gender ratio in an exposed population).[66-68]

B.   Semen analyses

More direct assessment of male reproductive capacity can be accomplished by focusing on the testis, the site of sperm cell production and the target organ for genetic damage. To establish the extent of toxicity to the testis, researchers can obtain a semen sample, measure the size of the testis, or take a testicular biopsy for histopathological examination. Standard semen analyses (including semen volume, sperm concentration, total sperm count, motility, and morphology) have been the primary research tools for studying the effects of toxicants on the male reproductive system. Occupational exposures to lead, DBCP, ethylene dibromide, and glycol ethers have been shown to affect sperm production in humans [1,36,51,69,70] and epidemiologic studies have successfully utilized semen quality as a marker of fertility.[29,65] Studies involving semen parameters are not without problems,[71-74] e.g., potential selection bias due to low compliance rates; inadvertent inclusion of vasectomized men; and substantial intra-individual variability in semen variables resulting in misclassification based on the static results of a single analysis. However, in contrast to longitudinal studies of individuals as well as clinical evaluations where more than one semen sample is required,[34] research has shown that in cross-sectional epidemiological investigations, a single semen sample from each participant generally is sufficient if obtained under defined conditions and according to a set protocol.[72,73,75]

Generally accepted normal ranges for the routine semen parameters established using World Health Organization (WHO) methods[76] and other well established criteria[77] are shown in Table 27.3. In epidemiologic studies conducted to investigate effects of an accidental exposure, it may be difficult to enroll an appropriate unexposed control group. In such cases, results obtained from exposed populations can be interpreted with respect to these reference values.

Only motile sperm are able to penetrate through cervical mucus, migrate through the female reproductive tract, penetrate the zona of the ova, and achieve fertilization.[34] In a routine semen analysis, overall quantitative motility is defined as the percentage of sperm that demonstrate any movement.[34,75] The forward progression of each spermatozoon is qualitatively graded in the following way: a = rapid progressive motility; b = slow or sluggish progressive motility; c= nonprogressive motility; and d = immotility. Motility and forward progression of spermatozoa analyzed visually by a technician is gradually being replaced by computer-assisted sperm analysis (CASA)

Table 27.3 Semen Analysis Reference Ranges for Normal Values

Volume	> 2.0 mL
Appearance	Whitish/Gray-yellow
Agglutination (scale 0 to 3)	0
Liquefaction	Within 30 minutes
Viscosity (scale 0 to 3)	0
pH	7.2 to 7.8
Sperm density	> 20 million/mL
Total count	> 40 million/mL
Motility (@ 37°C)	> 50%
Progressive Motility	> 50%
WHO Morphology	> 50% normal forms
Strict Kruger Morphology	> 14% normal forms
Viability (vital stain)	> 75% alive
Round cells	< 1.0 million/mL
White blood cells (peroxidase positive)	< 1.0 million/mL
Acrosome reaction assay	Delta ≥ 5: Positive

Source: Adapted from Bar-Chama, N. and Lamb, D.J., *Urol. Clin. N. Am.*, 21, 433–446, 1994; Schrader, S.H., in *Handbook of Human Toxicology*, CRC Press, Boca Raton, FL, 1997.

systems. CASA can provide useful information on the pattern and vigor of motion of sperm cells, including curvilinear velocity, straight-line velocity, linearity, and amplitude of lateral head displacement.[51,75]

Over the past 30 years, several classification schemes have been proposed for the morphological assessment of normal and abnormal appearing sperm. Variations in sperm size and shape are not distinct entities but, rather, represent a continuum. This provides a challenge within and especially among laboratories to establish a reliable system for morphological classification.[51] Because debate continues regarding the most valid methodology for morphology assessment[78] and to allow comparison with previous studies of reproductive toxicity which utilized the WHO semen analysis guidelines, some labs score all specimens by the strict Kruger and WHO criteria. New methods have been developed that use transmission and scanning electron microscopy to evaluate the ultrastructural morphology of sperm organelles.[79] Significant ultrastructural abnormalities have been reported among infertile as compared to fertile men[80] and in radiation-exposed salvage workers from Chernobyl.[81] With recent advances in computerized image analysis, methods for objective assessment of sperm head size and shape have been introduced. The andrology laboratory at NIOSH has pioneered a protocol for assessing sperm head morphometry.[51,82] Individual sperm heads are outlined using a digitizing tablet; the software used allows for calculations of area, perimeter, length, width, width/length ratio, and $4\Pi(area)/perimeter^2$ (Pi factor). However, serious impediments remain in achieving agreement among different analysis systems, therefore, comparisons across systems should be avoided.

The viability and motility of spermatozoa typically reflect seminal plasma quality.[51] Alterations in sperm viability, as measured by eosin stain exclusion or by hypo-osmotic swelling,[83] and alterations in sperm motility parameters[76] suggest a problem with the accessory sex glands. The accessory sex glands, which include the epididymis, prostate, and seminal vesicle, may be targeted by toxicants.[51] Ethylene dibromide is one substance that has been shown to affect the accessory sex glands following occupational exposure.[70] Biochemical analysis of seminal plasma provides insights into glandular function by evaluating marker chemicals secreted by each respective gland.[51] For example, the epididymis is represented by glycerylphosphorylcholine (GPC), the seminal vesicles by fructose, and the prostate gland by zinc. Measures of semen pH and volume provide additional general information on the nature of seminal plasma, reflecting posttesticular effects. A toxicant or its metabolite may act directly on accessory sex glands to alter the quantity or quality of their secretions. Alternatively, the toxicant may enter the seminal plasma and affect the sperm or may be carried to the site of fertilization by the sperm and affect the ova or conceptus. The presence of toxicants or their metabolites in seminal plasma can be analyzed using atomic absorption spectrophotometry or gas chromatography/mass spectrometry.

As much as any other factor, uncertainty in the results of studies addressing threats to male reproductive health stems from debate about the definition of normal semen quality and whether or not expected fluctuations in semen parameters are distinguishable from diminished reproductive capacity resulting from hazardous exposures.[73,74,84–86] It has proved difficult to resolve questions about the validity of using routine semen analyses to assess human fertility.[13,27,29] Which semen variables are the most sensitive with respect to perturbation by toxicant exposure and which are the most predictive of human fertility? Can threshold levels associated with impaired fertility be defined? Are shifts in sperm quantity and quality within populations related to measurable decreases in normal live births? Some relevant information about these questions has been provided by animal models.[47] However, because they rely on observations of the group as a whole, they have not been as useful in elucidating individual susceptibility, therefore impeding our ability to apply results from these models to humans.

The uncertainties associated with traditional semen measures have led to the recent development of biomarkers of sperm function and genetic integrity. It is hoped that these assays may prove to be more sensitive and more specific reflections of toxicant-induced effects in individuals.[34,87] However, issues related to the validity, availability, and practicality of the sperm function assays and the tests of genetic integrity described below have limited their integration into epidemiologic studies.[34,52] These concerns include the level of expertise needed to run some of the more complex assays, the time and expense involved in performing the tests, the difficulties establishing standardization and quality control of the assays, and the general doubts about the significance of isolated functional or genetic abnormalities in individual sperm cells.

C. Biomarkers of sperm function

Fertilization requires a series of intricate biochemical events that begins with sperm capacitation, followed by binding to the zona pellucida of the ovum and acrosomal discharge, binding to the oolemma, and finally, penetration into the ooplasm.[34,87] Abnormalities in any of the biochemical reactions by which sperm access and penetrate the ovum may be a source of infertility. There is considerable interest in determining the utility of including assessment of sperm functioning in epidemiologic studies of reproductive toxicity.[52] Although a variety of assays for evaluating sperm function have been developed recently, no single test is capable of evaluating all the steps involved in fertilization.[34] Certain sperm function assays may reflect toxicant effects at more that one site, for example, direct gonadotoxicity affecting spermatogenesis plus posttesticular effects on accessory sex gland secretions. A combination of tests can complement each other in providing a comprehensive evaluation of sperm functions. These include the penetration of sperm through cervical mucus (or viscous fluids simulating cervical mucus), the penetration of sperm into a zona-free hamster egg (sperm penetration assay or SPA), and the binding of sperm to the zona pellucida from a human ovum (hemi-zona assay or HZA). The acrosome reaction has been studied extensively as a predictor of fertilization success because it is a stable parameter of sperm function, which reflects the ability of the spermatozoa to capacitate independently of oocyte quality.[88] The acrosome, a membrane-bound organelle covering the anterior two thirds of the sperm head, contains numerous enzymes whose release is required for penetration of the zona pellucida. It is hypothesized that this release of enzymes is induced by one or more of the zona pellucida glycoproteins. In men with otherwise normal semen analyses, it has been shown that the failure of a significant proportion of sperm to undergo the acrosome reaction when appropriately stimulated is associated with lower *in vitro* fertilization (IVF) rates.[88] In addition, the acrosome reaction has a higher predictive value than standard semen analyses, including standard sperm head morphology.[88] Normal acrosomal morphology has been correlated with IVF success independently of acrosome activity and normal sperm head morphology.[89]

D. Biomarkers of genetic integrity or damage

Epidemiologic studies of large populations have demonstrated increased risk of certain congenital anomalies associated with various paternal occupations.[90] Given the low background population frequency of even the most common anomalies, such studies require a population base of thousands of pregnancies in order to have a reasonable probability of detecting an increased risk. Numerous case-control studies of childhood cancers have found significant associations with paternal occupations and exposures,[91] but between-study variation with respect to case populations, control groups, and methods of data collection and analysis makes it difficult to interpret the findings.

There has been considerable interest in developing more direct methods for use in epidemiologic studies to detect genetic damage in human germ cells that result from exposure to toxic agents.[92–94] If genomic abnormalities are found, they may help to explain some of the subfertility, the increased risk for spontaneous abortions, and the congenital anomalies and childhood cancers in offspring associated with exposure to particular toxicants.

Toxicants may cause genetic damage in various capacities, including disruption of the meiotic chromosome segregation (aneuploidy), fragmentation of the DNA, individual genetic mutations, disruption of the DNA structure (chromatin integrity), and production of DNA adducts. Assessing various parameters in human germ cells is important for understanding whether a particular genetic alteration can affect the next generation and to ascertain the level of toxicant exposures associated with specific germ cell end points.

Chromosomal abnormalities occur relatively frequently in human germ cells and are primarily of two types: numerical and structural. Karyotyping studies have shown that while oocytes demonstrate a higher frequency of numerical chromosomal abnormalities, human sperm demonstrate a higher frequency of structural abnormalities with numerical abnormalities being less frequent.[35] Approximately 1–2% of human sperm have an abnormal number of chromosomes, and an estimated 10% carry structural chromosome aberrations.[95,96]

Efficient technology for examining chromosomal abnormalities in sperm has only been developed recently, thereby increasing the availability of several promising biomarkers of genetic integrity. Each of the available methods has advantages and disadvantages that must be considered.[94] Though certain techniques (e.g., fluorescence *in situ* hybridization) may be more rapid, inexpensive, and technically straightforward, other more technically demanding and time-consuming assays (e.g., sperm karyotyping) can provide precise and detailed information.

The Comet assay detects genetic fragmentation by depicting DNA migrating out of the cell nucleus during electrophoresis.[97] Cells with undamaged DNA appear as intact heads without tails after specified electrophoresis times. DNA that has been fragmented will contain numerous strand breaks and will, therefore, migrate further than normal intact DNA. When visualized microscopically, the migrating DNA resembles the tail of a comet. After staining with ethidium bromide, the migrating DNA is quantified by measuring the intensity and extent of the fluorescence pattern. Increasing the duration of electrophoresis may enable detection of extremely low levels of DNA fragmentation. In addition, measuring the fluorescent intensity following DNA migration provides quantitative geometric measurements of the area and density of the dispersion DNA damage. The Comet assay has the significant advantage of being able to assess DNA fragmentation in individual cells.

The Comet assay has been used to measure DNA damage in individual human lymphocytes using relatively low doses of ionizing radiation and chemical genotoxins.[97] Refinements to the methods are being made to extend its applicability to other environmental exposures and lifestyle factors such as smoking[98] and other target tissues including sperm cells.[99–102] A small study of human semen samples, for example, detected DNA fragmentation using the Comet assay with DBCP, two estrogens (β-estradiol and the phytoestrogen daidzein) and 1,2-epoxybutene (a metabolite of 1,3-butadiene).[102]

Aneuploidy, i.e., abnormal chromosome number, can be detected using fluorescence *in situ* hybridization (FISH) with chromosome-specific DNA probes.[103] Multiple probes can be employed to evaluate numerous chromosomes in a single cell.[104] The chromosomes usually evaluated when assessing sperm with FISH are the sex chromosomes as these appear to be most at risk for nondisjunction, suggesting that there is a chromosome-specific variation in nondisjunction frequencies.[105] FISH has been used to assess factors that may induce sperm aneuploidy in humans such as advanced maternal and paternal age, cancer chemotherapy, and radiation.[106–108] The techniques show promise for assessing lifestyle factors like tobacco, caffeine, and alcohol[109,110] as well as environmental exposures including seasonal air pollution, pesticides, and heavy metals.[111–114]

Following completion of spermatogenesis, sperm undergo extensive differentiation and maturation. During spermiogenesis, DNA is tightly compacted and complexed to protamines. This chromatin structure is important for protection of the DNA and has a significant role in early human postfertilization events and embryo development. Flow cytometric techniques have been developed to evaluate the chromatin structure of sperm in order to correlate the findings to fertility and as a biomarker of exposure to reproductive toxicants. The sperm chromatin structure assay (SCSA) measures the resistance of sperm DNA to *in situ* denaturation (separating double stranded DNA into single strands) under thermal or chemical stress.[115] SCSA assesses flow cytogram-generated staining patterns, measuring a shift from green (native DNA) to red (denatured DNA) fluorescence in properly stained sperm chromatin. This shift is seen under conditions of stress to the sperm, such as low pH, and has been shown to correlate with toxic chemical exposures, drug exposures, and diseases.[116–120] Limited numbers of studies in animal (bovine) and human semen have suggested a relationship between sperm chromatin structure and fertility.[121–123] Occupational exposure to lead[124] and seasonal exposure to air pollution[56] have been associated with decreased sperm chromatin stability.

Unlike the cytogenetic assays (Comet and FISH), individual sperm are not evaluated with SCSA but with thousands of cells providing a representation of the whole ejaculate. There is a low variability of SCSA within individuals, with intraclass coefficients ranging between 67 and 90 in healthy volunteers.[115] Evaluation of semen samples collected in the same individual from two separate months showed highly repeatable results. This assay appears to be sensitive to early stages of chromatin alterations and is a potentially important method to assay for early events of toxicant-induced chromosome damage.

The cytogenetic and chromatin structure assays provide independent assessments of sperm quality that may or may not correlate well with other semen parameters. In infertility clinic populations, sperm density, total count, and morphology have shown low to moderate correlation with SCSA values.[115] However, in many studies in animals and humans, poor quality sperm chromatin structure was highly indicative of male subfertility.[123] Therefore, these assays may be of use in subfertile men who otherwise have normal semen parameters.[56]

Efforts are also being made to develop biochemical markers of sperm DNA damage. Reactive oxygen species (ROS) are a group of potentially destructive molecules implicated in the oxidative damage of biological structures. These ROS, including the superoxide anion, the hydroxyl radical, and hydrogen peroxide, may either be produced endogenously through cellular pathways of the mitochondria and lysosome or induced exogenously in reaction to environmental assaults. Ultraviolet and x-ray radiation and oxidatively reactive compounds, such as those found in cigarette smoke, alcohol, and air pollution, have all been shown to induce the formation of harmful ROS.[125] Over the past decade, concern has been raised after numerous studies reported the reactivity of ROS to DNA nucleotides and suggested the potential for ROS to generate genetic mutations that may evolve to cancer or birth defects if germ cells are damaged.[126,127] 8-hydroxydeoxyguanosine (8-OHdG) is one of many products of oxidative DNA damage and is currently used to evaluate ROS damage, including the analysis of 8-OHdG levels in the semen. DNA is enzymatically digested to excise damaged nucleosides and then analyzed with high-performance liquid chromatography (HPLC) to quantify 8-OHdG levels. Shen et al. used this assay to correlate a dramatic increase of 8-OHdG in semen of men exposed to cigarette smoke.[127] In addition to damaging DNA, ROS are also known to oxidize cellular membrane fatty acid components. Sperm cells are especially sensitive to this lipid peroxidation because of the increased density of unsaturated fatty acids needed for sperm membrane fluidity, and the decreased intracellular space available in sperm heads for antioxidant protection.[128] Oxidatively damaged lipids remain in the plasma membrane and may be assayed through a biochemical technique that converts lipid peroxides into detectable malondialdehyde.[129] These assays allow analysis and characterization of oxidative stressors that may affect male reproductive ability and may help determine effective preventative antioxidant interventions, if indicated.[130]

DNA adducts are the complexes formed between a toxicant or its metabolites and DNA. The presence of these adducts affects DNA synthesis and repair, and may induce genetic mutations that cause cancer or other adverse outcomes.[131] A common source of adduct-generating compounds that has been widely studied is polycyclic aromatic hydrocarbons (PAHs).[132] PAHs are environmentally ubiquitous compounds formed from the industrial manufacture and combustion of organic compounds found in coal, tars, petroleum oils, and cigarettes. PAHs do not form DNA adducts innately but must first be metabolically activated by cellular p450-dependent monooxygenases. Arene oxides, quinones, diol epoxides, and other PAH metabolites have all been discovered to form DNA adducts. Adducts can be detected through ^{32}P-radiolabeling, but HPLC seems to be more effective at determining the presence of DNA adducts.

In summary, biomarkers of genetic damage are increasingly used in the search to understand adverse reproductive health outcomes. These assays provide promising and sensitive approaches for investigating germinal and potentially heritable effects of exposures to toxicants, for identifying genetic polymorphisms which make an individual more susceptible to adverse reproductive effects (i.e., gene-environment interactions), and for confirming epidemiologic observations on smaller numbers of individuals.

E. Reproductive hormone profiles

The profile recommended by NIOSH to evaluate endocrine dysfunction associated with reproductive toxicity consists of serum concentrations of follicle-stimulating hormone (FSH), luteinizing hormone (LH), testosterone, and prolactin. Because of the pulsatile secretion of LH, testosterone and to a much lesser extent FSH, as well as the variability in the evaluation of reproductive hormones, it is recommended that three blood samples be drawn at set intervals in the early morning and the results pooled or averaged for clinical assessment.[133,134] In epidemiologic field studies, however, multiple blood samples are impractical and may significantly decrease participation rates.[135] Schrader et al. determined serum concentration of FSH, LH, testosterone, and prolactin as part of a longitudinal study of workers.[135] They assessed the reliability of the measurements over time and compared the results from a single sample to the average from 3 blood samples drawn 20 min apart; samples were drawn between 8 a.m. and 8 p.m. on 3 occasions 3 months apart. The precision for these hormone measurements was similar although there was some decrease in the intraclass correlation coefficient for LH and prolactin. The measurements from samples drawn 20 minutes apart were highly correlated, and the major sources of variation occurred across individuals and over time (samples drawn three months apart); therefore, multiple measurements at short intervals on the same day do not increase precision. Alternatively, LH and FSH can be measured in urine, providing indices of gonadotropin levels that are relatively unaffected by pulsatile secretion. However, if an exposure can affect hepatic metabolism of sex steroid hormones,[136] urinary measures of excreted testosterone metabolite (androsterone) or estradiol metabolite (estrone-3-glucuronide) are not recommended. There are currently no assays available to measure prolactin in urine.

Future assessment of reproductive hormones may extend to inhibin, activin and follistatin, polypeptides that are secreted primarily by the gonads and act on the pituitary to increase (activin) or decrease (inhibin and follistatin) FSH synthesis and secretion. Within the gonads, these peptides regulate steroid hormone synthesis and may also directly affect spermatogenesis. Ongoing studies are investigating the utility of serum inhibin-B level as an important marker of Sertoli cell function and *in utero* developmental toxicity.[137,138]

F.   Sexual function

Sexual function is attained through the integrated activities of the testes, the accessory sex glands, the endocrine control systems, and the neurological, behavioral, and psychological components of reproduction that are controlled by the central nervous system.[51] Assessments of libido, erection, ejaculation, and orgasm are difficult to make under normal conditions; therefore, detecting decrements associated with exposure to hazardous agents is challenging. Few objective measures have been developed for these outcomes, and sexual dysfunction is often attributed to and affected by psychological or physiological factors.[139] There is little literature on occupational exposures causing sexual dysfunction in men, however, there are suggestions that lead, carbon disulfide, stilbene, and cadmium can affect sexual function.[51] Therefore, ongoing research is underway on the utility of objective means of evaluating sexual function, for example, monitors that quantify the frequency and quality of nocturnal erections.[51]

IV. Conclusion

The determinants of the various endpoints of male reproductive health and dysfunction — including infertility, sexual dysfunction, sperm characteristics, adverse pregnancy outcomes, and reproductive system anomalies and cancers — are poorly understood. There is limited information about the mechanisms and the impact on male reproductive function of specific environmental and occupational exposures.[39] It is therefore important to identify the following: the exposures with the highest priority for study, the appropriate biomarkers of exposure and effect, and the effective means of preventing or controlling exposures to reproductive toxicants.[37] There is growing evidence that the male reproductive system can be influenced by exposure to hazardous substances during development; therefore, biomarkers that reflect exposures during pre-conception periods should be considered in human and *in vivo* experimental animal studies.[140] It is important that researchers consider the added value of integrating some of the biomarkers outlined in this paper, including tests of sperm function and genetic damage, into future epidemiologic studies of the impact of environmental and occupational exposures on male reproductive health.

References

1. Whorton, D. et al., Infertility in male pesticide workers, *Lancet*, 1259–1261, December 17, 1977.
2. Milby, T.H. and Whorton, D., Epidemiological assessment of occupationally related, chemically induced sperm count suppression, *JOM*, 22, 77–82, 1980.
3. Auger, J. et al., Decline in sperm quality among fertile men in Paris during the past 20 years, *New Engl. J. Med.*, 332, 281–285, 1995.
4. Bostofte, E., Serup, J., and Rebbe, H., Has the fertility of Danish men declined through the years in terms of semen quality? A comparison of semen qualities between 1952 and 1972, *Int. J. Fertil.*, 28, 91–95, 1983.
5. Bendvold, E., Semen quality in Norwegian men over a 20-year period, *Int. J. Fertil.*, 34, 401–404, 1989.
6. Carlson, E. et al., Evidence for decreasing quality of semen during past 50 years, *BMJ*, 305, 609–613, 1992.
7. Colborn, T., Dumanoski, D., and Myers, J.P., *Our Stolen Future*, Dutton Press, New York, 1996.
8. Irvine, D.S., Falling sperm quality, *BMJ*, 309, 476, 1994.
9. James, W.H., Secular trend in reported sperm count, *Andrologia*, 12, 381–388, 1980.
10. Swan, S.H., Elkin, E.P., and Fenster, L., Have sperm densities declined? A reanalysis of global trend data, *Environ. Health Perspect.*, 105, 1228–1232, 1997.
11. Fisch, H. et al., Semen analyses in 1,283 men from the United States over a 25-year period: no decline in quality, *Fertil. Steril.*, 65, 1009–1020, 1996.
12. Berling, S. and Wolner-Hanssen, P., No evidence of deteriorating semen quality among men in infertile relationships during the last decade: a study of males from Southern Sweden, *Hum. Reprod.*, 12, 1002–1005, 1997.
13. Rasmussen, P.E. et al., No evidence for decreasing semen quality in four birth cohorts of 1,055 Danish men born between 1950 and 1970, *Fertil. Steril.*, 68, 1059–1064, 1997.
14. Fisch, H., Ikeguchi, E.F., and Goluboff, E.T., Worldwide variations in sperm counts, *Urology*, 48, 909–911, 1996.
15. Bergstrom, R. et al., Increase in testicular cancer incidence in six European countries: a birth cohort phenomenon, *J. Natl. Cancer Inst.*, 88, 727–733, 1996.
16. Giwercman, A. et al., Evidence for increasing incidence of abnormalities of the human testis: a review, *Environ. Health Perspect.*, 101 (Suppl 2), 65–71, 1993.
17. Paulozzi, L.J., Erickson, D., and Jackson, R.J., Hypospadias trends in two U.S. surveillance systems, *Pediatrics*, 10, 831–834, 1997.
18. Paulozzi, L.J., International trends in rates of hypospadias and cryptorchidism, *Environ. Health Perspect.*, 107, 297–302, 1999.
19. Aho, M. et al., Is the incidence of hypospadias increasing? Analysis of Finnish hospital discharge data 1970–1994, *Environ. Health Perspect.*, 108, 463–465, 2000.
20. Sharpe, R.M. and Skakkebaek, N.E., Are oestrogens involved in falling sperm counts and disorders of the male reproductive tract? *Lancet*, 341, 1392–1395, 1993.
21. Toppari, J. et al., Male reproductive health and environmental xenosetrogens, *Environ. Health Perspect.*, 104 (Suppl 4), 741–803, 1996.
22. Savitiz, D.A. et al., Assessment of reproductive hazards and birth defects in communities near hazardous chemical sites, Part I: birth defects and developmental disorders, *Reprod. Toxicol.*, 11, 223–230, 1997.

23. Scialli, A.R. et al., Assessment of reproductive hazards and birth defects in communities near hazardous chemical sites, Part II: female reproductive disorders, *Reprod. Toxicol.*, 11, 231–242, 1997.

24. Wyrobek, A.J. et al., Assessment of reproductive hazards and birth defects in communities near hazardous chemical sites, III: guidelines for field studies of male reproductive disorders. *Reprod. Toxicol.*, 11, 243–259, 1997.

25. Olshan, A.F. and Mattison, D.R., Eds., *Male-Mediated Developmental Toxicity*, Plenum Press, New York, 1994.

26. Paul, M., Ed., *Occupational and Environmental Reproductive Hazards: A Guide for Clinicians*, Williams and Wilkens, Baltimore, 1993.

27. Lipshcultz, L.I., The debate continues: The continuing debate over the possible decline in semen quality, *Fertil. Steril.*, 65, 909–911, 1996.

28. Swan, S.H. and Elkin, E.P., Declining sperm quality: can the past inform the present? *Bioessays*, 21, 614–621, 1999.

29. Larsen, L. et al., Computer assisted semen analysis parameters as predictors for fertility of men from the general population: The Danish First Pregnancy Planner Study Team, *Hum. Reprod.*, 15, 1562–1567, 2000.

30. Bonde, J.P. et al., The European Asclepios study on occupational hazards to male reproductive capability: objectives, designs and populations, *Scand. J. Work Environ. Health*, 25 (Suppl 1), 49–61, 1999.

31. Mbizvo, M., Sentinel surveillance on waiting time to pregnancy and semen quality: a multicentre collaborative study, abstract presented at the Hazardous Substances and Male Reproductive Health Conference, New York, 1998.

32. Swan, S.H., Is semen quality influenced by environmental factors? What can be learned from international collaboration? abstract presented at the Hazardous Substances and Male Reproductive Health Conference, New York, 1998.

33. Westhoff, C., The epidemiology of infertility, in *Reproductive and Perinatal Epidemiology*, Kiely, M., Ed., CRC Press, Boca Raton, FL, 1991, pp. 43–61.

34. Bar-Chama, N. and Lamb, D.J., Evaluation of sperm function: what is available in the modern andrology laboratory, *Urol. Clin. N. Am.*, 21, 433–446, 1994.

35. Moosani, N. et al., Chromosomal analysis of sperm from men with idiopathic infertility using sperm karyotyping and fluorescence *in situ* hybridization, *Fertil. Steril.*, 64, 811–817, 1995.

36. Tas, S., Lauwerys, R., and Lison, D., Occupational hazards for the male reproductive system, *Crit. Rev. Toxicol.*, 26, 261–307, 1996.

37. Moorman, W.J. et al., Prioritization of NTP reproductive toxicants for field studies, *Reprod. Toxicol.*, 14, 293–301, 2000.

38. NIOSH, The National Occupational Research Agenda, NIOSH 96-115, Cincinnati, National Institute for Occupational Safety and Health, 1996.

39. Moline, J.M. et al., Exposure to hazardous substances and male reproductive health: a research framework, *Environ. Health Perspect.*, 108, 803–813, 2000.

40. Golden, A.L., Moline, J.M., and Bar-Chama, N., Male reproduction and environmental and occupational exposures: a review of epidemiologic methods, *Salud. Publica Mex.*, 41 (Suppl 2), S93–S105, 1999.

41. Schrader, S.M. et al., Laboratory methods for assessing human semen in epidemiologic studies: a consensus report, *Reprod. Toxicol.*, 6, 275–291, 1992.

42. Claudio, L., Bearer, C.F., and Wallinga, D., Assessment of the U.S. Environmental Protection Agency methods for identification of hazards to developing organisms, Part I: the reproduction and fertility testing guidelines, *Am. J. Ind. Med.*, 35, 543–553, 1999.

43. Claudio, L., Bearer, C.F., and Wallinga, D., Assessment of the U.S. Environmental Protection Agency methods for identification of hazards to developing organisms, Part I: the developmental toxicity testing guideline, *Am. J. Ind. Med.*, 35, 554–563, 1999.
44. Gray, L.E. et al., The value of mechanistic studies in laboratory animals for the prediction of reproductive effects in wildlife: endocrine effects on mammalian sexual differentiation, *Environ. Toxicol. Chem.*, 17, 109–118, 1998.
45. Ulbrich, B. and Palmer, A.K., Detection of effects on male reproduction: a literature survey, *J. Am. Coll. Toxicol.*, 14, 293–327, 1995.
46. Slott, V.L. et al., Synchronous assessment of sperm motility and fertilizing ability in the hamster following treatment with alpha-chlorohydrin, *J. Androl.*, 16, 523–535, 1995.
47. Chapin, R.E., Sloane, R.A., and Haseman, J.K., The relationships among reproductive endpoints in Swiss mice, using the reproductive assessment by continuous breeding database, *Fundam. Appl. Toxicol.*, 38, 129–142, 1997.
48. Gray, L.E. et al., Correlation of ejaculated sperm numbers with fertility in the rat, *Toxicologist*, 12, 433, 1992.
49. Chapin, R.E. and Sloane, R.A., Reproductive assessment by continuous breeding: evolving study design and summaries of ninety-one studies, *Environ. Health Perspect.*, 105 (Suppl 1), 199–395, 1997.
50. Schrader, S.M. and Kesner, J.S., Male reproductive toxicology, in *Occupational and Environmental Reproductive Hazards. A Guide for Clinicians*, Paul, M., Ed., Williams and Wilkens, Baltimore, 1993, pp. 3–17.
51. Schrader, S.M., Male reproductive toxicity, in *Handbook of Human Toxicology*, Massaro, E.J., Ed., CRC Press, Boca Raton, FL, 1997, pp. 962–980.
52. Scialli, A.R. and Lemasters, G.W., Epidemiologic aspects of reproductive toxicology, in *Reproductive Toxicology*, 2nd ed., Witorsch, R.J., Ed., Raven Press, New York, 1995, pp. 241–263.
53. Bonde, J.P., Giwercman, A., and Ernst, E., Identifying environmental risk to male reproductive function by occupational sperm studies: logistics and design options, *Occup. Environ. Med.*, 53, 511–519, 1996.
54. Joffe, M. et al., A time-to-pregnancy questionnaire designed for long-term recall: validity in Oxford, England, *J. Epidemiol. Comm. Health*, 49, 314–349, 1995.
55. Fikree, F.F., Gray, R.H., and Shah, F., Can men be trusted? a comparison of pregnancy histories reported by husbands and wives, *Am. J. Epidemiol.*, 138, 237–242, 1993.
56. Selevan, S.G. et al., Semen quality and reproductive health of young Czech men exposed to seasonal air pollution, *Environ. Health Perspect.*, 108, 887–894, 2000.
57. Zinaman, M.J., et al., Semen quality and human fertility: A prospective study with healthy couples, *J. Androl.*, 21, 145–153, 2000.
58. Baird, D.D., Wilcox, A.J., and Weinberg, C.R., Use of time to pregnancy to study environmental exposures, *Am. J. Epidemiol.*, 124, 470–480, 1986.
59. Joffe, M., Feasibility of studying subfertility using retrospective self reports, *J. Epidemiol. Comm. Health*, 43, 268–274, 1989.
60. Ford, W.C.L. et al., Increasing paternal age is associated with delayed conception in a large population of fertile couples: Evidence for declining fecundity in older men, The ALSPAC Study Team, Avon Longitudinal Study of Pregnancy and Childhood, *Hum. Reprod.*, 15, 1703–1708, 2000.

61. Jensen, T.K. et al., Selection bias in determining the age dependence of waiting time to pregnancy, *Am. J. Epidemiol.*, 152, 565–572, 2000.
62. Joffe, M. and Barnes, I., Do parental factors affect male and female fertility? *Epidemiology*, 11, 700–705, 2000.
63. Joffe, M. and Li, Z., Male and female factors in fertility, *Am. J. Epidemiol.*, 140, 921–929, 1994.
64. Juul, S., Keiding, N., and Tvede, M., Retrospectively sampled time-to-pregnancy data may make age-decreasing fecundity look increasing. European Infertility and Subfecundity Study Group, *Epidemiology*, 11, 717–719, 2000.
65. Fisch, H. et al., The relationship of sperm counts to birth rates: a population-based study, *J. Urol.*, 157, 840–843, 1997.
66. Davis, D.L., Gottlieb, M.B., and Stampnitzky, J.R., Reduced ratio of male to female births in several industrial countries, *JAMA*, 279, 1018–1023, 1998.
67. Akre, O. et al., Testicular nonseminoma and seminoma in relation to prenatal characteristics, *J. Natl. Cancer Inst.*, 88, 883–889, 1996.
68. Akre, O. et al., Risk factor patterns for cryptorchidism and hypospadias, *Epidemiology*, 10, 364–369, 1999.
69. Whorton, D. and Foliart, D., DBCP: eleven years later, *Reprod. Toxicol.*, 2, 155–161, 1988.
70. Schrader, S.M., Turner, T.W., and Ratcliffe, J.M., The effects of ethylene dibromide on semen quality: A comparison of short-term and chronic exposure, *Reprod. Toxicol.*, 2, 191–198, 1988.
71. Selevan, S.G., Epidemiology, in *Occupational and Environmental Reproductive Hazards. A Guide for Clinicians*, Paul, M., Ed., Williams and Wilkens, Baltimore, 1993, pp. 100–110.
72. Schrader, S.M. et al., Longitudinal study of semen quality of unexposed workers, Part I: study overview, *Reprod. Toxicol.*, 2, 183–190, 1988.
73. Schenker, M.B. et al., Prospective surveillance of semen quality in the workplace, *J. Occup. Med.*, 30, 336–344, 1988.
74. Larsen, S.B., Abell, A., and Bonde, J.P., Selection bias in occupational sperm studies, *Am. J. Epidemiol.*, 147, 681–685, 1998.
75. Schrader, S.M., Turner, T.W., and Simon, S.D., Longitudinal study of semen quality of unexposed workers: Sperm motility characteristics, *J. Androl.*, 12, 126–131, 1991.
76. World Health Organization, *WHO Laboratory Manual for the Examination of Human Sperm and Semen-Cervical Mucus Interaction*, 3rd ed., The Press Syndicate of the University of Cambridge, Cambridge, 1992.
77. Kruger, T.F. et al., Sperm morphologic features as a prognostic factor in *in vitro* fertilization, *Fertil. Steril.*, 46, 1118–1130, 1986.
78. Morgentaler, A. et al., Sperm morphology and *in vitro* fertilization outcome: a direct comparison of World Health Organization and strict criteria methodologies, *Fertil. Steril.*, 64, 1177–1182, 1995.
79. Bartoov, B. et al., Ultrastructural studies in morphological assessment of human spermatozoa, *Int. J. Androl.*, 5, 81–95, 1982.
80. Lipitz, S. et al., Sperm head ultramorphology and chromatin stability of males with unexplained infertility who fail to fertilize normal human ova *in vitro*, *Andrologia*, 24, 261–269, 1992.
81. Fischbein, A. et al., Ultramorphological sperm characteristics in the risk assessment of health effects after radiation exposure among salvage workers in Chernobyl, *Environ. Health Perspect.*, 105 (Suppl 6), 1445–1449, 1997.

82. Schrader, S.M., Turner, T.W., and Simon, S.D., Longitudinal study of semen quality of unexposed workers: sperm head morphometry, *J. Androl.*, 11, 32–39, 1990.

83. Schrader, S.M. et al., Sperm viability: A comparison of analytical methods, *Andrologia*, 18, 530–538, 1986.

84. Bigelow, P.L. et al., Association of semen quality and occupational factors: comparison of case-control analysis and analysis of continuous variables, *Fertil. Steril.*, 69, 11–18, 1998.

85. Levine, R., Methods for detecting occupational causes of male infertility, *Scand. J. Work Environ. Health*, 9, 371–376, 1983.

86. Hatch, M. and Marcus, M., Occupational exposures and reproduction, in *Reproductive and Perinatal Epidemiology*, Kiely, M., Ed., CRC Press, Boca Raton, FL, 1993, pp. 131–142.

87. Green, D.P., Mammalian fertilization as a biological machine: a working model for adhesion and fusion of sperm and oocyte, *Hum. Reprod.*, 8, 91–96, 1993.

88. Calvo, L. et al., Acrosome reaction inducibility predicts fertilization success at *in vitro* fertilization, *Hum. Reprod.*, 9, 1880–1886, 1994.

89. Menkveld, R. et al., Acrosomal morphology as a novel criterion for male fertility diagnosis: Relation with acrosin activity, morphology (strict criteria), and fertilization *in vitro*, *Fertil. Steril.*, 65, 637–644, 1996.

90. Olshan, A.F. and Schnitzer, P.G., Paternal occupation and birth defects, in *Male-Mediated Developmental Toxicity*, Olshan, A.F. and Mattison, D.R., Eds., Plenum Press, New York, 1994, pp. 153–167.

91. Buckley, J., Male-mediated developmental toxicity: paternal exposures and childhood cancer, in *Male-Mediated Developmental Toxicity*, Olshan, A.F. and Mattison, D.R., Eds., Plenum Press, New York, 1994, pp. 169–175.

92. Wyrobek, A.J., Methods and concepts in detecting abnormal reproductive outcomes of paternal origin, *Reprod. Toxicol.*, 7, 3–16, 1993.

93. Anderson, D., Factors contributing to biomarker responses in exposed workers, *Mutat. Res.*, 428, 197–202, 1999.

94. Martin, R.H. et al., Analysis of human sperm karyotypes in testicular cancer patients before and after chemotherapy, *Cytogenet. Cell Genet.*, 78, 120–123, 1997.

95. Mikamo, K., Kamiguchi, Y., and Tateno, H., The interspecific *in vitro* fertilization system to measure human sperm chromosomal damage, *Prog. Clin. Biol. Res.*, 372, 531–542, 1991.

96. Martin, R.H., Ko, E., and Rademaker, A., Distribution of aneuploidy in human gametes: comparison between human sperm and oocytes, *Am. J. Med. Genet.*, 39, 321–331, 1991.

97. Singh, N.P. et al., A simple technique for quantization of low levels of DNA damage in individual cells, *Exper. Cell Res.*, 175, 184–191, 1988.

98. Betti, C. et al., Comparative studies by Comet test and SCE analysis in human lymphocytes from 200 healthy people, *Mutat. Res.*, 343, 201–207, 1995.

99. Hughes, C.M. et al., A comparison of baseline and induced DNA damage in human spermatozoa from fertile and infertile men using a modified Comet assay, *Mol. Human Reprod.*, 2, 613–619, 1996.

100. Anderson, D. et al., Somatic and germ cell effects in rats and mice after treatment with 1,3-butadiene and its metabolites 1,2-epoxybutene and 1,2,3,4-diepoxybutane, *Mutat. Res.*, 391, 233–242, 1997.

101. Brinkworth, M.H. et al., Genetic effects of 1,3-butadiene on the mouse testis, *Mutat. Res.*, 397, 67–75, 1998.
102. Anderson, D. et al., DNA integrity in human sperm, *Teratog. Carcinog. Mutagen.*, 17, 97–102, 1997.
103. Martin, R.H., Spriggs, E., and Rademaker, A.W., Multicolour fluorescence *in situ* hybridization analysis of aneuploidy and diploidy frequencies in 225,846 sperm from ten normal men, *Biol. Reprod.*, 54, 394–398, 1996.
104. Martini, E. et al., Analysis of unfertilized oocytes subjected to intracytoplasmic sperm injection using two rounds of fluorescence *in situ* hybridization and probes to five chromosomes, *Hum. Reprod.*, 12, 2011–2018, 1997.
105. Wyrobek, A.J. et al., Detection of sex chromosomal aneuploidies X-X, Y-Y, and X-Y in human sperm using two-chromosome fluorescence *in situ* hybridization, *Am. J. Med. Genet.*, 53, 1–7, 1994.
106. Kinakin, B., Rademaker, A., and Martin, R., Paternal age effect of YY aneuploidy in human sperm, as assessed by fluorescence *in situ* hybridization, *Cytogenet. Cell. Genet.*, 78, 116–119, 1997.
107. Martin, R.H. et al., The relationship between paternal age, sex ratios, and aneuploidy frequencies in human sperm, as assessed by multicolor FISH, *Am. J. Med. Genet.*, 57, 1393–1399, 1995.
108. Robbins, W.A. et al., Chemotherapy induces transient sex chromosomal and autosomal aneuploidy in human sperm, *Nat. Genet.*, 16, 74–78, 1997.
109. Robbins, W.A. et al., Use of fluorescence *in situ* hybridization (FISH) to assess the effects of smoking, caffeine, and alcohol on aneuploidy load in healthy men, *Environ. Mol. Mutagen.*, 30, 175–183, 1997.
110. Rubes, J. et al., Smoking cigarettes is associated with increased sperm disomy in teenage men, *Fertil. Steril.*, 70, 715–723, 1998.
111. Ribas, G. et al., Herbicide-induced DNA damage in human lymphocytes evaluated by the single cell gel electrophoresis (SCGE) assay, *Mutat. Res.*, 344, 41–54, 1995.
112. Marchetti, F. et al., Paternally inheritied chromosomal structure aberrations detected in mouse first-cleaving zygote metaphases by multicolour fluorescence *in situ* hybridization painting, *Chromosome Res.*, 4, 604–613, 1996.
113. Rupa, D.S., Schuler, M., and Eastmond, D.A., Detection of hyperploidy and breakage affecting the 1cen-1q12 region of cultured interphase human lymphocytes treated with various genotoxic agents, *Environ. Mol. Mutagen.*, 29, 161–167, 1997.
114. Robbins, W.A. et al., Air pollution and sperm aneuploidy in healthy young men, *Environ. Epidemiol. Toxicol.*, 1, 125–131, 1999.
115. Evenson, D.P. et al., Individuality of DNA denaturation patterns in human sperm as measured by the sperm chromatin structure assay, *Reprod. Toxicol.*, 5, 115–125, 1991.
116. Evenson, D.P., Flow cytometric analysis of male germ cell quality, in *Methods in Cell Biology,* Crissman, H. and Darzynkiewicz, D., Eds., Vol. 38, Academic Press, San Diego, CA, 1990, pp. 401–410.
117. Evenson, D.P., Baer, R.K., and Jost, L.K., Long-term effects of triethylenemelamine exposure on mouse testis cells and sperm chromatin structure assayed by flow cytometry, *Environ. Mol. Mutagen.*, 14. 79–89, 1989.
118. Evenson, D.P. et al., Toxicity of thiotepa on mouse spermatogenesis as determined by dual-parameter flow cytometry, *Toxicol. Appl. Pharmacol.*, 82, 151–163, 1986.

119. Evenson, D.P. et al., Male reproductive capacity may recover following drug treatment with the L-10 protocol for acute lymphocytic leukemia (ALL), *Cancer*, 53, 30–36, 1984.
120. Evenson, D.P. et al., Male germ cell analysis by flow cytometry: effects of cancer, chemotherapy and other factors on testicular function and sperm chromatin structure, in *Proceedings of the Conference on Clinical Cytometry*, Andreeff, M.A., Ed., New York Academy of Science, New York, 1986, pp. 350–367.
121. Ballachey, B.E., Hohenboken, W.D., and Evenson, D.P., Heterogeneity of sperm nuclear chromatin structure and its relationship to fertility of bulls, *Biol. Reprod.*, 36, 915–925, 1987.
122. Ballachey, B.E., Evenson, D.P., and Saacke, R.G., The sperm chromatin structure assay: relationship with alternate tests of semen quality and heterospermic performance of bulls, *J. Androl.*, 9, 109–115, 1988.
123. Evenson, D.P., Darzynkiewicz, Z., Melamed, M.R., Relations of mammalian sperm chromatin heterogeneity to fertility, *Science* 240, 1131–1133, 1980.
124. Wildt, K., Eliasson, R., and Berlin, M., Effects of occupational exposure to lead on sperm and semen, in *Reproductive and Developmental Toxicity of Metals*, Clarkson, J.W., Nordberg, G.F., Sager, P.R., Eds., Plenum Press, New York, 1983, pp. 279–300.
125. Moller, P., Wallin, H., and Knudsen, L., Oxidative stress associated with exercise, psychological stress, and lifestyle factors, *Chemico-Biological Interactions*, 102, 17–36, 1996.
126. Fraga, C. et al., Ascorbic acid protects against endogenous oxidative DNA damage in human sperm, *Proc. Natl. Acad. Sci. USA*, 88, 11003–11006, 1991.
127. Shen, H.M. et al., Detection of oxidative DNA damage in human sperm and the association with cigarette smoking, *Reproductive Technology*, 11, 675–680, 1997.
128. Aitken, R.J., A free radical theory of male infertility, *Reprod. Fert. Dev.*, 6, 19–24, 1994.
129. Aitken, R.J., Harkiss, D., and Buckingham, D.W., Analysis of lipid peroxidation mechanisms in human spermatozoa, *Mol. Reprod. Dev.*, 35, 302–315, 1993.
130. Geva, E. et al., The effect of antioxidant treatment on human spermatozoa and fertilization rate in an *in vitro* fertilization program, *Fertil. Steril.*, 66, 430–434, 1996.
131. Santella, R.M., DNA damage as an intermediate biomarker in intervention studies, *Proc. Soc. Exp. Biol. Med.*, 216, 166–171, 1997.
132. Angerer, J., Mannschreck, C., and Gündel, J., Biological monitoring and biochemical effect monitoring of exposure to polycyclic aromatic hydrocarbons, *Int. Arch. Occup. Environ. Health*, 70, 365–377, 1997.
133. Santen, R.J. and Bardin, C.W., Episodic luteinizing hormone secretion in man: pulse analysis, clinical interpretation, physiologic mechanisms, *J. Clin. Invest.*, 52, 2617–2628, 1973.
134. Sokol, R.Z., Endocrine evaluations in the assessment of male reproductive hazards, *Reprod. Toxicol.*, 2, 217–222, 1988.
135. Schrader, S.M. et al., Measuring male reproductive hormones for occupational field studies, *JOM*, 35, 574–576, 1993.
136. Apostoli, P. et al., Steroid hormone sulphation in lead workers, *Br. J. Ind. Med.*, 46, 204–208, 1989.
137. Halvorson, L.M. and DeCherney, A.H., Inhibin, activin, and follistatin in reproductive medicine, *Fertil. Steril.*, 65, 459–469, 1996.

138. Jensen, T.K. et al., Inhibin-B as a serum marker of spermatogenesis: correlation to differences in sperm concentration and follicle-stimulating hormone levels. A study of 349 Danish men, *J. Clin. Endocrinol. Metab.*, 82, 4059–4063, 1997.

139. Arujo, A.B. et al., Relation between psychosocial risk factors and incident erectile dysfunction: prospective results from the Massachusetts male aging study, *Am. J. Epidemiol.*, 152, 533–541, 2000.

140. Shepard, T.H., Fantel, A.G., and Mirkes, P.E., Developmental toxicology: prenatal period, in *Occupational and Environmental Reproductive Hazards: A Guide for Clinicians*, Paul, M., Ed., Williams and Wilkens, Baltimore, 1993, pp. 37–51.

chapter twenty-eight

Peripheral benzodiazepine receptors: molecular biomarkers of neurotoxicity

Tomás R. Guilarte

Contents

1-56670-596-7/02/$0.00+$1.50
© 2002 by CRC Press LLC

Abstract   There is a need for the validation and application of biomarkers of neurotoxicity. The use of the peripheral benzodiazepine receptor (PBR), a glial-specific protein, has shown great promise to assess inflammation and neurodegeneration resulting from a number of brain insults. In this chapter, we present work describing its use in the assessment of brain injury following chemical-induced neurotoxicity. The advantage of the PBR as a biomarker of neurotoxicity is that it can visualize brain injury in the living brain using state-of-the-art, noninvasive imaging techniques.

I. Introduction

Human exposure to naturally occurring or anthropogenic chemicals contributes to the incidence of neurological disease. To estimate and minimize the human risk of neurological disease from chemical exposures, it is important to identify whether a chemical, or class of chemicals, produces neurotoxicity, and its regional specificity and mechanism of action. Biomarkers of neurotoxicity permit evaluation of exposure to an agent (dose) and the vulnerability (susceptibility) of specific brain structures and cell populations, such as neurons and glial cells, to damage (effect). Our approach to the development of a biomarker of neurotoxicity focuses on the peripheral benzodiazepine receptor (PBR), a glial-specific protein. The rationale for this strategy is that reactive gliosis is one of the earliest and most widespread response of the central nervous system to any type of injury. Quantification of a widespread response is important when there is a paucity of knowledge about neuronal or glial targets that may be damaged by chemical exposures. The goal of this chapter is to provide information on the validation and application of the PBR as a molecular biomarker of neurotoxicity. The novel aspect of this approach is that the PBR can be used as an *in vitro* screening method and an *in vivo* biomarker to visualize injury to the living brain using noninvasive imaging techniques. The latter is advantageous because of its potential use as a high-throughput screening method, allowing a reduction in the number of experimental animals by assessing the same animal over time. Further, this method can be used in human populations or groups of occupationally exposed individuals to assess brain damage from chemical exposure or from neurodegenerative disease. The validation and application of this molecular biomarker of neurotoxicity will be useful to researchers in academia, industry and regulatory agencies to assess the risk of neurological damage by exposures to specific chemicals, classes of chemicals or combinations of chemical exposures.

II. Need for molecular biomarkers of neurotoxicity

It has been estimated that of the nearly 70,000 man-made chemicals in use today, only a few have been tested for neurotoxicity.[1] The determination of the neurotoxic potential of specific chemicals or classes of chemicals will help to predict and minimize human neurological health risks. The development and validation of a biomarker of neurotoxicity is complicated by the molecular and cellular diversity of the mammalian brain and the multitude of targets that chemicals can interact with to produce damage. The development and use of biomarkers of neurotoxicity are needed to identify brain regions and neuronal populations specifically affected by exposures to environmental neurotoxicants. A chemically induced injury to the nervous system is often characterized by alterations in the expression of neuronal and/or glial markers (proteins) at single or multiple sites. Neuronal markers such as cellular organelles, structural proteins, enzymes, receptors, or transporters, expressed either in specific neuronal populations or common to all neurons, have been used as neurotoxic endpoints.[2,3] A limitation of using neuron-specific proteins as markers of neurotoxicity are the frequent lack of information about the site or population of neurons damaged by a specific chemical. Considering the multitude of neuronal phenotypes, screening of multiple neuronal marker proteins is often required to identify the site of damage.[3] Therefore, this approach makes the screening of even a single chemical an extremely expensive and time-consuming exercise. Further, in some cases the marker protein is not sensitive to subtle damage in which cell loss is not expressed.[4]

To be able to detect a neurotoxic endpoint, it is important that the biomarker not only be sensitive, but be able to measure damage to any brain cell type (i.e., neurons and glial cells). As a result of this important characteristic, our approach to the development of biomarkers of neurotoxicity has focused on glial-specific proteins.

III. Glial proteins as markers of neurotoxicity

The rationale for the use of glial proteins as markers of neurotoxicity is based on the fact that reactive gliosis is a stereotypic response of nervous tissue to injury, regardless of cell type, site of damage, or the means by which damage was produced.[5-8] This becomes most important if there is no *a priori* knowledge about the neurotoxic potential or target for a chemical. The assessment of glial-specific proteins as markers of neurotoxicity is not a new idea, and it has been explored by a number of investigators.[3,6-8] The most widely used and accepted approach has been to examine the levels of the astrocyte-specific protein, glial fibrillary acidic protein (GFAP). A large body of experimental evidence indicates that the initial response of the brain to damage (i.e., neurons and glial cells) is the proliferation and migration of microglia and the hypertrophy of astrocytes.[5,6,9] This response has been termed reactive gliosis. Damage to all neuronal and glial cell types examined to date has been shown to elicit reactive gliosis.[10] Reactive gliosis has been assessed by the staining or quantitative analysis of the astrocytic-localized protein denoted glial fibrillary acidic protein (GFAP).[4,11] Increased levels of GFAP have been measured after a number of CNS insults, ranging from physical damage to the damage resulting from the exposure to known environmental toxicants such as trimethyltin (TMT).[2,12] Recently, the assessment of GFAP levels in brain has been proposed as a biomarker for screening the adverse effects of chemicals on the central nervous system.[13,14] Microglia-specific markers have also been used to assess microgliosis by immunohistochemical methods.[15-18] In fact, it appears that microglia respond almost immediately following a brain insult and prior to astrocytic activation.[18-20] However, the microglial response appears to be short-lived, while the astrocytic response is longer lasting. Therefore, it would be important to assess microglia- and astrocytic-specific markers to assess brain damage.

Current methods available to assess reactive gliosis represent important advances in the development of biomarkers for detecting chemically induced or other types of brain damage; however, they have limitations. For example, the visualization of GFAP staining in brain tissue by immunohistochemistry is not quantitative, and it is unlikely that GFAP-immunohistochemistry cell counting will be of value in the quantitative assessment of neurotoxic insult. This is based on the fact that a greater amount of GFAP per astrocyte, rather than a greater number of astrocytes, appears to be the predominant response to chemically induced damage.[6,21,22] To overcome this limitation, an immunoassay has been developed in order to measure the levels of GFAP in dissected brain tissue.[23] This assay is sensitive and has been automated, but it lacks the anatomical resolution of the immunohistochemistry technique and it requires skilled microdissection of brain tissue. A major limitation with currently available methods that require the use of postmortem tissue is that they can not be applied for the assessment of brain injury *in vivo*.

We have focused our efforts on the peripheral benzodiazepine receptor (PBR), a glial specific protein. The novel aspect of the PBR approach is that it can overcome many of the limitations of presently available methods, and has the added advantage of being able to be used for *in vivo* assessment of neurotoxicity. The latter is possible because of the availability of a highly specific and selective ligand that can be labeled with a variety of radioisotopes for *in vitro* as well as *in vivo* imaging studies (See Positron Emission Tomography of Peripheral Benzodiazepine Receptor Expression in Human Neurological Disease below). Therefore, the PBR method offers for the first time a noninvasive, quantitative, high resolution, imaging approach to assess neurotoxic endpoints resulting from chemical-induced or other types of brain damage.

IV. Peripheral benzodiazepine receptors: what are they?

There are two types of benzodiazepine receptors. The central type is associated with the GABAa receptor complex and mediates the anxiolytic and anticonvulsive properties of benzodiazepines (e.g., diazepam, also known as valium) in the central nervous system.[24] Central type benzodiazepine receptors are known to be present in high concentrations in the brain, but not in peripheral organs. The second type of benzodiazepine receptors is the peripheral type, which is present at higher levels in peripheral tissues such as the lungs, adrenal glands, liver, kidneys, blood cells, and placenta, than in the brain.[25–27] The PBR differs from the central type in pharmacology, subcellular distribution and anatomical localization. The differences in pharmacology between the central- and peripheral-type benzodiazepine receptors were first identified with clonazepam, which binds to the central-type receptor with nanomolar affinity but exhibits low affinity for the PBR.[28–30] Studies on the subcellular distribution of the PBR indicate that this receptor is associated with the outer membrane of mitochondria.[31–33] Recent evidence raises the possibility that a small fraction of PBR may be localized in the plasma membrane of peripheral tissues.[27] Consistent with its localization to the mitochondrial outer membrane, activation of the PBR is the initial and rate-limiting step in the transport of cholesterol from the outer to the inner membrane of mitochondria. Thus, the PBR appears to play a key role in neurosteroid synthesis.[34–36] In addition to its important role in steroidogenesis, the PBR is implicated in cell growth and differentiation and in the immune response.[26]

Peripheral benzodiazepine receptors are present in low concentrations in the brain neuropil. However, high concentrations are expressed in certain structures within the brain such as the choroid plexus and ependymal cells of the ventricles.[37,38] Of importance, PBR levels in the brain are exclusively associated with glial cells.[25–27] The PBR has been characterized by radioligand binding assays and receptor autoradiography using a benzodiazepine [³H]-Ro 5–4864 or the isoquinoline carboxamide [³H]-PK11195.[28,39,40] These radioligands exhibit nanomolar affinity for the PBR and have extremely low affinity for the central-type benzodiazepine receptor. Studies have shown that [³H]-PK11195 has a higher selectivity for the PBR than [³H]-Ro 5–4864,[28] and is the preferred ligand in most published studies.

A.   Cellular basis and functional significance of injury-induced elevation of peripheral benzodiazepine receptors in the central nervous system

The PBR has been used as a marker of gliosis due to its preferential localization in glial cells.[39,41,42] The binding of [³H]-PK11195 to PBR is markedly increased in the brains of experimental animals following ischemic,[41,43] physical,[42,44] or chemically induced[39,45–49] damage. Further, is has been used to assess inflammation and degeneration in neurological disease (See Positron Emission Tomography of Peripheral Benzodiazepine Receptor Expression in Human Neurological Disease below). Levels of [³H]-PK11195 binding are selectively increased in brain structures undergoing inflammation or degeneration. In some cases, increased levels of PBR are also measured in secondary areas of brain injury resulting from the primary insult. Although brain PBR are known to be localized in astrocytes and microglia, it has been unclear what contribution each cell type makes to the increase in PBR levels measured following brain damage. Studies using an ischemic model of brain injury have suggested that increased PBR levels colocalize with activated microglia rather than reactive astrocytes.[41,43] Other studies using the model neurotoxicant trimethyltin (TMT) have shown that microglial activation occurs prior to reactive astrocytosis.[20] However, increased expression of GFAP is a prominent and long-lasting response in TMT-induced and other types of insults.[10,20,46] Consistent with the latter study, we have recently shown that PBR expression is present in microglia and astrocytes and follow the same profile of activation.[49]

The functional significance of increased PBR levels in glial cells following neuronal injury is still unclear. Microglia are known to migrate and proliferate in response to many types of brain injury.[15,50] It is possible that increased PBR levels are linked to both of these functions. Recent evidence suggests that PBR ligands can influence the rate of DNA synthesis and the chemotactic potential of tumor cell lines.[51] Further work is necessary to determine if similar effects are operational in activated microglia, and whether the PBR plays a role in the activation process.

The transport of cholesterol from the outer to the inner membrane of mitochondria has been demonstrated to be the rate-limiting step of steroidogenesis.[52–54] The PBR, being primarily a cholesterol transport protein, may be elevated in injured brain tissue to provide trophic support for either glial cell activation or neuronal recovery through the production of neurosteroids. Neurosteroids have been shown to increase neuronal survival in culture and following various forms of neuronal injury.[55,56] Additionally, they have been suggested as an autocrine regulator of astrocyte reactivity.[57,58] These trophic and regulatory mechanisms may explain the persistence of elevated PBR levels in the injured brain. The consequences of altered PBR expression and central nervous system steroid production in neuronal survival and recovery are not fully appreciated. Future investigations should provide valuable insights into the functional significance of these cellular reactions to brain injury.

B. Quantitative receptor autoradiography of peripheral benzodiazepine receptors

The use of quantitative receptor autoradiography techniques has provided a valuable tool to visualize the distribution of PBR in the normal and injured brain. This method has been described in detail.[46–49] As indicated previously, the PBR-specific radioligand [^{3}H]-PK11195 has been the most widely used in studies performing quantitative receptor autoradiography. [^{3}H]-PK11195 is commercially available only in the racemic form. A limiting factor in using racemic [^{3}H]-PK11195 for measuring PBR expression is the high levels of nonspecific binding that accounts for approximately 40% of total binding in rat brain slices. Recent advances in radiochemistry have shown that the *R*-enantiomer has a higher affinity for the PBR than the *S*-enantiomer and the use of this pharmacologically active enantiomer has emerged in the literature.[59] Therefore, the use of [^{3}H]-*R*-PK11195 reduces the amount of nonspecific binding to a low level. Figure 28.1 shows studies from our laboratory comparing the levels of nonspecific binding of racemic (*R,S*) and the phar-

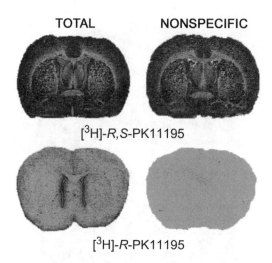

Figure 28.1 (See color insert following page 392.) Comparison of total and nonspecific binding for the racemic (R,S) and pharmacologically active R enantiomer of [³H]-PK11195. Total and nonspecific binding were performed in adjacent coronal rat brain slices at the level of the corpus striatum. Red represents high levels of binding; yellow-green, intermediate levels; and blue, low levels. The pharmacologically active R enantiomer of [³H]-PK11195 produces a dramatic decrease in nonspecific binding levels.

macologically active R-enantiomer. The figure shows [³H]-PK11195 autoradiograms of rat brain slices using the racemic [³H]-R,S-PK11195 and the pharmacologically active enantiomer [³H]-R-PK11195. The autoradiograms generated using the racemic radioligand have a greater percentage of their total binding at nonspecific sites as visualized in Figure 28.1. The functional effect of having high levels of nonspecific binding is that it reduces the practical observational range of specific binding at low levels of total binding. This is particularly important in measuring PBR levels in brain areas where subtle damage is expressed following neurotoxicant exposure. Thus, the use of the pharmacologically active R-enantiomer of [³H]-PK11195 improves the signal-to-noise ratio, increasing the ability to detect small changes in receptor expression.

C. Chemical-induced neurotoxicity: peripheral benzodiazepine receptors mark the spot

The use of PBR autoradiography has gained a great deal of popularity to assess chemical-induced as well as other types of brain injury. In our laboratory, we have used PBR expression as a molecular biomarker of neurotoxicity using a variety of model neurotoxicants. These include trimethyltin,[46,49] domoic acid,[47] MPTP,[48] methamphetamine,[60] 6-hydroxydopamine, and most recently cuprizone, a copper chelator that induces demyelination in the central nervous system.[61] Figure 28.2 shows PBR autoradiography using

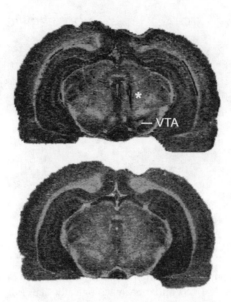

Figure 28.2 (See color insert following page 392.) Total and nonspecific binding of [³H]-(R,S)-PK11195 to coronal rat brain sections at the level of the ventral tegmental area (VTA). In this study, adult rats were stereotactically injected with the dopaminergic neurotoxicant 6-hydrodopamine into the VTA. The dark band to the left of the star (*) represents increased binding of [³H]-(R,S)-PK11195 to peripheral benzodiazepine receptors associated with the physical trauma of inserting a cannula guide for stereotactic injection. The dark area pointed by the line represents increased levels of [³H]-(R,S)-PK11195 binding resulting from the 6-hydroxydopamine-induced degeneration of dopaminergic neurons at the site of injection. Red represents high levels of binding; yellow-green, intermediate levels; and blue, low levels.

racemic [³H]-PK11195 to localize the sites of damage produced by stereotaxic injection of the dopaminergic neurotoxicant 6-hydroxydopamine into the ventral tegmental area of the brain. The injection of 6-hydroxydopamine results in the nearly complete degeneration of dopaminergic innervation on the ipsilateral side as assessed by a number of neuronal and behavioral enpoints.[62] Autoradiography of PBR levels using racemic [³H]-PK11195 was able to pinpoint the location of brain damage induced by the 6-hydroxydopamine injection as well as the physical damage produced by the cannula guide (Figure 28.2).

Our work has shown that the temporal pattern and the degree of PBR expression following chemical-induced brain injury are a direct reflection of the glial response and the type of damage produced by neurotoxicant exposure. For example, PBR expression following a single injection of TMT is not significantly increased until 3 days after exposure, but once elevated at 7 days, it remains increased in affected areas for a long period of time.[46] This is a reflection of the slow but progressive neuronal cell loss that TMT produces in limbic structures such as the hippocampus.[63] We have also shown that the early increase in PBR expression is associated with microglia activation, while the later response appears to be primarily related to astrocytes.[49] A similar effect as with TMT was observed with domoic acid, another neurotoxicant that damages limbic structures.[47] In this case, PBR levels are increased prior to any indication of seizure-induced neuronal cell loss. On the other hand, the PBR response is increased much more rapidly in the striatum and substantia nigra of animals exposed to the dopaminergic neurotoxicant MPTP, and the response is dose-dependent.[48] MPTP is a neurotoxicant that produces degeneration of dopaminergic terminal fields in the striatum and cell bodies in the substantia nigra. The PBR response to MPTP neurotoxicity appears to persist in the dorsal aspects of the striatum and in the substantia nigra, but it is transient in the ventral aspects of the striatum, possibly reflecting different degrees of sensitivity. These findings are consistent with a dorso-ventral gradient in the astrocytic response to MPTP damage in the corpus striatum.[64]

We are currently assessing the PBR response resulting from exposure to cuprizone. Cuprizone is a copper chelator that has been used to produce a model of demyelination in the central nervous system.[61] The corpus collosum and cerebellar peduncles are two brain regions that prominently express demyelination following cuprizone ingestion in mice.[65] Preliminary studies in our laboratory have shown that PBR levels are increased in these and other brain structures in a time dependent manner consistent with previous reports (see Figure 28.3).[65] We are currently assessing the cell type responsible for the increased levels of [³H]-R-PK11195 to PBR in this model of demyelination.

V. Positron emission tomography: imaging peripheral benzodiazepine receptors in the living brain

Advances in positron emission tomography (PET) have made possible noninvasive visualization and measurement of specific biochemical processes in the living brain with exquisite resolution. A detailed technical description of this technique is beyond the scope of this chapter.[66,67] Briefly, PET is based on the use of short-lived (minutes), positron-emitting radioisotopes (carbon-11, fluorine-18) for labeling organic molecules or drugs which are part of specific biochemical reactions or interact with specific recognition sites in the brain such as receptors or enzymes. Following the intravenous injection of a radioactively labeled ligand, the distribution of the radioactivity in the brain can be measured by coincidence detection of geometrically opposite gamma rays (0.511 KeV) that are emitted when the positron radiation emitted decays. The gamma rays are detected in coincidence by a ring of detectors in the PET scanner and the distribution of the radioactivity is computer reconstructed to form an image. The image is a direct reflection of the amount and spatial distribution of the interaction of the radiolabeled ligand with its biological target. Figure 28.4 is a PET image representing the normal distri-

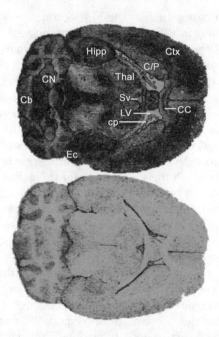

Figure 28.3 (See color insert following page 392.) Total binding of [³H]-*R*-PK11195 to peripheral benzodiazepine receptors in horizontal rat brain sections of cuprizone treated (top) animals (0.2% in diet for 4 weeks) and controls (bottom). Note the high levels of binding in cuprizone-treated brain in the corpus collosum (CC) and deep cerebellar nuclei (CN). Increased binding was also present in other brain structures such as the hippocampus (Hipp), cerebral cortex (Ctx), caudate/putamen (C/P), entorhinal cortex (Ec) and thalamus (Thal). High level of [³H]-*R*-PK11195 to peripheral benzodiazepine receptors is normally found in the choroid plexus (cp) and ventricles (3v, third ventricle). The cerebellum (Cb) is noted as an anatomical landmark. Red represents high levels of binding; yellow-green, intermediate levels; and blue, low levels.

bution of the PBR selective ligand [¹¹C]-*R*-PK11195 and the dopamine transporter selected ligand [¹¹C]-WIN 35,428 in a normal nonhuman primate brain. PBR expression is found at low concentrations and homogeneously distributed throughout the brain neuropil indicative of resident glial cells. On the other hand, the distribution of [¹¹C]-WIN 35,428 is concentrated in the caudate/putamen, a brain area highly enriched in high affinity dopaminergic transporters. We are currently using these imaging techniques to assess brain damage resulting from specific chemical exposures in nonhuman primates.

A. Positron emission tomography of peripheral benzodiazepine receptor expression in human neurological disease

The use of PBR-PET has received increased attention in recent years. Although PBR-PET studies were performed in the late 1980s when PET

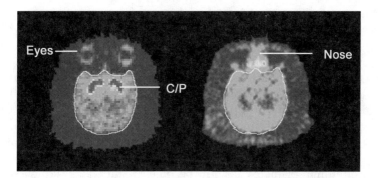

Figure 28.4 (See color insert following page 392.) PET studies of [¹¹C]-R-PK11195 to peripheral benzodiazepine receptors (right) and [¹¹C]-WIN 35,428 to dopamine transporters (left) in a normal nonhuman primate (baboon) brain (transaxial view). Note the highly concentrated levels of dopamine transporters in the caudate/putamen (C/P). A much more homogeneous distribution of peripheral benzodiazepine receptors in the brain neuropil is observed in the same animal. The brain is delineated by the white boundaries. Red represents high levels of binding; yellow-green, intermediate levels; and blue, low levels.

scanners were becoming available at research institutions, few studies were performed. The lack of studies was the result of poor image quality due to the use of racemic PK11195, poor spatial resolution of PET scanners, inherently low levels of PBR in the brain and the lack of appropriate mathematical models to accurately quantify PBR expression. The limited number of studies performed was related to the detection of gliomas, brain tumors that express high levels of PBR,[68,69] and the damage produced by stroke.[70] More recently, PBR-PET has been used in a number of human neurological diseases because advances in all of the areas that previously contributed to poor image quality have been made. For example, current PET scanners have better spatial resolution and greater axial sampling, the pharmacologically active (R)-PK11195 enantiomer is available and advanced mathematical models are currently used to accurately quantify brain radioactivity.

The use of PBR-PET has now been described in human studies of multiple sclerosis,[71,72] Rasmussen's encephalitis,[73] cerebral vasculitis associated with refractory epilepsy[74] and improved imaging of ischemic stroke.[75] From the above mentioned studies, it appears that the PBR is a useful marker of inflammation and ongoing gliosis and, in an indirect way, a marker of brain injury. For example, in multiple sclerosis PBR-PET was able to identify active lesions from nonactive lesions as defined by magnetic resonance imaging.[71,72] Further, PBR-PET was able to identify additional areas of brain damage beyond the lesion sites. In summary, it appears that PBR-PET provides a molecular marker of disease activity that can be monitored in the living human brain. Thus, it can be potentially useful to follow disease progression as well as to assess the effectiveness of therapeutic interventions.

B. *Future application of in vivo imaging of peripheral*
    benzodiazepine receptors in rodent and nonhuman primate
    models of disease

1. The use of quantitative receptor autoradiography has been a valuable tool to monitor the temporal and anatomical changes in [^{3}H]-R-PK11195 binding to PBR in rodent models of neurotoxicity. One of the limitations of this technique is that it requires postmortem brain tissue. The possibility of imaging biochemical reactions or recognition sites in the brain of living animals such as rodents and small nonhuman primates has now become a reality. This is due to the fact that unprecedented advances have taken place in the development of dedicated small animal PET scanners.[76,77] There are currently an emerging number of small animal scanners to perform PET studies in rodents and small nonhuman primates that offer spatial resolution in the order of 2–3 mm. Further, the resolution of the next generation of small animal scanners is in the order of 1 mm or less. These advances make it feasible to perform PET studies in the brain of living mice and rats as well as small monkeys. There are a number of advantages to this approach:

2. Small animal imaging allows investigators to monitor molecular changes in the living brain as a result of experimental treatments in real time, and can be measured in the same animal from development to aging.

3. The ability to monitor the animal in a longitudinal fashion reduces the number of animals needed to perform a particular study, since the animals do not have to be euthanized at specific time points during treatment to harvest brain tissue. Further, the animal serves as its own control, thus reducing biological variability.

4. Animal imaging allows the investigator to follow molecular changes in the brain in parallel with other cellular or behavioral outcomes that can be measured in the live animal.

5. Animal imaging saves having to terminate valuable rodent models such as transgenics, knockouts, or expensive and valuable nonhuman primates.

Animal imaging makes it possible to perform multiple studies per day in the same animal. The advances that can be by made by small animal PET imaging in combination with advances in genomics and proteomics, is only limited by the imagination of the scientists using it.

VI. Summary

Understanding glial cell responses under pathological conditions provides important information about marker proteins used to identify neurotoxic endpoints. Detection of changes in the expression of marker proteins based on immunohistochemical methods has inherent limitations and requires postmortem tissue. Validation of the PBR as a biomarker of neurotoxicity surmounts these limitations. The PBR is a glial protein that has been proven to be sensitive and specific in detecting sites of inflammation and brain damage. Studies from our laboratory and those of others support continued investigations on the use of the PBR as a biomarker of neurotoxicity. Considerable data has been generated in the validation of the PBR method in detecting chemical-induced neurotoxicity in rodent models. Ongoing studies are validating this methodology in nonhuman primates for subsequent studies in humans that are potentially exposed to environmental chemicals in the occupational setting. The ability to label the PBR-specific ligand PK11195 with positron-emitting radioisotopes makes possible the use of noninvasive PET scanning to study the PBR in the living human brain. The development of dedicated high-resolution PET scanners for studies in small animals such as rodents and small nonhuman primates will greatly facilitate the application of the PBR as an *in vivo* molecular biomarker of neurotoxicity.

Acknowledgments

My sincere thanks and gratitude go to my colleagues, present and previous doctoral and postdoctoral students and the research staff in my laboratory at The Johns Hopkins University Bloomberg School of Public Health. Their hard work and dedication has made many of these studies a reality. Further, the support of the National Institute of Environmental Health Sciences has been instrumental in providing the financial resources in advancing this work (grant number ES07062).

References

1. Tilson, H.A., Neurotoxicology in the 1990s, *Neurotoxicol. Teratol.*, 12, 293–300, 1990.
2. O'Callaghan, J.P., Neurotypic and gliotypic proteins as biochemical markers of neurotoxicity, *Neurotox. Teratol.*, 10, 445–452, 1988.
3. O'Callaghan, J.P., Jensen, K.F., and Miller, D.B., Quantitative aspects of drug- and toxicant-induced astrogliosis, *Neurochem. Int.*, 26, 115–124, 1995.
4. Brock, T.O. and O'Callaghan, J.P., Quantitative changes in the synaptic vesicle proteins, synapsin I and p38, and the astrocytic specific protein, glial fibrillary acidic protein, are associated with chemically induced injury to the rat central nervous system, *J. Neurosci.*, 7, 931–942, 1987.
5. Lindsay, R.M., Reactive gliosis, in *Astrocytes: Cell Biology and Pathology of Astrocytes*, Federoff, S. and Vernadakis A, Eds., Vol. 3, Academic Press, New York, 1986, pp. 231–262.
6. Norton, W.T. et al., Quantitative aspects of reactive gliosis: a review, *Neurochem. Res.*, 17, 877–885, 1992.
7. Riva-Depaty, I. et al., Contribution of peripheral macrophages and microglia to the cellular reaction after mechanical or neurotoxin-induced lesions of the rat brain, *Exp. Neurol.*, 128, 77–87, 1994.
8. Amat, J.A. et al., Phenotypic diversity and kinetics of proliferating microglia and astrocytes following cortical stab wounds, *Glia*, 16, 368–382, 1996.

9. Reier, P.J., Eng, L.F., and Jakeman, L., Reactive gliosis and axonal outgrowth in the injured CNS: is gliosis really an impediment to regeneration? in *Reactive Gliosis and Axonal Outgrowth*, A.R.Liss, New York, 1988, pp. 1–27.

10. O'Callaghan, J.P., Quantitative features of reactive gliosis following toxicant-induced damage of the CNS, *Ann. NY Acad. Sci.*, 679, 195–210, 1993.

11. Eng, L.F., Immunocytochemistry of the glial fibrillary acidic protein, in *Progress in Neurobiology*, Zimmerman, H.N., Eds., Vol. 5, Raven Press, New York, 1983, pp. 19–39.

12. O'Callaghan, J.P. and Miller, D.B., Nervous-system specific proteins as biochemical indicators of neurotoxicity, *TIPS*, 388–390, September 1983.

13. O'Callaghan, J.P., Assessment of neurotoxicity: Use of glial fibrillary acidic protein as a biomarker, *Biomed. Env. Sci.*, 4, 197–206, 1991.

14. Stone, R. New marker for nerve damage, *Science*, 259, 1541, 1993.

15. Kreutzberg, G.W., Microglia: a sensor for pathological events in the CNS, *Trends Neurosci.*, 19, 312–318, 1996.

16. Davis, E.J., Foster, T.D., and Thomas, W.E., Cellular forms and functions of brain microglia, *Brain Res. Bull.*, 34, 73–78, 1994.

17. Streit, W.J., An improved staining method for rat microglia cells using the lectin from *Griffonia simplicifolia* (GSA I-B4), *J. Histochem. Cytochem.*, 38, 1683–1686, 1990.

18. Fix, A.S. et al., Integrated evaluation of central nervous system lesions: stains for neurons, astrocytes, and microglia reveal the spatial and temporal features of MK-801-induced neuronal necrosis in the rat cerebral cortex, *Toxicol. Pathol.*, 24, 291–304, 1996.

19. Taniwaki, Y. et al., Microglia activation by epileptic activities through the propagation pathway of kainic acid-induced hippocampal seizures in the rat, *Neurosci. Lett.*, 217, 29–32, 1996.

20. McCann, M.J. et al., Differential activation of microglia and astrocytes following trimethyltin-induced neurodegeneration, *Neurosci.*, 72, 273–281, 1996.

21. O'Callaghan, J.P., Assessment of neurotoxicity using assays of neuron- and glia-localized proteins: chronology and critique, in *Neurotoxicology*, Tilson, H. and Mitchell, C., Eds., Raven Press, New York, 1992, pp. 83–100.

22. O'Callaghan, J.P., Quantitative features of reactive gliosis following toxicant-induced damage of the CNS, *Ann. New York Acad. Sci.*, 679, 195–210, 1993.

23. O'Callaghan, J.P., Quantification of glial fibrillary acidic protein: Comparison of slot-immunobinding assay to a novel sandwich ELISA, *Neurotox. Teratol.*, 13, 275–281, 1991.

24. Baestrup, C. and Squires, R.F., Benzodiazepine receptors in rat brain, *Nature*, 266, 732–734.

25. Gavish, M. et al., Biochemical, physiological, and pathological aspects of the peripheral benzodiazepine receptor, *J. Neurochem.*, 58, 1589–1601, 1992.

26. Zisterer, D.M. and Williams, D.C., Peripheral-type benzodiazepine receptors, *Gen. Pharm.*, 29, 305–314, 1997.

27. Woods, M.J. and Williams, D.C., Multiple forms and locations for the peripheral-type benzodiazepine receptor, *Biochem. Pharm.*, 52, 1805–1814, 1996.

28. Awad, M. and Gavish, M., Binding of [^{3}H]-RO 5–4864 and [^{3}H]-PK11195 to cerebral cortex and peripheral tissue of various species: species differences and heterogeneity in peripheral benzodiazepine binding sites, *J. Neurochem.*, 49, 1407–1414, 1987.

29. Bender, A.S. and Hertz, L., Pharmacological characteristics of diazepam receptors in neurons and astrocytes in primary cultures, *J. Neurosci Res.*, 18, 366–372, 1987.
30. Langer, S.Z. and Arbilla, S., Imidazopyridines as a tool for the characterization of benzodiazepine receptors: a proposal for the pharmacological classification as omega receptor subtypes, *Pharm. Biochem. Beh.*, 29, 763–766, 1988.
31. Anholt, R.R.H. et al., The peripheral benzodiazepine receptor localization to the mitochondrial outer membrane, *J. Biol. Chem.*, 261, 576–83, 1986.
32. Antkiewicz-Michaluk, L., Guidotti, A., and Krueger, K.E., Molecular characterization and mitochondrial density of a recognition site for peripheral-type benzodiazepine ligands, *Mol. Pharm.*, 34, 272–278, 1988.
33. Hirsch, J.D. et al., Characterization of ligand binding to mitochondrial benzodiazepine receptors, *Mol. Pharm.*, 34, 164–172, 1988.
34. Papadopoulos, V., Amri, H., Boujrad, N., Peripheral benzodiazepine receptor in cholesterol transport and steroidogenesis, *Steroids*, 62, 21–28, 1997.
35. Papadopoulos, V., Nowzari, F.B., and Krueger, K.E., Hormone stimulated steroidogenesis is coupled to mitochondrial benzodiazepine receptors, *J. Biol. Chem.*, 266, 3682–3687, 1991.
36. Krueger, K.E. and Papadopoulos, V., Mitochondrial benzodiazepine receptors and the regulation of steroid biosynthesis, *Ann. Rev. Pharmacol. Toxicol.*, 32, 211–237, 1992.
37. Richards, J.G. and Möhler, H., Benzodiazepine receptors, *Neuropharmacol.*, 23, 233–242, 1984.
38. Price, G.W. et al., *In vivo* binding to peripheral benzodiazepine binding sites in the lesioned rat brain: Comparison between [³H]-PK11195 and [¹⁸F]-PK14105 as makers for neuronal damage, *J. Neurochem.*, 55, 175–185, 1990.
39. Benavides, J. et al., Peripheral type benzodiazepine binding sites are a sensitive indirect index of neuronal damage, *Brain. Res.*, 421, 167–172, 1987.
40. Dubois, A. et al., Imaging of primary and remote ischemic and excitotoxic brain lesions: An autoradiographic study of peripheral type benzodiazepine binding sites in the rat and cat, *Brain Res.*, 445, 77–90, 1988.
41. Stephenson, D.T. et al., Peripheral benzodiazepine receptors are colocalized with activated microglia following transient global forebrain ischemia in the rat, *J. Neurosci.*, 15, 5263–5274, 1995.
42. Gehlert, D.R. et al., Increased expression of peripheral benzodiazepine receptors in the facial nucleus following motor neuron axotomy, *Neurochem. Int.*, 31, 705–713, 1997.
43. Myers, R., et al. Macrophage and astrocyte populations in relation to [³H]-PK11195 binding in rat cerebral cortex following a local ischaemic lesion, *J. Cer. Blood Flow Met.*, 11, 314–322, 1991.
44. Banati, R.B., Myers, R., and Kreutzberg, G.W., PK (peripheral benzodiazepine) binding sites in the CNS indicate early and discrete brain lesions: Microautoradiography detection of [³H]-PK11195 binding to activated microglia, *J. Neurocytol.*, 26, 77–82, 1997.
45. Bendotti, C. et al., Does GFAP mRNA and mitochondrial benzodiazepine receptor binding detect serotonergic neuronal degeneration in rat? *Brain Res. Bull.*, 34, 389–394, 1994.
46. Guilarte, T.R. et al., Enhanced expression of peripheral benzodiazepine receptors in trimethyltin-exposed rat brain: A biomarker of neurotoxicity, *Neurotoxicol.*, 16, 441–450, 1995.

47. Kuhlmann, A.C. and Guilarte, T.R., The peripheral benzodiazepine receptor is a sensitive indicator of domoic acid neurotoxicity, *Brian Res.*, 751, 281–288, 1997.

48. Kuhlmann, A.C. and Guilarte, T.R., Regional and temporal expression of the peripheral benzodiazepine receptor in MPTP neurotoxicity, *Toxicol. Sci.*, 48, 107–116, 1999.

49. Kuhlmann, A.C. and Guilarte, T.R., Cellular and subcellular localization of peripheral benzodiazepine receptors after trimethyltin neurotoxicity, *J. Neurochem.*, 74, 1694–1704, 2000.

50. Streit, W.J., Graeber, M.B., and Kreutzberg, G.W., Functional plasticity of microglia: a review, *Glia*, 1, 301–307, 1988.

51. Hardwick, M. et al., Peripheral-type benzodiazepine receptor (PBR) in human breast tissue: correlation of breast cancer cell aggressive phenotype with PBR expression, nuclear localization and PBR-mediated cell proliferation and nuclear transport of cholesterol, *Cancer Res.*, 59, 831–842, 1999.

52. Krueger, K.E. and Papadopoulos, V., Peripheral-type benzodiazepine receptors mediate translocation of cholesterol from outer to inner mitochondrial membranes in adrenocortical cells, *J. Biol. Chem.*, 265, 15015–15022, 1990.

53. Papadopoulos, V. et al., Targeted disruption of the peripheral-type benzodiazepine receptor gene inhibits steroidogenesis in the R2C leydig tumor cell line, *J. Biol. Chem.*, 272, 32129–32135, 1997.

54. Li, H. and Papadopoulos, V., Peripheral-type benzodiazepine receptor function in cholesterol transport: identification of a putative cholesterol recognition/interaction amino acid sequence and consensus pattern, *Endocrinol.*, 139, 4991–4997, 1998.

55. Baulieu, E.E., Neurosteroids: a novel function of the brain, *Psychoneuroendocrinol.*, 23, 963–987, 1998.

56. Lacor, P. et al., Regulation of the expression of peripheral benzodiazepine receptors and their endogenous ligands during sciatic nerve degeneration and regeneration: a role for PBR in neurosteroidogenesis, *Brain Res.*, 815, 70–80, 1999.

57. Jung-Testas, I. et al., Demonstration of steroid hormone receptors and steroid action in primary cultures of rat glial cells, *J. Steroid Biochem. Mol. Biol.*, 41, 621–631, 1992.

58. Del Cerro, S., Garcia-Estrada, J., and Garcia-Segura, L., Neuroactive steroids regulate astroglia morphology in hippocampal cultures from adult rats, *Glia*, 14, 65–71, 1995.

59. Shah, F. et al., Synthesis of the enantiomers of [N-methyl-^{11}C]-PK11195 and comparison of their behaviors as radioligands for PK binding sites in rats, *Nucl. Med. Biol.*, 21, 573–581, 1994.

60. Howard, A.S. et al., Peripheral benzodiazepine receptors are transiently increased after methamphetamine administration, *Toxicol. Sci.*, 54(1), 22, 2000.

61. Matsushima, G.K. and Morell, P., The neurotoxicant, cuprizone, as a model to study demyelination ad remyelination in the central nervous system, *Brain Pathol.*, 11, 107–116, 2001.

62. May, C.H. et al., Intrastriatal infusion of Lisuride: a potential treatment for Parkinson's disease? Behavioral and autoradiographic studies in 6-OHDA lesioned rats, *Neurodegeneration*, 3, 305–313, 1994.

63. Balaban, C.D., O'Callaghan, J.P., and Billingsley, M.L., Trimethyltin-induced neuronal damage in the rat brain: comparative studies using silver degeneration stains, immunocytochemistry and immunoassay for neurotypic and gliotypic proteins, *Neuroscience*, 26, 337–361, 1988.

64. Francis, J.W. et al., Neuroglial responses to the dopaminergic neurotoxicant 1-methyl-4-phenyl-1,2,3,6-tetrahydropyridine in mouse striatum, *Neurotoxicol. Teratol.*, 17, 7–12, 1995.
65. Kesterson, J.W. and Carlton, W.W., Histopathologic and enzyme histochemical observations of the cuprizone-induced brain edema, *Exp. Mol. Pathol.*, 15, 82–96, 1971.
66. Phelps, M.E., PET: the merging of biology and imaging into molecular imaging, *J. Nucl. Med.*, 41, 661–681, 2000.
67. Myers, R. et al., Use of two- and three-dimensional PET and [^{11}C]-R-PK11195 to image focal and regional brain pathology, in *Quantitative Functional Brain Imaging with Positron Emission*, Carson, R.E., Daube-Witherspoon, M.E., and Herscovitch, P., Eds., Academic Press, New York, , 1998, chap. 29, pp. 195-200.
68. Junck, L. et al., PET imaging of human glioma with ligands for the peripheral benzodiazepine site, *Ann. Neurol.*, 26, 752–758, 1989.
69. Pappata, S. et al., PET study of carbon-11-PK11195 binding to peripheral type benzodiazepine sites in glioblastoma: a case report, *J. Nucl. Med.*, 32, 1608–1610, 1991.
70. Ramsay, S.C. et al., Monitoring by PET of macrophage accumulation in brain after ischaemic stroke, *Lancet*, 339, 1054–1055, 1992.
71. Vowinckel, E. et al., PK11195 binding to the peripheral benzodiazepine receptor as a marker of microglia activation in multiple sclerosis and experimental autoimmune encephalomyelitis, *J. Neurosci. Res.*, 50, 345–353, 1997.
72. Banati, R.B. et al., *Brain*, 123, 2321–2337, 2000.
73. Banati, R.B. et al., [^{11}C]R-PK11195 positron emission tomography imaging of activated microglia *in vivo* in Rasmussen's encephalitis, *Neurology*, 53, 2199–2203, 1999.
74. Goerres, G.W. et al., Imaging cerebral vasculitis in refractory epilepsy using [^{11}C]R-PK11195 positron emission tomography, *Am. J. Radiol.*, 176, 1016–1018, 2001.
75. Pappata, S. et al., Thalamic microglia activation in ischemic stroke detected *in vivo* by PET and [^{11}C]PK11195, *Neurology*, 55, 1052–1054, 2000.
76. Tanaka, E., Animal Scanners, in *Principles of Nuclear Medicine*, Wagner, H.N., Jr., Szabo, S., and Buchanan, J., Eds., WB Saunders Co. Philadelphia, 1995, chap. 20, pp. 378–384.
77. Chatziioannou, A.F. et al., Performance evaluation of microPET: a high resolution leutetium oxyorthosilicate PET scanner for animal imaging, *J. Nucl. Med.*, 40, 1164–1175, 1999.

section V

Biomarkers and chemical toxicants

chapter twenty-nine

Biomarkers of organ injury from chemical exposure: concurrent inflammation as a determinant of susceptibility

Robert A. Roth and Patricia E. Ganey

Contents

Abstract   There are a number of factors that can determine the susceptibility of an individual to toxic responses from chemical exposure. Among these is a concurrent inflammatory response. We all experience episodes of inflammation from infections and various diseases; in addition, inflammagens such as endotoxin (lipopolysaccharide or LPS) reach the circulation after translocating from the gastrointestinal tract and may precipitate modest inflammatory responses. The picture that emerges from human and animal studies is that exposure to noninjurious amounts of LPS and possibly other inflammagens is commonplace and episodic. Accumulated evidence from experimental animal studies indicates that modest LPS exposure leads to an inflammatory response that can enhance the toxic effects of numerous xenobiotic agents, including drugs, food-borne toxins and environmental contaminants such as cadmium, mercury, polychlorinated biphenyls, organic solvents, and ozone. Inflammatory mechanisms are complex, involving activation of several cell types and the release of numerous inflammatory mediators that may precipitate overt injury in tissues homeostatically altered by exposure to an otherwise nontoxic dose of a xenobiotic agent. To develop appropriate biomarkers, it will be necessary to understand the mechanisms by which underlying inflammation augments toxic responses.

I. Introduction and perspective

Biomarkers of organ injury can be categorized into three types: markers of exposure, markers of effect, and markers of susceptibility.[1] Biomarkers of exposure typically are surrogate measures of internal dose or dose to target tissue for a xenobiotic agent. Markers of effect include pathophysiological, cellular, and molecular events that represent the toxic effects of an agent. By contrast, biomarkers of susceptibility are useful in defining which individuals in a population are most likely to respond with toxicity upon exposure to a xenobiotic agent or mixture.

This commentary is a brief exploration of the hypothesis that underlying inflammation is a potentially important determinant of individual susceptibility to toxic agents. To put this in perspective, the reader should bear in mind that there are numerous factors that can contribute to differences among individuals in their sensitivity to xenobiotic agents (Figure 29.1). These include variations in age, gender, xenobiotic metabolism, immunologic responses, reserve and repair capacity of tissues, absorption, coexisting disease, coexposure to additional xenobiotic agents, and nutritional status, as well as underlying inflammation. This large number of determinants is sure to coexist and exert influences in concert in any individual. Moreover, many are likely to vary with time, so that a snapshot of an individual's susceptibility, if one could be taken, would not necessarily be indicative of past or future susceptibility. The study of genetic factors that influence these determinants has become an important focus in toxicology in the past several years. An example is the discovery of polymorphisms in metabolic pathways that govern the toxicity of xenobiotic agents. It should be kept in mind, however, that environmental factors have the potential to exert important influences on most of these determinants (Figure 29.1).

II. Occurrence of inflammatory episodes

We all experience episodes of inflammation that expose one or more organs to biologically active inflammatory factors. The most obvious etiologic agents are bacteria and viruses, exposure to which usually precipitates inflammation as a response aimed at ridding the host of such pathogens. Specific immune reactions against antigens also typically culminate in an

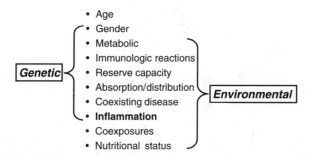

Figure 29.1 Most determinants of individual susceptibility, including concurrent inflammation, can be influenced by genetic and environmental factors.

acute or chronic inflammatory response. In addition, there are many diseases or conditions that are associated with tissue inflammation, among which are arthritis, hepatitis, atherosclerosis, asthma, and others. Finally, studies over the last two decades indicate that we commonly experience episodic exposure to bacterial endotoxin, a potent initiator of inflammation.

Endotoxin comprises proteins in association with lipopolysaccharide (LPS), a major constituent of the outer cell wall of Gram negative bacteria. Such bacteria normally inhabit the intestinal tract in numbers that exceed the number of eukaryotic cells in the human body. When these bacteria divide or are injured, large amounts of LPS are released in the intestinal lumen. Although it was once thought that the gastrointestinal (GI) mucosa was a perfect barrier to LPS, research over the last two decades has indicated that LPS can translocate from the GI lumen into the blood, and thereby the liver and perhaps other organs become exposed. The degree to which this happens in an unstressed GI tract remains a matter of debate: some investigators have detected plasma levels of LPS in healthy, unstressed people, whereas others have not.[2-4] Nevertheless, it is clear that numerous conditions can lead to an increase in plasma LPS concentration, probably as a result of enhanced GI translocation. These conditions include disease, surgery, trauma or ischemia that affects the GI tract, acute or chronic liver diseases, Reye's syndrome, and acute or chronic alcohol consumption.[5] Even the process of anesthetizing a patient has been reported to increase plasma LPS concentration significantly.[6] In experimental animals, alterations in diet have led to enhanced plasma LPS, as has exposure to certain xenobiotic agents.[7-10] In one study of long distance runners, participants averaged greater plasma LPS levels after than before a race, and the runners whose GI tracts were clearly stressed by the run experienced the greatest increases.[11] In addition to exposure from the GI route, considerable LPS exposure can occur via inhalation. This occurs especially in occupations in which LPS-laden dusts or aerosols are generated, but exposure in the home environment is also common.[12-16]

More research is needed to ascertain the frequency and magnitude of systemic LPS exposure in the human population. However, the picture that emerges from these clinical and experimental studies is that mild endotox-

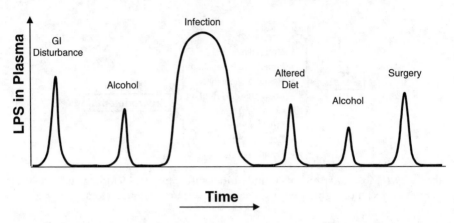

Figure 29.2 Current evidence suggests that exposure to LPS and probably other agents that induce inflammation is commonplace and episodic. In the case of LPS, many factors can increase translocation of LPS from the intestinal lumen into the blood. Such enhanced exposure may initiate modest inflammatory responses that can alter tissue homeostasis during periods of exposure.

emia in people is commonplace and episodic (Figure 29.2). The frequency and magnitude of increases in plasma LPS likely vary considerably among individuals and even within an individual, depending on lifestyle, disease, and other factors. Resultant exposure of organs to LPS is unlikely to be of sufficient magnitude to cause overt tissue injury, but such episodes may nevertheless precipitate a mild inflammatory response in tissues.

III. Underlying inflammation and sensitivity to toxicants

To the knowledge of the authors, there has been no systematic study in people of the relationship between inflammation and toxic response to xenobiotic agents. However, studies in experimental animals clearly suggest that modest inflammation can render individuals more sensitive to the toxic effects of a wide variety of xenobiotic agents.

We have recently studied the effects of modest inflammation imposed by a small dose of LPS on the acute hepatotoxicity of selected, food-borne toxicants in rats. For example, small, nontoxic doses of LPS convert a nontoxic dose of allyl alcohol, a chemical used in the manufacture of food flavorings and other agents, into one that produced marked periportal injury.[17] People and animals are exposed to aflatoxin B_1 as a fungal metabolite that contaminates grain and nut products. Administration of a small, nontoxic dose of LPS markedly enhanced periportal parenchymal cell necrosis and bile duct injury in rats exposed to aflatoxin.[18] Acute aflatoxicosis in humans is thought to result from aflatoxin B_1 doses that are similar to those that cause liver injury in rats.[19] Given this species similarity, the results in rodents raise the possibility that coexisting inflammation from LPS exposure or other causes may in part determine the severity of aflatoxicosis in human cases. In addition to causing acute hepatotoxicity, aflatoxin B_1 is a human and animal carcinogen. Several epidemiological studies have found that people with hepatitis have a greater risk of developing hepatocellular carcinomas from dietary aflatoxin B_1.[20-23] Since a defining feature of hepatitis is an hepatic inflammatory response, it may be that concurrent inflammation predisposes people to the carcinogenic effects of this fungal toxin.

Monocrotaline is one of many toxic pyrrolizidine alkaloids found in a wide variety of plant species, some of which are used in nutritional supplements and herbal teas. Exposure of people to these alkaloids has resulted in venoocclusive disease of the liver. Similarly, rats given large doses of monocrotaline experience acute, centrilobular hepatocellular injury accompanied by hemorrhage and destruction of endothelium in sinusoids and central veins. Small doses of monocrotaline and LPS, which are noninjurious when given by themselves, result in pronounced liver injury when given together.[24] In the case of this plant toxin, the nature of the hepatic lesions suggests that exposure to each of the agents potentiates the hepatotoxicity of the other. That is, the cotreatment results in a pyrrolizidine-like centrilobular lesion and a midzonal lesion like that which occurs after large, hepatotoxic doses of LPS. Dose/response studies using plasma alanine aminotransferase activity as a marker of hepatotoxicity indicated that LPS cotreatment reduced the threshold for liver injury from monocrotaline to between 13–33% of that from monocrotaline given alone. Herbal supplements containing pyrrolizidine alkaloids or other toxins are consumed in part to combat illnesses of various sorts, some of which may be associated with inflammation. This raises the possibility that people who consume these alkaloids during an episode of inflammation may be putting themselves at risk of an untoward response.

Accumulated evidence from several laboratories indicates that LPS exposure can enhance the effects of numerous hepatotoxicants. These include drugs, food-borne toxins, and environmental contaminants such as cadmium, solvents, and polychlorinated biphenyls (Table 29.1).[5] Although for some of these it is not known with certainty that inflammation is the cause of the increased sensitivity, this seems likely based on the known effects of LPS. Thus, the ability of underlying inflammation to augment hepatotoxicity is not specific to a chemical class or to chemicals that act by a specific mecha-

Table 29.1 Chemical-Induced Liver Injuries Enhanced by LPS Exposure

CCl_4	Halothane
Ethanol	Ethionine
D-Galactosamine	Chlorpromazine
Frog virus 3 protein	PCBs
T-2 Toxin	Allyl alcohol
Vomitoxin	Monocrotaline
Cadmium	Aflatoxin B_1

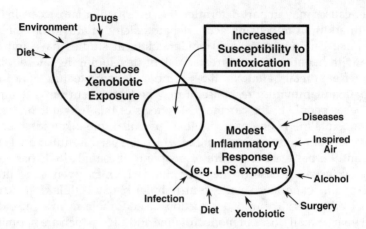

Figure 29.3 Inflammatory responses occurring concurrent with exposure to xenobiotic agents may render individuals particularly sensitive to toxicity.

nism. Moreover, the ability of concurrent inflammation to influence toxicity applies to extrahepatic organs as well. For example, LPS augments ozone-induced mucous cell metaplasia in the respiratory tract[25] and potentiates mercuric chloride nephrotoxicity.[26] Similarly, LPS acts synergistically with the trichothecene fungal toxin, deoxynivalenol, to cause B-cell apoptosis in lymphoid tissue.[27] Accordingly, the ability of LPS and probably other inflammagens to enhance toxic responses is likely a phenomenon applicable to a wide variety of xenobiotic agents that affect many tissues and act by various mechanisms. Such results have led to the idea that human exposure to LPS and other agents (e.g., pathogens) in the environment that precipitate a modest inflammatory response may act as a determinant of susceptibility to environmental and other toxic, xenobiotic agents (Figure 29.3). That is, people undergoing an inflammatory response may represent a subset of the population that is likely to respond to xenobiotic exposure with a toxic outcome.

IV. Mechanisms and hypotheses

To develop biomarkers with enough selectivity to be useful, it is often necessary to understand mechanisms by which an agent produces its toxic effects and factors that determine individual sensitivity. Much remains unknown about how LPS and other inflammagens exacerbate toxic responses. It seems likely that cells, soluble mediators and other signaling molecules activated or expressed during the inflammatory response underlie the heightened sensitivity. Inflammagens such as LPS cause tissue injury on their own in large doses, and it seems reasonable to expect that some of the same factors are at work during LPS augmentation of xenobiotic toxicity. Accordingly, clues as to mechanisms might be found in what is known about how large doses of LPS incite injury.

The pathogenesis of LPS-induced hepatocellular necrosis involves a chain or network of inflammatory events that appears to begin by binding of LPS to toll-like receptors (e.g., TLR4) within the plasma membranes of certain hepatic cells.[28,29] Binding of LPS to these receptors is facilitated by CD14 in cell membranes and LPS-binding protein in plasma.[30] Interaction of LPS with cellular receptors precipitates a complex network of events that, at large LPS doses, culminates in hepatocellular necrosis. One proposed scenario of these events is depicted in Figure 29.4. In the liver, binding of LPS to receptors on Kupffer cells results in cell activation, which in turn leads to expression of proinflammatory genes and the release of mediators including cytokines such as tumor necrosis factor-alpha (TNF)and interleukin-1, reactive oxygen species, proteases, and lipid metabolites such as platelet activating factor and prostaglandin D_2 (PGD_2).[31–35] Complement and coagulation systems in the plasma are also activated by LPS exposure through direct and indirect mechanisms.[36] Several of these mediators may have direct effects on parenchymal cells although overt cell injury appears to involve additional, downstream events. These include activation of other cells such as sinusoidal endothelium, stellate cells and platelets. The process of cell activation involves transcription factors such as nuclear factor-κB (NFκB), activation of which contributes to the expression of genes, resulting in products that amplify and extend the inflammatory response.[37–39] Stellate cell activation can lead to sinusoidal contraction, which promotes arrest of neutrophils in the microvasculature.[40] Activation of endothelial cells in sinusoids results in the expression of adhesion molecules that appear to be involved in transmigration of neutrophils, placing them in contact with parenchymal cells and activating them to release lysosomal proteases and other agents.[41,42] These neutrophil-derived agents, especially toxic proteases such as cathepsin G and elastase, appear to be important for hepatocellular necrosis during experimental endotoxemia and other acute, inflammatory liver conditions.[43,44]

It should be emphasized that the particular sequence described above and depicted in Figure 29.4 is a simplification, and the roles of numerous players remain a topic of considerable investigation and debate. Nevertheless, it is clear that inflammatory injury to liver and other tissues requires numerous events, cells, and mediators. Some of the same inflammatory factors may be critical to the amplification of toxic responses to xenobiotic agents by concurrent inflammation. Accordingly, recognition and understanding of these factors may provide clues as to how modest inflammation exacerbates toxic responses.

With this in mind we and others have begun to explore factors that are critical to the amplification of toxic responses by small doses of LPS in animals. The interaction between galactosamine and LPS is the most extensively studied amplification model to date. In this model, many of the factors that are key players in liver injury from large doses of LPS also play critical roles when smaller LPS doses are coupled with galactosamine administration to rodents. These factors include Kupffer cell activation, release of TNF and

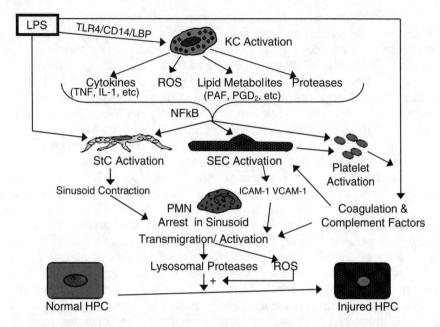

Figure 29.4 The initial event in endotoxin hepatotoxicity is binding of LPS to toll-like receptors and activation of Kupffer cells and perhaps other inflammatory cells. Various cellular and soluble mediators of inflammation are involved in downstream events critical to the ultimate injury to parenchymal cells of the liver. Many positive and negative feedback mechanisms that enhance or dampen the inflammatory response are not pictured. (Abbreviations: KC, Kupffer cell; StC, stellate cell; SEC, sinusoidal endothelial cell; PMN, polymorphonuclear leukocyte [neutrophil]; HPC, hepatic parenchymal cell. Other abbreviations are in text.)

reactive oxygen species, NFκB activation, expression of adhesion molecules on activated sinusoidal endothelium, and arrest, transmigration, and activation of blood neutrophils.[45-52]

LPS augmentation models involving other xenobiotic agents are less extensively studied but have provided some interesting contrasts. In a rat model of potentiation of allyl alcohol hepatotoxicity by LPS, inactivation of Kupffer cells prevented liver injury, suggesting that these cells are critical players.[17] Interestingly, however, a neutralizing antibody to TNF given to rats using a regimen that protects against liver injury from a large dose of LPS failed to provide protecton in this model.[53] This suggests that some other Kupffer cell-derived product may promote liver injury. A good candidate is PGD$_2$, which is formed by cyclooxygenase 2 (COX 2) and other enzymes in Kupffer cells and has been shown to render livers sensitive to injury.[54,55] Recent results indicate that inhibition of COX 2 reduces the potentiation of allyl alcohol hepatotoxicity by LPS, and addition of PGD$_2$ to parenchymal cells *in vitro* enhances the cytotoxic effect of allyl alcohol.[62]

In contrast to allyl alcohol, the augmentation of aflatoxin B_1 hepatotoxicity by LPS is prevented by TNF neutralization, suggesting a critical role for this cytokine.[56] In this model, neutrophil depletion prior to administration of aflatoxin B_1/LPS prevented parenchymal cell injury but did not affect injury to bile duct epithelial cells.[57] This result suggests that two different mechanisms, one neutrophil-dependent and the other not, are involved in the augmentation of aflatoxin B_1 hepatotoxicity by LPS and that the inflammatory factors that are critical to the response may depend on the target cell. A comprehensive review of the influence of LPS and other inflammagens on numerous other toxicants is beyond the scope of this short commentary. It will suffice to say that much remains to be learned about how inflammation enhances toxic responses and whether the critical inflammatory factors differ depending on the mechanism by which the xenobiotic agent acts. A general working hypothesis is shown in Figure 29.5. Numerous pathogens and other initiators of the inflammatory response such as LPS activate various cells. Likely to be important among these are macrophages and other inflammatory cells that accumulate in tissue and become activated to express and release a host of soluble mediators that directly or indirectly influence parenchymal cells that have been homeostatically altered by exposure to an environmental chemical or other xenobiotic agent. These influences might include gene expression, intracellular signal alteration, oxidative stress, membrane alterations, etc., that can push the homeostatically altered cell into a pathway leading to cell death. In addition, there is potential for inflammatory factors to influence metabolism of the xenobiotic agent (Figure 29.5). Cytokines produced during the inflammatory response are well known to control expression of genes encoding various isoforms of cytochromes P450.[58-61] For most cytochrome P450 isoforms, proinflammatory cytokines depress gene expression. Accordingly, xenobiotic agents that are inactivated by such isoforms may become more toxic during an inflammatory condition.

V. Biomarker development: research needs

Support for the hypothesis that concurrent inflammation represents a determinant of susceptibility to chemical intoxication currently rests on a growing database derived from experimental animal studies, a portion of which is reviewed above. One need is to evaluate the merit of this hypothesis in human and wildlife populations. To the knowledge of the authors, epidemiological or clinical studies of inflammation as a predisposing factor for toxic responses of any kind with any chemical are presently lacking.

The development of biomarkers of susceptibility related to the inflammatory response as a potentially important determinant is another issue. Choosing the most appropriate biomarkers will necessitate a better understanding of the mechanisms by which inflammation enhances toxic responses of various sorts. Which biomarkers of inflammation best predict increased sensitivity to chemical toxicity remains to be addressed. Are there

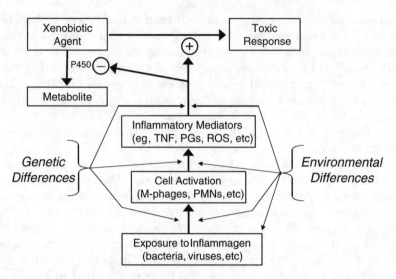

Figure 29.5 Inflammation can either influence the metabolism of a xenobiotic agent or modify its toxic action on cells to enhance tissue injury. This occurs through activation of cells and production of inflammatory mediators, processes which can be influenced by environmental and genetic factors.

biomarkers that are selective for various target organs? Results to date suggest that toxicities from xenobiotics that act by different mechanisms may be influenced by different inflammatory mediators. If so, will one biomarker of inflammation provide an adequate, general predictor of sensitivity, or will a cadre of them be needed? Many inflammatory factors appear and disappear rapidly during an inflammatory response, and the temporal relationship between onset of inflammation and toxicant exposure may strongly influence the magnitude of the toxic response. These qualities present additional challenges for biomarker development in this area.

To answer questions such as these will require substantial effort. At the next level, one might consider that environmental and genetic factors influence inflammatory responses (Figure 29.5). Drug metabolism polymorphisms have been a popular area of investigation as related to genetic determinants of susceptibility to toxic chemicals. Polymorphisms in genes that control the magnitude of inflammatory responses might prove to be quite important as well. It has only recently been recognized that differences in genes encoding the toll-like receptors have a major influence on biologic responses to inflammagens such as LPS. Similarly, polymorphisms in many genes involved the inflammatory cascade may influence individual susceptibility.

VI. Conclusion

There is substantial evidence that systemic exposure to inflammagens such as LPS is episodic and commonplace in people and that the magnitude of exposure can be influenced by a variety of conditions. Numerous studies in experimental animals indicate that exposure to LPS can markedly enhance the toxicities of a wide variety of xenobiotic agents. Together, these observations suggest that concurrent inflammation should be considered as a potentially important determinant of susceptibility to intoxication from environmental and other chemicals. However, additional study will be needed to understand the applicability to human and wildlife populations and to know which inflammatory factors and mechanisms are important so that appropriate biomarkers may be developed.

Acknowledgments

The authors are grateful to NIEHS for support of their research (grants ES04139 and ES08789) and to Mr. Ammar Hindash for his aid in preparing the manuscript.

References

1. Subcommittee on Pulmonary Toxicology, *Biologic Markers in Pulmonary Toxicology*, National Academy Press, Washington, D.C., 1989, pp. 2–3.
2. Nolan, J.P., Endotoxin, reticuloendothelial function, and liver injury, *Hepatology*, 1, 458–465, 1981.
3. Jacob, A.I. et al., Endotoxin and bacteria in portal blood, *Gastroenterology*, 72, 1268–1270, 1977.
4. Prytz, H. et al., Portal venous and systemic endotoxaemia in patients without liver disease and systemic endotoxaemia in patients with cirrhosis, *Scand. J. Gastroenterol.*, 11, 857–863, 1976.
5. Roth, R.A. et al., Is exposure to bacterial endotoxin a determinant of susceptibility to intoxication from xenobiotic agents? *Toxicol. Appl. Pharmacol.*, 147, 300–311, 1997.
6. Berger, D. et al., Time scale of interleukin-6, myeloid related proteins (MRP), C reactive protein (CRP), and endotoxin plasma levels during the postoperative acute phase reaction, *Shock*, 7, 422–426, 1997.
7. Spaeth, G. et al., Bulk prevents bacterial translocation induced by oral administration of total parenteral solution, *J. Parenter. Enteral. Nutr.*, 14, 442–447, 1990.
8. Rutenburg, A.M. et al., The role of intestinal bacteria in the development of dietary cirrhosis in rats, *J. Exp. Med.*, 106, 1–14, 1957.
9. Deitch, E.A., Microbial gastrointestinal translocation, in *Surgical Infections*, Fry, D.E. Ed., Little, Brown and Company, Boston, 1995, pp. 705–715.
10. Hurley, J.C., Antibiotic-induced release of endotoxin: a therapeutic paradox, *Drug Saf.*, 12, 183–195, 1995.
11. Brock-Utne, J.G. et al., Endotoxaemia in exhausted runners after a long-distance race, *S. Afr. Med J.*, 73, 533–536, 1988.
12. Rylander, R. and Haglind, P., Airborne toxins and humidifier disease, *Clin. Allergy*, 14, 109–112, 1984.
13. Dosman, J.A. et al., Chronic bronchitis and decreased forced expiratory flow rates in lifetime nonsmoking grain workers, *Am. Rev. Respir. Dis.*, 121, 11–16, 1980.

14. Pernis, B. et al., The role of bacterial endotoxins in occupational diseases caused by inhaling vegetable dusts, *Br. J. Indust. Med.*, 18, 120–129, 1961.
15. Mattsby-Baltzer, I. et al., Microbial growth and accumulation in industrial metal-working fluids, *Appl. Environ. Microbiol.*, 55, 2681–2689, 1989.
16. Peterson, R.D., Wicklund, P.E., and Good, R.A., Endotoxin activity of house dust extracts, *J. Allergy*, 35, 134–142, 1964.
17. Sneed, R.A. et al., Bacterial endotoxin enhances the hepatotoxicty of allyl alcohol, *Toxicol. Appl. Pharmacol.*, 144, 77–87, 1997.
18. Barton, C.C. et al., Bacterial lipopolysaccharide exposure augments aflatoxin B_1-induced liver injury, *Toxicol Sci.*, 55, 444–452, 2000.
19. Chao, T.C., Maxwell, S.M., and Wong, S.Y., An outbreak of aflatoxicosis and boric acid poisoning in Malaysia: a clinicopathological study, *J. Pathol.*, 164, 225–233, 1991.
20. Groopman, J.D. et al., Molecular epidemiology of aflatoxin exposures: validation of aflatoxin-N7-guanine levels in urine as a biomarker in experimental rat models and humans, *Environ. Health Perspect.*, 99, 107–113, 1993.
21. Jacobson, L.P. et al., Oltipraz chemoprevention trial in Qidong, People's Republic of China: study design and clinical outcomes, *Cancer Epidemiol. Biomarkers. Prev.*, 6, 257–265, 1997.
22. Qian, G.S. et al., A follow-up study of urinary markers of aflatoxin exposure and liver cancer risk in Shanghai, People's Republic of China, *Cancer Epidemiol. Biomarkers. Prev.*, 3, 3–10, 1994.
23. Ross, R.K. et al., Urinary aflatoxin biomarkers and risk of hepatocellular carcinoma, *Lancet*, 339, 943–946, 1992.
24. Yee, S.B. et al., Synergistic hepatotoxicity from coexposure to bacterial endotoxin and the pyrrolizidine alkaloid monocrotaline, *Toxicol. Appl. Pharmacol.*, 166, 173–185, 2000.
25. Fanucchi, M.V., Hotchkiss, J.A., and Harkema, J.R., Endotoxin potentiates ozone-induced mucous cell metaplasia in rat nasal epithelium, *Toxicol. Appl. Pharmacol.*, 152, 1–9, 1998.
26. Rumbeiha, W.K. et al., Augmentation of mercury-induced nephrotoxicity by endotoxin in the mouse, *Toxicology*, 151, 103–116, 2000.
27. Zhou, H.R. et al., Lipopolysaccharide and the trichothecene vomitoxin (deoxynivalenol) synergistically induce apoptosis in murine lymphoid organs, *Toxicol. Sci.*, 53, 253–263, 2000.
28. Means, T.K., Golenbock, D.T., and Fenton, M.J., The biology of toll-like receptors, *Cytokine Growth Factor Rev.*, 11, 219–232, 2000.
29. Beutler, B., Tlr4: central component of the sole mammalian LPS sensor, *Curr. Opin. Immunol.*, 12, 20–26, 2000.
30. Landmann, R., Muller, B., and Zimmerli, W., CD14, new aspects of ligand and signal diversity, *Microbes Infect.*, 2, 295–304, 2000.
31. Hewett, J.A. and Roth, R.A., Hepatic and extrahepatic pathobiology of bacterial lipopolysaccharides, *Pharmacol. Rev.*, 45, 382–411, 1993.
32. Burrell, R., Human responses to bacterial endotoxin, *Circ. Shock*, 43, 137–153, 1994.
33. Hardie, E.M. and Kruse-Elliott, K., Endotoxic shock, Part I: a review of causes, *J. Vet. Intern. Med.*, 4, 258–266, 1990.
34. Watson, R.W., Redmond, H.P., and Bouchier-Hayes, D., Role of endotoxin in mononuclear phagocyte-mediated inflammatory responses, *J. Leukoc. Biol.*, 56, 95–103, 1994.

35. Jaeschke, H., Reactive oxygen and mechanisms of inflammatory liver injury, *J. Gastroenterol. Hepatol.*, 15, 718–724, 2000.
36. Pearson, J.M. et al., The thrombin inhibitor, hirudin, attenuates lipopolysaccharide-induced liver injury in the rat, *J. Pharmacol. Exp. Ther.*, 278, 378–383, 1996.
37. Abraham, E., NF-kappaB activation, *Crit. Care Med.*, 28, N100–N104, 2000.
38. Wu, J. and Zern, M.A., NF-kappa B, liposomes and pathogenesis of hepatic injury and fibrosis, *Front Biosci.*, 4, D520-D527, 1999.
39. Christman, J.W., Lancaster, L.H., and Blackwell, T.S., Nuclear factor kappa B: A pivotal role in the systemic inflammatory response syndrome and new target for therapy, *Intensive Care Med.*, 24, 1131–1138, 1998.
40. Jaeschke, H. et al., Mechanisms of inflammatory liver injury: adhesion molecules and cytotoxicity of neutrophils, *Toxicol. Appl. Pharmacol.*, 139, 213–226, 1996.
41. Essani, N.A. et al., Transcriptional activation of vascular cell adhesion molecule-1 gene *in vivo* and its role in the pathophysiology of neutrophil-induced liver injury in murine endotoxin shock, *J. Immunol.*, 158, 5941–5948, 1997.
42. Essani, N.A., Fisher, M.A., and Jaeschke, H., Inhibition of NF-kappa B activation by dimethyl sulfoxide correlates with suppression of TNF-alpha formation, reduced ICAM-1 gene transcription, and protection against endotoxin-induced liver injury, *Shock*, 7, 90–96, 1997.
43. Ho, J.S. et al., Identification of factors from rat neutrophils responsible for cytotoxicity to isolated hepatocytes, *J. Leukoc. Biol.*, 59, 716–724, 1996.
44. Sauer, A. et al., Endotoxin-inducible granulocyte-mediated hepatocytotoxicity requires adhesion and serine protease release, *J. Leukoc. Biol.*, 60, 633–643, 1996.
45. Xu, H. et al., Leukocytosis and resistance to septic shock in intercellular adhesion molecule 1-deficient mice, *J. Exp. Med.*, 180, 95–109, 1994.
46. Nowak, M. et al., LPS-induced liver injury in D-galactosamine-sensitized mice requires secreted TNF-alpha and the TNF-p55 receptor, *Am. J. Physiol. Regul. Integr. Comp. Physiol.*, 278, R1202-R1209, 2000.
47. Takayama, F., Egashira, T., and Yamanaka, Y., NO contribution to lipopolysaccharide-induced hepatic damage in galactosamine-sensitized mice, *J. Toxicol. Sci.*, 24, 69–75, 1999.
48. Stachlewitz, R.F. et al., Glycine and uridine prevent D-galactosamine hepatotoxicity in the rat: role of Kupffer cells, *Hepatology*, 29, 737–745, 1999.
49. Jaeschke, H. et al., Glutathione peroxidase-deficient mice are more susceptible to neutrophil-mediated hepatic parenchymal cell injury during endotoxemia: importance of an intracellular oxidant stress, *Hepatology*, 29, 443–450, 1999.
50. Jaeschke, H. et al., Activation of caspase 3 (CPP32)-like proteases is essential for TNF-alpha-induced hepatic parenchymal cell apoptosis and neutrophil-mediated necrosis in a murine endotoxin shock model, *J. Immunol.*, 160, 3480–3486, 1998.
51. Chosay, J.G. et al., Neutrophil margination and extravasation in sinusoids and venules of liver during endotoxin-induced injury, *Am. J. Physiol.*, 272, G1195-G1200, 1997.
52. Komatsu, Y. et al., Role of platelet-activating factor in pathogenesis of galactosamine-lipopolysaccharide-induced liver injury, *Dig. Dis. Sci.*, 41, 1030–1037, 1996.
53. Sneed, R.A. et al., Pentoxifylline attenuates bacterial lipopolysaccharide-induced enhancement of allyl alcohol hepatotoxicity, *Toxicol. Sci.*, 56, 203–210, 2000.

54. Brouwer, A. et al., Production of eicosanoids and cytokines by Kupffer cells from young and old rats stimulated by endotoxin, *Clin. Sci.*, 88, 211–217, 1995.

55. Puschel, G.P. et al., Increase in prostanoid formation in rat liver macrophages (Kupffer cells) by human anaphylatoxin C3a, *Hepatology*, 18, 1516–1521, 1993.

56. Barton, C.C. et al., Endotoxin potentiates aflatoxin B_1-induced hepatotoxicity through a TNF-dependent mechanism, *Hepatology*, 33, 66–73, 2001.

57. Barton, C.C., Ganey, P.E., and Roth, R.A., Lipopolysaccharide augments aflatoxin B_1-induced liver injury through neutrophil-dependent and independent mechanisms, *Toxicol. Sci.*, 58, 208–215, 2000.

58. Liu, J. et al., Endotoxin pretreatment protects against the hepatotoxicity of acetaminophen and carbon tetrachloride: role of cytochrome P450 suppression, *Toxicology*, 147, 167–176, 2000.

59. Monshouwer, M. and Witkamp, R.F., Cytochromes and cytokines: changes in drug disposition in animals during an acute phase response, *Vet. Q.*, 22, 17–20, 2000.

60. Poloyac, S.M. et al., The effect of endotoxin administration on the pharmacokinetics of chlorzoxazone in humans, *Clin. Pharmacol. Ther.*, 66, 554–562, 1999.

61. Siewert, E. et al., Hepatic cytochrome P450 down-regulation during aseptic inflammation in the mouse is interleukin 6 dependent, *Hepatology*, 32, 49–55, 2000.

62. Ganey, P.E. et al., Involvement of cyclooxygenase-2 in the potentiation of allyl alcohol-induced liver injury by bacterial lipopolysaccharide, *Toxicol. Appl. Pharmacol.*, 174, 113–121, 2001.

chapter thirty

Modulation of GAP junctional communication by "epigenetic" toxicants: a shared mechanism in teratogenesis, carcinogenesis, atherogenesis, immunomodulation, reproductive- and neurotoxicities

James E. Trosko, Chia-Cheng Chang, and Brad Upham

Contents

Abstract Homeostasis in multicellular organisms, needed to regulate cell proliferation, differentiation, apoptosis, and adaptive responses of terminally differentiated cells, is mediated by an integration of extra-, intra- and gap junctional-intercellular communication (GJIC) mechanisms. The working hypothesis of this presentation is that with the disruption of normal GJIC, the following occurs:

1. During embryogenesis and fetal development can lead to either embryo lethality or birth defects
2. After initiation of stems cells can lead to the promotion of atherosclerotic plaques or to tumor promotion
3. During maturation of germ or lympho-reticular cells can lead to reproductive and immune toxicities
4. During development or functioning of the brain can lead to neurological dysfunctions

Most of the environmental toxicants, (e.g., PCBs, PBBs, 2,4,5-T, vomatoxin, methylmercury, cadmium, DDT, peroxisome proliferators, TCCD.), which can be teratogens, immune-modulators, tumor promoters, and reproductive- and neurotoxicants, work by epigenetic mechanisms and are able to inhibit GJIC reversibly, with threshold-like dose responses. Epigenetic mechanisms work by altering the expression of the genetic information at the transcriptional, translational, or posttranslational levels.

GJIC has been linked to the control of cell proliferation, differentiation, apoptosis and adaptive responses of differentiated cells. Experimental *in vitro* evidence, using a variety GJIC assays, has predicted the toxicities of many of these nongenotoxicants. Recent knock-out mice for several gap junction genes have demonstrated the critical role of GJIC in many of these disease states. Five human syndromes with inherited mutated gap junction genes for cardiovascular, deafness, cataract, neurological and epithelial tissue disorders have been identified. Recently, a method to measure GJIC *in vivo* has been developed and has been shown to link PCP's tumor promoting ability in rat liver to its ability to inhibit GJIC in rat liver *in vivo*. These assays to measure GJIC can be used to do the following:

1. Screen for this class of toxicants
2. Detect agents which ameliorate their toxicities
3. Assess the efficacies of remediation of toxic chemicals
4. Screen for chemopreventive agents
5. Study the mechanism by which these epigenetic chemicals work

In summary, the role that nonmutagenic environmental toxicants play in contributing to multiple disease states must include the mechanisms (cellular, biochemical, molecular) by which they trigger various signal transduction, intracellular communication mechanisms to modulate GJIC. Depending on the cell type and specific signal transduction mechanism that modulate

GJIC by the environmental toxicant, cells will either be induced to proliferate, differentiate, apoptose, or adaptively respond in an abnormal manner. Epigenetic toxicology, then, involves modulation of GJIC which could, depending on the developmental state of the organism, time of exposure, duration of exposure, cell and tissue type being exposed, and the concentration of the chemical, lead to many diverse chronic disease states.

I. Introduction

> Those researching the cancer problem will be practicing a dramatically different type of science than we have experience over the past 25 years. Surely much of this change will be apparent at the technical level. But ultimately, the more fundamental change will be conceptual.[1]

Mesmerized by the dazzling advances in biomedical research via molecular biology, all traditional disciplinary fields of biomedicine have been transformed by the techniques of molecular biology. Even the interdisciplinary field of toxicology, once driven by examining the toxic effects of chemicals at the whole animal, physiological or tissue levels, is being overtaken by molecular/biochemical studies of these toxic chemicals. While, in principle, these detailed studies have provided some useful insights, they have not fully provided the kinds of answers promised. Will the sequencing of the complete human genome answer the question as to how all diseases are caused, thereby giving us the correct sequence of DNA to prevent or cure the disease? In all likelihood, it will not for the reason that the higher organism, such as a rat or human being, is the result of a cybernetic interaction between a hierarchy of molecular, biochemical, organelle, cellular, tissue, organ, organ systems, having unique genetic components constantly interacting with environmental factors.[2] In other words, the whole is greater than the sum of its parts. One's linear sequence of nucleotides in their DNA is only one part of one's state of health or disease.

The Hanahan and Weinberg quote is meant to focus on the main thesis of this short review of mechanisms of toxicity. The word toxicity or even the field of toxicology means different things to different investigators. Originally, toxicity meant something is capable having a poisonous effect on an organism and toxicology meant a science to understand the method by which a toxicant or toxin might bring about the poisonous effect. Early methods to study the toxicity of chemicals on animals (birth defects, cancer, reproductive, neurological, behavioral effects, etc.), while providing some useful information on lethal doses, biologically toxic doses, could not provide a mechanistic understanding of the causes of these whole animal effects. With the advent of more reductionalistic approaches, cell-free, *in vitro*, biochemical, and molecular approaches were introduced into the field of toxicology to the point where many studies have disconnected the molecular levels with the whole animal levels.

It might be argued that the recent introduction of transgenic and knock-out, knock-in animals have reunited the reductionalistic approaches with the holistic approaches.

Again, there is some truth in this statement; however, philosophically and experimentally, more effort will be needed to make the connection. Recall the amazement when the predicted *p53* knock-out mouse survived and did not live up to the prediction; because *p53* was so vital to live, an organism not having this gene should not have survived. What is missing today in understanding the connection between the genetic information at the DNA level and the state of health or disease at the whole animal level is the complex interaction of the quality and quantity of that genetic information with the environmental factors at each level and how the different levels effects those above and below it.

The working hypothesis of this challenge to the prevailing paradigm which shapes most of the subdisciplinary fields of toxicology is that the modulation of GJIC can contribute to a common cellular mechanism to teratogenesis, carcinogenesis, atherogenesis, immunomodulation, and repro-ductive- and neurotoxicities.

II. Disrupted homeostasis as a shared toxicological mechanisms in many disease states

In a multicellular organism, such as a human being (at adulthood having approximately 100 trillion cells), the fertilized egg containing the total genetic information for the organism, must orchestrate the developmental process by regulating five cellular processes: cell proliferation, cell differentiation; apoptosis, adaptive functions of the terminally differentiated cells, and senescence of the cells. This includes generating pluripotent stem cells from the fertilized egg (toti-potent stem cell). These pluri-potent stem cells can, by either asymmetric cell division, give rise to one daughter which maintains stemness for further growth and repair and one daughter that can symmet-rically divide to expand the lineage of that particular differentiated cell type. The terminally differentiated cell, by becoming so highly specialized, appar-ently loses its ability to proliferate under most normal conditions Recent experiments on cloning with differentiated nucleic might belie that state-ment.[3] Herein lies one of the first concepts ignored in toxicology, namely exposure of a tissue/organ to a toxicant will not affect each cell type equally, i.e., pluripotent stem cell, progenitor cell, terminally differentiated cell equally. Normally, if a chemical is metabolized prior to becoming toxic, only cells with drug metabolizing capability will be affected.

Another factor in understanding how homeostasis during embryo, fetal, neonatal, adolescent, sexual, mature, and aging developmental processes occurs in the individual. From a conceptual framework, an integration of extra-, intra- and intercellular communication mechanisms, including cell-matrix and cell-adhesion communication mechanisms, constitutes the process of homeostasis.[2] Hormones, growth factors, cytokines, and neurotrans-mitters (extracellular signals) trigger a variety of intracellular signal trans-duction systems (Ca^{++}, protein kinases, ceramides, transcription factors, altered redox states, etc.), which in turn, modulates gene expression at the posttranslational, translational, or transcriptional levels as well as modulates the ability of cells to communicate via gap junctions.[4] This homeostatic concatenation of ionic/molecular information flow is what regulates whether a cell (stem; progenitor; terminally differentiated) proliferates, dif-ferentiates, apoptoses, adaptively responds, or senesces.

This delicate process must occur during development and maturation to allow the right amount of proliferation, the right kind and amount of differentiation, and appropriate apoptosis and adaptive responses in order that the individual is not born with a birth defect, can sexually mature, function neurologically/behaviorally in order to reproduce and function as a socially-contributing individual. Interference with this delicate homeostatic process can lead to many disease states, including embryonic, fetal neonatal death, birth defects, cancer, atherosclerosis, and reproductive and neurolog-ical dysfunctions.[5,6]

At the cellular level, mechanisms that can contribute to disruptions of homeostasis include mutagenesis, cell death (either necrosis or apop-tosis), and alterations in gene expression at the transcriptional, transla-tional, and posttranslational levels (epigenetic toxicology). Both mutations heredited via the germ line (e.g., xeroderma pigmentosum[7]) or induced in some pluripotent stem cell (e.g., mutated proto-oncogenes or tumor suppressor genes in carcinogenesis) can and does contribute to both genetic diseases and several somatic diseases (cancers, atherogenesis).[8] Massive cell death due to chemically-induced necrosis can also contribute to embryo/fetal death or birth defects as well as compensatory hyperpla-sia influencing the promotion phase of carcinogenesis[9] or atherogenesis.[10] Abnormal apoptosis, either induced or blocked has been shown to be correlated with many diseases.[11]

However, what has been generally ignored in the field of toxicology is the mechanism of chemically induced epigenetic alterations of gene expres-sion, without the induction of DNA damage or mutation induction in pluri-potent stem cells, at the transcriptional, translational, or posttranslational levels.[6,12] Nongenotoxic chemicals, such as TCDD, PCBs, PBBs, peroxisome proliferators, DDT, dieldrin, aldrin, perchlorophenol, drugs (e.g., phenobar-bital, thalidomide or forskolin), dietary ingredients (e.g., unsaturated fatty acids), green tea components, retinoids, carotenoids, etc., can either be down-or up-regulated GJIC at noncytotoxic levels.[13] Either by interfering with or mimicking the endogenous signaling extracellular molecules, these epige-netic toxicants can trigger intracellular signaling systems to either increase or decrease GJIC, which in turn, can alter gene expression. Finally, this could lead to abnormal induction of cell proliferation, differentiation, apoptosis, adaptive responses, or the senescence of cells.

III. GAP junctional communication: "where the toxic rubber hits the road to diseases"

In single cell organisms, survival is dependent on the ability of the organism to proliferate. When the physical environment induces changes that limits the ability of the single cell organism to proliferate, the species will die out if it were not for spontaneous or induced mutations existing within the population that can adapt to the new environmental conditions. Nutrient availability and temperature ranges are two major factors controlling cell proliferation in single cell organisms. During the transition from the single cell organism to the multicellular organism, new phenotypic functions appeared:

1. Growth control within the multicellular organism
2. Cellular differentiation that gave the organism higher levels of adaptive, survival abilities
3. Programmed cell death to allow sequential adaptive tissues during development
4. Formation of nested pluripotent stem cells
5. Adaptively responsive, terminally differentiated cells

Among the new genes that accompanied this set of new phenotypic traits not found in single cell organisms were the family of evolutionarily conserved genes coding for the gap junction proteins (connexin genes).[14]

These genes coded the proteins that self-organized into membrane-bound, heximeric hemichannels that linked up with the neighboring hemichannel, forming a means for ions and small molecular weight molecules to diffuse directly from cytoplasm of one cell to the cytoplasm of the coupled cell.[15] The cellular consequent of cellular coupling is to synchronize electrotonic activity in excitable tissues and to metabolically equilibrate non-excitable cells in tissues for pattern formation, growth control (contact inhibition), control of differentiation, and apoptosis.[16]

When Loewenstein discovered that normal cells, which control cell growth and normally differentiate, had functional gap junctions while cancer cells, which are characterized by the lack of growth control, cannot terminally differentiate and have abnormal apoptotic ability, do not have functional gap junctional intercellular communication,[17] the first link between the fundamental role that gap junctions played in homeostatic control of cellular functions and a disease (cancer) was made. Later, when nongenotoxic chemicals, such as the tumor-promoting phorbol esters, or DDT, were shown to inhibit GJIC reversibly, a mechanistic insight that demonstrated the importance of a complexity or higher phenomenon, that is an emergent or complexity[18] phenotype that goes beyond a reductionalistic approach to sequence only the DNA that codes for the connexin. When antibodies directed to the gap junction were applied to developing embryos and brought about birth defects,[19] as well as when various nongenotoxic chemicals, known to be reproductive-, immuno- and/or neurotoxicants were shown to modulate GJIC reversibly, additional evidence linked GJIC to other disease states. More recently, inherited mutated connexin genes in human beings were linked to the Charcot-Marie Tooth syndrome,[20] heart malformations,[21] nonsyndromic deafness,[22] cataracts,[23] and mucoepithelial dysplasia.[24] In addition, knock-out connexin mice have been associated with a wide variety of disease states, including fetal death, neonatal death, female reproductive disorder,[14] Charcot-Marie Tooth syndrome, and liver cancer predisposition.[25] These later knockouts clearly point out that many reductionalistic predictions, based only on molecular understanding but discounting other higher order functions, such as compensatory genes and functions in multicellular organisms, could not account for all the resulting phenotypes.

IV. GAP junctional intercellular communication: the biological rosetta stone in toxicology

The working hypothesis on which this brief analysis has been done should reflect the original quotation by Hanahan and Weinberg, as well as the phrase, "a biological Rosetta Stone." In the former case, what has been present is more of a integrating concept, than a list of detailed molecular/biochemical/cellular/whole animal experiments. Clearly, none of the given examples can reveal the concept that a fundamental integration of signaling mechanisms (extracellular, intracellular, GJIC, plus extracellular matrix-cellular and cell-cell adhesion) controls, in a homeostatic manner, the regulation of cell division, differentiation, apoptosis, and adaptive responses of stem, progenitor and terminally differentiated cells. On the other hand, reductionalistic approaches have ignored that biological systems are a cybernetic hierarchy of interacting systems that can be influenced by genetic and environmental interactions. The concept illustrates the old saw that the whole is greater than the sum of its parts.

The biological Rosetta Stone idea suggests that by looking at many different languages or different symbols representing a common experience (metaphor for the different disciplines examining, reductionalistically, carcinogenesis), one can sometimes infer the meaning of an unknown observation in one field (i.e., carcinogenesis) from similarities in two or more fields (e.g., teratogenesis, reproductive- and neurotoxicities) and can also infer the value of the process to do interdisciplinary examinations.

The concept being proposed is that during the evolution of multicellular organisms, the problem of controlling, in a homeostatic manner, the diverse cellular functions that must occur during embryonic, fetal, neonatal development, sexual maturation, and aging was solved by integrating the aforementioned signaling mechanisms to generate the correct ionic/molecular information for these diverse cellular options to occur, simultaneously, in one individual. Most cells in a multicellular organ do not exist as single cells but as coupled (electronically or metabolically) syncistium. We are not just 100 trillion individual cells but are a highly organized unit of cells. Normal genes, normal development, normal physiology, interacting with appropriate environmental factors, allow for growth factors, neurotransmitters, and hormones to trigger specific signal transducing systems in particular cells to do one of the potential cellular functions to occur by modulating GJIC. If the bottom line cellular communication mechanism, GJIC, is not modulated by these endogenous extracellular and intracellular signaling mechanisms, cells either do not uncouple or do not couple. The ground-state of the biological system is quiescent.

On the other hand, if the endogenous, extracellular, and intracellular communication mechanisms are interfered with at inappropriate times, nonadaptive cellular/whole animal consequences can develop. Modulation of embryo-fetal development could lead to the death of the organism or teratogenesis. Transient and chronic down regulation of GJIC by exogenous chemicals or by stable activation of oncogenes can bring about the proliferation of an initiated cell (tumor promotion or plaque formation in the artery). Up regulation of GJIC in gap junction defective cancer cells by chemicals can lead to anticancer therapy. Interference with differentiation in reproductive tissue (e.g., follicular cells and ova or between Sertoli and spermatogonia) can lead to reproductive dysfunction, while interference of GJIC between glial cells and neurons could lead to neurotoxicities. In other words, since GJIC exists in all tissues/organs and since GJIC can be modulated by normal episodic levels of endogenous extracellular signals, a means exists to regulate development, maturation, wound repair and adaptive functions. On the other hand, abnormal amounts/timing of either endogenous or exogenous extracellular signals could interfere with normal processes leading to a variety of disease states. Rosenkranz et al.,[26] seemed to confirm this idea when they concluded: "It was also surprising that irrespective of inhibition of GJIC, the observed prevalence of molecules that have the potential for jointly inducing allergic contact dermatitis, ocular irritation, sensory irritation, and respiratory hypersensitivity [teratogenesis and tumor promotion] is much greater than expected… again suggesting a commonality between the phenomena. These suggest a commonality in mechanisms that is worthy of further study."

In summary, this integrating concept was best described by Van P. Potter when he stated, "The biochemistry of cancer is a problem that obligates the investigator to combine the reductionalistic approaches of the molecular biologists with the holistic requirements of hierarchies within the organism. The cancer problem is not merely a cell problem, it is a problem of cell interaction, not only within tissues, but also with distal cells in other tissues. But in stressing the whole organism, we must also remember that the integration of normal cells with the welfare of the whole organism is brought about entirely by molecular messages acting on molecular receptors".[27]

Acknowledgments

The authors wish to acknowledge our appreciation for the excellent word processing skills of Mrs. Robbyn Davenport. In addition, the manuscript was written while the authors were supported in part by a grant from the NIEHS Basic Science Research Program [2 P42 ES04911].

References

1. Hanahan, D. and Weinberg, R.A., The hallmarks of cancer, *Cell*, 100, 57–70, 2000.
2. Trosko, J.E., Hierarchical and cybernetic nature of biological systems and their relevance to homeostatic adaptation to low level exposures to oxidative stress-inducing agents, *Environ. Health Perspect.*, 106, 331–339, 1998.
3. Wilmut, I. et al., Viable offspring derived from fetal and adult mammalian cells, *Nature*, 385, 810–813, 1997.
4. Trosko, J.E. and Ruch, R.J., Cell–cell communication in carcinogenesis, *Frontiers in Bioscience*, 3, 208–236, 1998.
5. Trosko, J.E., Chang, C.C., and Madhukar, B.V., *In vitro* analysis of modulators of intercellular communication: implications for biologically based risk assessment models for chemical exposures, *Toxicol. In Vitro*, 4, 635–643, 1990.
6. Trosko, J.E. et al., Epigenetic toxicology as toxicant-induced changes in intracellular signalling leading to altered gap junctional intercellular communication, *Toxicol. Letters*, 102–103, 71–78, 1998.
7. Cleaver, J.E., Xeroderma pigmentosum: genetic and environmental influences in skin carcinogenesis, *Int. J. Dermatol.*, 17, 435–444, 1978.
8. Trosko, J.E. et al., Genetic predispositions to initiation or promotion phases in human carcinogenesis, in *Biomarkers, Genetics, and Cancer*, Anton-Guirgis, H. and Lynch, H.T., Eds., Van Nostrand Reinhold Company, New York, 1985, pp. 13–37.
9. Trosko, J.E., Chang, C.C., and Medcalf, A., Mechanisms of tumor promotion: potential role of intercellular communication, *Cancer Invest.*, 1, 511–526, 1983.
10. Majesky, M.W. et al., Focal smooth muscle proliferation in the arotic intima produced by an initiation-promotion sequence, *Proc. Natl. Acad. Sci. USA*, 82, 3450–3454, 1985.
11. Thompson, C.B., Apoptosis in the pathogenesis and treatment of disease, *Science*, 267, 1456–1462, 1995.
12. Trosko, J.E. and Chang, C.C., Nongenotoxic mechanisms in carcinogenesis: role of inhibited intercellular communication, in *Banbury Report 31: Carcinogen Risk Assessment: New Directions in the Qualitative and Quantitative Aspects*, Hart, R. and Hoerger, F.D., Eds., Cold Spring Harbor Laboratory Press, Cold Spring Harbor, New York, 1988, pp. 139–170.
13. Trosko, J.E. and Chang, C.C., Modulation of cell–cell communication in the cause and chemoprevention/chemotherapy of cancer, *BioFactors*, 12, 259–263, 2000.
14. Bruzzone, R., White, T.W., and Paul, D.L., Connections with connexins: the molecular basis of direct intercellular signaling, *Eur. J. Biochem.*, 238, 1–27, 1996.
15. Loewenstein, W.R., Junctional intercellular communication: The cell–cell membrane channel, *Physiol. Rev.*, 61, 829–913, 1981.
16. Simon, A.M. and Goodenough, D.A., Diverse functions of vertebrate gap junctions, *Trends Cell Biol.*, 8, 477–483, 1988.
17. Loewenstein, W.R., Permeability of membrane junctions, *Ann. NY Acad. Sci.*, 137, 441–472, 1966.

18. Weng, C., Bhalla, U.S., and Lyengar, R., Complexity in biological signaling, *Science*, 284, 92–95, 1999.
19. Warner, A.E., Guthrie, S.C., and Gilula, N.B., Antibodies to gap junctional proteins selectively disrupt junctional communication in the early amphibian embryo, *Nature*, 311, 127–131, 1984.
20. Bergoffen, J. et al., Connexin mutations in X-linked Charcot-Marie-Tooth disease, *Science*, 262, 2039–2042, 1993.
21. Britz-Cunningham, S.H. et al., Mutations of the connexin 43 gap junction gene in patients with heart malformations and defects of laterality, *N. Engl. J. Med.*, 332, 1323–1329, 1995.
22. Shiels, A. et al., A missense mutation in the human connexin-50 gene (GJA8) underlies autosomal dominant "zonular pulverulent" cataract, on chromosome 1q, *Am. J. Hum. Genet.*, 62, 526–532, 1998.
23. Kelsell, D.P. et al., Connexin26 mutations in hereditary nonsyndromic sensorineuronal deafness, *Nature*, 387, 80–83, 1997.
24. Witkop, C.J. et al., Hereditary mucoepithelial displasia: A disease apparently of desmosome and gap junction formation, *Am. J. Human. Genet.*, 31, 414–427, 1979.
25. Temme, A. et al., High incidence of spontaneous chemically-induced liver liver tumors in mice deficient for connexin32, *Curr. Biol.*, 7, 713–716, 1997.
26. Rosenkranz, H.R., Pollack, N., and Cunningham, A.R., Exploring the relationship between the inhibition of gap junctional intercellular communication and other biological phenomena, *Carcinogenesis*, 21, 1007–1011, 2000.
27. Potter, V.P., Phenotypic diversity in experimental hepatomas: the concept of partially blocked ontogeny, *Br. J. Cancer*, 38, 1–23, 1978.

chapter thirty-one

Exposure to POPs and other congeners

Stephen H. Safe

Contents

Abstract   Persistent organic pollutants (POPs) are typically lipophilic compounds that are resistant to environmental breakdown and preferentially bioconcentrate in fish, wildlife, and human tissues. Most POPs are halogenated hydrocarbons as are some of their phenolic metabolites, including polychlorinated biphenyls (PCBs), dibenzofuran (PCDFs), dibenzo-*p*-dioxins (PCDDs), hydroxy-PCBs, some brominated aromatics, and organochlorine pesticides such as DDT and its metabolite DDE. High-resolution analytical techniques can now detect low levels of individual POPs in wildlife, human adipose tissue, and serum, and concentrations of different classes of POPs or individual congeners are now routinely used as biomarkers of exposure and correlated with adverse human health effects in various cohort studies. For example, studies in several countries have correlated *in utero* or early postnatal exposure to total PCBs with neurodevelopmental deficits in offspring. However, there are many inconsistencies in these correlational studies, and these include the sample timing for PCB analysis (e.g., *in utero* cord blood, breast milk), analyte quantitation (e.g., individual congeners, total PCBs, TEQs), and differences in neurodevelopmental testing. Increased levels of POPs have been observed in case-control studies using other patient groups (e.g., breast cancer), and there has been considerable variability between studies in these correlations. Delineating the potential role of POPs in various diseases must take into account several factors including biological plausibility, time of exposure, and other factors that will be discussed.

I. Introduction

POPs encompass a structurally-diverse group of halogenated hydrocarbons and aromatic compounds that are routinely detected in extracts from most environmental matrices (Figure 31.1.).[1-4] These chemicals are synthetic industrial compounds or combustion by-products and include PCBs, PCDDs, PCDFs, other halogenated aromatics, and organochlorine pesticides and their breakdown products.[5-10] POPs are characterized by their resistance to chemical and biological breakdown and lipophilicity, and these characteristics result in their environmental persistence and bioaccumulation in the food chain.[1-4] POP residues are routinely detected in air, water, fish, wildlife, and humans, and often undergo atmospheric transport from regions of high POP production/use to nonindustrial areas. This accounts for detection of these compounds in the environment and in humans living in remote polar Arctic communities.[1-4,11] POPs, such as PCBs, 1,1,1-trichloro-2–2-bis(*p*-chlorophenyl)ethane (p,p'-DDT), and 1,1-dichloro-2,2-bis(*p*-chlorophenyl)ethylene (p,p'-DDE) were initially detected as environmental contaminants in the late 1960s to early 1970s; DDT is still used for insect control of malaria in some less developed regions. POPs have continued to be a major concern of regulatory agencies worldwide and regulations regarding the uses and ultimate disposition of these compounds have been promulgated in many countries/regions. The success of these regulatory efforts is reflected in results of a recent study from Sweden which documents human breast milk levels of several classes of POPs from 1967–1997.[4] The results show that the most common POPs found in the environment have been decreasing dramatically with a >90 percent decline observed for DDE/DDT levels (Figure 31.2). POPs have been declining in most regions[4,12-16] with the major exception of region-specific hot spots where some wildlife populations may still have elevated levels of regional/local contaminants or dietary patterns involving high consumption of fish.[17] Interestingly, in the Swedish survey, levels of brominated diphenylethers, a minor POP category (1 percent of total POPs in human milk), have been increasing, and it will be important to determine the sources of these compounds and restrict their release into the environment.

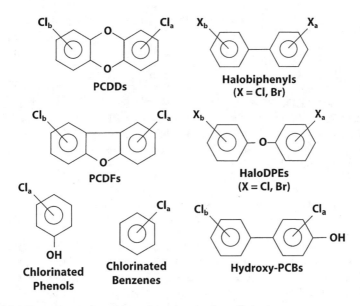

Figure 31.1 Structures of some persistent organic pollutants.

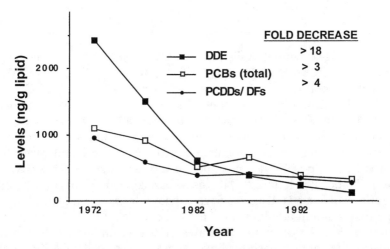

Figure 31.2 Levels of DDE, PCBs, PCDDs, and PCDFs in Swedish breast milk from 1972 to 1997. (From Norén, K. and Meironyté, D., *Chemosphere*, 40, 1111–1123, 2000.)

II. POP toxicology

A.   AHR-active compounds

The toxicology of POPs has been extensively investigated in laboratory animal models and initial studies primarily focused on those compounds that induce toxicity through the aryl hydrocarbon receptor (AHR).[18–23] 2,3,7,8-tetrachlorodibenzo-*p*-dioxin (TCDD) is the prototypical AHR agonist and

Table 31.1 TEFs for PCB, PCDD, and PCDF Congeners

Congener	TEF Value	Congener	No.	TEF Value
PCDDs		**Non-ortho PCBs**		
2,3,7,8-TCDD	1	3,3',4,4'-tetraCB	77	0.0001
1,2,3,7,8-PnCDD	1	3,4,4',5-tetraCB	81	0.0001
1,2,3,4,7,8-HxCDD	0.1	3,3',4,4',5-pentaCB	126	0.1
1,2,3,6,7,8-HxCDD	0.1	3,3',4,4',5,5'-hexaCB	169	0.01
1,2,3,7,8,9-HxCDD	0.1			
1,2,3,4,6,7,8-HpCDD	0.01			
OCDD	0.0001			
PCDFs		**Mono-ortho PCBs**		
2,3,7,8-TCDF	0.1	2,3,3',4,4'-pentaCB	105	0.0001
1,2,3,7,8-PnCDF	0.05	2,3,4,4',5-pentaCB	114	0.0005
2,3,4,7,8-PnCDF	0.5	2,3',4,4',5-pentaCB	118	0.0001
1,2,3,4,7,8-HxCDF	0.1	2',3,4,4',5-pentaCB	123	0.0001
1,2,3,6,7,8-HxCDF	0.1	2,3,3',4,4',5-hexaCB	156	0.0005
1,2,3,7,8,9-HxCDF	0.1	2,3,3',4,4',5'-hexaCB	157	0.0005
2,3,4,6,7,8-HxCDF	0.1	2,2',4,4',5,5'-hexaCB	167	0.00001
1,2,3,4,6,7,8-HpCDF	0.01	2,2',3,3',4,4',5-heptaCB	189	0.0001
1,2,3,4,7,8,9-HpCDF	0.01			
OCDF	0.0001			

hazard/risk assessment of POPs has focused, in part, on those halogenated aromatics (i.e., PCBs, PCDDs, and PCDFs) that exhibit TCDD-like toxicity. The major compounds of concern are summarized in Table 31.1. TCDD and related AHR agonists induce common biologic and toxic responses, including induction of phase 1 and phase 2 drug-metabolizing enzymes, modulation of many other genes, endocrine disruption, a wasting syndrome, immunotoxicity, modulation of carcinogenesis (enhancement and protection), hepatoxicity and porphyria, reproductive and developmental toxicity, chloracne, and other skin lesions.[18-23] It was recognized that the major differences in the effects of most AHR agonists were their potencies, and their rank order competitive binding affinities for the AHR correlated with their toxic/biochemical potencies for many responses. The correlations led to the toxic equivalency factor approach for hazard/risk assessment for AHR agonists (e.g., Table 31.1) in POP mixtures where

$$TEQ = \Sigma \, [PCDF_i][TEF_i] + \Sigma \, [PCDD_i][TEF_i] + \Sigma \, [PCB_i][TEF_i] + \ldots$$

TEQ represents TCDD or toxic equivalents, which is equal to the summation of the concentrations of individual PCDF, PCDD, and PCB congeners times their relative potency (TEF) compared to TCDD.[24-31] Toxic equivalency factors (TEFs) have been developed for the important environmentally relevant AHR-active PCB/PCDD/PCDF congeners, and the assigned values are derived from experimental studies in which congener potencies relative to TCDD (assigned a TEF of 1.0) were determined. For many individual compounds, the ranges of experimentally derived TEFs may vary by over 100-fold for different responses; however, it has been shown in several studies that calculated TEQs for mixtures of PCDDs and PCDFs using the TEFs shown in Table 31.1 are similar to experimentally derived TEQs.[32-34] Thus, TEQs can be used to estimate the potential TCDD-like toxicity of PCDD/PCDF mixtures, and TEFs/TEQs are routinely used for hazard and risk assessment of these compounds. However, in environmental samples, PCDDs/PCDFs are usually present along with much higher concentrations of PCBs (e.g., Figure 31.2), and extensive toxicity studies with PCB mixtures alone or in combination with TCDD or other PCDDs/PCDFs result in additive and nonadditive responses that are species- and response-dependent.[35,36] For example, induction of CYP1A1-dependent hepatic microsomal enzyme activity by Aroclors 1242, 1254, and 1260 in immature male Wistar rats gave essentially additive responses for the major AHR-active mono- and di-*ortho* substituted congeners. In contrast, the calculated immunotoxic potencies for commercial Aroclors in mice (e.g., Σ [PCB]$_i$ × TEF$_i$) were significantly higher than the observed immunotoxic effects indicating that PCB congeners in the mixture were antagonistic.[37] Subsequent studies using weak AHR agonists, such as PCB #153 (2,2',4,4',5,5'-hexaCB) or commercial Aroclor mixtures, and potent AHR-active compounds, including PCB #126 (3,3',4,4',5-pentaCB) and TCDD, clearly demonstrated nonadditive (antagonistic) interactions. For example, PCB #153 or Aroclor mixtures inhibit TCDD- or PCB #126-induced cleft palate formation, hydronephrosis, porphyria and immunotoxicity in mice;[37-43] embryotoxicity in Japanese medaka;[44] chick embryo malformations, edema and liver lesions;[45-46] and induction of CYP1A1-dependent activity (or reporter gene activity) in rat, mouse and chick embryo-derived cell lines.[47-50] These results are consistent with weak AHR agonists (e.g., Aroclors/PCB #153) exhibiting partial AHR antagonist activity and the AHR antagonist/agonist ratios of >1000/1 required for these inhibitory interactions are similar to those observed in environmental samples. For example, the ratio of the PCBs (324 ng/g lipid)/PCB #126 (0.076 ng/g/lipid) or Σ PCDDs/PCDFs in Swedish breast milk samples taken in 1997 was >1000.[4] In addition, many phytochemicals, such as flavonoids, indole-3-carbinol and related compounds, and carotenoids, also exhibit weak AHR agonist and partial AHR antagonist activities[51-54] and are present at high concentrations in the diet. Thus, a comprehensive risk assessment of low-level dietary POP-TEQs in food should factor in dietary intakes of phytochemical-derived AHR agonists/antagonists, as well as PCDDs, PCDFs, and PCB. Unfortunately, this question has not been seriously considered by regulators and regulatory scientists, and this is due, in part, to the lack of data on these interactions.

B.   AHR-inactive compounds

Several hundred POPs have been identified in the environment and hazard assessment of these compounds alone or as mixtures is poorly understood.[55]

Organochlorine pesticides induce a diverse spectrum of biochemical and toxic responses including neurotoxicity.[56-58] o,p'-DDT is estrogenic in some assays and p,p'-DDE is an antiandrogen.[59,60] A growing number of biochemical and toxic responses have been reported for the AHR-inactive PCBs and/or their hydroxyPCB metabolites, including neurotoxicity, estrogenic and antiestrogenic activities, modulation of multiple enzyme activities, tumor promotion activity, induction of phase 1 and phase 2 drug metabolizing enzymes, modulation of thyroid hormone responses, reproductive and developmental toxicity and immunomodulatory effects.[61-75] At present, it is not possible to develop integrative processes (e.g., TEQs/TEFs) for risk assessment of these mixtures.

III. Environmental POPs

A. Induced responses in human population

Several accidents involving high-level exposure to PCDDs, PCBs, and PCDFs have demonstrated a number of toxic responses in these populations including chloracne and other severe dermatological lesions, immune dysfunctions, and neurodevelopmental deficits in offspring exposed to these compounds *in utero*/early postnatal.[76-81] A recent study reported on the medical condition of individuals 14 years after accidental exposure to PCB/PCDF-contaminated rice oil in Taiwan, and it was concluded that these individuals "reported more frequent medical problems including skin diseases, goiter, anemia, and joint and spine diseases."[82] Several industrial cohorts exposed to TCDD as a contaminant during chemical production (usually involving chloro-phenols/chlorophenoxy herbicides) have been examined for cancer and other diseases, and the results tend to be inconsistent between studies.[83-86] For example, some studies of occupationally exposed workers showed increased incidence of ischemic heart disease and diabetes, whereas in a large U.S. industrial cohort, the overall incidence of these diseases was not increased although there may be some exposure-dependent correlations. Most of the industrial cohorts show an overall increased incidence of cancer; however, the pattern of increased individual cancers among these highly exposed groups is inconsistent. It is remarkable that the focus of these studies has always been on relatively low-level exposure to TCDD which is usually a trace contaminant involving exposures to high levels of other chemicals, such as chlorophenols/chlorophenoxy herbicides. The inconsistent results on disease incidence between studies suggests that compounds other than TCDD may be causal agents.

Jacobson and coworkers first reported on the possible linkage of environmental exposures to PCBs and neurodevelopmental defects in the offspring.[57] Their studies showed correlation with cord-blood PCB levels and poorer performance on a number of tests that measure developmental function in infants and children. These studies were carried out on a Michigan cohort of mothers many of whom had eaten Lake Michigan fish in the early

1980s. Moreover, in 1996, it was reported that offspring who were more highly exposed to PCBs *in utero* exhibited lower memory and attention at age 11.[87] There were some differences between results of the Michigan cohort and a study in North Carolina,[88] and in 1997, Jacobson and Jacobson concluded that "In light of the general decline in PCB levels in environmental samples since the early 1980s, future studies are not likely to detect similar deficits unless extensive efforts are made to include sufficient numbers of more heavily exposed children."[57] Nevertheless, research has continued in this area and the results from different studies investigating low-level exposure to PCBs are clearly of interest but not always consistent.[89-95] Moreover, with the improvement of analytical techniques for POPs, various groups measure not only total PCBs by individual congeners, selected congeners, PCB-TEQs, PCDD/PCDF-TEQs, and these values are measured pre- and postnatally. With the multiplicity of timing and combinations of compounds, it is difficult to appreciate the significance of some correlations. In a Dutch study, it was reported that perinatal exposure to PCBs "might be associated with a greater susceptibility to infectious disease," but decreased susceptibility to allergic reactions in 42-month-old children.[48,83-94] In contrast, there were no correlations between 73 biological/immunological parameters and PCB or POP levels in a group of 11-month-old infants in a German study.[42,47,95] The health significance of these low level exposures to POPs and the relevance of timing (i.e., prenatal and postnatal) is an issue that needs to be resolved, and it will also be important to incorporate differences in exposure within groups and between groups, to address the current effects in light of the overall decline in POP levels from the 1960s to the present.

B. Incidence of breast cancer

The linkage between OC pesticide exposure and breast cancer has been controversial, and in 1992 and 1993, two articles appeared[96,97] showing that PCB and DDE levels in adipose tissue and serum, respectively, were higher in breast cancer patients versus control subjects. These observations initiated studies worldwide on the potential correlation between POPs[98-115] (particularly PCBs and DDE) and increased incidence of breast cancer in women. Moreover, with the advent of high-resolution POP analyses coupled with steadily increasing sensitivity for detection of individual congeners, a large number of papers on breast and other cancers have investigated correlations between increased disease incidence and higher levels of individual or mixtures of POPs. This author and others pointed out that the linkage between PCB/DDE and breast cancer has a low biological plausibility based on laboratory animal and human data.[116-118] High occupational exposures to PCBs or DDE are not associated with increased incidence of breast cancer in women and this is supported by laboratory animal studies.[119-122] Long-term feeding of DDE to rodents has not been linked to increased mammary tumorigenesis, and Aroclor 1254 and Aroclor 1260 (PCB mixtures) administered in the diet to female Sprague-Dawley rats decrease age-dependent spontaneous mammary tumor incidence.[121] Other studies show that DDE can inhibit or enhance estrogen-dependent tumorigenesis, whereas several PCB congeners are antiestrogenic and PCB #77 (3,3',4,4'-tetraCB) inhibits carcinogen-induced mammary tumor formation in female Sprague-Dawley rats.[122] Many of these antiestrogenic responses are associated with inhibitory AHR-estrogen receptor α (ERα) crosstalk that has been extensively investigated in this laboratory.[123, 124]

Results of several studies in laboratories/clinics throughout the world have not observed correlations between higher levels of PCB or DDE in breast cancer patients versus controls. Some reports show correlations with other OC pesticides; however, the potential causal linkage between these compounds with breast cancer is problematic. Studies with a Danish cohort showed that although serum levels of PCBs/DDE were similar in breast cancer patients versus controls, dieldrin was associated with a dose-dependent increased risk for breast cancer and mortality from this disease.[103,115] In contrast, Dorgan and coworkers showed a nonsignificant dose-related decrease in serum dieldrin levels in breast cancer patients from a Columbia, Missouri, cohort.[110] A few recent reports have attempted to find correlations with individual PCB congeners, and these results are inconsistent (Table 31.2). For example, Aronson and coworkers reported that for PCBs #105 (2,3,3',4,4'-pentaCB) and #118 (2,3',4,4',5-pentaCB), there was a significant dose-dependent trend for increased levels of these congeners in breast cancer patients in Ontario, Canada. In contrast, this correlation was not observed for PCB #118 in adipose tissue from cohorts in Connecticut or serum samples from Columbia, Missouri, and differences in odd ratios were observed for many other compounds in the three studies (Table 31.2). PCBs #105 and #118 are weak Ah receptor agonists that exhibit antiestrogenic activity,[123,124] and the correlations have questionable biological significance. Future studies undoubtedly will show correlations between levels of some individual POPs with increased incidence of breast cancer or other diseases; however, the continuing pursuit of these correlations should be tempered by consideration of biological plausibility and the decreasing levels of most POPs in the environment and human tissues.

IV. Summary

POPs have represented an important class of environmental contaminants and have been a continuing concern and challenge for regulators, scientists, and the public. Early action was taken on restricting or banning the use of PCBs and DDT, and the growing concern regarding the adverse human health effects of PCDDs/PCDFs and other AHR-active compounds has resulted in regulations that have significantly decreased environmental levels of these POPs. It is somewhat paradoxical that as levels of these compounds are decreasing, there has been an increasing concern regarding the potential adverse health effects of low-level environmental exposure to some POPs. Many of the effects of most concern are related to *in utero*/early

Table 31.2 Odds Ratios for Breast Cancer Incidence
in Individuals Containing Different Levels
of Specific PCB Congeners[110,112,113]

	Aronson et al.[112]		Zheng et al.[113]		Dorgan et al.[110]	
	OR			OR		OR
PCB 105						
≤ 4.1 µg/kg	1.0	—	—	—	—	—
4.2–6.1	1.16	—	—	—	—	—
6.2–12	2.03	—	—	—	—	—
≥ 13	3.07	—	—	—	—	—
PCB 118						
≤ 16 µg/kg	1.0	—	—	0–49 ng/g	1.0	
17–27	1.25	< 28.6 µg/kg	1.0	50–74	0.9	
28–49	1.88	28.6–59.6	0.8	75–109	1.7	
≥ 50	2.31	≥ 59.7	0.6	110–533	1.2	
PCB 138						
≤ 50 µg/kg	1.0	—	—	0–69 ng/g	1.0	
51–71	1.38	< 63.7	1.0	70–93	1.6	
72–112	1.55	63.7–103.9	0.8	94–124	1.6	
≥ 113	1.56	≥ 104	0.7	125–359	1.7	

postnatal exposures to these compounds that are subsequently manifested in the offspring as children or adults. The relevance of these effects to current low-level exposures to POPs is clearly of interest, and it will also be important to confirm persistent effects in high-level exposure groups (e.g., occupational and accidental exposures).

Acknowledgments

The financial assistance of the National Institutes of Health (ES04917 and ES09106), the Electric Power Research Institute, and the Texas Agricultural Experiment Station is gratefully acknowledged.

References

1. Vallack, H.W. et al., Controlling persistent organic pollutants: what next? *Environ. Toxicol. Pharmacol.*, 6, 143–175, 1998.
2. Kuehl, D.W. and Butterworth, B., A national study of chemical residues in fish, III: study results, *Chemosphere*, 29, 523–535, 1994.
3. Tanabe, S. et al., Highly toxic coplanar PCBs: occurrence, source, persistency and toxic implications to wildlife and humans, *Environ. Pollut.*, 47, 147–163, 1987.
4. Norén, K. and Meironyté, D., Certain organochlorine and organobromine contaminants in Swedish human milk in perspective of past 20–30 years, *Chemosphere*, 40, 1111–1123, 2000.

5. Marklund, S. et al., Emissions of polychlorinated compounds in combustion of biofuel, *Chemosphere*, 28, 1895–1904, 1994.
6. Fiedler, H., Sources of PCDD/PCDF and impact on the environment, *Chemosphere*, 32, 55–64, 1996.
7. Olie, K., Vermeulen, P.L., and Hutzinger, O., Chlorodibenzo-*p*-dioxins and chlorodibenzofurans are trace components of fly ash and flue, *Chemosphere*, 6, 455–459, 1977.
8. Hutzinger, O. and Fiedler, H., From source to exposure: some open questions, *Chemosphere*, 27, 121–129, 1993.
9. Christmann, W. et al., PCDD/PCDF and chlorinated phenols in wood preserving formulations for household use, *Chemosphere*, 18, 861–865, 1989.
10. Norstrom, A. et al., Analysis of some older Scandinavian formulations of 2,4-dichlorophenoxy acetic acid and 2,4,5-trichlorophenoxy acetic acid for contents of chlorinated dibenzo-p-dioxins and dibenzofurans, *Scand. J. Work. Environ. Health*, 5, 375–378, 1979.
11. Buckley, E.H., Accumulation of airborne polychlorinated biphenyls in foliage, *Science*, 216, 520–522, 1982.
12. Turle, R., Norstrom, R.J., and Collins, B., Comparison of PCB quantitation methods: re-analysis of archived specimens of herring gull eggs from the Great Lakes, *Chemosphere*, 22, 201–213, 1991.
13. Schmitt, C.J., Zajicek, J.L., and Peterman, P.H., National contaminant biomonitoring program: residues of organochlorine chemicals in U.S. freshwater fish, 1976–1984, *Arch. Environ. Contam. Toxicol.*, 19, 748–781, 1990.
14. Lake, C.A. et al., Contaminant levels in harbor seals from the northeastern United States, *Arch. Environ. Contam. Toxicol.*, 29, 128–134, 1995.
15. Williams, L.L. et al., Polychlorinated biphenyls and 2,3,7,8-tetrachlorodibenzo-*p*-dioxin equivalent in eggs of red-breasted mergansers near Green Bay, Wisconsin, USA, in 1977–78 and 1990, *Arch. Environ. Contam. Toxicol.*, 29, 52–60, 1995.
16. Tremblay, N.W. and Gilman, A.P., Human health, the Great Lakes, and environmental pollution: a 1994 perspective, *Environ. Health Perspect.*, 103, 3–5, 1995.
17. Lagueux, J. et al., Cytochrome P450 CYP1A1 enzyme activity and DNA adducts in placenta of women environmentally exposed to organochlorines, *Environ. Res.*, 80, 369–382, 1999.
18. Poland, A. and Knutson, J.C., 2,3,7,8-Tetrachlorodibenzo-*p*-dioxin and related halogenated aromatic hydrocarbons: Examinations of the mechanism of toxicity, *Annu. Rev. Pharmacol. Toxicol.*, 22, 517–554, 1982.
19. Whitlock, Jr., J.P., Mechanistic aspects of dioxin action, *Chem. Res. Toxicol.*, 6, 754–763, 1993.
20. Lucier, G.W., Portier, C.J., and Gallo, M.A., Receptor mechanisms and dose-response models for the effects of dioxins, *Environ. Health Perspect.*, 101, 36–44, 1993.
21. Swanson, H.I. and Bradfield, C.A., The Ah-receptor: genetics, structure and function, *Pharmacogenetics*, 3, 213–223, 1993.
22. Landers, J.P. and Bunce, N.J., The Ah receptor and the mechanism of dioxin toxicity, *Biochem.J.*, 276, 273–287, 1991.
23. Okey, A.B., Riddick, D.S., and Harper, P.A., The Ah receptor: mediator of the toxicity of 2,3,7,8-tetrachlorodibenzo-*p*-dioxin (TCDD) and related compounds, *Toxicol. Lett.*, 70, 1–22, 1994.

24. Ahlborg, U.G. et al., Impact of polychlorinated dibenzo-*p*-dioxins, dibenzo-furans, and biphenyls on human and environmental health with special emphasis on application of the toxic equivalence factor concept, *Eur. J. Pharmacol.*, 228, 179–199, 1992.

25. North Atlantic Treaty Organization (NATO), *Scientific Basis for the Development of International Toxicity Equivalency Factor (I-TEF), Method of Risk Assessment for Complex Mixtures of Dioxins and Related Compounds*, Committee on the Challenges of Modern Society, 178, 1988.

26. Safe, S., Polychlorinated biphenyls (PCBs), dibenzo-*p*-dioxins (PCDDs), dibenzofurans (PCDFs) and related compounds: environmental and mechanistic considerations which support the development of toxic equivalency factors (TEFs), *C. R.C. Crit. Rev. Toxicol.*, 21, 51–88, 1990.

27. Safe, S., Polychlorinated biphenyls (PCBs): environmental impact, biochemical and toxic responses, and implications for risk assessment, *C. R.C. Crit. Rev. Toxicol.*, 24, 87–149, 1994.

28. Birnbaum, L.S. and DeVito, M.J., Use of toxic equivalency factors for risk assessment for dioxins and related compounds, *Toxicology*, 105, 391–401, 1995.

29. van Leeuwen, F.X.R. et al., Dioxins: WHO's tolerable daily intake (TDI) revisited, *Chemosphere*, 40, 1095–1101, 2000.

30. Van den Berg, M. et al., Toxic equivalency factors (TEFs) for PCBs, PCDDs, PCDFs for humans and wildlife, *Environ. Health Perspect.*, 106, 775–792, 1998.

31. Ahlborg, U.G. et al., Toxic equivalency factors for dioxin-like PCBs, *Chemosphere*, 28, 1049–1067, 1994.

32. Eadon, G. et al., Calculation of 2,3,7,8-TCDD equivalent concentrations of complex environmental contaminant mixtures, *Environ. Health Perspect.*, 70, 221–227, 1986.

33. Schrenk, D. et al., Assessment of biological activities of mixtures of polychlorinated dibenzo-p-dioxins: Comparison between defined mixtures and their constituents, *Arch. Toxicol.*, 65, 114–118, 1991.

34. Stahl, B.U., Kettrup, A., and Rozman, K., Comparative toxicity of four chlorinated dibenzo-p-dioxins (CDDs) and their mixture, Part I: Acute toxicity and toxic equivalency factors (TEFs), *Arch. Toxicol.*, 66, 471–477, 1992.

35. Safe, S., Development, validation and problems with the TEF approach for risk assessment of dioxins and related compounds, *J. Animal Sci.*, 76, 134–141, 1998.

36. Safe, S., Limitations of the toxic equivalency factor approach for risk assessment of TCDD and related compounds, *Teratogen. Carcinogen. Mutagen.*, 17, 285–304, 1998.

37. Harper, N. et al., Immunosuppressive activity of polychlorinated biphenyl mixtures and congeners: non-additive (antagonistic) interactions, *Fund. Appl. Toxicol.*, 27, 131–139, 1995.

38. Bannister, R. et al., Aroclor 1254 as a 2,3,7,8-tetrachlorodibenzo-*p*-dioxin antagonist: effects on enzyme activity and immunotoxicity, *Toxicology*, 46, 29–42, 1987.

39. Haake, J.M. et al., Aroclor 1254 as an antagonist of the teratogenicity of 2,3,7,8-tetrachlorodibenzo-*p*-dioxin, *Toxicol. Lett.*, 38, 299–306, 1987.

40. Biegel, L. et al., 2,2′,4,4′,5,5′-hexachlorobiphenyl as a 2,3,7,8- tetrachlorodibenzo-*p*-dioxin antagonist in C57BL/6J mice, *Toxicol. Appl. Pharmacol.*, 97, 561–571, 1989.

41. Davis, D. and Safe, S., Dose-response immunotoxicities of commercial polychlorinated biphenyls (PCBs) and their interaction with 2,3,7,8- tetrachlorodibenzo-p-dioxin, *Toxicol. Lett.*, 48, 35–43, 1989.

42. Davis, D. and Safe, S., Immunosuppressive activities of polychlorinated biphenyls in C57BL/6 mice: structure-activity relationships as Ah receptor agonists and partial antagonists, *Toxicology*, 63, 97–111, 1990.
43. Morrissey, R.E. et al., Limited PCB antagonism of TCDD-induced malformations in mice, *Toxicol. Lett.*, 60, 19–25, 1992.
44. Harris, G.E. et al., Embryotoxicity of extracts from Lake Ontario rainbow trout (*Oncorhynchus mykiss*) to Japanese medaka (*Oryzias latipes*), *Environ. Toxicol. Chem.*, 13, 1393–1403, 1995.
45. Zhao, F. et al., Inhibition of pentachlorobiphenyl-induced fetal cleft palate and immunotoxicity in C57BL/6 mice by 2,2',4,4',5,5'- hexachlorobiphenyl, *Chemosphere.*, 34, 1605–1613, 1997.
46. Zhao, F. et al., Inhibition of 3,3',4,4',5-pentachlorobiphenyl-induced chicken embryotoxicity by 2,2',4,4',5,5'-hexachlorobiphenyl, *Fundam. Appl. Toxicol.*, 35, 1–8, 1997.
47. Keys, B., Piskorska-Pliszczynska, J., and Safe, S., Polychlorinated dibenzofurans as 2,3,7,8-TCDD antagonists: *in vitro* inhibition of monooxygenase enzyme induction, *Toxicology Letters.*, 31, 151–158, 1986.
48. Aarts, J.M.M.J.G. et al., Species-specific antagonism of Ah receptor action by 2,2',5,5'-tetrachloro- and 2,2',3,3',4,4'-hexachlorobiphenyl, *Eur. J. Pharmacol.*, 293, 463–474, 1995.
49. Bosveld, A.T.C. et al., Mixture interactions in the *in vitro* CYP1A1 induction bioassay using chicken embryo heptocytes, *Organohal. Cmpds*, 25, 309–312, 1995.
50. Tysklind, M. et al., Inhibition of ethoxyresorufin-*O*-deethylase (EROD) activity in mixtures of 2,3,7,8-tetrachlorodibenzo-*p*-dioxin and polychlorinated biphenyls, *Environ. Sci. Pollut. Res.*, 4, 211–216, 1995.
51. Chen, I., Safe, S., and Bjeldanes, L., Indole-3-carbinol and diindolylmethane as aryl hydrocarbon (Ah) receptor agonists and antagonists in T47D human breast cancer cells, *Biochem. Pharmacol.*, 51, 1069–1076, 1996.
52. Ciolino, H.P., Wang, T.T., and Yeh, G.C., Diosmin and diosmetin are agonists of the aryl hydrocarbon receptor that differentially affect cytochrome P450 1A1 activity, *Cancer Res.*, 58, 2754–2760, 1998.
53. Bjeldanes, L.F. et al., Aromatic hydrocarbon responsiveness-receptor agonists generated from indole-3-carbinol *in vitro* and *in vivo*: Comparisons with 2,3,7,8-tetrachlorodibenzo-*p*-dioxin, *Proc. Natl. Acad. Sci. USA*, 88, 9543–9547, 1991.
54. Gradelet, S. et al., Ah receptor-dependent CYP1A induction by two carotenoids, canthaxanthin and β-apo-8'-carotenal, with no affinity for the TCDD binding site, *Biochem. Pharmacol.*, 54, 307–315, 1997.
55. Brouwer, A. et al., Functional aspects of developmental toxicity of polyhalogenated aromatic hydrocarbons in experimental animals and human infants, *Eur. J. Pharmacol.*, 293, 1–40, 1995.
56. Tilson, H.A. and Kodavanti, P.R.S., The neurotoxicity of polychlorinated biphenyls, *Neurotoxicology.*, 19, 517–525, 1998.
57. Jacobson, J.L. and Jacobson, S.W., Teratogen update: polychlorinated biphenyls, *Teratology*, 55, 338–347, 1997.
58. Seegal, R.F., Epidemiological and laboratory evidence of PCB-induced neurotoxicity, *Crit. Rev. Toxicol.*, 26, 709–737, 1996.
59. Robinson, A.K. et al., The estrogenic activity of DDT: the *in vitro* induction of an estrogen-inducible protein by *o,p*-DDT, *Toxicol. Appl. Pharmacol.*, 76, 537–543, 1984.

60. Kelce, W.R. et al., Persistent DDT metabolite *p,p'*-DDE is a potent androgen receptor antagonist, *Nature*, 375, 581–585, 1995.
61. Wong, P.W. and Pessah, I.N., Noncoplanar PCB 95 alters microsomal calcium transport by an immunophilin FKBP1two-dimensional dependent mechanism, *Mol. Pharmacol.*, 51, 693–702, 1997.
62. Bergman, Å., Klasson-Wehler, E., and Kuroki, H., Selective retention of hydroxylated PCB metabolites in blood, *Environ. Health Perspect.*, 102, 464–469, 1994.
63. Nishihara, Y. and Utsumi, K., 4-Chloro-4'-biphenylol as an uncoupler and an inhibitor of mitochondrial oxidative phosphorylation, *Biochem. Pharmacol.*, 36, 3453–3457, 1987.
64. Korach, K.S. et al., Estrogen receptor-binding activity of polychlorinated hydroxybiphenyls: Conformationally restricted structural probes, *Mol. Pharmacol.*, 33, 120–126, 1988.
65. Brouwer, A. and Van den Berg, K.J., Binding of a metabolite of 3,4,3',4'-tetrachlorobiphenyl to transthyretin reduces serum vitamin A transport by inhibiting the formation of the protein complex carrying both retinol and thyroxin, *Toxicol. Appl. Pharmacol.*, 85, 301–312, 1986.
66. Lans, M.C. et al., Structure-dependent, competitive interaction of hydroxy-polychlorobiphenyls, -dibenzo-p-dioxins and -dibenzofurans with human transthyretin, *Chem. -Biol. Interact.*, 88, 7–21, 1993.
67. Morse, D.C. et al., Alterations in rat brain thyroid hormone status following pre- and postnatal exposure to polychlorinated biphenyls (Aroclor 1254), *Toxicol. Appl. Pharmacol.*, 136, 269–279, 1996.
68. Connor, K. et al., Hydroxylated polychlorinated biphenyls (PCBs) as estrogens and antiestrogens: Structure-activity relationships, *Toxicol. Appl. Pharmacol.*, 145, 111–123, 1997.
69. Kuiper, G.G. et al., Interaction of estrogenic chemicals and phytoestrogens with estrogen receptor β, *Endocrinology*, 139, 4252–4263, 1998.
70. Kramer, V.J. et al., Hydroxylated polychlorinated biphenyl metabolites are antiestrogenic in a stably transfected human breast adenocarcinoma (MCF-7) cell line, *Toxicol. Appl. Pharmacol.*, 144, 363–376, 1997.
71. Moore, M. et al., Antiestrogenic activity of hydroxylated polychlorinated biphenyl congeners identified in human serum, *Toxicol. and Appl. Pharmacol.*, 142, 160–168, 1997.
72. Kester, M.H.A. et al., Potent inhibition of estrogen sulfotransferase by hydroxylated PCB metabolites: a novel pathway explaining the estrogenic activity of PCBs, *Endocrinology*, 141, 1897–1900, 2000.
73. Connor, K. et al., Structure-dependent induction of CYB2B by polychlorinated biphenyl congeners in female Sprague-Dawley rats, *Biochem. Pharmacol.*, 50, 1913–1920, 1995.
74. Buchmann, A. et al., Polychlorinated biphenyls, classified as either phenobarbital- or 3-methylcholanthrene-type inducers of cytochrome P-450, are both heptic tumor promoters in diethylnitrosamine-initiated rats, *Cancer Lett.*, 32, 243–253, 1986.
75. Laib, R.M., Rose, N., and Bunn, H., Hepatocarcinogenicity of PCB congeners: I. Initiation and promotion of enzyme-altered rat liver foci by 2,2',4,5'-tetra- and 2,2',4,4',5,5'-hexachlorobiphenyl, *Toxicol. Environ. Chem.*, 34, 19–22, 1991.
76. Kuratsune, M., Yusho (with reference to Yu-Cheng), in *Halogenated Biphenyls, Terphenyls, Naphthalenes, Dibenzodioxins and Related Products*, Kimbrough, R.D. and Jensen, A.A., Eds., Elsevier, Amsterdam, 1989, pp. 381–400.

77. Rogan, W.J. et al., Congenital poisoning by polychlorinated biphenyls and their contaminants in Taiwan, *Science*, 241, 334–336, 1988.
78. Kuratsune, M., Epidemiologic study on Yusho, a poisoning caused by ingestion of rice oil contaminated with a commercial brand for PCBs, *Environ. Health Perspect.*, 1, 119–128, 1972.
79. Bertazzi, P.A. et al., Cancer incidence in a population accidentally exposed to 2,3,7,8-tetrachlorodibenzo-*p*-dioxin, *Epidemiology*, 4, 398–406, 1993.
80. Kimbrough, P.M., Polychlorinated biphenyls (PCBs) and human health: an update, C. R.C. *Crit. Rev. Toxicol.*, 25, 133–163, 1995.
81. Mocarelli, P. et al., Change in sex ratio with exposure to dioxin, *Lancet*, 348, 409, 1996.
82. Guo, Y.L. et al., Chloracne, goiter, arthritis, and anemia after polychlorinated biphenyl poisoning: 14-year follow-up of the Taiwan Yucheng cohort, *Environ. Health Perspect.*, 107, 715–719, 1999.
83. International Agency for Research on Cancer (IARC), Polychlorinated dibenzo-*para*-dioxins and polychlorinated dibenzofurans, *IARC Monographs on the Evaluation of the Evaluation of the Carcinogenic Risk of Chemicals to Humans*, World Health Organization, Lyon, France, 69, 1997.
84. Becher, C. et al., PCDDs, PCDFs, and PCBs in human milk from different parts of Norway and Lithuania, *J. Toxicol. Environ. Health*, 46, 133–148, 1995.
85. Hooiveld, M., Heederik, D., and Bueno de Mesquita, H.B., Preliminary results of the second follow-up of a Dutch cohort of workers occupationally exposed to phenoxy herbicides, chlorophenols and contaminants, *Organochlor. Cmpds.*, 30, 185–189, 1996.
86. Fingerhut, M.A. et al., Cancer mortality in workers exposed to 2,3,7,8-tetra-chlorodibenzo-*p*-dioxin, *N. Engl. J. Med.*, 324, 212–218, 1991.
87. Jacobson, J.L. and Jacobson, S.W., Intellectual impairment in children exposed to polychlorinated biphenyls *in utero*, *N. Engl. J. Med.*, 335, 783–789, 1996.
88. Gladen, B.C. and Rogan, W.J., Effects of perinatal polychlorinated biphenyls and dichlorodiphenyl dichloroethene on later development, *J. Pediatr.*, 119, 58–63, 1991.
89. Patandin, S. et al., Effects of environmental exposure to polychlorinated biphenyls and dioxins on cognitive abilities in Dutch children at 42 months of age, *J. Pediatr.*, 134, 33–41, 1999.
90. Huisman, M. et al., Perinatal exposure to polychlorinated biphenyls and dioxins and its effect on neonatal neurological development, *Early Hum. Dev.*, 41, 111–127, 1995.
91. Huisman, M. et al., Neurological condition in 18-month-old children perinatally exposed to polychlorinated biphenyls and dioxins, *Early Hum. Dev.*, 43, 165–176, 1995.
92. Koopman-Esseboom, C. et al., Effects of polychlorinated biphenyl/dioxin exposure and feeding type on infants' mental and psychomotor development, *Pediatrics*, 97, 700–706, 1996.
93. Winneke, G. et al., Developmental neurotoxicity of polychlorinated biphenyls (PCBs): cognitive and psychomotor functions in 7-month-old children, *Toxicol. Lett.*, 102–103, 423–428, 1998.
94. Weisglas-Kuperus, N. et al., Immunological effects of background exposure to polychlorinated biphenyls and dioxins in Dutch toddlers, *Organohal. Cmpds.*, 49, 84–86, 2000.

95. Abraham, K. et al., No measurable changes of biological parameters in breast-fed infants due to POP background exposure, *Organohal. Cmpds.*, 48, 143–144, 2000.
96. Falck, F. et al., Pesticides and polychlorinated biphenyl residues in human breast lipids and their relation to breast cancer, *Arch. Environ. Health*, 47, 143–146, 1992.
97. Wolff, M.S. et al., Blood levels of organochlorine residues and risk of breast cancer, *J. Natl. Cancer Inst.*, 85, 648–652, 1993.
98. Lopez-Carrillo, L. et al., Dichlorodiphenyltrichloroethane serum levels and breast cancer risk: a case-control study from Mexico, *Cancer Res.*, 57, 3728–3732, 1997.
99. Van't Veer, P. et al., DDT (dicophane) and postmenopausal breast cancer in Europe: case control study, *Br. J. Med.*, 315, 81–85, 1997.
100. Hunter, D.J. et al., Plasma organochlorine levels and the risk of breast cancer, *New Engl. J. Med.*, 337, 1253–1258, 1997.
101. Schecter, A. et al., Blood levels of DDT and breast cancer risk among women living in the north of Vietnam, *Arch. Environ. Contam. Toxicol.*, 33, 453–456, 1997.
102. Moysich, K.B. et al., Environmental organochlorine exposure and postmenopausal breast cancer risk, *Cancer Epidemiol. Biomarkers Prev.*, 7, 181–188, 1998.
103. Hoyer, A.P. et al., Organochlorine exposure and risk of breast cancer, *Lancet*, 352, 1816–1820, 1998.
104. Guttes, S. et al., Chlororganic pesticides and polychlorinated biphenyls in breast tissue of women with benign and malignant breast disease, *Arch. Environ. Contam. Toxicol.*, 35, 140–147, 1998.
105. Liljegren, G. et al., Case-control study on breast cancer and adipose tissue concentrations of congener specific polychlorinated biphenyls, DDE and hexachlorobenzene, *Eur. J. Cancer Prev.*, 7, 135–140, 1998.
106. Helzlsouer, K.J. et al., Serum concentrations of organochlorine compounds and the subsequent development of breast cancer, *Cancer Epidemiol. Biomarkers Prev.*, 8, 525–532, 1999.
107. Moysich, K.B. et al., Polychlorinated biphenyls, cytochrome P4501A1 polymorphism, and postmenopausal breast cancer risk, *Cancer Epidemiol. Biomarkers Prev.*, 8, 41–44, 1999.
108. Krieger, N. et al., Breast cancer and serum organochlorines: a prospective study among white, black, and Asian women, *J. Natl. Cancer Inst.*, 86, 589–599, 1994.
109. Zheng, T. et al., Environmental exposure to hexachlorobenzene (HCB) and risk of female breast cancer in Connecticut, *Cancer Epidemiol. Biomarkers Prev.*, 8, 407–411, 1999.
110. Dorgan, J.F. et al., Serum organochlorine pesticides and PCBs and breast cancer risk: results from a prospective analysis, *Cancer Causes Control*, 10, 1–11, 1999.
111. Demers, A. et al., Risk and aggressiveness of breast cancer in relation to plasma organochlorine concentrations, *Cancer Epidemiol. Biomarkers Prev.*, 9, 161–166, 2000.
112. Aronson, K.J. et al., Breast adipose tissue concentrations of polychlorinated biphenyls and other organochlorines and breast cancer risk, *Cancer Epidemiol. Biomarkers Prev.*, 9, 55–63, 2000.

113. Zheng, T. et al., Breast cancer risk associated with congeners of polychlorinated biphenyls, *Am. J. Epidemiol.*, 152, 50–58, 2000.

114. Zheng, T. et al., Risk of female breast cancer associated with serum polychlorinated biphenyls and 1,1-dichloro-2,2′-bis(*p*-chlorophenyl)ethylene, *Cancer Epidemiol. Biomarkers. Prev.*, 9, 167–174, 2000.

115. Hoyer, A.P. et al., Organochlorine exposure and breast cancer survival, *J. Clin. Epidemiol.*, 53, 323–330, 2000.

116. Safe, S., Environmental and dietary estrogens and human health: is there a problem? *Environ. Health Perspect.*, 103, 346–351, 1995.

117. Safe, S., Endocrine disruptors and human health: Is there a problem (An update), *Environ. Health Perspect.*, 108, 487–493, 2000.

118. Ahlborg, U.G. et al., Organochlorine compounds in relation to breast cancer, endometrial cancer, and endometriosis: an assessment of the biological and epidemiological evidence, *Crit. Rev. Toxicol.*, 25, 463–531, 1995.

119. Silinskas, K.C. and Okey, A.B., Protection by 1,1,1-trichloro-2,2-bis(p-chlorophenyl)ethane (DDT) against mammary tumors and leukemia during prolonged feeding of 7,1two-dimensionalimethylbenz(a)anthracene to female rats, *J. Natl. Cancer Inst.*, 55, 653–657, 1975.

120. Scribner, J.D. and Mottet, N.K., DDT acceleration of mammary gland tumors induced in the male Sprague-Dawley rat by 2-acetamidophenanthrene, *Carcinogenesis*, 2, 1235–1239, 1981.

121. Mayes, B.A. et al., Comparative carcinogenicity in Sprague-Dawley rats of the polychlorinated biphenyl mixtures Aroclors 1016, 1242, 1254, and 1260, *Toxicol. Sci.*, 41, 62–76, 1998.

122. Ramamoorthy, K. et al., 3,3′,4,,4′-Tetrachlorobiphenyl exhibits antiestrogenic and antitumorigenic activity in the rodent uterus and mammary and in human breast cancer cells, *Carcinogenesis*, 20, 115–123, 1999.

123. Safe, S., Modulation of gene expression and endocrine response pathways by 2,3,7,8-tetrachlorodibenzo-*p*-dioxin and related compounds, *Pharmacol. Therap.*, 67, 247–281, 1995.

124. Safe, S., Chemically-induced oestrogenic and anti-oestrogenic activity: structure-dependent effects and molecular mechanisms of action, in *Endocrine and Hormonal Toxicology*, Harvey, P.W., Rush, K.C., and Cockburn, A., Eds., John Wiley and Sons, New York, 1999, pp. 461–485.

chapter thirty-two

Polybrominated diphenyl ethers: route, dose, and kinetics of exposure to humans

Åke Bergman, Lars Hagmar, Peter Höglund, and Andreas Sjödin

Contents

Abstract Polybrominated diphenyl ethers (PBDEs) are produced commercially as a mixture of chemical compounds that are brominated to differing degrees (i.e., PentaBDE, OctaBDE or DecaBDE). PBDEs are used as flame retardants in textiles, rubber products, electronics, furnishings, and building materials. Sixty-seven thousand metric tons of PBDEs were manufactured in 1999 for use in semi-open systems. Because of their use as flame retardants, PBDEs are ubiquitous in the environment. Some foods in the human diet, mainly fish, accumulate a significant level of PBDEs. Fatty fish and mammals consumed by humans play a particularly important role in determining the

PBDE level in the tissues of individuals who are not occupationally exposed to these chemicals. In general, low levels (1–7 pmol/g lipid weight [l.w.]) of PBDEs are found in human plasma but the concentration of PBDEs in the milk of Swedish mothers has been increasing dramatically. The dominating PBDE congener, 2,2′,4,4′-tetraBDE is typically one to two orders of magnitude lower in concentration than the persistent PCB congener, 2,2′,4,4′,5,5′-hexachlorobiphenyl (CB-153). Higher brominated diphenyl ether congeners, i.e., hepta-BDE to decaBDE, have been detected at elevated concentrations in workers at electronics recycling plants and in engineers who install and repair computers. The most likely source of exposure is by inhalation of PBDEs adsorbed to particulates in the air. The concentration of PBDEs measured in the air in work environments supports this idea. Some PBDEs such as decabrominated diphenyl ether (BDE-209) have a short half-life in humans (approximately 7 days), indicating that this compound undergoes rapid elimination and/or biotransformation.

I. Introduction

Flammable materials such as textiles and wood are often treated with chemicals to reduce their flammability. Flame retardants have been in use for more than 2,000 years; however, chemical products for protecting modern polymeric materials used in textiles, construction, electronics, and furnishings were developed relatively recently.[1] Flame retardants (FRs) currently in use are the following: 1) inorganic chemicals such as aluminum hydroxide, antimony oxides and ammonium salts; 2) organophosphorus compounds such as phosphate esters; and 3) organohalogen compounds.[1,2] Organic nitrogen containing FRs are also used at low concentration for certain applications. More than 600,000 tons of FRs are presently being produced per year[2] of which more than 200,000 tons are brominated flame retardants (BFRs);[2,3] chlorinated FRs are produced at a lower level. Plastics such as polyvinyl chloride (PVC) are inherently flame resistant because of the chlorine in the polymer; however, these compounds are still treated with flame retardants.

BFRs may include as many as 75 different chemicals[1] but the major compounds in use are tetrabromobisphenol A (TBBPA) and its derivatives, polybrominated diphenyl ethers (PBDEs) with different degrees of bromination and hexabromocyclododecane (HBCDD). The 1999 production figures for these BFRs were recently released by the Bromine Science Environmental Forum.[3] The production levels were 120,000 tons TBBPA, 8,500 PentaBDE, 3,800 OctaBDE, 55,000 tons DecaBDE, and 16,000 tons HBCDD. The chemical structures of the major BFRs are shown in Figure 32.1. These chemicals are characterized by different physico-chemical properties. TBBPA has two phenolic groups and has acidic properties (pK_a=7.7 and 8.5).[4] TBBPA is a reactive BFR, mainly applied to plastics where it is covalently bound to the polymer. TBBPA is the major flame retardant used in printed circuit boards. The second largest class of BFRs is PBDEs, of which 209 congeners are theoretically possible. Commercially produced PentaBDE and OctaBDE products

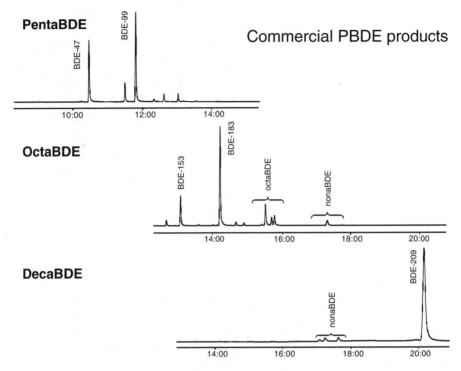

Figure 32.1 Chemical structures of tetrabromobisphenol A (TBBPA) and three neutral derivatives of TBBPA; examples of the three major commercial products of PBDEs and hexabromocyclododecane (HBCDD) are shown.

contain PBDE congeners with four to six and six to nine bromine atoms, respectively (Figure 32.2); BDE-209 is the principal chemical component in DecaBDE (Figure 32.2). Any single PBDE compound is much less complex than the commercially produced PBDE products.[5] PBDE congeners have been numbered according to the PCB numbers.[6] PBDEs are lipophilic neutral compounds. HBCDD is a lipophilic, neutral, cycloaliphatic compound which is a mixture of α-, β-, and γ-HBCDD (similar to hexachlorocyclohexane mixtures). Important characteristics of BFRs are their diversity, multiplicity, and complexity; this group of compounds has large differences in reactivity, lipophilicity, acidity, structure, and molecular weight.

This chapter discusses issues and concerns about human exposure to PBDEs. PBDEs have been reported in atmospheric samples from the U.K., Sweden,[7,8] and at remote locations like Alert and Dunai in the Canadian and Russian Arctic regions, respectively.[9] These data indicate that PBDEs are transported for long distances. 2,2',4,4'-tetrabromodiphenyl ether (BDE-47) is the major PBDE congener in environmental samples. PBDEs are also detected in sediment[10–12] and sewage sludge.[10,12] BDE-209, a perbrominated diphenyl ether, has also been detected in the environment. Particularly high concentrations of BDE-209, 40 to 1700 ng/g dry weight, were reported in estuaries

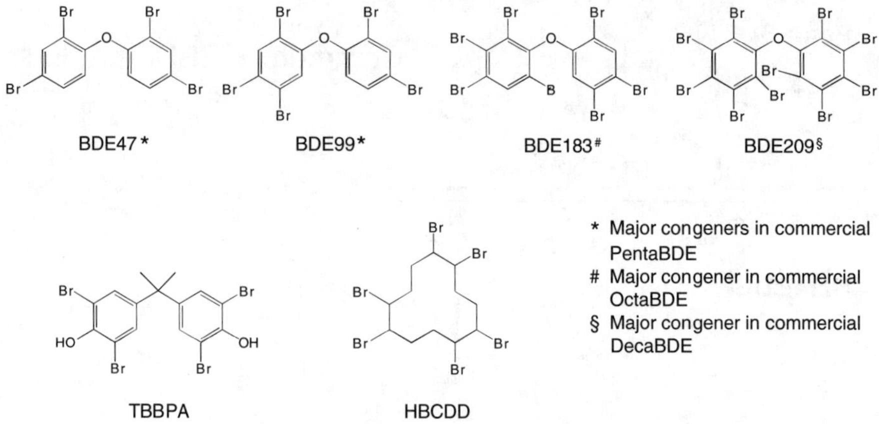

BDE47*   BDE99*   BDE183#   BDE209§

TBBPA   HBCDD

* Major congeners in commercial
  PentaBDE
\# Major congener in commercial
  OctaBDE
§ Major congener in commercial
  DecaBDE

Figure 32.2 Composition of three commercial PBDE products: PentaBDE, OctaBDE, and DecaBDE.

of the Liffey (Ireland), Schelde (Belgium/Germany), and Mersey and Humber (U.K.) rivers.[12] Other PBDEs were present at much lower concentration.

Most studies of PBDEs have focused on congeners with four to six bromine atoms. In addition, few studies measure BDE-209 because this congener is difficult to analyze and requires different instrumental conditions than lower brominated analogs (i.e., different GC column length and injection method).

II. Human exposure to polybrominated diphenyl ethers

A.   Routes of exposure

Humans are exposed to PBDEs in their diet or in the air by inhalation. The most commonly discussed route of exposure to persistent environmental contaminants such as PCBs and DDT/DDE is via dietary intake of fish and marine mammals that accumulate these compounds in their fatty tissue.[13] Dairy products are a less well recognized but possibly significant dietary source of human exposure to these compounds, e.g., dioxin exposure in the Netherlands[14]. The concentration of BFRs has been studied in wildlife, but few studies have examined their level in the human diet. A recent paper reported that the PBDEs in chicken consumed in the United States were 40 ng/g l.w. or lower,[15] which is 60 times lower than the average concentration in salmon from Lake Michigan.[16] Chickens grown in the vicinity of a PBDE manufacturer have a different PBDE congener pattern than other chickens. Table 32.1 lists the average concentration of PBDEs in several fish species.[16] The fish species shown in Table 32.1 are all consumed by humans. It is interesting to note that the concentration of BDE-47 in Lake Michigan steelhead trout (1.800 ng/g l.w., n=6) is not significantly different from the level of the persistent 2,2',4,4',5,5'-hexachlorobiphenyl (CB-153) (2.200 ng/g l.w.,

Table 32.1 Concentrations (ng/g lipid weight) of 2,2′,4,4′-Tetrabromodiphenyl Ether in Some Recently Sampled Fish Species from North America (the Great Lakes) and Sweden

Fish Species	Sampling Year	Sampling Site	Concentration	Reference
Herring	1999	Baltic Proper	18 (15–21)	35
(*Clupea harengus*)		Swedish West coast	9.8 (8.1–12)	35
Salmon	1995	Dalälven (river)	200 (100–410)	17
(*Salmo salar*)				
Coho salmon	1996	Lake Michigan	1600 (820–5200)	16
(*Oncorhynchus kisutch*)				
Steelhead trout	1996	Lake Michigan	1700	17
(*Oncorhynchus mykiss*)				
Lake trout	1998	Lake Ontario	560	20
(*Salvelinus namaycush*)				
Pike	1996	Lake Bolmen	3.0	21
(*Esox lucius*)		(Sweden)		

n=6) in the fish (including salmon) from Lake Michigan.[16,17] In fish, there are only two reports of higher PBDEs levels than these; the highest PBDE levels ever measured were in the early 1980s in pike in a Swedish river contaminated by PentaBDE[18] and a level of 57 000 ng/g l.w. was detected in a composite sample of carp in Virginia.[19] In general, PBDEs (mainly BDE-47) are found at a lower concentration than the most persistent PCB congeners, but that is not necessarily true everywhere. Also, it has been reported that the PBDE level is increasing in lake trout from Lake Ontario[20] while the concentration of BDE-47 seems to be leveling off in pike in Sweden.[21]

Humans are also exposed to PBDEs in the air. Ambient levels of PBDEs are low but indoor and/or occupational exposure to PBDEs may occur; for example, relatively high exposure has been documented for workers in electronics recycling and for computer service engineers.[24–27] PBDEs are mainly present in the particulate fraction of the air rather than in the semi-volatile fraction in a pumped air sample.[26] However, this varies with different chemical compounds: 30% of BDE-47 was in the semivolatile fraction, but <1% of BDE-209 was in the semivolatile fraction of air from a dismantling plant for electronics.[26]

B. Human levels

Several recent studies have determined the average PBDE level in humans in several countries. Table 32.2 shows a summary of PBDE levels in blood, milk, and several tissues. BDE-47, BDE-209, and a few other low brominated diphenyl ethers have been measured. BDE-209 is bioavailable and is detected in individuals not directly exposed to PBDE, i.e., cleaners at a hospital[24] and in U.S. blood donors[28]. The concentration of BDE-209 is highly variable and

Table 32.2   Concentrations (ng/g l.w.) of Polybrominated Diphenyl Ethers (PBDEs) in Human Milk and Blood

Gender F (Female) M (Male)	n	Sampling Year	Tissue	Geographical Area	Median Mean Pool	BDE47 ng/g l.w.	BDE47 Range ng/g l.w.	ΣPBDE	No. of PBDEs Analyzed	Reference
					Single Samples					
F	20	1990	milk	Sweden	pool	0.81	—	1.21	8	30
F	40	1997	milk	Sweden	pool	2.28		4.02	8	30
M	4	1994	adipose	Sweden	single		1.7–4.9	3.8–7.7	9	36
F	10	1992	milk	Canada	median	1.75	0.31–18.7	3.14		37
					mean	3.39			5.8	37
F	200	1981/1982	milk	Canada (wide)	pool			0.2	6	37
F	100	1992	milk	Canada (wide)	pool			16.2	6	37
F	11	1994–1998	milk	Finland	median	0.77				38
					mean	1.1	0.36–2.80	2.2	3	38
F	8	1985	blood	Germany	median	1.8		2.6	8	32
M	8	1985	blood	Germany	median	2.6		3.4	8	32
F	10	1999	blood	Germany	median	2.8		4.3	8	32
M	10	1999	blood	Germany	median	3.4		5.4	8	32
F	20	1997	blood	Sweden	median	1.6	<1–?	3.3	5	24
M	20[a]	1992	blood	Sweden	median	0.4	0.1–2.5			29
M	19[a]	1992	blood	Latvia	median	0.26	0.1–0.72			29
M	12[b]	1992	blood	Sweden	median	2.2	0.96–5.7			29
M	26[b]	1992	blood	Latvia	median	2.4	1.4–5.5			29
M + F	12 + 12	1998	blood	Japan	median	0.5	0.1–2.0	4.0		39
F	5	1995	breast adipose	USA (Calif.)			7–28			40

[a] Low fish consumer.
[b] High fish consumer.

was as high as 30 pmol/g l.w. in a few cases. The median level of BDE-47 in nonoccupationally exposed humans is 1.0–7.0 pmol/g l.w.

BDE-47 is the dominating PBDE congener in biota and its level in human samples is generally lower than the level of CB-153 and other persistent compounds. The level of CB-153 and BDE-47 were 220 and 0.4 ng/g l.w., respectively, in a cohort of Swedish nonfish-eating males; in contrast, the level of CB-153 and BDE-47 were 450 and 2.2 ng/g l.w., respectively, in males who ate 12–20 fish meals per month.[29] In measurements made in 1997, the level of CB-153 was 330 pmol/g l.w. in blood from Swedish female cleaners; the same study showed that the level of BDE-47 was 3.2 pmol/g l.w.[24] In 1988, the average concentrations of CB-153 and BDE-47 in U.S. blood donors were 360 and 1.3 pmol/g l.w., respectively.[28]

Few studies have examined the temporal trend of PBDE levels in humans. However, a study in Sweden showed that PBDEs are increasing by a factor of two every fifth year (Figure 32.3).[30] Similar trends were shown in studies with blood samples from Germans and milk samples in Swedes.[31,32] In contrast, the levels of PCB, DDT and dioxins[31] are decreasing in milk from Swedes.

Workplace-related exposure to PBDEs has been demonstrated in electronics[24] and computer technicians[27] but not in computer clerks.[24] Figure 32.4 shows median levels of several PBDE congeners associated with occupational exposure to PBDEs. The highest exposure was in persons dismantling electronics; a lower exposure was detected in computer technicians. There is a correlation between work hours and level of BDE-153 detected in computer technicians.[27] Higher brominated diphenyl ethers, 2,2′,4,4′,5,5′-hexaBDE (BDE-153), 2,2′,3,4,4′,5,5′-heptaBDE (BDE-183] and BDE-209 accu-

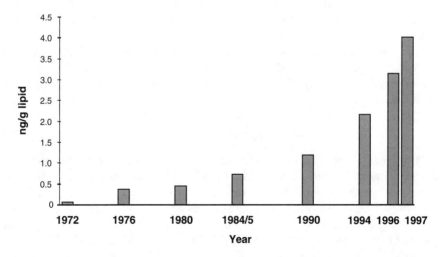

Figure 32.3 Time trend in exposure to PBDEs in human milk between 1972 and 1997. (The data were originally presented by Meironyté, D., Norén, K., and Bergman, Å., *J. Toxicol. Environ. Health,* 58, Part A, 101–113, 1999.)

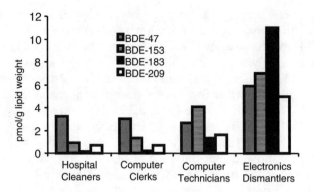

Figure 32.4 Concentration of selected PBDE congeners in electronics technicians, computer technicians, computer clerks, and cleaners. (From Sjödin, A. et al., *Environ. Health Perspect.*, 107, 643–648, 1999; Jakobsson, K. et al., *Chemosphere*, in press.)

mulate as a result of occupational exposure, while. BDE-47 does not accumulate as a result of occupational exposure.

C. Kinetics

All PBDE congeners are bioavailable, even those whose molecular mass is high such as BDE-209. BDE-209 is taken up slowly, but it does accumulate in fatty tissue. Some studies address the rate of decline of PBDE in the blood of individuals occupationally exposed to PBDEs by monitoring the change in PDBE level during vacation periods.[25] The blood concentration of BDE-209 decreased quite rapidly (half-life 6.8 days), but the level of BDE-183 decreased more slowly (half-life 85 days). Some metabolites of BDE-209 have been identified.[33] Data were insufficient to determine the half-lives of PBDE congeners with six or fewer bromine atoms (long half-lives; small amount of data).

BDE-47 is the major PBDE congener in human samples. BDE-47 seems to be highly persistent but there is evidence that it is metabolized in rats and mice.[34] Hydroxylated PBDEs were detected but they have not yet been characterized structurally.

III. Conclusion

A large number of brominated chemicals are used as flame retardants. Most PBDE congeners used as flame retardants are persistent compounds, with the exception that DecaBDE is a semipersistent compound in humans. PBDEs are ubiquitous environmental contaminants found in wildlife at high trophic levels of low ppm concentrations. PBDEs, including BDE-209 are in general present in human subjects at low ppb (<10 ppb) concentration. 2,2′,4,4′,5,5′-HexaBDE, 2,2′,3,4,4′,5,5′-heptaBDE and decaBDE are elevated in humans who work as electronic or computer technicians and are occupationally exposed. BDE-209 is bioavailable but it has a much shorter half-life in humans than BDE-183.

In general, the level of PBDEs in human samples is significantly lower than the levels of the most persistent PCB congeners. On the other hand, PBDE levels are increasing in some populations while the levels of other environmental contaminants including dioxin and PCBs are decreasing. Therefore, the implications and impact of environmental PBDEs on human and environmental health deserve further study.

Acknowledgments

We are grateful to Ulrika Örn for her kind help with figures and tables in this work. Financial support has been obtained from the Swedish Work Life Council, the Swedish EPA, the Medical Faculty at Lund University, and the Faculty of Natural Sciences and Mathematics at Stockholm University.

References

1. WHO Environmental Health Criteria 192, Flame Retardants: A General Introduction, International Program on Chemical Safety, World Health Organization, Geneva, 1997.
2. OECD, Selected brominated flame retardants, *Risk Reduction Monograph 3*, OECD, Environment Directorate, Paris, 1994.
3. BSEF 2000, Bromine Science Environmental Forum, www.bsef.com, 2001.
4. WHO, Environmental Health Criteria 172. Tetrabromobisphenol A and derivatives, International Program on Chemical Safety, WHO, Geneva, 1995.
5. Sjödin, A., Occupational and dietary exposure to organohalogen substances with special emphasis on polybrominated diphyl ethers. Thesis, Stockholm University, Stockholm, 2000.
6. Ballschmiter, K., Mennel, A. and Buyten, J., Long chain alkyl-polysiloxanes as non-polar stationary phases in capillary gas chromatography, *Fresenius J. Anal. Chem.*, 346, 396–402, 1993.
7. Peters, A.J., Coleman, P. and Jones, K.C., Organochlorine pesticides in UK air, *Organohal. Cmpds.*, 41, 447–450, 1999.
8. de Wit, C.A., Brominated flame retardants, *Swedish EPA Report* 5065, 41–43, 2000.
9. Alaee, M., Levels and trends of PBDEs in North American environment, Abstract from The 2nd International Workshop on Brominated Flame Retardants, Stockholm, 2001, pp. 117–120, www.kemi.se.
10. Nylund, K. et al., Analysis of some polyhalogenated organic pollutants in sediment and sewage sludge, *Chemosphere*, 24, 1721–1730, 1992.
11. Sellström, U. et al., Polybrominated diphenyl ethers and hexabromocyclododecane in sediment and fish from a Swedish river, *Environ. Toxicol. Chem.*, 17, 1065–1072, 1998.
12. Sellström, U. et al., Brominated flame retardants in sediments from European estuaries, the Baltic sea and in sewage sludge, *Organohal. Cmpds.*, 40, 383–386, 1999.
13. Asplund, L. et al., Polychlorinated biphenyls, 1,1,1-trichloro-2,2-bis(p-chlorophenyl)ethane (p,p'-DDT) and 1,1-dichloro-2,2-bis(p-chlorophenyl)-ethylene (p,p'-DDE) in human plasma related to fish consumption, *Arch. Environ. Health*, 49, 477–486, 1994.

14. Liem, A.K.D. and Theelen, R.M.C., in *Dioxins: Chemical Analysis, Exposure and Risk Assessment*, Thesis from Utrecht University, The Netherlands, 1997.
15. Huwe, J.K.; et al., Analysis of mono- to decabrominated diphenyl ethers in chickens at the part per billion level, *Chemosphere*, 46, 635–640, 2002.
16. Manchaster-Neesvig, J.B., Valters, K., and Sonzogny, W.C., Comparison of polybrominated diphenyl ethers (PBDEs) and polychlorinated biphenyls (PCBs) in Lake Michigan salmonids, *Environ. Sci. Technol.*, 35, 1072–1077, 2001.
17. Asplund, L. et al., Levels of polybrominated diphenyl ethers (PBDEs) in fish from the Great Lakes and Baltic Sea, *Organohal. Cmpds.*, 40, 351–354, 1999.
18. Andersson, Ö. and Blomkvist, G., Polybrominated aromatic pollutants found in fish in Sweden, *Chemosphere*, 10, 1051–1060, 1981.
19. Hale, R.C. et al., Comparison of brominated diphenyl ether fire retardant and organochlorine burdens in fish from Virginia rivers (USA), *Organohal. Cmpds.*, 47, 65–68, 2000.
20. Luross, J.M. et al., Spatial and temporal distribution of polybrominated diphenyl ethers and polybrominated biphenyls in lake trout from the Great Lakes, Abstract from The 2nd International Workshop on Brominated Flame Retardants, Stockholm, 2001, pp. 401–404, www.kemi.se.
21. Kierkegaard, A. et al., Temporal trends of a polybrominated diphenyl ether (PBDE), a methoxylated PBDE, and hexabromocyclododecane (HBCD) in Swedish biota, *Organohal. Cmpds.*, 40, 367–370, 1999.
22. Lindström, G. et al., Identification of 19 polybrominated diphenyl ethers (PBDEs) in long-finned pilot whale (*Globicephala melas*) from the Atlantic, *Arch. Environ. Contam. Toxicol.*, 36, 355–363, 1999.
23. Grandjean, P. et al., Relation of a seafood diet to mercury, selenium, arsenic, and polychlorinated biphenyl and other organochlorine concentrations in human milk, *Envir. Res.*, 71, 29–38, 1995.
24. Sjödin, A. et al., Flame retardant exposure: Polybrominated diphenyl ethers in blood from Swedish workers, *Environ. Health Perspect.*, 107, 643–648, 1999.
25. Hagmar, L. et al., Biological half-lives of polybrominated diphenyl ethers and tetrabromobisphenol A in exposed workers, *Organohal. Cmpds.*, 47, 198–201, 2000.
26. Sjödin, A. et al., Flame retardants in indoor air at an electronics recycling plant and at other work environments, *Environ. Sci. Technol.*, 35, 448–454, 2001.
27. Jakobsson, K.; et al. Exposure to polybrominated diphenyl ethers and tetrabromobisphenol A among computer technicians, *Chemosphere*, 46, 709–716, 2002.
28. Sjödin, A., Patterson., Jr., D.G., and Bergman, Å., Brominated flame retardants in serum from U.S. blood donors, *Environ. Sci. Technol.*, 35, 3830–3833, 2001.
29. Sjödin, A. et al., Influence of the consumption of fatty Baltic Sea fish on plasma levels of halogenated environmental contaminants in Latvian and Swedish men, *Environ. Health Perspect.*, 108, 1035–1041, 2000.
30. Meironyté, D., Norén, K., and Bergman, Å., Analysis of polybrominated diphenyl ethers in Swedish human milk: A time trend study, 1972–1997, *J. Toxicol. Environ. Health*, 58 Part A, 101–113, 1999.
31. Norén, K. and Meironyté, D., Certain organochlorine and organobromine contaminants in Swedish human milk in perspective of past 20–30 years, *Chemosphere*, 40, 1111–1123, 2000.
32. Schröter-Kermani, C. et al., The German environmental specimen bank: application in trend monitoring of polybrominated diphenyl ethers in human blood, *Organohal. Cmpds.*, 47, 49–52, 2000.

33. El Dareer, S.M. et al., Disposition of decabromobiphenyl ether in rats dosed intravenously or by feeding, *J. Toxicol. Environ. Health*, 22, 405–415, 1987.
34. Örn, U. and Klasson-Wehler, E., Metabolism of 2,2´,4,4´-tetrabromodiphenyl ether in rat and mouse, *Xenobiotica*, 28, 199–211, 1998.
35. Nylund, K. et al., Spatial distribution of some polybrominated diphelyl ethers and hexabromocyclododecane in herring (*Clupea harengus*) along the Swedish coast, Abstract from The 2nd International Workshop on Brominated Flame Retardants, Stockholm, 2001, pp. 349–352, www.kemi.se,.
36. Meironyté, D., Bergman, Å., and Norén, K., Polybrominated diphenyl ethers in Swedish human liver and adipose tissue, *Arch. Environ. Contam. Toxicol.*, 40, 564–570, 2001.
37. Ryan, J.J. and Patry, B., Determination of brominated diphenyl ethers (BDEs) and levels in Canadian human milks, *Organohal. Cmpds.*, 47, 57–60, 2000.
38. Strandman, T., Kostinen, J., and Vartiainen, T., Polybrominated diphenyl ethers (PBDEs) in placenta and human milk, *Organohal. Cmpds.*, 47, 61–64, 2000.
39. Nagayama, J., Tsuji, H., and Takasuga, T., Comparison between brominated flame retardants and dioxins or organochlorine compounds in blood levels of Japanese adults, *Organohal. Cmpds.*, 48, 27–30, 2000.
40. She, J. et al., Analysis of PBDEs in seal blubber and human breast adipose tissue samples, *Organohal. Cmpds.*, 47, 53–56, 2000

chapter thirty-three

Bioanalytical approaches for the detection of dioxin and related halogenated aromatic hydrocarbons

Michael S. Denison, Scott R. Nagy, Michael Ziccardi,
George C. Clark, Michael Chu, David J. Brown, Guomin Shan,
Yukio Sugawara, Shirley J. Gee, James Sanborn,
and Bruce D. Hammock

Contents

Abstract Proper epidemiological, risk assessment and exposure analysis of 2,3,7,8-tetrachlorodibenzo-p-dioxin (TCDD, dioxin) and related halogenated aromatic hydrocarbons (HAHs) requires accurate measurements of these chemicals in the species of interest and in various exposure matrices, i.e., biological, environmental, and food. High-resolution instrumental analysis techniques are established for these chemicals; however, these procedures are costly and time consuming, and as such, they are impractical for large-scale sampling studies, i.e., for epidemiological studies and assessment of areas with widespread contamination. Accordingly, we have developed two simple, rapid, sensitive, and inexpensive assays (a TCDD immunoassay and a novel recombinant cell bioassay called CALUX) for the detection and relative quantitation of TCDD and related HAHs in extracts of a variety of matrices. These assays can be used individually or in combination as relatively rapid prescreening tools to identify positive samples for subsequent instrumental analysis and to drive chemical fractionation of complex mixtures in order to identify novel chemicals or classes of chemicals which exert dioxin-like activity. The availability of these bioanalytical approaches should greatly facilitate large-scale sampling studies needed for the accurate assessment of exposure to this ubiquitous class of environmental contaminants.

I. Introduction

HAHs, such as polychlorinated dibenzo-p-dioxins (PCDDs), biphenyls (PCBs), and dibenzofurans (PCDFs) represent a large group of compounds which can produce toxic effects at concentrations which can occur in the environment. HAHs have been identified worldwide in a variety of wildlife, domestic and human tissues as well as in food, water, and soil samples. Because of their ubiquitous distribution, resistance to biological and chemical degradation, high toxicity, and potential for bioaccumulation/biomagnification, HAHs can have a significant impact on the health and well-being of humans and animals.[1,2] Exposure to and bioaccumulation of HAHs, including the prototypical and most potent member of this class, TCDD, have been observed to produce a wide variety of species-specific and tissue-specific toxic and biological effects, including lethality, reproductive dysfunction, birth defects, endocrine disruption, liver toxicity, impaired immune function, cancer, and alterations in gene expression.[1-3] Given these issues, the detection and quantitation of these chemicals in biological, environmental, and food samples is of paramount importance. In addition, epidemiological and risk assessment analysis of HAHs/TCDD in humans and animals requires that an accurate measurement of the internal level of exposure be made. However, HAHs are found not as individual congeners, but as complex HAH mixtures, of which the relative and absolute concentrations of individual congeners can vary dramatically. Consequently, one problem in the evaluation of risk to HAHs is the identification and quantitation of toxic/bioactive HAH congeners in biological samples. Although sophisticated cleanup procedures followed by high-resolution gas chromatography-mass spectrometry (GC/MS) can separate, identify, and quantitate individual PCDD, PCB, and PCDF congeners,[4,5] these procedures require highly sophisticated equipment and training, a large amount of sample for analysis, and are also costly and time consuming, particularly when samples may theoretically contain large numbers of different HAH isomers and congeners (up to 209 different PCB, 135 different PCDF and 75 different PCDD congeners). Additionally, the concentration of HAHs in a given sample provides only part of the information necessary to evaluate their potential for biological/toxicological effects in humans and animals. Therefore, from a human and animal health as well as an environmental standpoint, inexpensive and rapid bioassays capable of detecting and estimating the relative potency of complex mixtures of HAHs would be valuable. Accordingly, we have developed a rapid, inexpensive cell bioassay system that can be used with immunoassays in a two-tiered approach to detect the presence of these chemicals in a variety of biological and environmental matrices. These assays can be used individually or in combination as prescreening tools to allow identification of positive samples for subsequent analysis by GC/MS or other technologies, and they are amenable for large-scale analysis necessary for epidemiological studies. In addition, the bioassay can be used to drive chemical purification of complex mixtures to identify previously unknown components that have a similar mechanism of action as the HAHs and used in combination with chromatographic methods to support identification of HAHs in a sample.

II. Tier 1: CALUX bioassay

The first tier screening bioassay is based on the biochemical mechanism of action of TCDD (Figure 33.1) and related HAHs and involves the Ah receptor (AhR), a ligand-dependent transcription factor which can activate the expression of target genes specifically in response to these chemicals.[6] Following high-affinity binding of TCDD/HAH, the cytosolic ligand–AhR complex undergoes transformation during which it translocates into the nucleus, dissociates from its protein complex, and following its association with the Arnt (AhR nuclear translocator) protein, the AhR complex is converted into its high affinity DNA binding form.[7,8] The binding of the transformed heteromeric ligand–AhR–Arnt complex to its specific DNA recognition site, the dioxin responsive element (DRE), stimulates transcriptional activation of the adjacent gene.[7,9] Taking advantage of the ability of DREs to confer TCDD responsiveness upon an adjacent promoter and gene,[9] combined with the availability of several sensitive reporter genes, we constructed a recombinant TCDD/HAH-inducible expression vector which contain an easily measurable and extremely sensitive reporter gene (firefly luciferase) under HAH/AhR-inducible control of four DREs.[10] Cell lines that have been stably transfected (Figure 33.1) respond to TCDD and related HAHs with the induction of firefly luciferase in a dose- (Figure 33.2a), time-, and AhR-dependent manner.[10] Luciferase activity is readily quantitated by measurement of the amount of light produced in an enzymatic assay. This chemically activated luciferase expression (CALUX) cell bioassay system is inexpensive, rapid (approximately 5 h for complete analysis), sensitive (with a minimal detection limit of about 100–200 ppq TCDD), and has been optimized and streamlined such that the assay can be carried out in a 96-well microtiter format.[11] Because it has been optimized for 96-well formats, the same diluters, dispensers, and readers that are used for enzyme-linked immunosorbent assays (ELISA) can be used with the CALUX system. The system is adaptable in that other reporter systems can be used and also the general approach can be applied to a variety of promoter systems to detect not only HAHs but

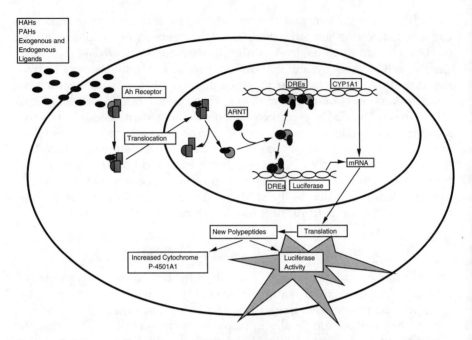

Figure 33.1 AhR-dependent molecular mechanism of TCDD action. The AhR stimulates expression of DRE-dependent endogenous genes as well as the stably integrated DRE-luciferase reporter gene.

estrogens, peroxisome proliferators, and a variety of other compounds with specific receptor systems.[12] In addition, we have recently demonstrated that the CALUX bioassay is also amenable for direct detection of TCDD and TCDD-like chemicals present in whole serum samples, without the need for solvent extraction.[11] Similar to samples which have been solvent extracted, induction of luciferase activity by TCDD-containing serum occurs in a time- and dose-dependent manner and this modification of the bioassay can detect as little as 5–10 ppt TCDD and/or TCDD-equivalents in a 50 µl aliquot of whole serum.[11] This modification of the CALUX assay not only has applications for the screening of samples where only small volumes of blood are available (e.g., infants and endangered and/or small species), but it will allow for large-scale screening of populations for epidemiology studies.

The utility of the CALUX bioassay for detection and relative quantitation of TCDD and related HAHs has been validated in several studies and they demonstrate a high degree of correlation between the CALUX bioassay and GC/MS results.[13–15] The results of an analysis of soil sample extracts by GC/MS and the CALUX bioassay are shown in Figure 33.2b. These data reveal an excellent correlation (R^2=0.932) between the soil extract TCDD equivalents (TEQs) estimated by GC/MS and those equivalents estimated using CALUX bioassay. The CALUX bioassay is simpler, faster and far less expensive than conventional chemical analysis techniques and perhaps more

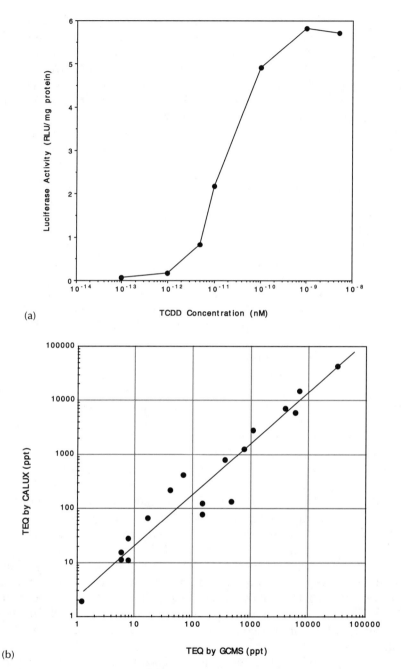

(a)

(b)

Figure 33.2 Tier I CALUX bioassay. (a) TCDD dose response curve for luciferase induction in stably transfected mouse hepatoma (H1L1.1c2) cells. (b) Relationship between TCDD equivalents (TEQs) determined by CALUX and GC/MS of soil sample extracts.

importantly, this assay provides a quantitative measure of the biological potency of the complex mixture, relative to the most potent ligand, TCDD. Given that the CALUX bioassay has the advantage of being able to detect a variety of TCDD-like HAHs and other AhR agonists, it is a perfect first-stage screen. Samples that are positive in this bioassay can undergo congener-specific HAH analysis using GC/MS. However, if confirmation of the presence and relative concentration of TCDD in CALUX-positive extracts is the goal, this can more readily and inexpensively be accomplished using our second tier dioxin selective immunoassay as described below.

III. Tier 2: TCDD immunoassay

Immunoassay, as a rapid, sensitive and selective bioanalytical method, has been widely used for environmental and biological monitoring of small molecular toxins. However, due to the lipophilic properties and the difficulty of the chemistry of dioxins, only a few attempts to detect TCDD by immunoassay have been reported,[16–19] and their sensitivities are not as high as is desirable. Based on careful hapten design and synthesis, a sensitive polyclonal antibody-based ELISA was developed in this laboratory,[20,21] which exhibited an I_{50} value of 12 pg/well (240 ng/L or 0.75 nM), with working range from 2–240 pg/well. Theoretically, an ELISA sensitivity is determined by antibody affinity. The ultimate detection limit of an assay is approximately 10–100 times lower than the K_d of antibody.[22] The K_d of antibody used in our study was measured by accelerator mass spectrometry[23] and found to be 0.1 nM suggesting that a more sensitive immunoassay can be achieved using this antibody. Thus, a series of new coating antigen haptens were designed and synthesized. After extensive screening, a highly sensitive assay was developed with coating Hapten II (Table 33.1). The I_{50} of new assay system was 1.8 pg/well (36 ng/L) (Figure 33.3a) with a lower detection limit of 0.2 pg/well (4.0 ng/L). Because of the restrictions on the use of TCDD, its toxicity, and the high cost of disposal, many laboratories avoid using TCDD as a primary standard. Thus, a TCDD surrogate standard, 2,3,7-trichloro-8-methyl dibenzo-p-dioxin (TMDD), was developed and used in our study.[24] TMDD and related compounds have similar polarity to TCDD and can be used as a secondary analytical standard in a variety of assays including the above bioassay as well as GC/MS. According to the cross-reactivity study[25]), most of the dioxins and PCDFs with a high TEF value (> 0.1) have strong or moderate cross-reactivities in this assay, which suggests that this assay might be a good indicator of toxicity of PCDDs and PCDFs in the test samples.

This immunoassay has been validated with extracts from fish and egg samples by GC/MS. A good agreement between GC/MS and ELISA TEQs was obtained from linear regression analysis (y=1.12x − 4.08, R^2=0.89) (Figure 33.3b), and no matrix effects were found for these extracts as prepared. A fairly good correlation between ELISA and TEF values was also observed with these samples (Y=0.78x + 6.37, R^2=0.90). Although there is an overesti-

Table 33.1 Structures of TCDD, Surrogate Standard TMDD and Dioxin Haptens

Compound	Structure
TCDD	
TMDD	
Hapten I (immunogen)	
Hapten II	

mation by ELISA in comparison to TEF values, the strong correlation between this ELISA and TEF values indicates that this assay can be used as a TEF screening method for dioxins and PCDFs on its own or sequentially to a more general screen based on the Ah receptor as a tier two method. The method has shown good correlation with spiked samples and with GC/MS results performed by this and other laboratories. For example, in one case, a blank, spiked recovery and positive soil samples were extracted with hexane and analyzed directly by ELISA after evaporation of the hexane with no further clean up. The ELISA data on the crude hexane extract agreed well with ELISA data on the same samples following a Florasil column indicating that for ELISA of soil samples the column was not needed. The ELISA data on the hexane extract agreed well with GC/MS data obtained after a 14-step clean-up and analysis procedure. These data suggest that this ELISA can be used as a rapid, inexpensive screen to detect dioxins in the soil samples.

The antibodies characterized above have a number of other analytical uses. For example, ELISA is of course easy to use as a postcolumn detection system for high-performance liquid chromatography,[26] and the CALUX assay also can be formatted to use in this manner. This approach provides either a confirmation experiment where one can say how much immunore-activity or CALUX activity has the same retention time as TCDD or as a method to tentatively identify other materials that are positive in the assay but are not TCDD. These and other antibodies can be used for immunoaf-finity purification prior to either bioassay or chromatographic analysis such as LC or GC/MS. A simple application of these antibodies is to add them to diluted aliquots of samples which have shown positive in either in CALUX or chromatographic assays. The antibodies will bind TCDD like chemicals

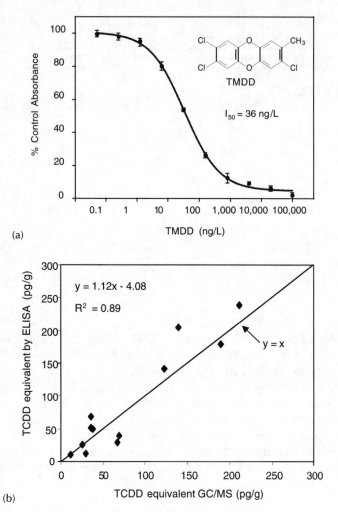

Figure 33.3 Tier II immunoassay. (a) TMDD immunoassay inhibition curve. (b) Relationship between TCDD equivalents (TEQs) determined by ELISA and GC/MS of fish and egg sample extracts.

and can be precipitated by a variety of techniques such as magnetic beads coated with goat anti-rabbit antibody or Protein A. If the positive response is removed one can say with high confidence that the assay was detecting a TCDD like chemical.

IV. Application of CALUX and TCDD immunoassay

The CALUX bioassay and the immunoassay provide a two tiered screening system that can be utilized as relatively rapid and inexpensive assays for the detection and relative quantitation of TCDD and TCDD-like chemicals in a wide range of samples and sample extracts (Figure 33.4). The CALUX assay

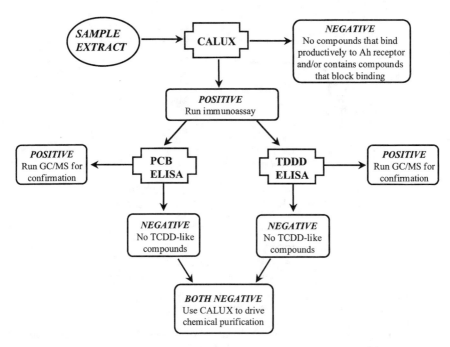

Figure 33.4 Overview of the two-tiered screening system for detection and relative quantitation of TCDD and TCDD-like chemicals in sample extracts.

can be used alone to eliminate samples that do not contain HAHs that bind to the Ah receptor. False negatives at this step could result from the presence of AhR antagonists in a sample; however, the presence of these inhibitory agents can readily be determined by measuring the ability of the sample to reduce the induction response of a known amount of TCDD. With a positive CALUX, the sample can be analyzed directly by chromatographic procedures or taken to a second tier where it is further evaluated by immunoassay. As shown in Figure 33.4, antibodies to TCDD, PCBs or other HAHs can be used to discriminate positives further in the CALUX assay. This second tier will put the sample in one of four categories: containing TCDD, containing PCBs that react with Ah receptor, containing TCDD and PCBs or containing an Ah receptor agonist that is neither TCDD or PCB. In samples where only TCDD or PCBs are known to be the primary contaminant either the CALUX or immunoassay can be used alone. Additionally, the CALUX and immunoassays can be used to prescreen large numbers of samples in order to identify those that should be subsequently analyzed by the more costly and time consuming GC/MS procedures. With proper controls, obtaining false negatives with the CALUX and ELISA assays is difficult. Thus, these techniques can be used to screen out large numbers of negative samples. Often the most time-consuming part of GC/MS and LC/MS analysis is cleaning an instrument. The CALUX and immunoassays provide a way to rank samples so that samples with suspected low levels of HAHs can be run on the instrument

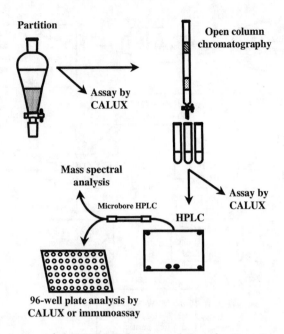

Figure 33.5 Overview of the CALUX bioassay-driven fractionation scheme for the purification and identification of novel AhR agonists.

when the sensitivity is high and samples suspected of high concentrations of HAHs can be either diluted or run subsequently.

Not only have these assays have been used to detect TCDD and related chemicals in variety of biological, environmental and food samples, but the CALUX bioassay provides an avenue to identify and characterize novel chemicals and classes of chemicals which can bind to and activate the AhR signal transduction pathway. Thus, a positive CALUX that is shown by subsequent immunoassay or immunoprecipitation assay (Figure 33.4) to be due to neither TCDD nor PCB must contain a novel AhR agonist. The identify of the activating chemical(s) can be obtained by using the CALUX bioassay to monitor a classical purification scheme based on differential extraction and column chromatography as presented in Figure 33.5.

Acknowledgments

This research was supported by NIEHS/EPA Superfund Basic Research Grant ES04699 and NIEHS SBIR grant R43ES80372.

References

1. Safe, S., Polychlorinated biphenyls (PCBs), dibenzo-p-dioxins (PCDDs), and related compounds: environmental and mechanistic considerations which support the development of toxic equivalency factors, *Crit. Rev. Toxicol.,* 21, 51–88, 1990.
2. Giesy, J.P., Ludwig, J.P., and Tillitt, D.E., Dioxins, dibenzofurans and wildlife, in *Dioxins and Health,* Schecter, A., Ed., Plenum Press, New York, 1994, pp. 249–307.
3. DeVito, M.J. and Birnbaum, L.S., Toxicology of dioxins and related chemicals, in *Dioxins and Health,* Schecter, A., Ed., Plenum Press, New York, 1994, pp. 139–162.
4. Ballschmiter, K., Rappe, C., and Buser, H.R., Chemical properties, analytical methods and environmental levels of PCBs, PCTs, PCNs and PBBs, in *Halogenated Biphenyls, Terphenyls, Naphthalenes, Dibenzodioxins and Related Products,* Kimbrough, R.D. and Jensen, J., Eds., Elsevier Science, Amsterdam, 1989, pp. 239–293.
5. Rappe, C. and Buser, H.R., Chemical and physical properties, analytical methods, sources and environmental levels of halogenated dibenzodioxins and dibenzofurans, in *Halogenated Biphenyls, Terphenyls, Naphthalenes, Dibenzodioxins and Related Products,* Kimbrough, R.D. and Jensen, J., Eds., Elsevier Science, Amsterdam, 1989, pp. 239–293.
6. Denison, M.S., Elferink, C.F., and Phelan, D., The Ah receptor signal transduction pathway, in *Toxicant-Receptor Interactions in the Modulation of Signal Transduction and Gene Expression,* Denison, M.S. and Helferich, W.G., Eds., Taylor and Francis, Bristol, PA, 1998, pp. 3–33.
7. Denison, M.S., Fisher, J.M., and Whitlock, Jr., J.P., Inducible, receptor-dependent protein-DNA interactions at a dioxin-responsive transcriptional enhancer, *Proc. Natl. Acad. Sci. USA,* 85, 2528–2532, 1988.
8. Hord, N.G. and Perdew, G.H., Physicochemical and immunocytochemical analysis of the aryl hydrocarbon receptor nuclear translocator: characterization of two monoclonal antibodies to the aryl hydrocarbon receptor nuclear translocator, *Molec. Pharmacol.,* 46, 618–624, 1994.
9. Denison, M.S., Fisher, J.M., and Whitlock, Jr., J.P., The DNA recognition site for the dioxin-Ah receptor complex: nucleotide sequence and functional analysis, *J. Biol. Chem.,* 263, 17721–17724, 1988.
10. Garrison, P.M. et al., Species-specific recombinant cell lines as bioassay systems for the detection of 2,3,7,8-tetrachlorodibenzo-p-dioxin-like chemicals, *Fund. Appl. Toxicol.,* 30, 194–203, 1996.
11. Ziccardi, M.H., Gardner, I.A., and Denison, M.S., Development and modification of a recombinant cell bioassay to directly detect halogenated and polycyclic aromatic hydrocarbons in serum, *Toxicol. Sci.,* 54, 183–193, 2000.
12. Rogers, J.M. and Denison, M.S., Recombinant cell bioassays for endocrine disruptors: development of a stably transfected human ovarian cell line for the detection of estrogenic and anti-estrogenic chemicals, *In Vitro Mol. Toxicol.,* 13, 67–82, 2000.
13. Murk, A.J. et al., Chemical-activated luciferase gene expression (CALUX): a novel *in vitro* bioassay for Ah receptor active compounds in sediments and pore water, *Fundam. Appl. Toxicol.,* 33, 149–160, 1996.

14. Hooper, K. et al., CALUX™ results correlate with GC/MS/MS data from Kazakhstan breast milk samples, *Organohal. Cmpds.*, 45, 236–239, 2000.
15. Van Overmeire, I. et al., A comparative study of GC-HRMS and CALUX™ TEQ determinations in food samples by the Belgian Federal Ministries of Public Health and Agriculture, *Organohal. Cmpds.*, 45, 196–199, 2000.
16. Stanker, L. et al., Monoclonal antibodies for dioxin: antibody characterization and assay development, *Toxicol.*, 45, 229–243, 1987.
17. Vanderlaan, M. et al., Improvement and application of an immunoassay for screening environmental samples for dioxin contamination, *Environ. Toxicol. Chem.*, 7, 859–870, 1988.
18. Langley, M.N. et al., Immunoprobes for polychlorinated dibenzodioxins: synthesis of immunogen and characterization of antibodies, *Food Agricul. Immunol.*, 4, 143–151, 1992.
19. Harrison, R.O. and Carlson, R.E., An immunoassay for TEQ screening of dioxin/furan samples: Current status of assay and applications development, *Chemosphere*, 334, 915–928, 1997.
20. Sanborn, J.R. et al., Hapten synthesis and antibody development for polychlorinated dibenzo-*p*-dioxin immunoassays, *J. Agric. Food Chem.*, 46, 2407–2416, 1988.
21. Sugawara, Y. et al., Development of a highly sensitive enzyme-linked immunosorbent assay based on polyclonal antibodies for the detection of polychlorinated dibenzo-*p*-dioxins, *Anal. Chem.*, 70, 1092–1099, 1998.
22. Jackson, T.M. and Ekins, R.P., Theoretical limitations on immunoassay sensitivity. Current practice and potential advantages of fluorescent Eu3+ chelates as non-radioisotopic tracers, *J. Immunol. Methods*, 87, 13–20, 1986.
23. Shan, G.M. et al., Isotope-labeled immunoassays without radiation waste, *Proc. Natl. Acad. Sci. USA*, 97, 2445–2449, 2000.
24. Shan, G. et al., Surrogates for dioxin analyses: analytical, ELISA and toxicological aspects, *Organohal. Cmpds.*, 45,188–191, 2000.
25. Shan, G. et al., A highly sensitive dioxin immunoassay and its application to soil and biota samples, *Anal. Chim. Acta.*, 444, 169–178, 2001.
26. Krämer, P.M., Li, Q.X., and Hammock, B.D., Integration of liquid chromatography with immunoassay: an approach combining the strengths of both methods, *J. AOAC Int.*, 77, 1275–1287, 1994.

chapter thirty-four

Application of biomarkers to population studies on food chain contaminants in Nunavik (Arctic Quebec, Canada)

Éric Dewailly, Pierre Ayotte, Marc Rhainds,
Suzanne Bruneau, Chris Furgal, Jacques Grondin,
Benoît Lévesque, Carole Blanchet, Daria Pereg,
and Gina Muckle

Contents

Abstract   Over the past 10–15 years, we have conducted several studies in the Inuit population of Nunavik (Arctic Quebec, Canada) to document exposure and investigate possible health effects related to contaminants found in the Arctic food chain. Early investigations focused on assessing biomarkers of exposure to traditional food chain contaminants such as organochlorines and heavy metals in various subgroups of the Inuit population (pregnant women and lactating mothers, children, and adults). As standards for additional compounds became available, we expanded our analyses to include toxaphene congeners, chlordane metabolites, and more recently, hydroxylated polychlorinated biphenyls. A first epidemiological study was conducted to investigate the effect of prenatal exposure to organochlorines on biomarkers of immune function and on the incidence of infectious diseases in infants during the first year of life. Studies have also been initiated to validate markers of early biological effects linked to developmental deficits and to prenatal exposure to organochlorines. Epidemiological studies are underway in Nunavik to investigate the effect of polychlorinated biphenyls (PCBs) and mercury exposure on child development and to corroborate our previous findings on immune system dysfunction. The inclusion of validated biomarkers in these studies may help in identifying relevant associations between exposure to contaminants and detrimental health effects in the Inuit population.

I. Introduction

Located on a vast territory of 563,515 km² north of 55th parallel, the Inuit population of Nunavik (Northern Quebec, Canada) is estimated at 8,970 people, distributed among 14 villages (Figure 34.1). Compared to the rest of Canada, the Inuit population is young. In 1991, 40% of the Inuit were 15 years of age and less, and 2% were 65 years old and over as compared to Canadian figures for the same year which were 20% and 11%, respectively.[1] Inuit are confronted with challenging environmental conditions such as extreme cold, and, historically, the abundance of arctic fauna has supported the survival of this population. Their traditional diet consists of marine mammals , e.g., white whale (beluga), seal, fish, and caribou, eaten fresh (raw or cooked) or dried, using the skin, blubber, liver and fat in different meals.

Persistent organic pollutants such as organochlorines (OCs) reach the Arctic through long-range oceanic and, above all, atmospheric transport.[2,3] OCs design a class of chemical compounds that comprises PCBs, dichlorodiphenyl trichloroethane (DDT) and its main metabolite dichlorodiphenyl dichloroethylene (DDE), hexachlorobenzene (HCB), chlordane-related compounds, hexachlorocyclohexane, and toxaphene among others. Because of their lipophilic nature, OCs tend to bioaccumulate in fatty tissues of organisms along the food chain, with relatively high concentrations being observed in lipid-rich tissues such as the blubber and liver of sea mammals. For example, in the Canadian Arctic, PCB concentrations average 1, 4, and

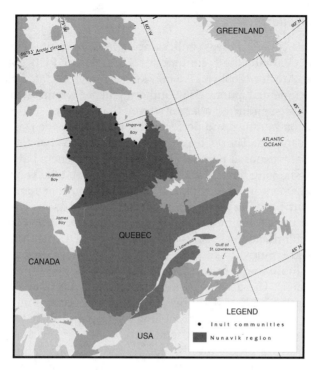

Figure 34.1 Localization of Nunavik (Arctic Quebec, Canada).

6 mg/kg in blubber of ringed seal, beluga, and narwhal, respectively.[4] With regard to heavy metals, mercury and lead are of greatest concern. Originating from anthropogenic (e.g., mining, smelting, fossil fuel burning, and waste incineration) and natural (local geology, volcanoes, and degassing in aquatic environments) sources, these contaminants have been found in all components of the Arctic ecosystem.[2,4] The source of mercury (geological or anthropogenic) in the Arctic food chain continues to be a subject of debate.[2]

II. Biomarkers of exposure to OCs and heavy metals

Three large surveys were conducted in Nunavik to assess exposure to OCs and heavy metals in various subgroups of the Inuit population:

1. A breast milk survey conducted in 1989–1990 among 107 Inuit women[5,6] with a pilot phase that took place in 1986 and included 24 participants.[7]
2. The Santé Quebec Inuit Health Survey performed in 1992 that included 492 adults from Nunavik.[8,9]
3. A cord-blood monitoring program that took place between 1993 and 1997 and included 480 newborns.[10]

A. Organochlorines

In 1986–87, we collected breast milk samples from 24 Inuit women living in Nunavik to evaluate their content for OCs. Mean PCB concentration (expressed as Aroclor 1260) was 3.6 mg/kg in breast milk fat,[7] 7 times higher than that documented among 536 Southern Quebec women (0.52 mg/kg).[11] In 1989–1990, we conducted a larger survey that included 107 women from all Nunavik communities and performed congener-specific PCB analyses of breast milk samples. The PCB profile was similar to that found in other environmentally exposed population, with PCB congeners IUPAC nos. 138, 153, and 180 showing the highest concentrations. The total concentration of ten PCB congeners in Inuit milk fat was similar to that found in beluga blubber (about 1 mg/kg lipids; 2.9 mg/kg expressed as Aroclor 1260), and the congener profile resembled that of the polar bear. Mean concentrations of the main chlorinated pesticides or metabolites (DDE, dieldrin, HCB, heptachlor epoxide, mirex) in Inuit breast milk were between 2 and 10 times greater than those found in samples previously collected from 50 Southern Quebec women.[6] These were the first results showing that Inuit mothers exhibited one the highest body burden known to occur from exposure to OC residues present in the environment, by virtue of their location at the highest trophic level of the arctic food web.

Concentrations of dioxin-like compounds (2,3,7,8-chlorosubstituted dibenzo-*p*-dioxins and dibenzofurans, nonortho substituted coplanar PCBs) were determined in a subset of 40 milk samples selected out of the 107 collected in the course of the survey mentioned above.[5] The mean total concentration, expressed as 2,3,7,8-tetrachlorodibenzo-*p*-dioxin equivalent (TEQ), was 51.3 ngTEQ/kg, compared to 23.1 ng/kg in 16 pooled samples from Caucasian women of the Quebec City region.

In the Santé Quebec Inuit Health Survey, PCBs, *p,p'*-DDE, *trans*-nonachlor, oxychlordane, and HCB were the organochlorines most frequently detected (100% of samples). Mean PCB concentration (expressed as Aroclor 1260) was 5.6 mg/kg lipids and mean DDE concentration, 2.3 mg/kg lipids (unpublished data). The parent chlordane compounds, α- and γ-chlordane, were detected respectively in 8% and 55% of samples, and low concentrations were found (0.003–0.006 mg/kg lipids) compared to their main metabolites *trans*-nonachlor and oxychlordane.[8] Additional analyses were performed to quantify two other chlordane metabolites, nonachlor-III and photoheptachlor. These compounds were found to be major chlordane metabolites, as mean concentrations determined in a subset of samples (n = 57) were respectively 0.141 and 0.011 mg/kg lipids.[12] *p,p'*-DDT was present in low concentrations (0.07 mg/kg) compared to its main metabolite *p,p'*-DDE. The DDT/DDE concentration ratio was about 30, indicating that exposure to DDT did not occur through direct contact with the technical mixture. Finally, two toxaphene congeners, parlar 26 (T2) and parlar 50 (T12) were detected in all eight samples analysed for these compounds, with mean concentrations averaging 0.07 and 0.10 mg/kg lipids, respectively.[8] These congeners are the most recalcitrant of all toxaphene congeners and were previously found in several species of the arctic food chain and in Inuit breast milk.[13] No correlation was observed between concentrations of PCB congeners and those of toxaphene congeners.[8]

Twenty pooled plasma samples were formed, each made of individual samples collected in the Santé Quebec Inuit Health Survey from the same age group, sex, and region of residence, to allow for dioxin-like compound determination. The mean total concentration of PCBs and dioxin-like compounds were respectively 4.1 mg/kg lipids and 184.2 ngTEQ/kg lipids, compared to 0.13 mg/kg lipids and 26.1 ngTEQ/kg lipids for three control pooled plasma samples from Southern Quebec. Total PCBs and dioxin-like compound concentrations were strongly correlated (r=0.98; p < 0.0001), increased with age, and were greater in men than in women. Body burdens of PCBs and dioxin-like compounds documented in this study were close to those which induced adverse health effects in laboratory animals.[14]

The prenatal period has been shown to be critical for the production of adverse developmental effects induced by PCB exposure. Therefore, an umbilical cord-blood monitoring program was initiated in 1993, first in Nunavik and then in the entire Canadian Arctic[15] and all circumpolar countries.[16] Mean PCB concentrations in umbilical cord plasma samples ranged from 0.3–2.0 µg/liter (expressed as Aroclor 1260) for newborns from various Aboriginal populations. The highest concentrations were found among the Inuit of Nunavik and Baffin Island (Nunavik, Canada). Comparative data obtained in 656 newborns from Southern Quebec indicated a mean PCB concentration of 0.51 µg/liter.[17] Several Inuit neonates displayed PCB concentrations that exceed the threshold concentration beyond which cognitive impairments are expected to occur.[10]

Through collaborative work with the National Wildlife Research Centre (Canadian Wildlife Services), hydroxylated PCB metabolites (OH-PCBs) and other chlorinated phenolic compounds were identified and their relative concentrations determined in whole blood from 13 male and 17 female Inuit that participated in the Santé Quebec Inuit Health survey and a pooled whole blood sample from Southern Quebec. Total OH-PCB concentrations were variable among the Inuit samples, ranging over two orders of magnitude (0.117 to 11.6 ng/g whole blood wet weight). These concentrations were equal to and up to 70 times those found for the Southern Quebec pooled whole blood sample. Geometric mean concentrations of total OH-PCBs were 1.73 and 1.01 ng/g whole blood for Inuit men and Inuit women, and 0.161 ng/g whole blood for the Southern population pool. Limited data are available for comparison, but levels of OH-PCBs in Inuit are higher than those previously reported in the literature for other populations. Pentachlorophenol (PCP) was the dominant phenolic compound in blood, constituting 46% (geometric mean) of the total quantitated chlorinated phenolic compounds. PCP concentrations in Inuit blood ranged from 0.558–7.77 ng/g on a wet weight basis.[18]

B.   Heavy metals

1.   Lead

Lead was detected in all 492 blood samples of participants to the Santé Quebec Inuit Health Survey. The arithmetic mean concentration was 0.49 µmol/liter and the geometric mean concentration was 0.42 µmol/liter. Concentrations ranged from 0.04–2.28 µmol/liter. An analysis of variance was performed using log transformed concentrations of lead as dependent variable and age, smoking, and waterfowl (goose and duck) consumption as independent variables. All three independent variables showed statistically significant associations and explained 30% percent of the variation.[9]

Cord-blood samples were collected from Nunavik Inuit newborns (cord-blood monitoring program) to document prenatal lead exposure and investigate the possible sources. Mean (geometric) concentration was 0.19 µmol/liter (95% CI: 0.18–0.20), compared to 0.076 µmol/liter (95% CI: 0.074–0.079) in 1109 Southern Quebec neonates.[17] Close to 7% of the 475 participating neonates had a cord-blood lead concentration equal or greater than the 0.48-µmol/liter level of intervention adopted by governmental agencies. An epidemiological investigation as well as determination of the lead stable isotope ratio (^{206}Pb/^{207}Pb) in cord-blood samples were conducted among 29 newborns with lead levels equal to or greater than 0.48 µmol/liter (high-exposure group). Each newborn from this group was paired with an Inuit newborn from the same community who had a low blood lead level and three controls previously sampled during a cord-blood survey carried out in Southern Quebec. Mean (geometric) cord-blood lead concentrations in Inuit neonates from the high- and low-exposure groups, and in control neonates (Southern Quebec) were respectively 0.62 µmol/liter (95% CI: 0.57–0.67; n = 29); 0.20 µmol/liter (95% CI: 0.17–0.23; n = 31) and 0.11 µmol/liter (95% CI: 0.10–0.11; n = 89). The average lead isotope ratio among all Inuit neonates was 1.195 (range: 1.166–1.230; 95% CI: 1.190–1.200) and was not different between high- and low-exposure groups. This ratio was similar to that found in some of the most popular brands of shotgun cartridges used by Nunavik hunters. Southern Quebec neonates showed a lower radiogenic ratio (1.166; range = 1.126–1.230; 95% CI: 1.163–1.168; Student t-test: p ≤ 0.0001) indicating a different source of exposure. We concluded that the ingestion of lead shot residues in game meat, most notably waterfowl, was likely responsible for the higher lead levels found in Nunavik Inuit neonates.[19]

2.   Mercury

Mercury was detected in 100% of the 492 samples collected in the Santé Quebec Inuit Health Survey. Mean (geometric) total mercury concentration was 79.6 nmol/liter among Inuit adults, ranging from 4 to 560 nmol/liter. Inorganic mercury was measured in a subset of 18 individuals with high total mercury values. Among these individuals, total mercury averaged 265 nmol/liter and inorganic mercury 48.7 nmol/liter, representing 18% of total mercury. Traditional food consumption estimated by the number of seal meat meals, was strongly associated with mercury exposure with concentrations varying from 153, 106, 97, and 72 nmol/liter in the daily, weekly, monthly, and never consumption class groups of seal meat meal frequency. Omega-3 fatty acids and mercury concentrations were strongly associated (r = 0.56, p < 0.001). Age, as well as seal and beluga consumption were the two independent variables showing statistically significant associations and explained 30% of the variation of individual blood levels of mercury in females.[9]

Among all aboriginal populations surveyed in the Canadian Arctic during the cord-blood monitoring program, the Inuit of Nunavik and Baffin Island exhibited the highest concentrations of mercury in blood, with mean concentrations of 70.8 nmol/liter and 52.0 nmol/liter, respectively. Mean (geometric) concentration of mercury in cord-blood samples from 1109 southern Quebec neonates was 4.82 nmol/liter (95% CI = 4.56, 5.08).[17] A significant proportion of Nunavik and Baffin Island Inuit newborns had concentrations above the critical threshold for neurological deficits. Variations in exposure levels resulted from different dietary habits in these Canadian subgroups.[10]

3. Cadmium

High concentrations of cadmium found in the liver and kidneys of caribous and sea mammals of the Canadian Arctic have led to the hypothesis that the Inuit might be highly exposed to cadmium though their diet. Data from the Santé Quebec Inuit Health Survey confirmed that blood cadmium of Inuit is indeed high by comparison to levels reported in other population surveys. Blood cadmium levels were strongly associated with the smoking status. Among nonsmokers, blood cadmium levels were comparable to those reported in nonsmokers elsewhere in the world and were not associated with dietary habits such as the consumption of sea mammals. In reference to international standards, blood cadmium concentrations were high enough among the Inuit to warrant a public health intervention.[20]

III. Biomarkers of effect

A. Biomarkers of immune system function

Our first epidemiological study investigated whether organochlorine exposure is associated with the incidence of infectious diseases in Inuit infants. The number of infectious disease episodes in 98 breast-fed and 73 bottle-fed infants was compiled during the first year of life. Concentrations of OCs were measured in early breast milk samples and used as surrogates to prenatal exposure levels. Biomarkers of immune system function (lymphocyte subsets, plasma immunoglobulins) were determined in venous blood samples collected from infants at 3, 7, and 12 months of age. Otitis media was the most frequent disease with 80.0% of breast-fed and 81.3% of bottle-fed infants experiencing at least one episode during the first year of life.

During the second follow-up period, the risk of otitis media increased with prenatal exposure to *p,p'*-DDE, HCB, and dieldrin. The relative risk (RR) for 4- to 7-month-old infants in the highest tertile of *p,p'*-DDE exposure as compared with infants in the lowest was 1.87 (95% CI: 1.07–3.26). The relative risk of otitis media over the entire first year of life also increased with prenatal exposure to *p,p'*-DDE (RR 1.52; 95% CI: 1.05–2.22) and HCB (RR 1.49; 95% CI: 1.10–2.03). Furthermore, the relative risk of recurrent otitis media (≥ three episodes) augmented with prenatal exposure to these compounds. No clinically relevant differences were noted between breast-fed and bottle-fed infants with regard to biomarkers of immune function and prenatal OC exposure was not associated with these biomarkers. These results suggest that prenatal OC exposure could be a risk factor for acute otitis media in Inuit infants.[21]

B.   Biomarkers of developmental effects

We investigated markers of early biological effects possibly related to OC exposure in Inuit women. CYP1A1-dependent enzyme activity (EROD) and DNA adducts were measured in placenta samples obtained from 22 Inuit women from Nunavik. These biomarkers were also assessed in 30 women from a Quebec urban center (Sept-Îles) as a reference group. Prenatal OC exposure was determined by measuring these compounds in umbilical cord plasma. Placental EROD activity and the amount of DNA adducts potentially induced by OC exposure were significantly higher in the Nunavik group than in the reference group. For both biomarkers, smoking was found to be an important confounding factor. OC exposure was significantly associated with EROD activity and DNA adduct levels when stratifying for self-declared smoking status. We concluded at that time that CYP1A1 enzyme induction and DNA adducts in placental tissue constitute useful biomarkers of early effects induced by environmental exposure to organochlorines.[22]

However, in the latter study, there were few Inuit women who did not smoke during pregnancy and the smoking status was not ascertained with a biomarker. We conducted a second study to determine if environmental exposure to PCBs induces placental CYP1A1 in Inuit women. Placenta and cord-blood samples were obtained from 35 Inuit women from Nunavik and 30 women from Sept-Ies (reference population). We previously validated the use of cotinine concentration in meconium as a marker of prenatal exposure to tobacco smoke.[23] PCB concentrations were measured in cord plasma, and CYP1A1 activity (EROD) was assessed in placenta. Despite the higher PCB exposure of the Inuit population, both groups showed similar EROD activities when the data were stratified according to the smoking status ascertained by the cotinine concentration in meconium. In the Nunavik population, EROD activity was correlated with 2,2',4,4',5,5'-hexachlorobiphenyl plasma concentration (a marker of exposure to the environmental PCB mixture), but a multivariate analysis failed to demonstrate a significant contribution of PCB exposure to placental CYP1A1 activity when tobacco smoking was included in the analysis. This study showed that low-level environmental PCB exposure does not influence CYP1A1 activity in the placenta, leaving tobacco smoking as a major modulating factor.[24]

IV. Ongoing epidemiological studies that include biomarker measurements

A.   Evaluation of preschool children

The purpose of this study is to examine the long-term consequences of prenatal exposure to PCBs and methylmercury on the development of the nervous system. This project was designed to study domains of effects overlooked in most previous studies that addressed this issue. Of particular interest is the impact of exposure on neurophysiological and neurological endpoints that could be related to learning difficulties and disabilities. Approximately 300 children for whom prenatal PCBs and methylmercury exposure was documented during the cord-blood monitoring program, have now reached the age of five. The final sample will include 100 Nunavik Inuit children, ages 5–6 years. Prenatal exposure to PCBs chlorinated pesticides and mercury were obtained at birth through cord-blood analyses. Current body burdens of OC and methylmercury are documented by peripheral blood plasma and hair analyses, respectively. This study focuses on neurological and neurophysiological endpoints. To be more specific, the Amiel-Tison and Stewart's neurological examination is performed. Fine neuromotor functions refer to the involuntary components of gross motor skills and to the organization of voluntary movements executed repeatedly. A selection of tests, which are more sensitive to detect subtle changes in the fine neuromotor function than the traditional neurobehavioral tests used in neurophysiological testing, guides our choice. As a result, new technologies from the field of cognitive neuroscience used with adults have been adapted for child testing and are used in this study. The electrophysiological methods (visual and cognitive evoked potentials) are objective tools, selected to evaluate the maturation and integrity of the central visual system, as well as attentional and memory processing.

B.   Child development cohort study

This project is designed to replicate and extend previous findings by studying a highly exposed cohort of infants, and using infant tests that have the potential to provide information regarding possible mechanisms of action. 300 Inuit infants from Nunavik and Greenland will be assessed at birth, 6 and 11 months of age. The impact of PCBs and methylmercury exposure on newborn's thyroid hormones, physical growth, immune system function, physical and neurological maturity, overall health, mental, psychomotor and neurobehavioral development, and visual and spatial information processing will be studied. This research will also provide the opportunity to perform long-time trend analyses of human exposure to environmental contaminants (OCs and heavy metals).

C. *Salluit study*

Between 1971 and 1978, Health and Welfare Canada (Medical Services Branch) carried out mercury exposure assessments in 350 Aboriginal communities in Canada. Salluit inhabitants in particular were found to have unexpectedly high levels; 17% of the test results were over 500 nmol/liter, whereas in the other Nunavik communities only 2% of the test results were over this concentration (concentrations higher than 500 nmol/liter are considered to cause health problems). Further investigation in Salluit was deemed "imperative"[25] to find the cause of these high levels of mercury. The objectives of this ongoing study are to evaluate time trends of mercury exposure (20 years), to develop biomarkers of oxidative damage, and to assess sensorimotor functions.

V. Conclusions

Several studies were carried out documenting exposure of the Inuit population of Nunavik to OCs and heavy metals. Biomarkers of exposure were used to reveal an unusually high body burden of several OCs, mercury and lead in this population. Earlier studies suggest that this level of exposure can adversely affect neurological development in infants and children. A significant proportion Inuit infants would appear to be potentially at risk for such adverse neurodevelopmental effects; however, they may be afforded some level of protection by the high levels of selenium and omega-3 fatty acids in the Nunavik diet.[26] Biomarkers of exposure and effect are being used in several ongoing epidemiological studies in Nunavik, which may improve understanding of health risks in the Nunavik and other Arctic populations.

References

1. Hodgins, S., *Health and What Affects It in Nunavik: How Is the Situation Changing?* Nunavik Regional Board of Health and Social Services, Kuujjuaq, 1997.
2. Barrie, L.A. et al., Arctic contaminants: sources, occurence and pathways, *Sci. Tot. Environ.*, 122, 1–74, 1992.
3. Macdonald, R.W. et al., Contaminants in the Canadian Arctic: 5 years of progress in understanding sources, occurrence and pathways, *Sci. Tot. Environ.*, 254, 93–234, 2000.
4. Muir, D.C.G. et al., Arctic marine ecosystem contamination, *Sci. Tot. Environ.*, 122, 75–134, 1992.
5. Dewailly, É. et al., Breast milk contamination by PCDDs, PCDFs and PCBs in Arctic Quebec: a preliminary assessment, *Chemosphere*, 25, 1245–1249, 1992.
6. Dewailly, É. et al., Inuit exposure to organochlorines through the aquatic food chain in Arctic Quebec, *Environ. Health Perspect.*, 101, 618–620, 1993.
7. Dewailly, É. et al., High levels of PCBs in breast milk of Inuit women from Arctic Quebec, *Bull. Environ. Contam. Toxicol.*, 43, 641–646, 1989.

8. Santé Quebec. *A Health Profile of the Inuit: Report of the Santé Quebec Health Survey Among the Inuit of Nunavik,* Vol. 1, Jetté, M., Ed., Ministère de la Santé et des Service sociaux, Gouvernement du Quebec, Montréal, 1994.

9. Dewailly, É. et al., Exposure of the Inuit population of Nunavik (Arctic Quebec) to lead and mercury, *Arch. Environ. Health.,* 56(4), 350–357, 2001.

10. Muckle, G., Dewailly, É., and Ayotte, P., Prenatal exposure of Canadian children to polychlorinated biphenyls and mercury, *Can.J. Public Health,* 89, S20–S25, 1998.

11. Dewailly, É. et al., Polychlorinated biphenyl (PCB) and dichlorodiphenyl dichloroethylene (DDE) concentrations in the breast milk of women in Quebec, *Am. J. Public Health,* 86, 1241–1246, 1996.

12 Zhu, J. et al., Persistent chlorinated cyclodiene compounds in ringed seal blubber, polar bear fat, and human plasma from Northern Quebec, Canada: identification and concentrations of photoheptachlor, *Environ. Sci. Technol.,* 29, 267–271, 1995.

13. Stern, G.A. et al., Isolation and identification of two major recalcitrant toxaphene congeners in aquatic biota, *Environ. Sci. Technol.,* 26, 1838–1840, 1992.

14. Ayotte, P. et al., PCBs and dioxin-like compounds in plasma of adult Inuit living in Nunavik (Arctic Quebec), *Chemosphere,* 34, 1459–1468, 1997.

15. Van Oostdam, J., et al., Human health implications of environmental contaminants in Arctic Canada: A review, *Sci. Tot. Environ.,* 230, 1–82, 1999.

16. Arctic Monitoring and Assessment Programme, *AMAP Assessment Report: Arctic Pollution Issues,* AMAP, Oslo, 1998.

17. Rhainds, M. et al., Lead, mercury, and organochlorine compound levels in cord blood in Quebec, Canada, *Arch. Environ. Health,* 54:40–47, 1999.

18. Sandau, C.D. et al., Analysis of hydroxylated metabolites of PCBs (OH-PCBs) and other chlorinated phenolic compounds in whole blood from Canadian Inuit, *Environ. Health Perspect.,* 108, 611–616, 2000.

19. Dewailly, É. et al., Lead shot as a source of lead poisoning in the Canadian Arctic, *Epidemiology,* 11 (4), S146, 2000.

20. Rey, M. et al., High blood cadmium levels are not associated with consumption of traditional food among the Inuit of Nunavik, *J. Toxicol. Environ. Health,* 51, 5–14, 1997.

21. Dewailly, É. et al., Susceptibility to infections and immune status in Inuit infants exposed to organochlorines, *Environ. Health Perspect.,* 108, 205–211, 2000.

22. Lagueux, J. et al., Cytochrome p450 CYP1A1 enzyme activity and DNA adducts in placenta of women environmentally exposed to organochlorines, *Environ. Res.,* 80, 369–382, 1999.

23. Pereg, D. et al., Cigarette smoking during pregnancy: comparison of biomarkers for inclusion in epidemiological studies, *Biomarkers,* 6(2), 161–173, 2001.

24. Pereg, D. et al., Environmental exposure to polychlorinated biphenyls and placental CYP1A1 activity in Inuit women from Northern Quebec, *Environ. Health Perspect.,* in press.

25. Wheatley, M.A. and Wheatley, B. The effect of eating habits on mercury levels among Inuit residents of Sugluk, P.Q., *Études/Inuit/Studies,* 5(1), 27–43, 1981.

26. Dewailly, É. et al., Weighing contaminant risks and nutrient benefits of country food in Nunavik, *Arct. Med. Res.,* 55, 13–19, 1996.

chapter thirty-five

University-community partnership for the study of environmental contamination at Akwesasne

David O. Carpenter, Alice Tarbell, Edward Fitzgerald,
Michael J. Kadlec, David O'Hehir, and Brian Bush

Contents

Abstract The Mohawk Nation at Akwesasne is a Native American community of about 12,000 persons who reside at the juncture of New York, Ontario, and Quebec. This community has been severely affected by local contamination with polychlorinated biphenyls (PCBs) as a result of PCB use and misuse by three local aluminum industries. The PCBs have contaminated the traditional fishing waters of the reserve, making the fish unsafe to eat. This has altered employment patterns and diet. For the past 14 years, scientists from the University at Albany and the New York State Department

1-56670-596-7/02/$0.00+$1.50
© 2002 by CRC Press LLC

of Health have worked with this community to determine levels of PCBs in body fluids, the relation of consumption of contaminated fish to PCB levels in individuals, and to determine the health effects of exposure to these xenobiotics. The local community body which has functioned as the partner in this research is the Akwesasne Task Force on the Environment. In this paper, we review the history of environmental problems at Akwesasne, the evidence that PCB levels are related to contaminated fish consumption, and some of the preliminary evidence for health effects of PCB exposure in this population. In addition, we describe how the university and the community partners have developed and maintained a collegial and trusting working relationship that benefits both groups and describe some of the principles that have evolved in terms of how an academic research community may optimally work with any community. While this particular community is different from many others, we believe that many of the lessons learned are applicable to any situation where the target of study is a human population in a specific community.

I. Introduction

Akwesasne (Land Where the Partridge Drums) is a 22,000-acre community of about 12,000 Mohawk Native Americans who live at the juncture of New York, Ontario, and Quebec (Figure 35.1). For generations, the Mohawk community has derived employment, food, and enjoyment from the St. Lawrence River and its fast flowing tributaries which descend from the Adirondack Mountains. However, because of the availability of cheap electricity as a result of construction of the Robert Moses Power Dam, in the late 1950s, three aluminum foundries were constructed just upstream from Akwesasne, operated by General Motors, ALCOA, and Reynolds Metals. All three used PCBs as hydraulic fluids, which leaked and, as a result, PCB-contaminated sludge from the wastewater treatment system leaked and drained into the rivers. Since PCBs are persistent, lipophilic, and bioaccumulate in the food chain, this resulted in significant levels in the local fish, which is the major protein source for this community for centuries.

The General Motors site, which abuts the Akwesasne Reserve, was listed as a National Priority List site in 1985, having on-site PCB concentrations in soils and sludge as high as 4 percent, with significant off-site migration. A cap was placed over the major landfill site in 1988, but only a small amount of contaminated soils and sediments have been removed to date. The Reynolds and ALCOA sites are upstream from Akwesasne. As a result of the PCB contamination the New York State Department of Environmental Conservation has declared 1,000 acres near the St. Lawrence, Grasse, and Raquette Rivers to be an inactive hazardous site.

From the beginning of this project, our study was a collaborative endeavor with the members of the Akwesasne community and has required hard work on our parts to generate mutual respects and trust. The initial contacts came through awareness of the Mohawks of an investigation of

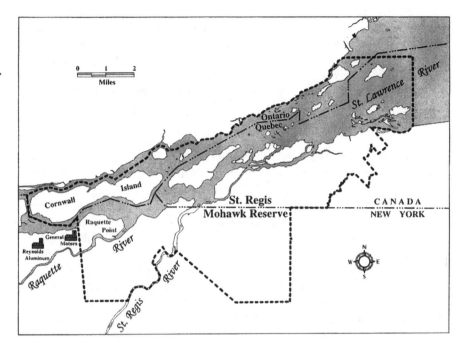

Figure 35.1 Map of Akwesasne showing the Reserve boundaries and the locations of the General Motors and Reynolds Metals foundry sites.

human milk PCB contamination at Oswego, NY, which had recently been published by the New York State Department of Health. They agreed to commence a pilot study of human milk collected from women at Akwesasne. This project faltered several times partly because of lack of trust by the Mohawks of government agencies. In addition, the Mohawks had been subjected in the past to more or less blundering research studies by academic institutions that were not sensitive to the needs and wishes of the community. When the community cooperated in these studies, no results were returned to the community and they had the perception that they had been used as guinea pigs for the benefit only of the academics. As a result, the community determined that would not happen again. While the Mohawks were concerned about the possible seriousness to their health, and particularly that of their children,[1] they also demanded a greater role in future studies and a greater degree of communication regarding the results. This paper describes the problem, some of the results of the studies, and the interactions between the academic community and the Mohawk Nation at Akwesasne.

As trust grew and the possibility of strong federal Superfund funding arose, a highly interactive multidisciplinary study of the deleterious effects of PCB on humans and wildlife, involving chemists, toxicologists, anthropologists, epidemiologists, behavioral psychologists, ecologists, and engineers was developed and funded for a period of 13 years through the Superfund Basic Research Program of the National Institute of Environmental Health

Sciences. The School of Public Health of the University at Albany was the lead organization and included faculty from Syracuse University and SUNY Oswego as well. Since the School of Public Health is a collaborative endeavor of the university and the New York State Department of Health, this mechanism allowed professional staff who had faculty appointments and resources of the Department of Health to be participants in the study.

II. Polychlorinated biphenyls

PCBs are a mixture of different chemical compounds. There are 209 possible congeners, depending upon how many chlorines are on the biphenyl ring and where they are positioned. PCBs were manufactured as mixtures with varying percentages of chlorination. The mixture used at the three aluminum plants near Akwesasne was Aroclor 1248, which contains 48 percent chlorine by weight. Aroclor 1248 contains more than 50 individual chlorobiphenyl compounds (congeners) (Figure 35.2 top), but the concentration of each ranges widely. No contemporary method of PCB analysis is capable of measuring every PCB congener. What is shown in Figure 35.2 (top) is the measurement of the amount of 79 different congeners, shown in the order in which they elute from the column, which is roughly related to the degree of chlorination. Each congener has distinct physiological actions on a variety of living organisms, including mammals (particularly humans), birds, and, several lower species such as fish, snails, and aquatic insects.[2] Epidemiological studies suggest that PCB exposure is associated with a number of different health effects including cancer, immune suppression, neurobehavioral abnormalities, and disruption of various endocrine systems, including the thyroid and sex steroids. While PCBs are widely distributed throughout foodstuffs, especially animal products, consumption of contaminated fish is a major source of human exposure. Thus, the contamination of local fish has the potential to cause all of the diseases listed above.

Since different PCB congeners have different biologic activities,[3] it is important that the individual congeners be measured, not just the total PCB concentration. While total PCB levels may provide a rough indication of exposure, it is the profile of individual congeners that will indicate which diseases are of increased risk in exposed people. Our laboratory has been a leader in development of routine techniques for determination of a large number of PCB congeners.[4,5] This analysis is performed on gas chromatographs equipped with electron capture detectors, using state-of-the-art capillary chromatography. Over the number of years of our project, many new developments in capillary chromatography have occurred, as well as the introduction of more modern gas chromatographs allowing for increased sensitivity as well as analysis of additional separations. Due to the evolution of this methodology the list of congeners identified and quantified has increased with time, which is why varying numbers of peaks are shown in the figures below. As the method was revised, the PCB numbers were reassigned based on the elution order of the congeners. For each revision, these

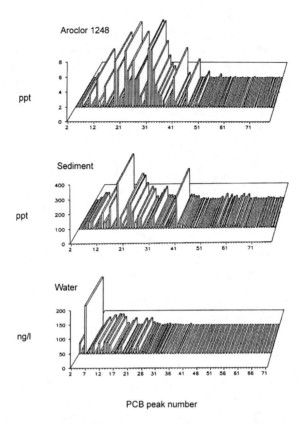

Figure 35.2 PCB congener profiles from an Aroclor 1248 standard (top), a sediment sample from Contaminant Cove (middle) and a water sample from Contaminant Cove (bottom). The Aroclor standard was 100 ppt, while the total PCB concentration for the sediment was 2.6 ppm and the water was 619.4 ng/l. The congener peak numbers reflect the order in which the peak come off the GC, which correlates roughly with increasing degrees of chlorination.

PCB numbers correlate to specific congeners which are identified in our previous publications.

Unfortunately, our ability to interpret all of the information obtained with congener-specific measurements is somewhat limited at present, since not all congeners have been studied by themselves. There are several added complications. The rates of clearance of PCBs from the body by liver metabolism varies greatly with different congeners, depending on how many and where the chlorines are located. Lower chlorinated congeners are, in general, more easily metabolized than are higher chlorinated ones. Those congeners that are most persistent in the human body are those that are present in the Aroclor mixtures and have six or more chlorines. It is clear that some adverse effects of PCBs are mediated by activation of the Ah receptor, an action similar to that of 2,3,7,8-dibenzo-p-dioxin,[6] and these are known to cause cancer and immune system suppression. Only the coplanar PCBs (congeners that do not have chlorines in the ortho positions, adjacent to the biphenyl rings) are effective activators of this receptor. In contrast, those congeners with chlorines in the ortho positions are not known to be carcinogenic, but cause neurobehavioral toxicity and endocrine disruptive effects.[3] Finally, various congeners differ in their relative volatility and water solubility. In general, the lower chlorinated congeners are more volatile and water soluble.

A.   PCBs in soils, sediments, and water

Soils, sediments, and water are all routes of possible human exposure to PCBs, either directly through unintended ingestion, inhalation of dust particles and volatile PCBs or through dermal absorption, or indirectly as a means of contamination of fish or other foodstuff. Figure 35.2 (middle) shows the congener pattern of PCBs in a sediment from Contaminant Cove, the small cove immediately adjacent to the General Motors Foundry Site, as well as the pattern present in water in the Cove (Figure 35.2 bottom). This particular sample had 2.6 ppm of PCBs, but it is apparent when comparing Figures 35.2A and 35.2B that the congener pattern is not identical. This is because anaerobic soil bacteria are capable of removing some of the chlorines in the meta and para positions around the molecule.[7,8] Thus with time the congener profile may change although the anaerobic bacteria are not capable of destroying the PCB molecule.[9]

Hwang et al.,[10] have reported results of PCB analysis of 119 surface soil samples from a variety of sites within and near Akwesasne. Most were not significantly higher than other sites in upstate New York, with an overall mean concentration of 0.061 ppm. Several samples from near the General Motors facility and on the Raquette River had higher levels, with the maximum being 0.886 ppm. Before the remediation, PCB levels in soil on the General Motors site ranged from less than 1 ppm to 380 ppm.

B.   PCB levels in fish, game, and vegetables

Much of the initial concern over PCBs centered on measurements in wildlife. Stone (unpublished) reported finding 835 ppm wet weight PCBs in snapping turtle fat from a turtle collected near to General Motors. Since the turtle is considered by the Mohawks to be a sacred animal, this had particular significance to the population. Masked shrews were found to have levels from 30–11,522 ppm (lipid basis) in 1985.

Table 35.1 gives levels of PCBs in local fish.[2] For reference, the FDA[11] prohibits the sale of fish which have PCB concentrations of greater than 2 ppm. It is clear from this table that many fish, especially bottom feeders like bullheads, are highly contaminated. The fish taken from near the General Motors site showed significantly higher levels of PCBs than did those from the St. Lawrence River although not all species of fish had the same degree of elevation. Equally important, the congener pattern of fish from Turtle Creek near General Motors is similar to that of Aroclor 1248, whereas those

Table 35.1 Mean Total PCB Levels in Standard Fillets of Fish Collected from the Vicinity of the General Motors Foundry and in the St. Lawrence River

Species	Near General Motors (Mean and Range in ppm)	St. Lawrence River and Tributaries (Mean and Range in ppm)
Brown bullhead	20.55 (<0.15–81.49)	1.82 (<0.15–10.73)
Northern pike	2.73 (0.48–5.12)	0.42 (<0.15–3.52)
Rock bass	1.04 (<0.15–4.02)	0.18 (0.15–0.86)
White sucker	6.39 (0.29–11.0)	0.17 (<0.15–0.63)
Yellow perch	3.41 (0.20–12.26)	0.61 (0.15–0.86)

from other areas differed. Rock bass, for which other fish are not the primary food source, showed low levels of PCBs, whereas eels, catfish, and carp had significantly elevated levels even in the St. Lawrence. In addition, there was considerable variation in levels among different fish of the same species, which presumably reflects what and where they ate. These studies resulted in strong advice from the Mohawk governing leaders and state and federal agencies not to consume fish from the river, especially for women of child-bearing age and children.

It has usually been assumed that the major source of the PCB body burden in fish reflects oral intake of either contaminated sediment or smaller fish. Figure 35.3 shows the congener specific profile of PCBs from a carp caught in Contaminant Cove, a small cove immediately adjacent to the General Motors site which drains into the St. Lawrence, in comparison with the congener profile of Aroclor 1248, the major PCB mixture used in this area. In these studies, we measured 68 different PCB peaks, each representing a single or, in a few cases, double or triple congeners. In general, the lower numbers reflect lower degrees of chlorination, but the exact composition of each peak is known from calibration studies. The congener profiles in these fish are not identical to those for Aroclor 1248. For the carp, which is a bottom feeder that ingests large quantities of sediment, there is an accumulation of

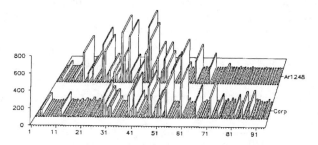

Figure 35.3 PCB congener profile for Aroclor 1248 and from a carp caught in Contaminant Cove in the spring. The fish shows an accumulation of some higher chlorinated congeners than those present in Aroclor 1248 in quantity. This may reflect an Aroclor 1254 pattern obtained from other parts of the river, or may reflect bioaccumulation of congeners present in Aroclor 1248 is a smaller relative proportion.

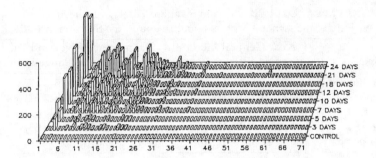

Figure 35.4 PCB accumulation in caged trout maintained within the water column in Contaminant Cove and fed commercial fish food for periods of time up to 24 days. Note the progressive accumulation of congeners having a profile similar to that within the water (Figure 35.2 bottom).

higher chlorinated congeners, probably reflecting the fact that the greater the degree of chlorination the more persistent the congener. In contrast, the perch shows a shift toward lower chlorinated congeners. Since perch are not bottom feeders but derive food from worms, insects, and plants, they usually do not have as high levels as the carp. The differences may result from a variety of factors, including bacterial dechlorination of PCBs in the sediments, selective metabolism of some congeners in the fish, and differential water solubility and volatilization of the congeners.

If all exposure to fish comes from food sources and sediments, then fish farming would be a possible mechanism of providing the traditional food stuff without PCB contamination. However, a possibility remains that fish can also absorb PCBs directly from the water column. To test this possibility, a study was done in which caged trout were placed in contaminated water and fed clean commercial fish food. Figure 35.4 shows the PCB congener profile in these fish as a function of the days of exposure to the water column. The PCBs in the fish reflect the congener pattern in the water (Figure 35.2, bottom), not that of the sediment (Figure 35.2, middle). Clearly, fish can absorb PCBs directly from the water and can bioconcentrate them such that after 24 days the levels in the fish exceed those in the water by several hundred-fold. Since it is the congeners with fewer chlorines that are more soluble in water, the pattern of contamination is different from that seen in sediments and soils and in wild fish living in contaminated water. Absorption from the water column may also contribute to the congener patterns seen in the yellow perch in Figure 35.2.

Fitzgerald et al.,[12] have reported PCB levels in drinking water and vegetables. They found the mean tap water concentration to be 16 ppt (range 10–42 ppt). Almost all of the vegetables studied had undetectable levels (detection limit=0.2 ppb), although a few samples of corn and tobacco from a garden near to the General Motors facility were found to contain PCB levels over 100 ppb. However, these low levels make clear the conclusion that the major source of exposure to Mohawks is consumption of contaminated fish.

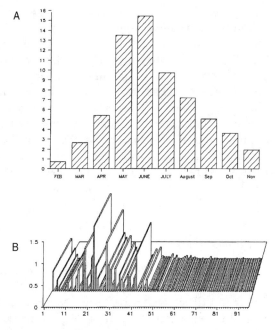

Figure 35.5 (A) PCBs in measured in air near Contaminant Cove at various months of the year. The air samplers collected volatile and air-borne particulates (B) Congener profile of the PCBs collected in June.

C. PCBs in air

In addition to ingestion and absorption through skin, a third possible source of PCB exposure for humans is inhalation. PCBs are, in general, not considered to be volatile. However, they are to a degree and, as is the case for water solubility, volatilization of PCBs is greater for lower chlorinated than higher chlorinated congeners.[13] In addition, soil particles containing bound PCBs can become airborne and can be inhaled or deposited at some distance from the original site. For more than one year, we obtained monthly air PCB concentrations from six sites at Akwesasne as a first step toward determination of whether inhalation is a significant source of PCB exposure in humans. The filters used collected particulate and volatile fractions, so these cannot be distinguished in these data. Figure 35.5 shows total PCB concentrations in the air near Contaminant Cove, the most contaminated area, as a function of time of year, and shows the congener profile in the air samples. Air concentrations varied with the seasons, being greatest in the spring and early summer, probably because this the time of year when contaminated sediments are exposed and drying, and then the dried sediments can create dust which is blown around by wind.

The congener profile of PCBs collected on the air filter is relatively similar to that for the sediment at Contaminant Cove (Figure 35.2, middle), which is adjacent to the sampling site. This suggests that particulate transport was significant. However, there is a relative enrichment of the lower chlorinated congeners. More lightly chlorinated congeners are relatively more volatile, but there is also some volatilization of higher chlorinated congeners.

It is known that PCBs evaporate with water,[13,14] and the observation that air levels were greater in May than in July and August when temperatures are warmer is consistent with evaporation from contaminated sediments exposed when the spring floods recede. These outdoor air levels are low relative to those recorded in indoor air in many older buildings and homes,[2] and it is likely that the levels in outdoor air do not pose any particular threat to humans, although this may not be the case for volatile PCBs from contaminated sources tracked inside. The other sites at Akwesasne from which air samples were obtained did not show significant elevation above background.[15]

D. PCBs in Akwesasne residents

Fitzgerald and colleagues[16-18] have reported PCB concentrations in blood from Akwesasne women (pre- and postpartum), men, and children. Figure 35.6 shows the average PCB congener patterns for 139 Mohawk men. The arithmetic mean total PCB concentration was 5.4 ppb, while the geometric mean was 3.3 ppb. The maximum concentration found was 31.7 ppb.[12] These values are certainly higher than the background U.S. average but are not exceptionally high in relation to some occupational cohorts.[2] All but one man reported eating local fish, with an average consumption of eight meals per month in the past and a decline to two meals per month during the year previous to study. This decline is likely due to the advisories issued by tribal leaders against local fish consumption. Fish consumption among Mohawk women also declined significantly between 1986 to 1992. In a study of 97 women interviewed within one month of postpartum, Fitzgerald et al.,[17] reported consumption of an average of 10.7 local fish meals per year among Mohawks who gave birth in 1986–1989, compared to 3.6 and 0.9, respectively, for women who gave birth in 1990 and 1991–92. Still, one third of Mohawk mothers ate local fish during pregnancy.[17]

Figure 35.7 shows mean breast milk PCB concentrations in Mohawk women at three time periods as compared to women from Women Infant and Child clinics in rural Washington and Schoharie counties in New York where there is no particular contamination of PCBs. Up through 1989, there was a significant elevation in breast milk PCB concentration in Mohawk women, but this difference disappeared after 1990. As a result of the advice by tribal elders in the mid 1980s, the Mohawks reduced fish consumption, and when they did eat local fish, trimmed fat and skin, which contain a disproportionate percentage of the contaminants. These observations show clearly that consumption of contaminated fish was a major source of PCB body burden in this population, and that following fish consumption advisories can be effective in reducing sources of exposure to infants.

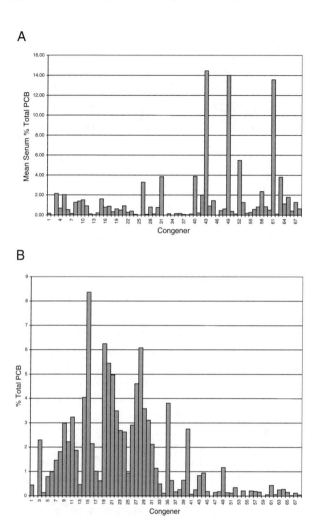

Figure 35.6 Congener profile of mean serum PCB concentrations from 139 Mohawk men (A) as compared with the congener profile of a yellow perch caught near the Reserve (B). In the Mohawk men the more heavily chlorinated congeners predominate, especially IUPAC #138 (peak 43), #153 (peak 49) and #180 (peak 61). These congeners are typically the most prevalent in all surveys of human serum because they resist metabolism by the cytochrome P-450 enzyme system. They are characteristic of Aroclors 1254 and 1260. These commercial PCB mixtures are ubiquitous pollutants and are commonly fund in the Lake Ontario and St. Lawrence River. However, the same serum also show some lower chlorinated congeners such as IUPAC #28 (peak 15), #74 (peak 26) and #99 (peak 31). These congeners are present in Aroclor 1248, used at Akwesasne. They are also found in the fish caught off-shore of these facilities (B). Use of statistical "fingerprinting" analysis demonstrated that the Mohawk men who ate the most local fish had a serum PCB congener pattern that more closely approximated that found in yellow perch caught near the local industrial facilities that did men who ate the least amount of fish.

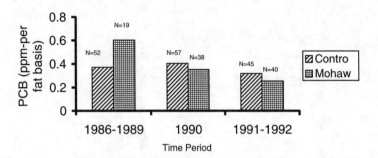

Figure 35.7 Breast milk concentrations of total PCBs among Mohawk mothers as compared to a non-Native control population drawn from Washington and Schoharie counties, New York, as a function of time. The decline in breast milk PCB concentrations reflects the decline on consumption of local contaminated fish.

III. Health effects of PCB exposure

These initial studies demonstrated clearly that Akwesasne Mohawks were exposed to PCBs, and that a major source of exposure was consumption of contaminated fish. We have continued a collaborative project with this population over the past 6 years, now investigating health effects in children born at the time when fish consumption was common, and in adults. The results of these studies have not yet been completed, since data collection has ended less than one year ago. However, we have preliminary results from two studies, both of which are in press.

One of the projects in our interdisciplinary effort was a study of adolescents between 10 and 16 years of age who were born during the period when consumption of contaminated fish was common. This study involved obtaining blood determinations of PCBs, thyroid hormones, sex steroid hormones, and routine blood chemistries as well as extensive interviews and psychological and cognitive tests. The first component of this study to be completed[19] has focused on the relationship between PCBs in serum and thyroid function. The PCB levels in these children are not excessive by most standards, ranging from undetectable to a maximum of 4.75 ppb in 117 children, with a mean of 1.82 ppb. However, the levels of thyroid stimulating hormone (TSH) were directly correlated with PCB levels, while the levels of free and total thyroxine were inversely correlated with PCB levels. Though the PCB and the thyroid indicators are within what would be considered to be normal levels, these observations clearly show that even low-level PCB exposure results in biochemical changes in the body.

We have also completed a portion of a study of the diagnoses of disease at the St. Regis Mohawk Health Service Clinic, at Akwesasne, using the computerized databases maintained by the Indian Health Service.[20] The first component, which has been completed, has looked only at disease incidence in this population as compared to the general U.S. population. Several diseases showed a much higher than expected frequency in the Mohawk population than expected, including hypothyroidism, diabetes, osteoporosis, and asthma.

Results as yet unpublished from our study of adult Mohawks, where we obtained information from interviews and determined serum concentrations of PCBs and routine clinical chemistries, showed that 20.3 percent of adults over 30 years of age have diabetes, and 88 percent of these people are aware of their disease. However, 15.2 percent have hypothyroidism (predominately women), but only 53 percent know they have this disease. These studies do not, of course, prove that these diseases are elevated because of environmental contamination with PCBs. However, diabetes has been shown to be elevated in U.S. servicepersons who handled dioxin-containing Agent Orange in the Vietnam War,[21] and while diabetes is frequent in Native populations. Since coplanar PCB congeners act like dioxin there is reason to pursue the correlation between exposure and disease here as well. The clear demonstration of the relationship between thyroid hormone levels and serum PCBs in the adolescents[19] and much animal work[22] is consistent with the possibility that hypothyroidism may be elevated in exposed individuals. Dr. Negoita is continuing this study with measurement of PCB levels in individuals with hypothyroidism and randomly selected controls.

IV. University-community relationships

Our collaborative study with the Mohawks at Akwesasne has lasted over 15 years and still continues. However, the above description of the PCB analytical results does not convey what is perhaps the most important lesson learned from this study, which has more to do with how an academic program works with a community, especially a minority community that has had less than optimal previous experiences with government organizations and in academic research programs of this type. Some time earlier, this community was approached by an academic group for study of the possible effects of fluoride contamination since aluminum foundries release a significant amount of fluoride. The community cooperated with the researchers, providing blood samples, giving interviews, and assisting with collection of environmental samples. Then they waited for the results, but they basically never heard again from this research team. This experience caused much distress in the community, which only added to the alienation they already felt toward the governments and academic institutions of Canada and the United States and the individuals who represent the dominant culture.

In response to this experience and other events that had dramatically affected life at Akwesasne (including especially construction of the Moses Dam on the St. Lawrence River with the subsequent appearance of the aluminum industries), the community formed a group known as the Akwesasne Task Force on the Environment (ATFE). This group consisted of members of the community with interest in, or knowledge of, the environmental problems facing the community, and it obtained recognition by the tribal authorities to be a gate keeper for any further research activities at the Akwesasne.

For these past 15 years, we have worked closely with the ATFE and have benefited and learned from it while we have continued studies that the ATFE accepted as benefiting the community. The guiding principles which the ATFE has established for such research collaborations are based on traditional Native American values plus a concern that the community be a full partner in any activity involving members of the community. The traditional Mohawk values of Peace, Good Mind, and Strength reflect the values of contentment but with the mental and physical resources to prevent being taken advantage of. Therefore, the ATFE identified three factors which were essential if the community were to collaborate with outside researchers. They are as follows:

> **Respect:** In order to develop a good research agreement, the researchers and the community must show respect for each other. Respect is generated by understanding each others' social, political and cultural structures. The researchers cannot assume that they believe in the same things or share the same goals and expectations. Communication must work both ways in order that a good research agreement is generated. The scientists must communicate in understandable ways what is the purpose of the studies to be done, and must demonstrate to the community that it is to their direct benefit to be a part of the study. The researchers must not treat the community participants as guinea pigs in a laboratory study but rather as partners in a mutually beneficial investigation. Importantly, the community must serve as the teacher to the researchers in aspects of their history, religion, and value systems. The researchers must learn cultural sensitivity, just as the members of the community must learn enough about the scientific questions being pursued so that they can determine whether it is in their best interest to be participants. Definitions and assumptions must be clarified and questioned by each side. The community and the researchers must listen to each other with an open mind and with respect. Consensus and a mediation process must be used to develop procedures which can be honored by the researchers and the community.
>
> **Equity:** Resources should be shared by the researchers with the community in ways that are of direct benefit to the community and yet contribute to the research activities. Money is only one form of equity, but it is an important one. If the research is supported by a grant or contract, a part of the funds must go to the community. This can be in the form of hiring community members, providing equipment or resources to the community that can outlast the particular project, or providing training for community members that will be valuable after the project is finished. Community knowledge, networks, personnel, and political/social power are other forms of equity useful to the project. Each of these commodities has value and must be shared between the researchers and community if a good agreement

is to be formulated. It is necessary to continue to review equity distribution over the duration of an agreement.

Empowerment: This is the sharing of power and the results from a good research agreement developed by the community and the researcher(s). All the participants feel that their needs are being met and that their credibility is increasing. Partnership and responsibility continue to grow as increasing respect and equity enter the agreement. The application of the research as a useful instrument of the community is balanced with the researchers need for good science. Empowerment also means that the community and the researchers must share the authorship. Even though this is sometimes difficult, the increase in empowerment and credibility benefits the good research agreement. The provision of training in research methods to members of the community is also an important form of empowerment. A truly empowered community will ultimately no longer need research directed by persons external to the community since they will have built there own internal expertise.

When these principles are followed, research becomes a true collaboration among partners, both of which have their own strengths and contributions. The community becomes invested in the research, and the researchers have an investment in the community. By being a full partner in the definition of the problem, design of the studies, and interpretation of the conclusions, the community has more motivation to facilitate changes that will benefit them as individuals, whether it is changes in individual behavior or larger community actions. To achieve this ultimate goal may be hard work, but the trust generated and the friendships created are major rewards. However, even from the solely selfish interests of the researcher, this kind of partnership facilitates cooperation with the community, which is often a difficult thing to achieve.

Our research group has accepted and benefited from having worked with this empowered community. We have worked with ATFE in planning grant applications and research projects and have tailored our project plans to attend to the concerns of the community. All research papers from the research go to the Task Force for their information and approval. One member of the community has been assigned to each project as a special representative of the community. These persons visit our laboratories and are provided as much information about the research projects as possible. We have supported community members to attend policy discussions, held within our research team and at national conferences. We have held annual retreats at the Akwesasne, usually in the form of a 2–3-day meeting on Stanley Island, with 10 cabins in the middle of the St. Lawrence that does not have phones or soda machines and where at least a third of our 70–80 people had to use tents. These meetings were accompanied by a public meeting on the mainland, where questions from the greater community were answered and their concerns listened to. Last year, we had an art contest for

Mohawk children, with prizes for the best depictions of an environmental problem at Akwesasne.

This is not to imply that all of this has been easy. We have had some significant disagreements, including some that took many months to resolve. But we have tried to build our relationship upon the principles of research and trust and have worked and taken the time to find solutions that are satisfactory to the members of the community and the scientific staff. We have learned that there are at least four keys to successful working with communities. These include having respect for individuals and for the local culture, listening to community concerns and problems and then acting upon them, providing resources that directly benefit the community, and having the goal of empowering the community so that it does not depend upon outside forces to deal with local concerns.

References

1. Akwesasne Task Force on the Environment, Superfund clean-up at Akwesasne: a case study in environmental injustice, *Internat. J. Contemp. Sociol.* 34, 267–290, 1997.
2. Agency for Toxic Substances and Disease Registry (ATSDR), *Toxicological Profile for Polychlorinated Biphenyls*, ATSDR, Atlanta, GA, , 2000.
3. Hansen, L.G., *The Ortho Side of PCBs: Occurrence and Disposition*, Kluwer Academic Publishers, Boston, MA, 1999.
4. Bush, B., Snow, J., and Koblintz, R., Polychlorobiphenyl (PCB) congeners, p,p′-DDE, and hexachlorobenzene in maternal and fetal cord blood from mothers in upstate New York, *Arch. Environ. Contamin. Toxicol.*, 13, 517–527, 1984.
5. DeCaprio, A.P. et al., Routine analysis of 101 polychlorinated biphenyl congeners in human serum by parallel dual-column gas chromatography with electron capture detection, *J. Analty. Toxicol.*, 24, 403–420, 2000.
6. Safe, S.H., Polychlorinated biphenyls (PCBs): Environmental impact, biochemical and toxic responses, and implications for risk assessment, *Crit. Rev. Toxicol.*, 24, 87–149, 1994.
7. Liu, X. et al., An investigation of factors limiting the reductive dechlorination of polychlorinated biphenyls, *Environ. Toxicol. Chem.*, 15, 1738–1744, 1996.
8. Sokol, R.C., Bethoney, C.M., and Rhee, G.-Y., Reductive dechlorination of reexisting sediment polychlorinated biphenyls with long-term laboratory incubation, *Environ. Toxicol. Chem.*, 17, 982–987, 1998.
9. Kim, J. and Rhee, G.-Y., Population dynamics of polychlorinated biphenyl-dechlorinating microorganisms in contaminated sediments, *Appl. Environ. Microbiol.*, 63, 1771–1776, 1997.
10. Hwang, S.-A. et al., Assessing environmental exposure to PCBs among Mohawks at Akwesasne through the use of geostatistical methods, *Environ. Res.*, 80, S189–S199, 1999.
11. Food and Drug Administration (FDA), Polychlorinated biphenyls (PCBs): reduction of tolerance, *Federal Register*, 44, 38330–38340, 1979.
12. Fitzgerald, E.F. et al., Polychlorinated biphenyl (PCB) and dichlorobiphenyl dichloroethylene (DDE) exposure among Native American men from contaminated Great Lakes fish and wildlife, *Toxicol. Ind. Health*, 12, 361–368, 1996.

13. Chiarenzelli, J.R. et al., PCB volatile loss and the moisture content of sediment during drying, *Chemosphere*, 34, 2429–2436, 1997.
14. Bushart, S.P. et al., Volatilization of extensively dechlorinated polychlorinated biphenyls from historically contaminated sediments, *Environ. Toxicol. Chem.*, 17, 1927–1933, 1998.
15. Chiarenzelli, J. et al., Defining the sources of airborne polychlorinated biphenyls: evidence for the influence of microbially dechlorinated congeners from river sediment? *Can.J. Fish Aquat. Sci.*, 57, 86–94, 2000.
16. Fitzgerald, E.F. et al., Chemical Contaminants in the Milk of Mohawk Women from Akwesasne, New York State Department of Health, 1992.
17. Fitzgerald, E.F., Brix, K.A., and Huang, S.-A., Exposure to PCBs from hazardous waste among Mohawk women and infants at Akwesasne, U.S. Department of Health and Human Services, Agency for Toxic Substances and Disease Registry, Atlanta, GAGA, PB95–159935, 1995a.
18. Fitzgerald, E.F. et al., Fish PCB concentrations and consumption patterns among Mohawk women at Akwesasne, *J. Exp. Anal. Environ. Epidemiol.*, 5, 1–19, 1995b.
19. Schell, L.M. et al., Polychlorinated biphenyls and thyroid function in adolescents of the Mohawk Nation at Akwesasne, *Acta Medica Auxologica*, 2001, in press.
20. Negoita, S. et al., Chronic disease surveillance of St. Regis Mohawk Health Service patients, *J. Public Health Mgt. Practice*, 7, 84–91, 2001.
21. Henriksen, G.L. et al., Serum dioxin and diabetes mellitus in veterans of Operation Ranch Hand, *Epidemiology*, 8, 252–258, 1997.
22. McKinney, J.D. and Waller, C.L., Polychlorinated biphenyl as hormonally active structural analogues, *Environ. Health Perspect.*, 102, 290–297, 1994.

section VI

Nanotechniques and biomarkers

chapter thirty-six

Electronic biosensing

G.R. Facer, O.A. Saleh, D.A. Notterman, and L.L. Sohn

Contents

Abstract Analysis of fluid samples in the biosciences has traditionally relied on chemical and optical detection techniques.[1–3] While these techniques provide diverse and fundamental information, they have a number of disadvantages which prevent their universality. For instance, most samples must be chemically altered prior to analysis; photobleaching can place a time limit on optically probing fluorophore-tagged samples; and elaborate and expensive optical apparatus (external or integrated) is frequently required. Given these drawbacks, one poses the following question: "What alternatives to chemical and optical detection are there for bioanalysis?"

In this chapter, we suggest that electronic detection may be the answer to this question. As we will demonstrate, purely electronic techniques can probe a sample and its chemical environment directly without modification or amplification over a range of time scales. In addition, we will also demonstrate that electronic detection is ideally suited for micro total analytical systems (µ-TAS), or "lab-on-a-chip" microfluidic devices[4] as current micro- and nanofabrication techniques can produce compact, robust, and low-cost electronic sensors directly on a microfluidic chip, thus obviating the need for external detectors. We describe in detail three electronic detection techniques: capacitance cytometry,[5] dielectric spectroscopy,[6] and resistive sensing.[6a] We have developed these in our laboratory to demonstrate the power of elec-

1-56670-596-7/02/$0.00+$1.50
© 2002 by CRC Press LLC

tronic sensing in biological systems. While it is still in its infancy, electronic biosensing has already shown that it is well suited to many detection and monitoring applications in biology.

I. Capacitance cytometry

The electrical properties of biological cells are of great interest, as they can provide opportunities to develop novel, rapid assays for disease[7] and integrated hybrid chips for electronics.[8,9] Previous electrical studies of cells have focused on external macroscopic properties, such as cell membrane responses or volume, and have primarily reflected those of large ensembles of cells.[10,11] In this section, we describe capacitance cytometry, which we have recently developed and allows us to investigate some of the internal properties of individual cells.[5] Capacitance cytometry can quantify the DNA content of single eukaryotic cells from a diverse set of organisms, ranging from yeast to mammals. In addition, it can be employed as an assay for abnormal changes in DNA content, such as are frequently encountered in neoplastic cells. By monitoring the DNA content of populations of cells with this technique, one can produce a profile of their cell-cycle kinetics. Thus, capacitance cytometry may serve as a medical diagnostic tool whose low-detection limit of just one cell can identify the presence of malignancy in small quantities of tissue, and without special processing.

Though able to interrogate cells one by one, standard laser flow cytometry requires sample preparation such as sample staining or manipulation of cells. In contrast, our electronic technique requires no special preparation. Thus, capacitance cytometry has the potential to be simpler, faster, and less expensive than standard laser flow cytometry.

The fundamental basis of capacitance cytometry is an AC capacitance measurement. This extremely sensitive yet robust electronic technique allows one to probe the polarization response of a wide range of materials — organic and inorganic — to an external electric field. In the past, capacitance measurements have been used to identify and investigate a number of different materials in bulk.[12] More recently, it has been used to investigate ensembles of biological cells, determining cell size and cellular membrane capacitance, in order to assay cell-cycle progression[10] and to differentiate normal and malignant white blood cells.[11] In contrast, here we employ capacitance measurements as a means of detecting and quantifying the polarization response of DNA within the nucleus of single eukaryotic cells. Since DNA is a highly charged molecule, in an applied low-frequency AC electric field its polarization response, in combination with the motion of the surrounding counterions, can be substantial.[13] We measure this response as a change in total capacitance, ΔC_T, across a pair of microelectrodes as individual eukaryotic cells suspended in buffer solution flow one by one through a microfluidic channel (Figure 36.1). Unlike a Coulter counter, which measures displaced volume as cells or particles flow through a small orifice,[14] our integrated microfluidic chip measures the polarization response of a cell as it passes

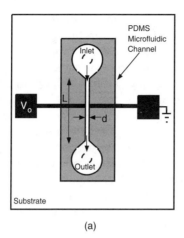

(a)

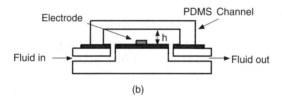

(b)

Figure 36.1 Schematic illustration of the integrated microfluidic device. (a) Top view shows the entire device, including electrode configuration, inlet and outlet holes for fluid, and the PDMS microfluidic channel. The electrodes are made of gold and are 50 μm wide. The distance d separating the electrodes is 30 μm. The width of the PDMS microfluidic channel is also d, the length L is 5 mm, and the height h is either 30 μm or 40 μm. (b) Side view along the vertical axis of the device shows a detailed view of fluid delivery. Fluid delivery is accomplished with a syringe pump at non-pulsatile rates ranging from 1 to 300 μL/hr. (From Sohn, L.L. et al., *Proc. Natl. Acad. Sci. USA*, 97, 10687–10690, 2000. With permission.)

through an electric field region. The data we obtain therefore relate, at least in part, to the charge distribution within the cell.

Electronic measurements in conductive solutions often lead to complications due to charge-screening effects at the electrode-fluid interface, i.e., electrode polarization. The measurements we report here are subject to these ionic effects, a fact that prevents our interpreting absolute capacitance values. However, since electrode polarization is localized to the electrode surface (to within the Debye screening length for the solution) and remains constant for a particular device geometry, ion concentration, and applied frequency, we can determine cellular properties by comparing changes in total capacitance values among different cells passing through our device.

The fabrication of our integrated microfluidic device is a multistage process. Photolithography is first used to fabricate a pair of 50 μm-wide, gold microelectrodes, constituting the sensor, onto a glass or quartz substrate. Figure 36.1a shows a schematic of the entire device. The distance separating the two electrodes is 30 µm, three times larger than the average diameter of the eukaryotic cells we examined.[1] Millimeter-sized holes are then drilled through the substrate on either side of the completed electrodes in order to provide an inlet and outlet for the fluid and cells.

Once the central device has been fabricated, we use soft lithography[15] to create a polydimethyl siloxane (PDMS) microfluidic channel. To avoid complications arising from a cell passing directly over only one electrode and to minimize the effects of electrode polarization, we chose the channel width to be the distance separating the two electrodes. In our experiments, we employed two different channel heights, h = 30 µm and h = 40 µm. The two gave quantitatively similar results, up to an overall scale factor (see below). Once we aligned and positioned the PDMS channel over the electrodes and holes (see Figure 36.1b), we used a syringe pump (KD Scientific Syringe Pump, Model KD2100) to deliver fluid to the completed device at nonpulsatile rates ranging from 1 µl/hr to 300 µl/hr.

We measure the capacitance of the completed device using a commercial capacitance bridge (Andeen Hagerling AH2500A 1kHz Ultra-Precision Capacitance Bridge). This bridge applies a voltage (V_{rms} = 250 mV) at a frequency of 1 kHz across the device.[2] By electrically shielding the device and controlling the temperature precisely (to within ± 0.05°C), we are able to achieve noise levels of ~5 aF when the microfluidic channel is dry and 0.1–2 fF when wet.

We have used our device to compare the DNA content of individual eukaryotic cells. Since the position of such a cell along the mitotic cell cycle is strictly related to DNA content — a cell in G_0/G_1-phase has 2N DNA content, a cell in G_2/M-phase has 4N DNA content, and a cell in S-phase has between 2N and 4N DNA content — we should be able to determine the phase of an individual cell. Because DNA is a highly charged molecule, we anticipate that it will produce a change in capacitance and this change should approximately scale with the DNA content of the cell, at least at frequencies up to 1 kHz. Thus, the device response ΔC_T to a cell in G_2/M-phase should be roughly twice that in G_0/G_1-phase, as the former has twice the DNA content (4N versus 2N DNA content); the response to a cell in S-phase should be between that of the G_0/G_1–G_2/M-phases; and, the response to a hyperdiploid cell (greater than 4N DNA content) should be greater than that of either a G_0/G_1-, S-, or G_2/M-phase cell.

Mouse myeloma cells (SP2/0), a malignant cell line, were grown in suspension to a density of approximately 10^5 cells/mL. The cells were then washed in phosphate-buffered saline (PBS) solution (pH 7.4), fixed in 75% ethanol at –20°C for a minimum of 24 h, washed again with PBS solution, treated with RNAase, and then washed and resuspended for storage in 75% ethanol. Standard analysis (FACScan flow cytometer, Becton Dickinson Immunocytometry Systems, San Jose, CA), following treatment with a nucleic acid probe (SYTOX® Green Nucleic Acid Stain, Molecular Probes, Eugene Oregon), showed that approximately 41% of the cells were in G_0/G_1-

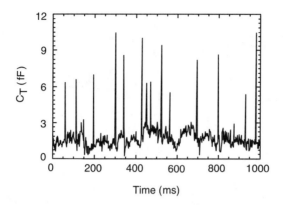

Figure 36.2 Device response over a course of 1000 ms to fixed mouse myeloma SP2/0 cells suspended in 75% ethanol and 25% phosphate buffered saline solution at 10°C. Distinct peaks are present in the data; each peak corresponds to a single cell flowing past the electrodes. The slight difference in peak widths is an artifact of the time-resolution limit of the data acquisition. The channel height of the device was 30 μm. (From Sohn, L.L. et al., *Proc. Natl. Acad. Sci. USA*, 97, 10687–10690, 2000. With permission.)

phase, 40% in S-phase, 18% in G_2/M phase of the cell cycle, and <1% were hyperdiploid.

For any given experimental run, microliters of fixed cells at a concentration of 10^5 cells/mL were injected into our device at a rate of 1 μL/hr. Using cells tagged with a fluorescent probe (SYTOX® Green Nucleic Acid Stain) we visually confirmed that, at this dilute concentration, cells flowed one by one through the microfluidic channel at an average cell velocity of ~250 μm/sec.[3] The Reynolds number was estimated to be Re~10,[2] thus ensuring that flow in the channel was laminar.

Figure 36.2 is a representative response we obtained when cells passed through a device having a channel height of 30 μm. As shown in the figure, we observe a series of sharp peaks whose amplitudes ΔC_T range from ~3 fF to ~12 fF. The individual peaks are separated by time intervals ranging from 40–100 ms. Optical observations during similar measurement runs confirm that each peak does indeed correspond to a single cell flowing past the electrodes.

A central analysis technique in flow cytometry is the DNA histogram, which provides a visual representation of the number of cells as a function of DNA content and, therefore, the proportion of cells in each phase of the cell cycle. Figure 36.3 is a histogram resulting from our capacitance measurements. As shown, there are two distinct populations of SP2/0 cells: one corresponding to 2N DNA content centered at 12.3 fF and the other corresponding to 4N DNA content centered at 23.0 fF. Based upon this capacitance histogram, we judge that approximately 48% of the cells are in G_0/G_1 phase, 30% S phase, 22% G_2/M phase, and <1% hyperdiploid. This distribution is comparable to that achieved with standard laser flow cytometry (Figure 36.3 inset).

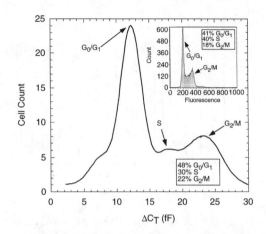

Figure 36.3 Frequency histogram of the SP2/0 cells obtained with a device of $h =$ 30 μm, as compared to that obtained by conventional laser flow cytometry. The ungated histogram shows two major peaks, one centered at 12.3 fF, corresponding to G_0/G_1-phase, and one centered at 23.0 fF, corresponding to G_2/M-phase. The distribution of cells at capacitances less than 10 fF correspond to hypodiploid cells; the distribution of cells at capacitances greater than 27 fF are due to hyperdiploid cells. Based on the histogram obtained, we judged that approximately 48% are in G_0/G_1-phase, 30% in S phase, and 22% in G_2/M-phase. This cell cycle distribution is comparable to that obtained by conventional flow cytometry. Inset: Histogram obtained via conventional flow cytometry. The data have been gated and do not include hypo- and hyperdiploid cells. Two peaks at fluorescence channels 190 and 380 correspond to G_0/G_1- and G_2/M-phases, respectively. (From Sohn, L.L. et al., *Proc. Natl. Acad. Sci. USA*, 97, 10687–10690, 2000. With permission.)

The histogram shown in Figure 36.3 strongly suggests that our device is able to differentiate cells in different phases of the cell cycle. We can safely rule out the possibility that the measured differences in capacitance are due to cells flowing past the electrodes at different channel positions with respect to the electrodes, as we have optically confirmed that the cells flow in the center of the channel and directly between the electrodes. Since flow in the channel is laminar, we neither expect nor observe lateral motion of cells across the channel width. Over 60 devices have been tested and showed similar quantitative results, thus excluding irregularities of device fabrication.

To confirm experimentally that we are indeed differentiating cells based on their DNA content and not by size or volume (G_0/G_1 cells have half the DNA content of G_2/M cells, and are also smaller), we have also measured and compared avian red blood cells (Accurate Chemical and Scientific Corporation, Westbury, NY) to mammalian (sheep) red blood cells (Sigma Chemical Company, St. Louis, MO), both fixed with glutaraldehyde. Whereas avian red blood cells possess 2N DNA and are therefore in G_0/G_1-phase, mammalian red blood cells (the same 6–7 μm size as avian cells) contain no DNA. We observed capacitance peaks when avian cells flowed through our device, but we observed no significant peaks when interrogating the mam-

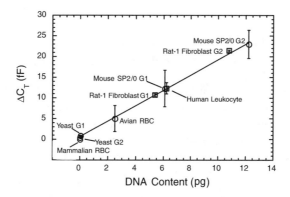

Figure 36.4 Change in capacitance ΔC_T vs. DNA content of mouse SP2/0, yeast, avian, and mammalian red blood cells. As shown, there is a linear relationship between ΔC_T and DNA content at 1 kHz frequency. Open circles correspond to data taken with a device whose channel height was 30 μm; open squares correspond to data taken with a device whose channel height was 40 μm. The 40-μm data were scaled by the ratio of the ΔC_T's obtained for mouse SP2/0 cells measured with 30-μm- and 40-μm-high channel devices (see Reference 4). All data were obtained at a temperature T = 10°C and in phosphate-buffered saline solution. (From Sohn, L.L. et al., *Proc. Natl. Acad. Sci. USA*, 97, 10687–10690, 2000. With permission.)

malian red blood cells, even after a series of experimental runs and measurements with a number of different devices. This confirms that we are indeed measuring DNA content rather than cell size or volume.

The avian red blood cells we measured have an average capacitance change, ΔC_T, of 5.0 fF. Significantly, avian red blood cells have less DNA content than SP2/0 cells and produce a smaller signal. Indeed, the ratio of observed signals of the two different types of cells (5.0 fF to 12.3 fF) is in remarkable quantitative agreement with the ratio of their DNA content (2.5 pg for *Gallus domesticus* vs. 6.1 pg for *Mus muscularis*).[16]

To determine the exact relationship between capacitance ΔC_T and DNA content, we plot ΔC_T and DNA content for the different cell types measured. As shown in Figure 36.4, a strong linear dependence exists between the two at a frequency of 1 kHz.[4] This suggests that there may be a species-independent relationship between the DNA content of eukaryotic cells and the resulting change in capacitance as these cells transit in a low-frequency electric field. Since other cellular constituents may scale with DNA content (such as nuclear histones), we cannot be certain that the entirety of the capacitance signal is derived from DNA. However, the relationship between DNA content and ΔC_T holds across cells of the four species (yeast, mouse, rat, and human) that we sampled. Since it is unlikely that all of these species have the same stoichiometric relationship between DNA and other nuclear and cytoplasmic constituents, the most likely explanation for the linear relationship between DNA content and capacitance signal is that the latter is strictly a function of the former. On an operational basis, at least, the relationship appears to be sufficiently durable to allow prediction of DNA content on the basis of ΔC_T.

Indeed, by using capacitance cytometry, we are able to detect progressive alterations in DNA content. Thus, Rat-1 rodent fibroblast cells were synchronized in the G_0/G_1-phase of the cell cycle by placing them in serum-depleted media (containing 0.1% fetal bovine serum or FBS) for 72 h. Subsequently, these cells were again permitted to grow in a serum-replete media (containing 10% FBS). Aliquots of cells were harvested from the depleted media and at intervals following the re-addition of serum. Measurement of DNA content was performed by capacitance cytometry (using a 40-μm-high channel) and by standard laser flow cytometry. A comparison of the histograms derived from capacitance cytometry and flow cytometry (Figure 36.5) indicates that cells cultured in the depleted media (t = 0 h) are synchronized at a single value of ΔC_T (centered at 3.2 fF in the histogram), which represents G_0/G_1-phase. Twelve hours following the addition of serum, the DNA content of some cells has increased as they enter the S-phase of the cell cycle. By 21 h, most of the cells are in the S-phase, and contain an amount of DNA between G_0/G_1- and G_2/M-phase. By 30 hours, many of these cells have transited G_2/M-phase (centered at 6.4 fF), and by 48 h, the cell population once more has the appearance of an asynchronously-growing population.

In summary, we describe a new method of directly determining the DNA content of single eukaryotic cells, capacitance cytometry, that uses microelectrodes to measure changes in capacitance as cell-bearing fluid flows through a micron-sized channel. Of particular interest, we demonstrate a tight and linear relationship between the capacitance and the DNA content of a cell, and we show that this relationship is not species-dependent (among yeast, mouse, rat, and human). That this relationship does not respect interspecies boundaries implies that the measurement, ΔC_T is dependent upon DNA content *per se* rather than upon an associated cellular constituent that scales with the content of DNA.

DNA content analysis is a core technique in examining cellular physiology. We have demonstrated that our integrated microfluidic device can replicate the DNA histograms of standard laser flow cytometry. The potential applications of this simple and economical device are numerous, ranging from the experimental enumeration of DNA content in biological model systems to the determination of aneuploidy and proportion of S-phase in clinical tumor samples. In addition to advantages in cost, size, robustness, and complexity, sample preparation for the device is quite simple, compared to laser-based flow cytometry. In the experiments reported here, cell samples were fixed in ethanol or glutaraldehyde, but such treatment is not necessary, thus opening the possibility of monitoring DNA content in living tissue, such as peripheral blood, sputum, or cerebrospinal fluid. This might be advantageous in tumor cell detection and in real-time monitoring of the effects of pharmacological agents on cell cycle and cell death. Capacitance cytometry is also uniquely suited to application in microchip-based cell sorting; current efforts in this area make use of external optical detectors.[17,18]

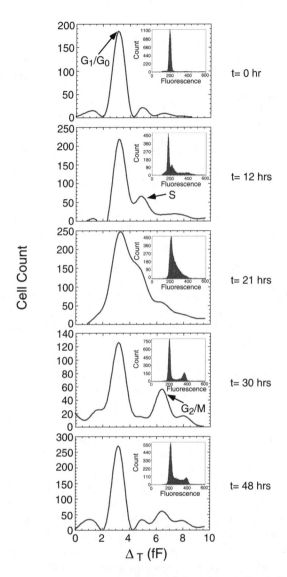

Figure 36.5 DNA Progression of Rat-1, Rodent Fibroblast Cells. Cells were G_0/G_1-arrested (t = 0 hr) and then allowed to progress through one mitotic cell cycle in synchrony. At t = 12 h, the cells are beginning to enter S-phase and at t = 21 hrs, they have fully entered this phase. At t = 30 h, the cells have entered G_2/M-phase. This is shown by the secondary peak at $\Delta C_T = 6.4$ fF. At t = 48 hrs, the cells have completed one mitotic cell cycle and are once again in G_0/G_1-phase. The G_0/G_1-, S-, and G_2/M-phases are indicated with arrows. The data shown were taken at T = 10°C with a microfluidic channel whose height was 40 μm. Standard laser flow cytometry data for the same population of cells is shown in the insets for comparison. (From Sohn, L.L. et al., *Proc. Natl. Acad. Sci. USA*, 97, 10687–10690, 2000. With permission.)

Table 36.1 Some Responses to Electric Fields in Fluid Samples, with
Characteristic Frequency Ranges

Physical Mechanism	Characteristic Frequency (Hz)
Ionic screening at electrodes	$< 10^{-3} - 10^9$
Counterion screening around cells and macromolecules	$1 - 10^9$
Charge buildup at interfaces (membranes, etc.)	$10^3 - 10^8$
Electrorotation	$10^3 - 10^8$
Deformations of macromolecules	$> 10^6$
Rotations of solvent molecules	$10^9 - 10^{11}$ (H_2O)

The use of capacitance cytometry with these microchip-based cell sorters would further reduce the cost, size, and complexity of these devices.

II. Dielectric spectroscopy

Dielectric spectroscopy is another example of electronic detection. As opposed to capacitance cytometry, which measures capacitance at a 1 kHz frequency, dielectric spectroscopy measures permittivity as a function of frequency (from 40 Hz to 50 GHz). This direct, nondestructive, and sensitive technique can probe a system at various length scales, from centimeters to microns, with sample volumes as small as femtoliters. It, thus, provides information about the species present and the chemical environment.

In the past, a variety of research groups have applied dielectric spectroscopy across an extremely broad range of frequencies: from millihertz to tens of gigahertz. Interpreting the results of these experiments, however, has proved far from trivial. One reason for this is the number of physical processes which contribute to dielectric measurements in fluid samples. Table 36.1 below gives a partial illustration of the diverse interwoven phenomena involved.

Despite this complexity, many of the different contributions to permittivity data are well understood. For example, a simple screening theory after Cole and Cole[19] define general features in the dielectric spectra, all relating to polarization relaxations. These features are generally classified as α-, β-, or γ- dispersions. α-dispersion is the permittivity enhancement by rearrangements of small ions, including screening at the fluid interface. β-dispersion arises from distortions of cellular membranes and macromolecules. γ-dispersion is due to rotations and deformations of small, polar molecules or groups (frequently, the solvent). Access to a broad frequency range is imperative with biological samples, due to their chemical diversity.[20,21] In solution with total ionic strengths = 0.1 M, α-dispersion extends up to = 1 GHz, while β-dispersions extend from = 1 kHz[22] up to the relaxational modes of macromolecules in the infrared (THz) and beyond.

Dielectric spectra from α-, β-, and γ-dispersions have a common form[23,24]: at low frequencies, the polarization is able to closely follow the applied electric

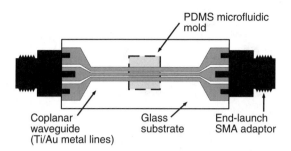

Figure 36.6 CPW device, showing the Ti/Au wave guide (not to scale) and microfluidic sample containment. Across the central portion, the inner line width is 40 μm, outer line widths 380 μm, and the inner-outer separation is 7 μm. Total substrate length is 34 mm. (From Facer, G.R., Notterman, D.A., and Sohn, L.L., *Appl. Phys. Lett.*, 78, 996–998, 2001. With permission.

field (relative permittivity $\varepsilon = \varepsilon_{lf}$); at high frequencies, applied excitations oscillate too fast for the charges to response ($\varepsilon = \varepsilon_{HF}$). Generally, $\varepsilon_{LF} >> \varepsilon_{HF}$.

We have developed a coplanar waveguide (CPW) on chip which allows us to perform dielectric spectroscopy on samples confined to a microfluidic channel or well across nearly nine orders of magnitude in frequency, from 40~Hz to 26.5~GHz.[6] Because coupling to the sample is capacitive, our CPWs allow measurements from DC to microwave frequencies, without the need for surface functionalization or chemical binding.[25] A wide range of species can, therefore, be analyzed rapidly and directly. An added feature of our CPW is that its planar geometry allows for straightforward integration with microfluidic devices.[5,26,27] Below, we discuss the fabrication of the CPW devices, and the low frequency to microwave spectra we have obtained for biomolecular solutions and cell suspensions.

A schematic diagram of our CPW devices is shown in Figure 36.6. They are symmetric metal transmission lines comprised of a 40-μm-wide central strip bordered by two grounded 380-μm-wide conductors. Each metal region is an evaporated Ti/Au 50 Å/500 Å base topped with an electrodeposited gold layer (total Au thickness 1~μm). The substrate is glass, and connection is via end-launch SMA adaptors. Capacitive coupling to the fluid is achieved by encapsulating the metal lines in 1000 Å of PECVD-grown silicon nitride. Silicone poly(dimethylsiloxane), PDMS, confines the fluid.

At frequencies below 100 MHz, the relative permittivity is obtained from the impedance Z via $\varepsilon = 1/j\omega Z C_o$, where C_o is the capacitance through the sample volume when empty and is typically ~10 fF. Z data are obtained with a Hewlett-Packard 4294A impedance analyzer (excitation amplitude 500~mV). We have confirmed that the data are free of nonlinear conductive effects. Microwave data (= 45 MHz) are phase-sensitive transmission and reflection coefficients (S-parameters) at the adaptors, obtained with a Hewlett-Packard 8510C vector network analyzer.

We have examined a variety of samples, including solutions of hemoglobin (derived from washed and lysed human red blood cells) and live *E.*

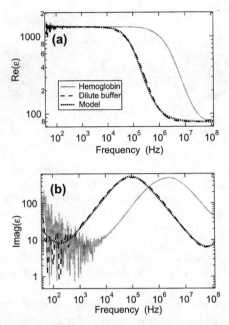

Figure 36.7 Relative permittivity data: real (a) and imaginary (b) components. Solid traces are from hemoglobin (100 µg/mL), dashed traces for Tris buffer (1 mM, pH 8), and dotted curves are Cole-Cole calculations as per Equation (36.1). (From Facer, G.R., Notterman, D.A., and Sohn, L.L., *Appl. Phys. Lett.*, 78, 996–998, 2001. With permission.)

coli suspensions. The concentration of hemoglobin is 100 µg/mL in 0.25~M Tris buffer (pH 8), and that of DNA is 500 µg/mL, in 10~mM Tris and 1~mM EDTA (pH 8) buffer. *E. coli* aresuspended in 85% 0.1 M $CaCl_2$/15% glycerol. For our measurements, we employed molded microfluidic channels and simpler enclosed wells. Results are consistent (within a scaling factor for the fluid -CPW overlap length) for sample volumes ranging from = 3 pL to = 2 0 µL. For the following discussions, we present data from capped 10 µL wells.

Figure 36.7 shows ε from 40 Hz to 110 MHz, for hemoglobin, dilute Tris buffer (concentration 1 mM, pH 8), and a Cole-Cole[19] model calculation relating ε to the angular frequency ω.

$$\varepsilon = \varepsilon_{HF} + \frac{\varepsilon_{LF} - \varepsilon_{HF}}{1 + (j\omega\tau)} - j\frac{\sigma_{LF}}{\omega} \qquad (36.1)$$

Here $\varepsilon_{LF} - \varepsilon_{HF}$ is the dielectric increment, τ is a characteristic time constant, $\alpha = 1$ defines the sharpness of the transition, and σ_{LF} is the DC conductivity. For the calculation in Figure 36.7, $\varepsilon_{LF} - \varepsilon_{HF} = 1340$, $\tau = 1.70$ µs, $\alpha = 0.91$, and $\sigma_{LF} = 40$ nS. A small series resistance (90W) is included in the model to fit high-frequency loss within the CPW.

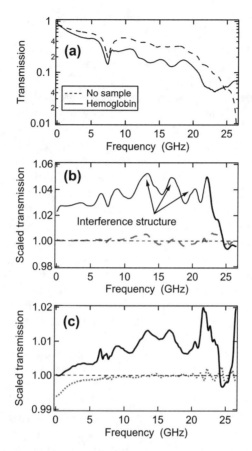

Figure 36.8 Microwave transmission data. (a) Raw data, for the cases of no sample (dotted) and a 100-μg/mL hemoglobin solution (solid). (b) Normalized data (using the respective buffers) for 100 μg/mL hemoglobin (solid trace) and 500-μg/mL phage λ-DNA (dashed), showing the difference in their microwave response. (c) Solid trace is the (buffer-normalized) response of *E. coli*, and the dotted trace is that of the Tris buffer from the hemoglobin solution (normalized using deionized H_2O). (From Facer, G.R., Notterman, D.A., and Sohn, L.L., *Appl. Phys. Lett.*, 78, 996–998, 2001. With permission.)

The spectra in Figure 36.7 show two features. First, the dielectric increment of the high-frequency transition is a constant of the measurement geometry. Second, and in contrast, the ε_{LF} to ε_{HF} transition frequency is directly proportional to the total ionic strength of the solution. As shown, the dispersion model Equation 36.1 describes the data well.

Figure 36.8 shows transmission data from 45~MHz to 26.5~GHz. In Figure 36.8a, raw transmission and reflection are shown for two control cases: a dry sample setup and deionized water. Figures 36.8b and 36.8c contain transmission data sets for hemoglobin, DNA, and live *E. coli* which have been normalized with respect to their corresponding buffers. Figure 36.8c also shows (dotted trace) transmission data from the buffer used for hemo-globin measurements, normalized using deionized water data. This, in par-ticular, demonstrates that even at high salt concentrations (0.25~M Tris-HCl), the microwave effects of buffer salts are limited to a monotonic decrease in transmission below 10~GHz.

Three descriptive notes should be made regarding the data. First, peri-odic peak and trough features (such as those marked by arrows in Figure 36.8b) are interference effects due to reflections at the SMA adaptors and the fluid itself. Second, the SMA adaptors impose the high-frequency cutoff at 26.5~GHz. Finally, reproducibility of the microwave data has been verified for three CPW devices, using several successive fluidic assemblies on each. Only the interference structure changes slightly from device to device.

The most striking aspect of the microwave data is that the transmission through the hemoglobin and bacteria specimens is higher than that through their respective buffer samples. In addition, the response due to 100 μg/mL of hemoglobin is far stronger than that for DNA even though the DNA is more concentrated (500 μg/mL). Furthermore, the hemoglobin exhibits increased transmission across a frequency range from <100 MHz to 25 GHz, which is unique among the samples measured to date (by contrast, the onset of increased transmission in the bacteria data is at ~1 GHz). The increases in transmission are not correlated to any change in reflection, indicating that there is a decrease in power dissipation within the sample. Finally, the breadth of the response implies no resonant process is at play (as is also the case for the *E. coli* data). We must, therefore, conclude that the increased transmission represents an increase in the transparency of the medium to microwaves, i.e., that these specimens are better dielectrics than water alone at this frequency. The fact that this frequency range coincides with the γ-dispersion transition in water (implying high dissipation) is most likely a contributing factor to the success of detection.

Other samples measured, for which data are not shown here (G. R. Facer, D. A. Notterman, and L. L. Sohn, unpublished), include collagen, bovine serum albumin, and ribonucleic acid solutions. These macromolecule solu-tions exhibit behavior highly similar to that of the DNA in Figure 36.8b (i.e., with the 10–20 GHz interference features present) and not to that of the buffer solution. This raises the possibility that the strength and shape of the interference features are more sensitive to the presence of macromolecules and their counterion clouds than just to simple salts. Again, it is reasonable to conclude that this frequency range is significant due to the γ-dispersion of water. The reason for the strength of transmission enhancement by hemo-globin, compared to that by nucleic acids or other proteins, is yet to be confirmed, but we hypothesize that it associated with the activity of the central heme complex.

In summary, we have developed CPW devices which allow us to perform dielectric spectroscopy on small volumes of biological samples confined with a microfluidic channel or well. These devices yield permittivity spectra across an exceptionally broad range of frequencies: from 40 Hz to 26.5 GHz thus far. Neither chemical treatment nor surface activation is required. By combining transmission line design with robust thin-film insulation, sensitivity to sample properties can be achieved in low- and high-frequency regimes within a single device.

III. Resistive sensing in artificial nanopores

While we have demonstrated in the previous two sections that AC measurements are effective in detecting and quantifying biological systems in solution, we show here that simple DC measurements are just as effective, especially when quantitative measurements of the size and concentration of nanoscale particles are critical for studies of colloidal and macromolecular solutions. Traditionally, sizing is accomplished through ultracentrifugation, chromatography, gel electrophoresis,[28] or dynamic light scattering.[29] We have developed[6a] an alternative sizing method based on the Coulter technique of particle sensing.[30] Coulter counters typically consist of two reservoirs of particle-laden solution separated by a membrane and connected by a single pore through that membrane. By monitoring changes in the electrical current through the pore as individual particles pass from one reservoir to another, a Coulter counter can measure the size of particles whose dimensions are on the order of the pore dimensions. Though this method has long been used to characterize cells several microns in diameter,[31,32] its relative simplicity has led to many efforts to employ it to detect nanoscale particles[33-36] including viruses.[37]

Here, we describe the first working realization of a Coulter counter on a microchip. Our device, fabricated on top of a quartz substrate using standard microfabrication techniques, utilizes a four-point measurement of the current through the pore. We are able to control the pore dimensions precisely, which we can easily measure using optical and atomic force microscopies. Knowing the exact pore dimensions allows us to predict quantitatively the response of the device to various sized particles. We have fabricated pores with lateral dimensions between 400 nm and 1 µm and used them to detect latex colloidal particles as small as 87 nm in diameter. Furthermore, we demonstrate the ability of the device to detect ~500 nm diameter colloids with a resolution of ±10 nm. The device has numerous applications in sizing and separating nanoscale particles in solution and is easily integrated with other on-chip analysis systems.

Our device, shown in Figure 36.9a, is fabricated in multiple stages. Each stage consists of lithographic pattern generation, followed by pattern transfer onto a quartz substrate using either reactive ion etching (RIE) or metal deposition and lift-off. The first stage is the fabrication of the pore. A line is patterned on the substrate using either photolithography (PL) for line widths = 1 µm, or electron-beam lithography (EBL) for line widths between 100 and 500 nm, and then etched into the quartz using a CHF_3 Ri.e., The substrate subsequently undergoes a second stage of PL and RIE to define two reservoirs that are 3.5 µm deep, separated by 10 µm, and connected to each other by the previously defined channel. The length of the pore is defined in this

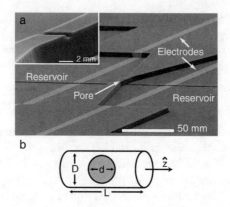

Figure 36.9 (a) Scanning electron micrograph of our microchip Coulter counter. The 3.5-m deep reservoirs and the inner Ti/Pt electrodes, which control the voltage applied to the pore but pass not current, are only partially shown. The outer electrodes, which inject current into the solution, are not visible in this image. The inset shows a magnified view of this device's pore, which has dimensions 5.1 x 1.5 x 1.0 $\mu m.^3$ (b) A schematic diagram of a spherical particle of diameter d in a pore of diameter D and length L. (From Saleh, O.A. and Sohn, L.L., *Rev. Sci. Inst.*, 72, 4449–4451, 2001.)

second stage by the separation between the two reservoirs. The final stage consists of patterning four electrodes across the reservoirs, followed by two depositions of 50 Å/250 Å Ti/Pt in an electron-beam evaporator with the sample positioned 45 degrees from normal to the flux of metal to ensure that the electrodes are continuous down both walls of the reservoirs.

The device is sealed on top with a silicone-coated (Sylgard 184, Dow Corning Corp.) glass coverslip before each measurement. Prior to sealing, the silicone and the substrate are oxidized in a dc plasma to insure the hydrophilicity[38] of the reservoir and pore and to strengthen the seal[39] to the quartz substrate. After each measurement, the coverslip is removed and discarded and the substrate is cleaned by chemical and ultrasonic methods.[5] Thus, each device can be reused many times.

We have measured solutions of negatively charged (carboxyl-coated) latex colloids (Interfacial Dynamics, Inc.), whose diameters range from 87 nm to 640 nm. All colloids were suspended in a solution of 5x concentrated TBE buffer with a resistivity of 390 Ω-cm and pH 8.2. To reduce adhesion of the colloids to the reservoir and pore walls, we added 0.05% v/v of the surfactant Tween 20 to every solution. The colloidal suspensions were diluted significantly from stock concentrations to avoid jamming of colloids in the pore; typical final concentrations were ~108 particles/mL. The pore and reservoirs were filled with solution via capillary action.

The sensitivity of a Coulter counter relies upon the relative sizes of the pore and the particle to be measured. The resistance of a pore R_p increases by δR_p when a particle enters since the particle displaces conducting fluid. δR_p can be estimated[32] for a pore aligned along the z-axis (see Figure 36.10b) by

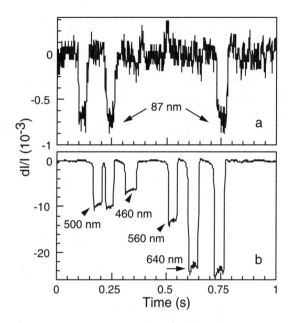

Figure 36.10 Relative changes in baseline current $\delta I / I$ vs. time for (a) monodisperse solution of 87 nm diameter latex colloids measured with an EBL-defined pore of length 8.3 μm and cross section 0.16 μm,[2] and (b) a polydisperse solution of latex colloids with diameters 5460 nm, 500 nm, 560 nm, and 640 nm measured with a PL-defined pore of length 9.5 μm and cross section 1.2 μm.[2] Each downward current pulse represents an individual particle entering the pore. The four distinct pulse heights in (b) correspond as labeled to the four different colloid diameters. (From Saleh, O.A. and Sohn, L.L., *Rev. Sci. Inst.*, 72, 4449–4451, 2001.)

$$\delta R_p = \rho \int \frac{dz}{A(z)} - R_p \tag{36.2}$$

where $A(z)$ represents the successive cross sections of the pore containing a particle, and ρ is the resistivity of the solution. For a spherical particle of diameter d in a pore of diameter D and length L, the relative change in resistance is

$$\frac{\delta R_p}{R_p} = \frac{D}{L} \left[\frac{\arcsin(d/D)}{(1-(d/D)^{1/2}} - \frac{d}{D} \right] \tag{36.3}$$

Equations. 36.2 and 36.3 assume that the current density is uniform across the pore and, thus, is not applicable for cases where the cross section $A(z)$ varies quickly, i.e., when $d \ll D$. For that particular case, Deblois and Bean[33] formulated an equation for δR_p based on an approximate solution to the Laplace equation:

$$\frac{\delta R_p}{R_p} = \frac{d^3}{LD^2}\left[\frac{D^2}{2L^2} + \frac{1}{\sqrt{1+(D/L)^2}}\right]F\left(\frac{d^3}{D^3}\right) \qquad (36.4)$$

where $F(d^3/D^3)$ is a numerical factor that accounts for the bulging of the electric field lines into the pore wall. When employing Equation 36.4 to predict resistance changes, we find an effective value for D by equating the cross-sectional area of our square pore with that of a circular pore.

If R_p is the dominant resistance of the measurement circuit, then relative changes in the current I are equal in magnitude to the relative changes in the resistance, $|\delta I/I| = |\delta R_p/R_p|$, and Equations 36.3 and 36.4 can be directly compared to measured current changes. This comparison is disallowed if R_p is similar in magnitude to other series resistances, such as the electrode/fluid interfacial resistance, R_{eff}, or the resistance R_u of the reservoir fluid between the inner electrodes and the pore. We completely remove R_{eff} from the electrical circuit by performing a four-point measurement of the current (see Figure 36.9a). We minimize R_u by placing the inner electrodes close to the pore (50 μm away on either side), and by designing the reservoir with a cross section much larger than that of the pore. For a pore of dimensions 10.5 μm by 1.04 μm² we measured R_p = 36 MΩ, in good agreement with the 39 MΩ value predicted by the pore geometry and the solution resistivity. This confirms that we have removed R_u and R_{eff} from the circuit.

Figure 36.10 shows representative data resulting from measuring a monodisperse solution of colloids 87 nm in diameter with an EBL-defined pore (Figure 36.10a), and from measuring a polydisperse solution containing colloids of diameters 460 nm, 500 nm, 560 nm, and 640 nm with a PL-defined pore (Figure 36.10b). Each downward current pulse in Figure 36.10, corresponds to a single colloid passing through the pore.[6] For the data shown, 0.4 V was applied to the pore. In other runs, the applied voltage was varied between 0.1 and 1 V to test the electrophoretic response of the colloids. We found that the width of the downward current pulses varied approximately as the inverse of the applied voltage, as is expected for simple electrophoretic motion.

Figure 36.11 shows a histogram of ~3000 events measured for the polydisperse solution. The histogram shows a clear separation between the pore's response to the differently-sized colloids. The peak widths in Figure 36.11 represent the resolution of this device, which we find to be ±10 nm in diameter for the measured colloids. This precision approaches the intrinsic variations in colloid diameter of 2–4%, as given by the manufacturer. In this run, the maximum throughput was 3 colloids/s, a rate easily achievable for all of our samples. Event rates are limited by the low concentrations needed to avoid jamming.

We used a device whose pore size was 10.5 μm by 1.04 μm² to measure colloids ranging from 190 nm to 640 nm in diameter. Figure 36.12 shows the comparison between the measured mean pulse heights and those predicted by Equations. 36.3 and 36.4. As shown, there is excellent agreement between the measured and calculated values, with the measured error insignificant

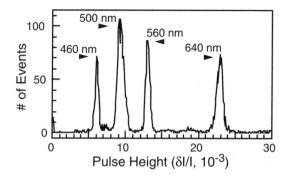

Figure 36.11 A histogram of pulse heights resulting from measuring the polydisperse solution shown in Figure 37.10b. The resolution for this particular device ±10 nm in diameter for the particles measured. (From Saleh, O.A. and Sohn, L.L., *Rev. Sci. Inst.*, 72, 4449–4451, 2001.)

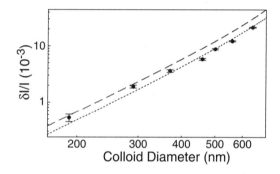

Figure 36.12 Comparison of measure $\delta I/I$ values (circles) to those predicted by Equation 36.3 (dotted line) and Equation 36.4 (dashed line). The measured data were taken over several runs on a single PL-defined pore of length 10.6 μm and cross section 1.04 μm.[2] Error bars for the larger colloid sizes are obscured by the size of the plotted point. AS the colloid diameter increases, there is a transition from agreement with Equation 36.4 to Equation 36.3. This reflects the fact that the derivation of Equation 36.4 assumes the colloid diameter d is much less than the pore diameters D; conversely Equation 36.3 relies on an assumption that holds only as d approaches D, and breaks down for smaller colloids. (From Saleh, O.A. and Sohn, L.L., *Rev. Sci. Inst.*, 72, 4449–4451, 2001.)

compared to the range of pulse heights. In addition, the measurements more closely follow Equation 36.4 for small d and Equation 36.3 for larger D, as was anticipated in the derivation of those equations.

In summary, we have demonstrated a microchip Coulter counter that behaves in a quantitatively predictable manner, and detects particles as small as 87 nm in diameter. The ease of device fabrication allows us to decrease the pore size, and thus decrease the detectable particle size. For example, by extending our use of EBL, we can produce pore diameters of <100 nm. Assuming a pore length of 2 μm and a current sensitivity similar to that already achieved, such a pore could detect particles <30 nm in diameter, the length scale of many biological macromolecules. The strength of our device in size-based separation could then be utilized to study the fractionation of mixtures of large proteins or DNA molecules[6a] (O. A. Saleh and L. L. Sohn, submitted for U. S. Patent 2000).

IV. Summary

A wide range of microfluidic chips has recently been developed for detection of biological systems.[4] A fundamental component of such chips is the detector, and in many applications, multiple detectors are required. Currently, optical detection is the most commonly employed detection scheme. As an example, Mathies et al.,[4] have developed an array capillary electrophoresis system that uses an off-chip confocal scanning system. As in many microchip applications, this optical detection system is off-chip. Only Burns et al.,[40] has successfully incorporated an optical detector on-chip; however, this achievement has not been with ease nor has it been demonstrated to be easily expandable to parallel arrays of optical detectors.

Equally significant, many analytes of interest are neither inherently fluorescent nor easily tagged with artificial fluorophores. Consequently, a number of groups are investigating an electrochemical detector consisting of multiple electrodes that are located inside microchannels for measuring electrochemical potentials.[4] Like the optical detection scheme, this detector is also limiting because the electrodes erode with time.

Given these limitations in analyte detection for existing microfluidic chips, it is clear that an alternative, cost-effective technology is needed. In this chapter, we have suggested that electronic techniques can measure directly the unique intrinsic properties of analytes. The ability to detect the solid-state properties of biological samples is advantageous to other forms of microchip sensing as little sample preparation, such as those required by optical detection, is needed. The electronic sensors we have presented here are all based on current integrated circuit technology. As such, fabricating arrays of sensors on a microfluidic chip can be easily and cheaply accomplished. The cost per chip is estimated to be $0.50; the total cost of a system (with interfacing electronic and computer) ~$3000. Overall, these chips are compact and rugged, thus making on-site or in-the-field deployment more than feasible.

References

1. Nie, S. and Zare, R.N., Optical detection of single molecules, *Annu. Rev. Biophys. Biomol. Struct.*, 26, 567–596, 1997.
2. Weiss, S., Fluorescence spectroscopy of single molecules, *Science*, 283, 1676–1683, 1999.
3. MacBeath, G. and Schreiber, S.L., Printing proteins as microarrays for high-throughput function determination, *Science*, 289, 1760–1763, 2000.

4. Jed Harrison, D. and van den Berg, A., Eds., *Micro Total Analysis Systems '98*, Banff, Canada, 13–16 October, 1998.

5. Sohn, L.L. et et al., Capacitance cytometry: measuring biological cells one by one, *Proc. Natl. Acad. Sci. USA*, 97, 10687–10690, 2000.

6. Facer, G.R., Notterman, D.A., and Sohn, L.L., Dielectric spectroscopy for bioanalysis: from 40 Hz to 26.5 GHz in a microfabricated waved guide, *Appl. Phys. Lett.*, 78, 996–998, 2001.

6a. Saleh, O.A. and Sohn, L.L., Quantitative sensing of nanoscale colloids using a microship Coulter counter, *Rev. Sci. Inst.*, 72, 4449–4451, 2001.

7. Ayliffe, H.E., Frazier, A.B., and Rabbitt, R.D., Electric impedance spectroscopy using microchannels with integrated metal electrodes, *IEEE J. Microelectromech. Sys.*, 8, 50–57, 1999; Huang, Y. et al., The removal of human breast cancer cells from hematopoietic CD34(+) stem cells by dielectrophoretic field-flow-fractionation, *J. Hematother. Stem Cell Res.*, 8, 481–490, 2000.

8. Fromherz, P. et al., Membrane transistor with giant lipid vesicle touching a silicon chip, *Appl. Phys.* A 69, 571–576, 1999; Vassanelli, S. and Fromherz, P., Neurons from rat brain coupled to transistors, *Appl. Phys.*, A 65, 85–88, 1997; and Maher, M.P. et al., The neurochip: a new multielectrode device for stimulating and recording from cultured neurons. *J. Neurosci. Meth.*, 87, 45–56, 1999.

9. Kawana, A.. in *Nanofabrication and Biosystems*, Hoch, H.C., Jelinski, L.W., and Craighead, H.G., Eds., Cambridge University Press, New York, 1996, pp. 258–276; Jung, D.R. et al., Cell-based sensor microelectrode array characterized by imaging x-ray photoelectron spectroscopy, scanning electron microscopy, impedance measurements, and extracellular recordings, *J. Vac. Sci. Technol.*, A 16, 1183–1188, 1998.

10. Asami, K., Gheorghiu, E., and Yonezawa, T., Real-time monitoring of yeast cell division by dielectric spectroscopy, *Biophys. J.*, 76, 3345–3348, 1999.

11. Polevaya, Y. et al., Time domain dielectric spectroscopy study of human cells, Part II: normal and malignant white blood cells, *Biochim. Biophys. Acta*, 15, 257–71, 1999.

12. Pethig, R., *Dielectric and Electronic Properties of Biological Materials*, John Wiley and Sons, Ltd., New York, 1979.

13. Takashima, S., *J. Mol. Biol.*, 7, 455–467, 1963; Bone, S. and Small, C.A., Dielectric studies of ion fluctuations and chain bending in native DNA, *Biochim. Biophys. Acta*, 1260, 85–93, 1995; and Yang, Y. et al., Dielectric response of triplex DNA in ionic solution from simulation, *Biophys. J.*, 69, 1519–1527, 1995.

14. Yen, A., *Flow Cytometry: Advanced Research and Clinical Applications*, CRC Press, New York, 1989.

15. Xia, Y., Kim, E., and Whitesides, G.M., Micromolding in capillaries: applications in microfabrication, *Chem. Mater.*, 8, 1558–1567, 1996.

16. Tiersch, T.R. and Wachtel, S.S., On the evolution of genome size of birds, *J. Hered.*, 82, 363–368, 1991; Greilhuber, J., Volleth, M., and Loidl, J., Genome size of man and animals relative to the plant Allium cepa, *J. Genet. Cytol.*, 25, 554–560, 1983.

17. Fu, A.Y. et al., A microfabricated fluorescence-activated cell sorter, *Nature Biotech.*, 17, 1109–1111, 1999.

18. Schrum, D.P. et al., Microchip flow cytometry using electrokinetic focusing, *Anal. Chem.*, 71, 4173–4177, 1999.

19. Cole, K.S. and Cole, R.H., Dispersion and absorption in dielectrics, *J. Chem. Phys.*, 9, 341, 1941.

20. Onaral, B., Sun, H.H., and Schwan, H.P., Electrical properties of bioelectrodes, *IEEE Trans. Biomed. Eng.*, 31, 827, 1984.
21. Cirkel, P.A., van der Ploeg, J.P.M., and Koper, G.J.M., Branching and percolation in lecithin wormlike micelles studied by dielectric spectroscopy, *Physica A.*, 235, 269–278, 1997.
22. Gimsa, J. and Wachner, D., A unified resistor-capacitor modes for impedance, dielectrophoresis, electrorotation, and induced transmembrane potential, *Biophys. J.*, 75, 1107–1116, 1998.
23. Schwan, H.P. and Takashima, S., Electrical conduction and dielectric behaviour in biological systems, in *Encyclopedia of Applied Physics*, Vol. 5, VCH, New York, 1993, pp. 177–200.
24. Raicu, V., Dielectric dispersion of biological matter: model combining Debye-type and "universal" responses. *Phys. Rev.*, E 60, 4677–4680, 1999.
25. Hefti, J., Pan, A., and Kumar, A., Sensitive detection method of dielectric dispersions in aqueous-based, surface-bound macromolecular structures using microwave spectroscopy, *Appl. Phys. Lett.*, 75, 1802–1804, 1999.
26. Cooper, J.M., Towards electronic petri dishes and picolitre-scale single-cell technologies, *Trends Biotechnol.*, 17, 226–230, 1999.
27. Duffy, D.C. et al., Rapid prototyping of microfluidic systems in poly(dimethylsiloxane), *Anal. Chem.*, 70, 4974–4984, 1998.
28. Alberts, B. et al., *Molecular Biology of the Cell*, Garland Publishing, Inc., New York, 1994.
29. Russel, W.B., Saville, D.A., and Schowalter, W.R., *Colloidal Dispersions*, Cambridge University Press, New York, 1989.
30. Coulter, W.H., U.S. Patent No. 2,656,508, issued 20 Oct. 1953.
31. Kubitschek, H.E., Electronic counting and sizing of bacteria, *Nature*, 182, 234–235, 1958.
32. Gregg, E.C. and David Steidley, K., Electrical counting and sizing of mammalian cells in suspension, *Biophys.J.*, 5, 393–405, 1965.
33. DeBlois, R.W. and Bean, C.P., Counting and sizing of submicron particles by the resistive pulse technique, *Rev. Sci. Instrum.*, 41, 909–913, 1970.
34. Koch, M., Evans, A.G.R., and Brunnschweiler, A., Design and fabrication of a micromachined coulter counter, *J. Micromech. Microeng.*, 9, 159–161, 1999.
35. Sun, L. and Crooks, R.M., Fabrication and characterization of single pores for modeling mass transport across porous membranes, *Langmuir*, 15, 738–741, 1999.
36. Kobayashi, Y. and Martin, C.R., Toward a molecular Coulter counter type device, *J. of Electroanalytical Chemistry*, 431, 29–33, 1997.
37. DeBlois, R.W., Bean, C.P., and Wesley, R.K.A., Electrokinetic measurements with submicron particles and pores by resistive pulse technique, *J. Colloid Interface Sci.*, 61, 323–335, 1977.
38. Fakes, D.W. et al., The surface analysis of a plasma modified contact-lens surface by SSIMS, *Surf. Interface Anal.*, 13, 233–236, 1988.
39. Chaudhury, M.K. and Whitesides, G.M., Direct measurement of interfacial interactions between semispherical lenses and flat sheets of poly(dimethylsiloxane) and their chemical derivatives, *Langmuir*, 7, 1013–1025, 1991.
40. Burns, M.A. et al., An integrated nanoliter DNA analysis, *Science*, 282, 484–487, 1998.

chapter thirty-seven

Two-photon fluorescence microscopy: a review of recent advances in deep-tissue imaging

Chen-Yuan Dong, Lily Hsu, Ki-Hean Kim, Christof Buehler, Peter D. Kaplan, Thomas D. Hancewicz, and Peter T. C. So

Contents

I. Introduction

Noninvasive optical imaging of living tissues in three dimensions offers new research opportunities in biology, toxicology, and medicine. Two-photon microscopy is an important recent invention that makes it possible to image living tissue.[1] Three-dimensional tissue imaging based on two-photon excitation has been applied in research on cell growth during tissue engineer-

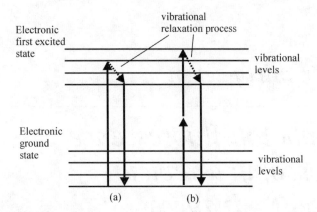

Figure 37.1 Jablonski diagrams for (a) one-photon and (b) two-photon excitation. One-photon excitation occurs through the absorption of a single photon. The two-photon process occurs through the simultaneous absorption of two lower energy photons. After either excitation process, the fluorophore relaxes to the lowest energy level of the first excited electronic state. The subsequent fluorescence emission processes is independent of the mode of excitation.

ing,[2–4] chemical transport across the dermal epithelium,[5] and carcinogenesis in transgenic animals.[6]

Two-photon microscopy is a fluorescence imaging technique (Figure 37.1). Conventional one-photon fluorescence excitation results from the absorption of a single ultraviolet (UV) or visible photon that results in the electronic transition of a fluorescent molecule from the ground to the excited state. After vibrational relaxation, the fluorescent molecule returns to the ground state with the emission of a lower energy photon. For two-photon excitation, the excitation transition results from the fluorescent molecule simultaneously absorbing two photons. Energy conservation dictates that the two absorbed photons have energies that are approximately half the energy difference between the ground and the excited states. These photons are typically in the infrared (IR) spectral range. After vibrational relaxation, the fluorescent molecule resides in the same excited state independent of the excitation process. Therefore, the emission spectrum and the fluorescent lifetime are the same regardless whether the molecule is excited via one- or two-photon process. Multiphoton excitation is a more general term that describes the process in which the fluorescent molecule is excited by the absorption of two or more photons.

Two-photon microscopy has several key features. First, the excitation volume of two-photon microscopy is confined to a sub-femtoliter focal volume greatly reducing off-focus background. Two-photon microscopy has inherent three-dimensional sectioning capability. The typical photo excitation volume has a radial dimension of 0.3 μm and an axial dimension of 0.8 μm. Second, since the excitation is localized, photo excitation induced damage is also limited to the same small volume. In contrast, one-photon confocal microscopy, another common three-dimensional optical imaging method that localizes the observation volume using a confocal pinhole, often produces greater specimen photodamage since the excitation volume is extended.[7] Third, two-photon microscopy allows the imaging of near-UV fluorescent molecules in tissues using IR light that is absorbed and scattered significantly less than UV or visible photons used in one-photon microscopy. The tissue absorption extinction coefficient in the UV range can be two to three orders of magnitude higher than in the IR region. Rayleigh scattering cross-section is proportional to the inverse fourth power of the wavelength and is significantly reduced at longer wavelengths.[8] Therefore, two-photon excitation has greater depth penetration. In a recent study comparing imaging characteristics between one-photon and two-photon microscopy, it was shown that the two-photon technique can resolve structures at least twice as deep as the confocal approach.[9] Finally, the wide spectral separation between the two-photon excitation and fluorescence emission wavelengths allows high sensitivity detection. Unlike one-photon excitation where the excitation wavelengths are separated by around 50 nm, the multiphoton excitation and emission wavelength is separated by about 200 nm or more from the fluorescence. The broader spectral separation ensures more sensitive detection of fluorescence emission while minimizes scattered excitation light contamination. Further, fluorescent spectroscopy studies of specimen biochemistry are easier because this spectral separation allow the entire emission spectrum to be measured.

This chapter provides an overview of two-photon imaging technology. We will subsequently review a number of recent advancements in this technology including two-photon video-rate imaging, resolution enhancement based on maximum likelihood deconvolution, and tissue biochemical assays based on two-photon spectroscopy.

II. Basic two-photon microscopy instrumentation

There are two key components in a multiphoton instrument. First, an ultrafast femtosecond laser source is essential for efficient two-photon excitation. Although two-photon excitation by picosecond and continuous wave lasers have been demonstrated,[10,11] the narrow pulse width (around 100 fs) of femtosecond lasers ensures high temporal concentration of excitation photons needed for efficient two-photon excitation. The second major component is a scanning microscope system. High-quality objective lens ensures diffraction-limited focusing of the excitation light. The spatial confinement of excitation photons further increases photon flux and enhances nonlinear photon absorption probability.

A typical two-photon microscope design is shown in Figure 37.2. Titanium Sapphire laser is one of the most convenient femtosecond light sources for two-photon microscopy (Mira 900, Coherent Inc., Mountain View, CA). The excitation light from the laser is guided towards an x-y galvanometric

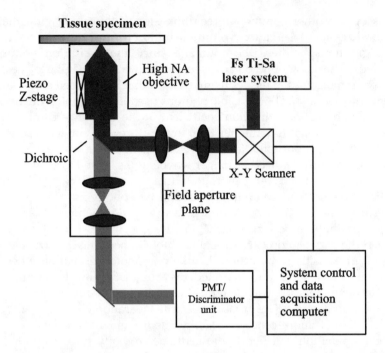

Figure 37.2 A schematic of two-photon fluorescence microscope design.

mirror scanner system (Model 6350, Cambridge Technology, Cambridge, MA). Raster scanning the laser beam using the galvanometric scanner allows two-dimensional imaging. The laser beam is reflected by a dichoric mirror toward the microscope. Additional optics for beam expansion is used to ensure overfilling the back aperture of the microscope objective. The uniform filling of the back aperture is necessary for diffraction-limited focusing at the specimen. Fluorescence is generated at the focal point and is collected by the same objective. Axial scanning of the specimen is achieved by a piezo-driven, objective positioner (P721 PIFOC®, Physik Instrumente (PI), Germany). three-dimensional imaging is thus achieved by the combination of x-y scanner mirrors and axial piezo positioner.

A wide range of high-quality objectives are available for two-photon imaging. High numerical aperture (NA) objectives are often chosen to minimize the excitation volume and to optimize equal to image resolution. Oil-immersion objectives have higher numerical apertures and are efficient for thin specimens with thickness less than 10 µm. For thicker tissue specimens, water-immersion objectives are often used since they have refractive index better matched to many tissue specimens. Index matching minimizes image and signal deterioration due to spherical aberration.

The fluorescence signal passes through the dichroic mirror and additional barrier filters before reaching the detector. Either analog detection mode or single-photon counting detection mode can be used. Single-photon counting detection is illustrated in Figure 37.2. Individual photon burst is detected by high-sensitivity, low-noise photomultiplier tubes (R7200, Hamamatsu, Bridgewater, NJ) and separated from detector noise using a discriminator circuit. The digitized signal is recorded by the data acquisition computer. The data acquisition computer is responsible for correlating the photon signal with positions of the x-y scanner and the axial piezo positioner.

III. Deep tissue imaging based on two-photon microscopy

Although two-photon excitation allows microscopic resolution in living tissues with unprecedented depth, the performance of this technique is still limited due to complex, heterogeneous optical properties of tissue samples. For example, two-photon imaging of corneal structures, which is fairly transparent, can be accomplished down to a depth of hundreds of microns based on tissue autofluroescence.[12] Two-photon method has been used to image blood flow in neuronal tissues, which are slightly more opaque, using exogenous contrast agents down to a depth of over 500 μm.[13] On the other hand, the skin is a much more complicated optical system. Optical coherence tomography has been used to measure the scattering coefficients and refractive index of skin. Both properties are found to change significantly from the surface stratum corneum down to the upper dermis.[14,15] In the stratum corneum, the refractive index is around 1.5 and the scattering coefficient is between 1–1.5 mm^{-1}. In lower epidermis, the index of refraction decreases while the scattering coefficients increases; the refractive index is around 1.34 and the scattering coefficient is between 4–7 mm^{-1} in the basal layer. In the upper dermal layers, the refractive index and scattering coefficient are about 1.41 and 5–8 mm^{-1}. The refractive index variation prevents easy index matching and results in significant spherical aberration. Two-photon imaging of skin based on auto-fluorescence can only penetrate to a depth of about 200 μm.[16,17] In two-photon microscopy, specimen absorption, scattering, and varying refractive indices all contribute to limit the useful penetration depths two-photon microscopy by restricting the reach of the incident photons and interfere with the collection of the fluorescence emission.

Since its invention a decade ago, two-photon fluorescence microscopy has been found applications in many diverse tissue imaging applications. This technology has been used in the study of tissue physiology,[12,16,18] neurobiology,[19–24] and embryology.[25–28]

The use of two-photon microscopy for tissue imaging is well demonstrated in the study of dermal physiology based on autofluorescence. With two-photon microscopy, skin cellular and extracellular structures down to the dermal layer can be imaged nondestructively. Figure 37.3 shows the stratum corneum, the basal layer, and the upper dermis imaged based on tissue endogenous fluorescence. The observed structural features correspond well to known histology data.

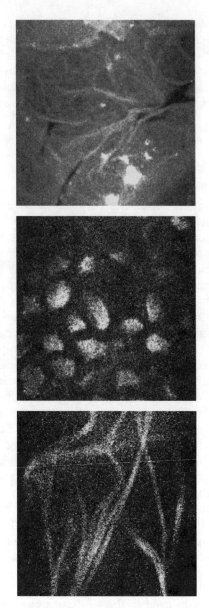

Figure 37.3 Two-photon auto-fluorescence images of human skin. Top: stratum corneum; middle: basal layer; and bottom: fibrous dermal layer. (Zeiss Fluar 100× oil.)

IV. Recent advances in two-photon microscopy

A. Video-rate two-photon microscopy

Typical two-photon microscopes have a frame rate on the order of a few
seconds. Even though this scan rate is sufficient in many cases, it is inade-

quate to study some biological process with fast kinetics such as calcium signaling.[29] For the clinical applications, faster scan rate allows efficient imaging of a macroscopic area of the tissue specimen which is needed for thorough disease diagnosis. High-speed imaging also minimizes image degradation due to physiological motions with frequencies on the order of 1 Hz. Finally, high-speed two-photon imaging allows the implementation of three-dimensional tissue image cytometry.[30] Three-dimensional image cytometry allows *in situ* sampling the cellular properties inside tissues. High-speed imaging ensures that a sufficiently large cell population can be sampled to provide statistically accurate measure of cellular properties. The potential of two-photon three-dimensional image cytometry has been demonstrated in three-dimensional cell cultures where rare cells can be identified at ratio as low as 1 in 10^5.

One of the first video-rate two-photon microscopic systems is based on the line-scanning approach.[31,32] Resonance mirror and multifocus scanning have also been implemented.[29,33,34] Another possibility in achieving video-rate two-photon microscopy is by using a rotating polygonal mirror to replace one scanning axis.[35]

The major limitation of video-rate two-photon microscope is the inevitable reduction in signal-to-noise level. As the scan rate is increased, the number of excitation laser pulses available for excitation per pixel decreases. For a video-rate microscope (30 frames per second) generating 256 x 256 pixel images, the pixel residence time is 500 ns. There are approximately 40 laser pulses available for excitation during this time using a typical titanium-sapphire laser operating at 80 MHz repetition frequency. Within the residence time in a pixel, each fluorescent molecule at a given pixel can emit at most 10 photons without excitation saturation. Therefore, the low signal photon flux is a key constraint for video-rate two-photon imaging. Possible solutions to circumvent this problem include increasing the laser repetition frequency, circularizing of the laser polarization, and using an analog detection scheme.[35] The most promising approach is the parallelization of excitation and detection as implemented in multifocal multiphoton microscopes.[33] As this technology being further developed, the utility of video-rate microscopes has been demonstrated in the study of cellular calcium signaling[29] and tissue physiology. Collagen/elastin fiber structures in the dermal layer of *ex vivo* human skin can be clearly visualized based on autofluorescence at a frame rate of 12 Hz (Figure 37.4).

B.   *Enhancing image resolution based on maximum likelihood deconvolution*

In addition to the improving image acquisition speed, it is also important to optimize the image resolution and signal level for deep tissue applications. Though two-photon microscopy has good resolution and its resolution is limited to 200–300 nm due to diffraction effect, numerical deconvolution methods allow further image resolution enhancement. Details of cellular and

Figure 37.4 (See color insert following page 392.) A video-rate image of the elastin fiber structures in the dermal layer (Zeiss Fluar 40× oil.)

tissue structures with features smaller than the diffraction limit can sometimes be visualized based on numerical deconvolution. However, it should be noted that the resolution improvement beyond diffraction limit based on numerical deconvolution is rather modest due to signal-to-noise limitation. In deep tissue imaging, image resolution is often degraded due to spherical aberration and scattering effects. Numerical deconvolution allows some restoration of these images by partially negating aberration and scattering effects. Further, deconvolution based on maximum likelihood approach allows image restoration but also extracts the point spread function, a numerical quantification of the degree of image degradation in the specimen. The application of maximum likelihood blind deconvolution method to image heterogeneous, complex tissues allows a measure of its optical properties.

We have demonstrated the use of numerical deconvolution to further improve two-photon microscope images. Digital images were deconvoluted based on a maximum likelihood algorithm (Autodeblur™, AutoQuant Imaging, Inc., Watervliet, NY). Two-photon images were acquired of bovine pulmonary endothelial cells, labeled with BODIPY FL phallacidin (green F-actin) and DAPI (blue nucleus) (F-14780, Molecular Probes, Eugene, OR). A raw, unprocessed two-photon image and one after numerical deconvolution are shown in Figure 37.5. The deconvoluted image clearly has additional structural details. The most significant improvement is in the axial sections. Prior to deconvolution, the cross sections of the actin fiber bundles are elongated ovals due to the lower resolution of the two-photon microscope along the axial direction. This lower axial resolution does not allow individual fibers to be distinguished. The deconvolution process equalizes the resolutions along the axial and the radial directions and restores the cross-sections of the actin fiber bundles to an expected circular geometry allowing individual fibers to be better resolved. The application of deconvolution method to two-photon tissue images were illustrated using data from *ex vivo* human skin. In Figure 37.6a shows the raw and deconvoluted images of the

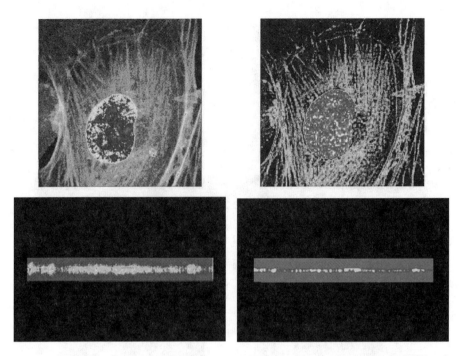

Figure 37.5 (See color insert following page 392.) Two-photon fluorescence (left) and blind deconvoluted (right) images of a bovine pulmonary endothelial cell (green F-actin, blue nucleus). Top: lateral view; bottom: axial section. (Zeiss Fluar 100× oil.)

basal layer of an *ex vivo* human skin. Figure 37.6b contains the raw and deconvoluted images from the dermal layer of the skin. In the basal layer case, deconvolution allows the observation of melanin granules that are not visible in the raw images. The individual elastin fibers are better resolved in the dermal layer. The benefit of using deconvolution to further improve the resolution of two-photon tissue images is being explored.

C.   Two-photon spectral characterization of tissue biochemistry

Two-photon microscopy imaging of endogenous fluorescent species or exogenous fluorescent probes has been shown to be a powerful method for the quantification of tissue structure and biochemistry. Autofluorescence is observed ubiquitously in many tissue types; the presence and the distribution of these endogenous fluorescent species provide tissue functional information related to its physiological and pathological states. Fluorescence spectroscopy has been used to characterize different tissue types such as from the colon, lung, cervix, and skin.[36,37] In addition to tissue type characterization, spectroscopy has been used to monitor tissue physiological states.[38,39] The differences in tissue excitation spectra have been used to for disease diagnosis such as distinguishing between malignant and normal tissues.[40–43] Tissue spectroscopy has also been used to study the effects of

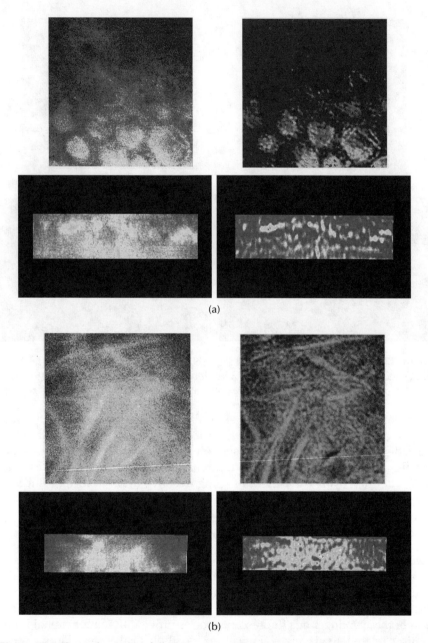

Figure 37.6 (See color insert following page 392.) Image restoration of two-photon images of *ex vivo* human skin using maximum likelihood approach. (a) auto-fluorescence (left) and blind deconvoluted (right) images of human basal layer. Top: lateral view; bottom: axial section. (Zeiss Fluar 40× oil.) (b) auto-fluorescence (left) and blind deconvoluted (right) images of human basal layer. Top: lateral view; bottom: axial section. (Zeiss Fluar 40× oil.)

aging and photoaging.[44] The use of exogenous probes have been applied to monitor a variety of tissue and cellular biochemical states such the concentration of metabolites (e.g., pH, calcium, zinc),[29] the fluidity of lipid membrane,[45] and the distribution of oxygen.[46]

While autofluorescence is observed ubiquitously in many tissue types, the identities and distributions of these fluorescent molecules have not been completely characterized. The different fluorescent species are expected to have different fluorescence excitation and emission spectra. We have applied fluorescence excitation spectroscopy to identify autofluorescence biochemical species in *ex vivo* human skin. Total fluorescence emission intensity from 350 to 550 nm is measured at each pixel as a function of excitation wavelength from 720 to 920 nm. Spectral data are analyzed by self-modeling curve resolution (SMCR) approaches to extract spectroscopic components from these two-photon images.

The data to be analyzed are four-dimensional that include three spatial dimensions and an excitation wavelength dimension (x, y, z, λ). Ignoring possible correlation between pixels, the four dimensional data set is converted into two dimensions (n, λ) for further analysis (two-way factor analysis) where n is the index of the pixels in a three-dimensional data set. We assume no correlation between pixels in the SMCR analysis. Therefore, the presence of a correlation of biochemical species distributions with physiological structural features in the tissue serves as an independent verification of this approach's accuracy.

Two-way factor analysis assumes a bilinear data structure and attempts to separate the original data matrix of spectra, **D**, into two sub-matrices, **C** and **S**

$$D = CS^T + E \qquad 37.1$$

where, C is a matrix of coefficients related to the real concentration profiles (scores), S is a matrix of vectors related to the real spectra (factors), E is a matrix of spectral residuals, and T is the transpose operator.

The equation above describes the standard abstract principal component analysis solution. Alternative least square analysis with appropriate constraints is used to optimize the data fit by minimizing the error matrix. From the SMCR analysis, major factors (biochemical components) contributing autofluorescence in the human skin are identified. The SMCR analysis provides us with the excitation spectrum of the individual component and the three-dimensional distribution of each component which can be re-expressed as a depth distribution profile for that component. The depth distribution profile describes the distribution of each component within the volume of skin as a function of depth. The distribution of each chemical component can be compared to the known histological structure of the skin.

In the analysis of skin excitation data, several autofluorescent components are identified by SMCR analysis. One of the components identified corresponds to elastin fibers in the dermis (Figure 37.7a). An image recon-

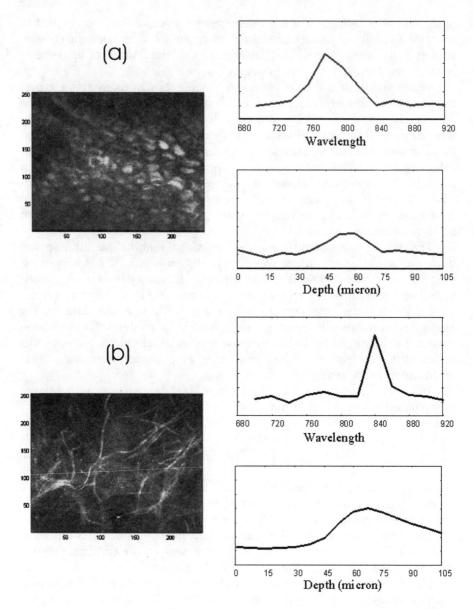

Figure 37.7 Two independent spectral components isolated in *ex vivo* human skin based on SMCR: (a) a spectral component corresponds to elastin fibers in the dermis and (b) a spectral component corresponds to melanin (or a fluorophore that colocalizes with melanin) in the epidermal-dermal junction. For (a) and (b): (left) A two-dimensional image of the concentration distribution of the spectral component (right, top) The spectrum of this component. (right, bottom) The depth distribution profile of this component.

structed from the three-dimensional distribution of this component shows the well-defined morphology of elastin fibers in the dermis. The depth distribution profile shows that this component is not seen in the epidermis but only in the dermis. This observation is also consistent with skin physiology. Another component identified represents melanin or a biochemical component that colocalizes with melanin (Figure 37.7b). The reconstructed images from the distribution of this component have the typical morphology of the melanin caps at the basal layer. The distribution of this component is found to maximize at a depth between 40 to 60 μm corresponding to the epidermal–dermal junction of the skin as expected.

V. Conclusions

A decade after the invention of two-photon microscopy, this technique has become one of the most important techniques in modern optical microscopy. Two-photon microscopy has three-dimensional resolved imaging capability, enhanced depth penetration, and is compatible with study tissue structure and biochemistry *in vivo*. Recent advances in high-speed imaging, resolution enhancement based on numerical deconvolution, and spectroscopic resolution allow further progress in tissue diagnosis. Three-dimensional image cytometry based on high-speed imaging measures cellular properties in tissues with high statistical precision through sampling a large cellular population. Numerical deconvolution methods provide higher resolution imaging of finer structures in cells and extracellular matrix. Finally, the combination of two-photon spectroscopy and microscopy is a powerful method to simultaneous assay tissue function and structure.

References

1. Denk, W., Strickler, J.H., and Webb, W.W., Two-photon laser scanning fluorescence microscopy, *Science*, 248, 73–76, 1990.
2. Agarwal, A. et al., Collagen remodeling by human lung fibroblasts in three-dimensional gels, *Am. J. Respir. Crit. Care Med.*, 159, A199, 1999.
3. Agarwal, A. et al., Two-photon scanning microscopy of epithelial cell-modulated collagen density in three-dimensional gels, *FASEB J.*, 14, A445, 2000.
4. Agarwal, A. et al., Two-photon laser scanning microscopy of epithelial cell-modulated collagen density in engineered human lung tissue, *Tissue Eng.*, 7, 191–202, 2001.
5. Yu, B. et al., *In vitro* visualization and quantification of oleic acid induced changes in transdermal transport using two-photon fluorescence microscopy, *J. Invest. Dermatol.*, 117, 16–25, 2001.
6. Brown, E.B. et al., *In vivo* measurement of gene expression, angiogenesis and physiological function in tumors using multiphoton laser scanning microscopy, *Nat. Med.*, 7, 864–868, 2001.
7. Pawley, J.B., Ed. *Handbook of Confocal Microscopy*, Plenum, New York, 1995.
8. Jackson, J.D., *Classical Electrodynamics*, John Wiley & Sons, New York, 1975.

9. Centonze, V.E. and J.G. White, Multiphoton excitation provides optical sections from deeper within scattering specimens than confocal imaging, *Biophys. J.*, 75, 2015–2024, 1998.

10. Booth, M.J. and S.W. Hell, Continuous wave excitation two-photon fluorescence microscopy exemplified with the 647-nm ArKr laser line, *J. Microsc.*, 190, 298–304, 1998.

11. Hell, S.W., Booth, M., and Wilms, S., Two-Photon Near- and Far-Field Fluorescence Microscopy with Continuous-Wave Excitation, *Opt. Lett.*, 23, 1238–1240, 1998.

12. Piston, D.W., Masters, B.R., and Webb, W.W., Three-dimensionally resolved NAD(P)H cellular metabolic redox imaging of the *in situ* cornea with two-photon excitation laser scanning microscopy, *J. Microsc.*, 178, 20–27, 1995.

13. Kleinfeld, D. et al., Fluctuations and stimulus-induced changes in blood flow observed in individual capillaries in layers 2 through 4 of rat neocortex, *Proc. Natl. Acad. Sci. USA*, 95, 15741–15746, 1998.

14. Tearney, G.J. et al., Determination of the refractive-index of highly scattering human tissue by optical coherence tomography, *Opt. Lett.*, 20, 2258–2260, 1995.

15. Knuttel, A. and Boehlau-Godau, M., Spatially confined and temporally resolved refractive index and scattering evaluation in human skin performed with optical coherence tomography, *J. Biomed. Opt.*, 5, 83–92, 2000.

16. Masters, B.R., So, P.T., and Gratton, E., Multiphoton excitation fluorescence microscopy and spectroscopy of *in vivo* human skin, *Biophys.J.*, 72, 2405–2412, 1997.

17. So, P.T.C., Kim, H., and Kochevar, I.E., Two-photon deep tissue *ex vivo* imaging of mouse dermal and subcutaneous structures, *Opt. Exp.*, 3, 339–350, 1998.

18. Bennett, B.D. et al., Quantitative subcellular imaging of glucose metabolism within intact pancreatic islets, *J. Biol. Chem.*, 271, 3647–3651, 1996.

19. Denk, W., Two-photon scanning photochemical microscopy: mapping ligand-gated ion channel distributions, *Proc. Natl. Acad. Sci. USA*, 91, 6629–6633, 1994.

20. Yuste, R. and Denk, W., Dendritic spines as basic functional units of neuronal integration, *Nature*, 375, 682–684, 1995.

21. Yuste, R. et al., Mechanisms of calcium influx into hippocampal spines: Heterogeneity among spines, coincidence detection by NMDA receptors, and optical quantal analysis, *J. Neurosci.*, 19, 1976–1987, 1999.

22. Svoboda, K., Tank, D.W., and Denk, W., Direct measurement of coupling between dendritic spines and shafts, *Science*, 272, 716–719, 1996.

23. Svoboda, K. et al., *In vivo* dendritic calcium dynamics in neocortical pyramidal neurons, *Nature*, 385, 161–165, 1997.

24. Svoboda, K. et al., Spread of dendritic excitation in layer 2/3 pyramidal neurons in rat barrel cortex *in vivo*, *Nat. Neurosci.*, 2, 65–73, 1999.

25. Mohler, W.A. et al., Dynamics and ultrastructure of developmental cell fusions in the Caenorhabditis elegans hypodermis, *Curr. Biol.*, 8, 1087–1090, 1998.

26. Mohler, W.A. and White, J.G., Stereo-4-D reconstruction and animation from living fluorescent specimens, *Biotechniques*, 24, 1006–1010, 1012, 1998.

27. Squirrell, J.M. et al., Long-term two-photon fluorescence imaging of mammalian embryos without compromising viability, *Nat. Biotechnol.*, 17, 763–767, 1999.

28. Summers, R.G. et al., The orientation of first cleavage in the sea urchin embryo, Lytechinus variegatus, does not specify the axes of bilateral symmetry, *Dev. Biol.*, 175, 177–183, 1996.

29. Fan, G.Y. et al., Video-rate scanning two-photon excitation fluorescence microscopy and ratio imaging with cameleons, *Biophys. J.*, 76, 2412–2420, 1999.
30. Kim, K.H. et al., Three-dimensional image cytometry based on a high-speed two-photon scanning microscope, *SPIE Proc.*, 4262, 238–246, 2001.
31. Guild, J.B. and Webb, W.W., Line scanning microscopy with two-photon fluorescence excitation, *Biophys. J.*, 68, 290a, 1995.
32. Brakenhoff, G.J. et al., Real-time two-photon confocal microscopy using a femtosecond, amplified Ti:sapphire system, *J. Microsc.*, 181, 253–259, 1996.
33. Bewersdorf, J., Pick, R., and Hell, S.W., Mulitfocal multiphoton microscopy, *Opt. Lett.*, 23, 655–657, 1998.
34. Buist, A.H. et al., Real time two-photon absorption microscopy using multipoint excitation, *J. Microsc.*, 192, 217–226, 1998.
35. Kim, K.H., Buehler, C., and So, P.T.C., High-speed, two-photon scanning microscope, *Appl. Opt.*, 38, 6004–6009, 1999.
36. Nilsson, A.M.K. et al., Near infrared diffuse reflection and laser-induced fluorescence spectroscopy for myocardial tissue characterization, *Spectrochimica Acta Part A*, 53, 1901–1912, 1997.
37. Zeng, H. et al., Spectroscopic and microscopic characteristics of human skin autofluorescence emission, *Photochem. Photobiol.*, 61, 639–645, 1995.
38. Cothren, R.M. et al., Gastrointentinal tissue diagnostic by laser-induced fluorescence spectroscopy at endoscopy, *Gastrointest. Endosc.*, 36, 105–111, 1990.
39. Richards-Kortum, R. et al., Spectroscopic diagnosis of colonic dysplasia. *Photochem. Photobiol.*, 53, 777–786, 1991.
40. Glasgold, R. et al., Tissue autofluorescence as an intermediate endpoint in NMBA-induced esophageal carcinogenesis, *Cancer Lett.*, 82, 33–41, 1994.
41. Schantz, S.P. et al., Native cellular fluorescence and its application to canacer prevention, *Environ. Health Perspect.*, 105, 941–944, 1997.
42. Licha, K. et al., Hydrophilic cyanine dyes as contrast agents for near-infrared tumor imaging: synthesis, photophysical properties and spectroscopic *in vivo* characterization, *Photochem. Photobiol.*, 72, 392–398, 2000.
43. Beauvoit, B. and Chance, B., Time-resolved spectroscopy of mitochondria, cells, and tissues under normal and pathological conditions, *Mol. Cell. Biochem.*, 184, 445–455, 1998.
44. Kollias, N. et al., Endogenous skin fluorescence includes bands that may serve as quantitative markers of aging and photoaging, *J. Invest. Derm.*, 111, 776–780, 1998.
45. Yu, W. et al., Fluorescence generalized polarization of cell membranes: a two-photon scanning microscopy approach, *Biophys. J.*, 70, 626–636, 1996.
46. Helmlinger, G. et al., Interstitial pH and pO2 gradients in solid tumors *in vivo*: high-resolution measurements reveal a lack of correlation, *Nat. Med.*, 3, 177–182, 1997.

Index

A

Accelerator mass spectrometry
    DNA damage measured using, 161
    polycyclic aromatic hydrocarbons–DNA
        adducts detected using, 146
Acrylonitrile, 372
Activin, 401
Adenomatous polyposis coli, 175
Aflatoxin
    biomarkers of exposure, 10
    characteristics of, 307–308
    hepatitis B virus and, 310–311
    molecular epidemiology of, 308–310
    p53 tumor suppressor gene mutations,
        311–314
Aflatoxin B1
    description of, 309, 435
    ligation-mediated polymerase chain
        reaction mapping of, 150
    liver cancer and, 150, 435, 439
Aflatoxin–albumin adduct levels, 311, 315
Aflatoxin-nucleic acid adduct, 310
Ah receptor
    background, 358
    cell proliferation regulated by, 359–360
    description of, 358
    gene transcription, 358–359
    ligand-activated, 359
    polychlorinated biphenyls mediated by,
        511
    signal transduction pathway of, 358–360
Akwesasne
    background, 508
    history of, 508
    industrial plants in, 508
    location of, 509
    polychlorinated biphenyl exposure in
        air, 515–516
        drinking water, 514
        fish, 512–514
        health effects, 518–519
        health problems associated with, 510
        humans, 516–518
        overview of, 510
        sediments, 512
        soil, 512
        sources of, 508
        vegetables, 514
        water, 512
    Task Force on the Environment,
        519–522
    university-community relationships,
        519–522
Allele
    cancer susceptibility, 74–75
    CYP2D6*1, 46
    definition of, 40
    DNA repair genes, 70
    genotypic distribution, 42
    PON1, 47
    PON1*3, 47
    rolling circle amplification for *in situ*
        discrimination of, 330
    XRCC1, 71–72
Antibodies, polycyclic aromatic
        hydrocarbons–DNA adducts
        detected using, 146
Anti-7β,8α-dihydroxy-9α,10α-epoxy-
        7,8,9,10-tetrahydrobenzo[a]-
        pyrene, *See* BPDE
Antiestrogen drugs
    raloxifene, 132–133
    tamoxifen, *See* Tamoxifen
    toremifene, 132
Antioxidants
    anticarcinogenic potential of, 366, 373
    ascorbate, 374–375

gene expression patterns for predicting
toxicity, 58
Toxicology
homeostatic disruptions, 449
variations in meaning of, 447
2,3,7-Trichloro-8-methyl dibenzo-p-dioxin,
488–489
Trimethyltin, 374, 413, 416
Tumor suppressor genes
description of, 320
p53, See p53
real-time polymerase chain reaction
measurement of, 327
Two-photon fluorescence microscopy
applications of, 553
deep tissue imaging using, 553–554
description of, 549–550
features of, 550–551
image resolution enhancements,
555–557
instrumentation for, 551–553
microscope design, 551–552
numerical deconvolution and, 556
one-photon microscopy vs., 550–551
resolution of, 555–556
schematic diagram of, 550
self-modeling curve resolution
approaches, 559–560
tissue biochemistry characterized by,
557–561
video-rate, 554–555

U

Ultraviolet light radiation
carcinogenesis associated with, 373–374
cataracts caused by, 372–373
photoproduct removal, 164
SupF system for studying mutations
caused by, 188–193
Unscheduled DNA synthesis, 88
Urine
arsenic concentrations in
arsenic species, 250–251
characteristics of, 257

description of, 249–250
drinking water arsenic concentrations
correlated with, 251–258
exposure biomarker use of, 250
toenail arsenic concentrations
correlated with, 251–258
mercury excretion in, 228
Uroporphyrinogen decarboxylase, 49–50

V

Validation process, 12–13

X

XC3H-4, 343–344
Xenobiotic chemicals, 59
Xenopus laevis
allelic variants in, 343–352
CCCH zinc fingers, 344–345
description of, 339–340, 340
disadvantages of, 343
expressed sequence tags
alignment of, 346–351
description of, 340
polymorphisms in, 352
sequencing of, 340–343, 352
genomic sequencing efforts, 352
transcript arrays of, 347
XC3H-4, 343–344
Xeroderma pigmentosum, 66, 163
XRCC1, 71
XRCC3, 73

Z

Zooplankton
description of, 289
fish ingestion of, 291, 293
metal levels in, 289
size of, 293–295
structure of, 293–295
taxa-specific information, 291